Atom Chips

Edited by
Jakob Reichel and Vladan Vuletić

Related Titles

Stolze, J., Suter, D.

Quantum Computing
A Short Course from Theory to Experiment

2008
ISBN: 978-3-527-40787-3

Matta, C.F., Boyd, R.J. (eds.)

The Quantum Theory of Atoms in Molecules
From Solid State to DNA and Drug Design

2007
ISBN: 978-3-527-30748-7

Bruß, D., Leuchs, G. (eds.)

Lectures on Quantum Information

2007
ISBN: 978-3-527-40527-5

Vogel, W., Welsch, D.-G.

Quantum Optics

2006
ISBN: 978-3-527-40507-7

Bachor, H.-A., Ralph, T.C.

A Guide to Experiments in Quantum Optics

2004
ISBN: 978-3-527-40393-6

Weidemüller, M., Zimmermann, C. (eds.)

Interactions in Ultracold Gases
From Atoms to Molecules

2009
ISBN: 978-3-527-40750-7

Leuchs, G., Beth, T. (eds.)

Quantum Information Processing

2003
ISBN: 978-3-527-40371-4

Cohen-Tannoudji, C., Dupont-Roc, J., Grynberg, G.

Atom-Photon Interactions
Basic Processes and Applications

1998
ISBN: 978-0-471-29336-1

Atom Chips

Edited by
Jakob Reichel and Vladan Vuletić

WILEY-VCH Verlag GmbH & Co. KGaA

The Editors

Prof. Jakob Reichel
Laboratoire Kastler Brossel de l'E.N.S.
24, rue Lhomond
75231 Paris Cedex 05
Frankreich

Prof. Vladan Vuletić
Massachusetts Inst. of Technology,
Room 26-231
77, Massachusetts Avenue
Cambridge, MA 02139-4309
USA

■ All books published by Wiley-VCH are carefully produced. Nevertheless, authors, editors, and publisher do not warrant the information contained in these books, including this book, to be free of errors. Readers are advised to keep in mind that statements, data, illustrations, procedural details or other items may inadvertently be inaccurate.

Library of Congress Card No.: applied for

British Library Cataloguing-in-Publication Data:
A catalogue record for this book is available from the British Library.

Bibliographic information published by the Deutsche Nationalbibliothek
The Deutsche Nationalbibliothek lists this publication in the Deutsche Nationalbibliografie; detailed bibliographic data are available on the Internet at http://dnb.d-nb.de.

© 2011 WILEY-VCH Verlag GmbH & Co. KGaA, Boschstr. 12, 69469 Weinheim, Germany

All rights reserved (including those of translation into other languages). No part of this book may be reproduced in any form – by photoprinting, microfilm, or any other means – nor transmitted or translated into a machine language without written permission from the publishers. Registered names, trademarks, etc. used in this book, even when not specifically marked as such, are not to be considered unprotected by law.

Typesetting le-tex publishing services GmbH, Leipzig
Printing and Binding Fabulous Printers Pte Ltd, Singapore
Cover Design Adam Design, Weinheim

Printed in Singapore
Printed on acid-free paper

ISBN 978-3-527-40755-2

Contents

Preface *XV*

List of Contributors *XVII*

Part One Fundamentals *1*

1 From Magnetic Mirrors to Atom Chips *3*
Andrei Sidorov and Peter Hannaford
1.1 Introduction *3*
1.2 Historical Background *4*
1.3 Magnetic Mirrors for Cold Atoms *7*
1.3.1 Basic Principles *7*
1.3.2 Experimental Realization of Magnetic Mirrors *9*
1.3.2.1 Macroscopic Array of Rare-Earth Magnets of Alternating Polarity *9*
1.3.2.2 Micro-Fabricated Grooved Magnetic Mirrors *10*
1.3.2.3 Micro-Fabricated Array of Current-Carrying Conductors *11*
1.3.2.4 Magneto-Optical Recording of Magnetic Microstructures *12*
1.4 The Magnetic Film Atom Chip *13*
1.4.1 Background *13*
1.4.2 BEC on a Magnetic Film Atom Chip *14*
1.4.3 Spatially Resolved RF Spectroscopy to Probe Magnetic Film Topology *16*
1.4.4 Adiabatic Splitting of a BEC for Asymmetric Potential Sensing *19*
1.4.5 Spatially Inhomogeneous Phase Evolution of a Two-Component BEC *21*
1.4.6 BEC on Other Permanent-Magnet Atom Chips *22*
1.5 Permanent Magnetic Lattice on a Magnetic Film Atom Chip *23*
1.5.1 Background *23*
1.5.2 Basic Principles *24*
1.5.2.1 One-Dimensional Magnetic Lattice *24*
1.5.2.2 Two-Dimensional Magnetic Lattice *25*
1.5.2.3 Permanent 1D Magnet Lattice for Ultra-Cold Atoms *26*
1.5.2.4 Other Permanent Magnetic Lattices *28*

1.6	Summary and Conclusions 28
	References 29

2 Trapping and Manipulating Atoms on Chips 33
Jakob Reichel

2.1	Introduction 33
2.2	Overview of Trapping Techniques 34
2.3	Magnetic Traps for Neutral Atoms 35
2.3.1	Magnetic Interaction 35
2.3.2	Stability against Spin-Flip Losses 36
2.3.3	Quadrupole Traps 37
2.3.4	Ioffe–Pritchard Traps 37
2.3.5	Some General Properties of Magnetic Traps 38
2.4	The Design of Wire Patterns for Magnetic Potentials 39
2.4.1	Conductor Elements and Multipoles 39
2.4.2	Wire Guide 40
2.4.3	Conductor Cross ("Dimple" Trap) 41
2.4.4	"H", "Z", and "U" Traps 43
2.4.5	Finite Wire Dimensions 44
2.4.6	Maximum Confinement 46
2.4.6.1	Field Gradient 46
2.4.6.2	Field Curvature and Trap Frequency 46
2.4.7	Combining Elements: Arrays, Conveyors and Others 47
2.5	Real Wires: Roughness and Maximum Current 48
2.5.1	Effect of Wire Roughness 48
2.5.2	Heat Transport and Maximum Current 49
2.5.2.1	Wire–Substrate Interface 49
2.5.2.2	Heat Evacuation through the Substrate 51
2.6	Loading Techniques 51
2.6.1	Mirror-MOT 51
2.6.2	Magnetic Elevator 52
2.6.3	"Mode Matching" 52
2.7	Vacuum Cells 53
2.7.1	Traditional Cell 53
2.7.2	Compact Cell with Atom Chip Wall 55
2.8	Conclusion and Outlook 57
	References 58

3 Atom Chip Fabrication 61
Ron Folman, Philipp Treutlein and Jörg Schmiedmayer

3.1	Introduction 61
3.2	Fabrication Challenges 62
3.3	The Substrate 63
3.4	Lithography 65
3.4.1	Optical Lithography 65

3.4.2	Electron-Beam Lithography	67
3.5	Metallic Layers	68
3.5.1	Deposition and Etching	68
3.5.1.1	Electroplating	68
3.5.1.2	Evaporation and Lift-Off Metallization	70
3.5.1.3	Wet and Dry Etching	72
3.5.1.4	Designing Potentials by Postprocessing the Wires	73
3.5.2	Effects of Roughness and Homogeneity of the Fabricated Structures	74
3.5.3	Special Metals	76
3.5.3.1	Alloys	76
3.5.3.2	Superconductors	77
3.5.3.3	Semiconductors	79
3.5.4	Permanent Magnets	80
3.5.5	Metal Outlook	82
3.6	Additional Features	85
3.6.1	Planarization and Insulation	85
3.6.2	On-Chip Mirrors	87
3.6.3	Multi-Layer Chips	88
3.7	Current Densities and Tests	91
3.8	Photonics on Atom Chips	93
3.8.1	Fiber-Based Integrated Optics	93
3.8.1.1	SU8 – Holding Structures	93
3.8.1.2	Fiber-Based Fluorescence Detector	94
3.8.1.3	Fiber Cavities	95
3.8.2	Microlens and Cylindrical Lens	97
3.8.3	Microdisks and Microtoroids	98
3.8.4	Mounted and Fully Integrated Fabry–Pérots	99
3.8.5	Planar Optics	101
3.8.6	Photonics Outlook	102
3.9	Chip Dicing, Mounting, and Bonding	104
3.10	Further Integration and Portability	106
3.11	Conclusion and Outlook	109
	References	110

Part Two Ultracold Atoms near a Surface 119

4 Atoms at Micrometer Distances from a Macroscopic Body 121
Stefan Scheel and E.A. Hinds
4.1	Introduction	121
4.2	Principles of QED in Dielectrics	123
4.3	Relaxation Rates near a Surface	126
4.3.1	Spin Flips near a Dielectric or Metallic Surface	126
4.3.2	Spin Flips near a Superconductor	130
4.3.3	Transverse Spin Relaxation	132
4.3.4	Heating	133
4.3.5	Electric Dipole Coupling of Molecules to a Surface	134

4.4	Casimir–Polder Forces	138
4.5	Closing Remarks	144
	References	145

5 Interaction of Atoms, Ions, and Molecules with Surfaces 147
Carsten Henkel

5.1	Qualitative Overview	147
5.1.1	Electromagnetic Dipole Moments	148
5.1.2	Electromagnetic Field Strengths	149
5.1.3	Digression: Surface Green Functions	151
5.2	Interaction Potentials	153
5.2.1	Charges and Permanent Dipoles	153
5.2.2	Van der Waals Potential	154
5.2.3	Casimir–Polder Potential	155
5.2.4	Recent Developments	156
5.3	Surface-Induced Atomic Transitions	157
5.3.1	Visible Frequencies: Spontaneous Emission	158
5.3.2	Thermal Frequencies: Spin-Flips	159
5.3.3	Trap Heating	161
5.3.4	Atom Chips and Decoherence	162
5.4	Perspectives	165
	References	166

Part Three Coherence on Atom Chips 171

6 Diffraction and Interference of a Bose–Einstein Condensate Scattered from an Atom Chip-Based Magnetic Lattice 173
A. Günther, T.E. Judd, J. Fortágh and C. Zimmermann

6.1	Introduction	173
6.2	Experimental Setup	174
6.2.1	The BEC Apparatus	174
6.2.2	The Magnetic Lattice Chip	177
6.3	The Magnetic Lattice Potential	178
6.3.1	Infinite Lattice	178
6.3.2	Finite Size Effects	181
6.3.3	The Double Meander Potential	182
6.4	Diffraction and Interference	184
6.4.1	Diffraction Scheme	184
6.4.2	Theoretical Model for the Interaction	185
6.4.3	Diffraction in the Raman–Nath Regime	189
6.4.4	Evolution of the Wave Function after the Lattice Interaction	190
6.5	Ballistic Expansion and Phase Imprinting	194
6.6	Experimental Results	195
6.7	Effect of Atomic Interactions	202
6.7.1	Modeling BEC Surface Diffraction	202
6.7.2	Density Profile Dynamics	203

6.7.3	Phase Modification by Inter-Atomic Interactions	204
6.7.4	Comparison of the Interacting Theory with Experiment	205
6.7.5	Locating the Low-Interaction Regime	206
6.8	Conclusion	207
	References	208

7 Interferometry with Bose–Einstein Condensates on Atom Chips *211*

Thorsten Schumm, Stephanie Manz, Robert Bücker, David A. Smith and Jörg Schmiedmayer

7.1	Introduction	211
7.2	Atom Chip BEC Splitters Based on Static Fields	213
7.2.1	Transverse Splitting	213
7.2.1.1	The Two-Wire Splitter	214
7.2.1.2	The Five-Wire Splitter	216
7.2.1.3	The Y Splitter	218
7.2.2	Longitudinal Splitting	221
7.2.3	Electrostatic Splitter	222
7.3	Atom Chip BEC Splitters Based on Dressed Adiabatic Potentials	224
7.3.1	Dressed Adiabatic State Potentials	225
7.3.2	A BEC Splitter Based on Dressed Adiabatic State Potentials	228
7.3.3	Beyond the Rotating-Wave Approximation	230
7.3.4	Implementation on an Atom Chip	231
7.3.5	Advantages of RF-Induced Splitters over Static Splitters	232
7.4	Matter–Wave Interferometry with Bose–Einstein Condensates	234
7.4.1	Theoretical Aspects	234
7.4.2	Experimental Realizations	238
7.4.2.1	Coherent Splitting on Atom Chips	239
7.4.2.2	Interference of Independent Condensates	240
7.4.2.3	Phase Dynamics of Split Condensates	240
7.4.2.4	Merging of Split Condensates	245
7.5	Interferometry with 1D quasi condensates	246
7.5.1	Coherently Split 1D BECs: Coherence Dynamics	247
7.5.1.1	Decoherence of Uncoupled 1D Systems	249
7.5.1.2	Coherence Dynamics for Coupled 1D Condensates	251
7.5.2	Independent 1D BECs: Noise Statistics of Interference Amplitude	252
7.5.2.1	Average Interference Amplitude Square	253
7.5.2.2	Full Counting Statistics of Interference Amplitude	255
7.6	Summary and Outlook	257
	References	259

8 Microchip-Based Trapped-Atom Clocks *265*

Vladan Vuletić, Ian D. Leroux and Monika H. Schleier-Smith

8.1	Basic Principles	265
8.2	Atomic-Fountain versus Trapped-Atom Clocks	265
8.3	Optical-Transition Clocks versus Microwave Clocks	267

8.4	Clocks with Magnetically Trapped Atoms: Fundamental Limits to Performance *267*
8.5	Clocks with Magnetically Trapped Atoms: Experimental Demonstrations *271*
8.6	Readout in Trapped-Atom Clocks *274*
8.7	Spin Squeezing *277*
	References *278*

9 Quantum Information Processing with Atom Chips *283*
Philipp Treutlein, Antonio Negretti and Tommaso Calarco

9.1	Introduction *283*
9.2	Ingredients for QIP with Atom Chips *284*
9.3	Qubit States with Long Coherence Lifetime *285*
9.4	Qubit Rotations (Single-Qubit Gates) *288*
9.5	Single-Qubit Readout (Single-Atom Detection) *290*
9.6	Single-Qubit Preparation (Single-Atom Preparation) *291*
9.7	Conditional Dynamics (Two-Qubit Gates) *291*
9.7.1	Internal-State Qubits and Collisional Interactions *292*
9.7.2	Motional-State Qubits and Collisional Interactions *298*
9.7.3	Alternative Chip-Specific Approaches to Entanglement Generation *300*
9.7.4	Cavity-QED-Based Schemes *300*
9.7.5	Quantum Gate Schemes that Can Be Adapted from Other Contexts *301*
9.8	Hybrid Approaches to QIP on a Chip *303*
9.8.1	Hybrid Approaches to Entanglement Generation *303*
9.8.2	Interfacing Atoms (Storage/Processing Qubits) with Photons (Flying Qubits) *304*
9.8.3	Quantum Information Technology for Precision Measurement and Other Applications *304*
9.9	Conclusion and Outlook *305*
	References *305*

Part Four New Directions *309*

10 Cryogenic Atom Chips *311*
Gilles Nogues, Adrian Lupaşcu, Andreas Emmert, Michel Brune, Jean-Michel Raimond and Serge Haroche

10.1	Introduction *311*
10.2	Superconducting Atom Chip Setup: Similarities and Differences with Conventional Atom Chips *312*
10.2.1	Experimental Considerations *312*
10.2.1.1	Chip Fabrication and Wiring *312*
10.2.1.2	The Cryogenic Cell *314*
10.2.2	Trapping and Cooling: First Results *316*
10.2.2.1	Magnetic Trap *316*
10.2.2.2	Forced Evaporation and Quantum Degeneracy *317*

10.3	Perspectives for Cryogenic Atom Chips: A New Realm of Investigations *319*	
10.3.1	Probing the Superconducting Film Current Distribution *319*	
10.3.2	Integration of Atom Chips with Superconducting Circuit Elements *321*	
10.3.2.1	Coupling with a Superconducting Qubit *321*	
10.3.2.2	Coupling with a Superconducting Resonator: On-Chip CQED *322*	
10.3.3	Atom Chips for Circular Rydberg States *325*	
10.4	Conclusion *328*	
	References *329*	
11	**Atom Chips and One-Dimensional Bose Gases** *331*	
	I. Bouchoule, N.J. van Druten and C.I. Westbrook	
11.1	Introduction *331*	
11.2	Regimes of One-Dimensional Gases *332*	
11.2.1	Strongly versus Weakly Interacting Regimes *334*	
11.2.2	Nearly Ideal Gas Regime *335*	
11.2.3	Quasi-Condensate Regime *338*	
11.2.3.1	Density Fluctuations *340*	
11.2.3.2	Phase Fluctuations *341*	
11.2.4	Exact Thermodynamics *342*	
11.3	1D Gases in the Real World *345*	
11.3.1	Transverse Trapping and Nearly 1D Bose Gases *345*	
11.3.2	Applying 1D Thermodynamics to a 3D Trapped Gas *347*	
11.3.3	Longitudinal Trapping *347*	
11.3.3.1	Local Density Approximation *348*	
11.3.3.2	Validity of the Local Density Approximation *349*	
11.3.4	3D Physics versus 1D Physics *349*	
11.4	Experiments *351*	
11.4.1	Failure of the Hartree–Fock Model *352*	
11.4.2	Yang–Yang Analysis *353*	
11.4.3	Measurements of Density Fluctuations *355*	
11.4.3.1	A Local Density Analysis *355*	
11.4.3.2	Ideal Gas Regime: Observation of Bunching *356*	
11.4.3.3	Quasi-Condensate Regime: Saturation of Atom Number Fluctuations *358*	
11.5	Conclusion *359*	
	References *360*	
12	**Fermions on Atom Chips** *365*	
	Marcius H.T. Extavour, Lindsay J. LeBlanc, Jason McKeever, Alma B. Bardon, Seth Aubin, Stefan Myrskog, Thorsten Schumm and Joseph H. Thywissen	
12.1	Introduction *365*	
12.2	Theory of Ideal Fermi Gases *366*	
12.2.1	Thermodynamics *366*	
12.2.2	Density Distribution *368*	

12.2.3	Crossover to Fermi Degeneracy 370
12.3	The Atom Chip 371
12.3.1	Chip Construction and Wire Pattern 372
12.3.2	Electrical and Mechanical Connections 372
12.3.3	The Z-Wire Magnetic Trap 373
12.4	Loading the Microtrap 373
12.4.1	Laser Cooling and Magnetic Transport to the Chip 374
12.4.2	Loading Bosons and Fermions onto the Atom Chip 374
12.4.3	Effective Trap Volume 375
12.4.4	A Full Tank of Atoms: Maximum Trapped Atom Number 376
12.4.5	Effect of Geometry on Loaded Atom Number 377
12.5	Rapid Sympathetic Cooling of a K-Rb Mixture 377
12.5.1	Forced Sympathetic RF Evaporation 378
12.5.2	K-Rb Cross-Thermalization 379
12.5.3	Density-Dependent Loss 380
12.5.4	Required Temperature 380
12.5.5	Experimental Signatures of Fermi Degeneracy 381
12.6	Species-Selective RF Manipulation 382
12.6.1	Sympathetic RF Evaporation 383
12.6.2	Species-Selective Double Wells 385
12.7	Fermions in an Optical Dipole Trap near an Atom Chip 387
12.7.1	Optical Trap Setup 388
12.7.2	Loading the Optical Trap 388
12.7.3	Microwave and RF Manipulation 389
12.8	Discussion and Future Outlook 390
	References 391
13	**Micro-Fabricated Chip Traps for Ions** 395
	J.M. Amini, J. Britton, D. Leibfried and D.J. Wineland
13.1	Introduction 395
13.2	Radio-Frequency Ion Traps 396
13.2.1	Motion of Ions in a Spatially Inhomogeneous RF Field 396
13.2.2	Electrode Geometries for Linear Quadrupole Traps 398
13.3	Design Considerations for Paul Traps 399
13.3.1	Doppler Cooling 399
13.3.2	Micromotion 401
13.3.3	Exposed Dielectric 402
13.3.4	Loading Ions 403
13.3.5	Electrical Connections 404
13.3.6	Motional Heating 405
13.4	Measuring Heating Rates 406
13.5	Multiple Trapping Zones 407
13.6	Trap Modeling 408
13.6.1	Modeling 3D Geometries 408
13.6.2	Analytic Solutions for Surface-Electrode Traps 409

13.7	Trap Examples	*411*
13.8	Future	*415*
	References	*417*
	Index	*421*

Preface

This book intends to give both an introduction and an in-depth review of the beautiful physics being done with atom chips. Topics range from the manipulation of single atoms to the quantum entanglement between many atoms, and from interferometry with atomic matter waves to studies of fundamental atom–surface interactions.

For about three decades researchers have used magnetic and electric fields from DC to optical frequencies to confine neutral atoms for a variety of experiments and applications. The term *atom chip* has come to designate setups where microscopic or micro-fabricated structures, typically confined to a surface, generate three-dimensional trapping fields in the vicinity of the surface.

At its inception, the atom chip was regarded primarily as a tool to conveniently generate electromagnetic fields varying on a small length scale, and as such is related to early prototypes of magnetic mirrors. In fact, the attainment of Bose–Einstein condensation on a chip in 2001 in Tübingen and Munich was the first landmark that brought atom chips to the attention of the physics community at large. Since then, a growing number of research groups has adopted microchips as a convenient and fast method for the creation of Bose–Einstein condensates (BECs), and now also degenerate Fermi gases.

The strongly confining, complex, multi-parameter potentials that can be realized with atom chips have enabled experimentalists to explore new situations. For example, studies of one-dimensional quantum gases are benefitting from extremely elongated single traps that can be generated on atom chips, and BECs have been diffracted from specifically designed magnetic lattices realized on the chip surface.

However, atom chips are not merely devices to form atom traps by a combination of conductors and insulators on a surface. Atom chips promise rich functionality and integrability, and possibly nano-scale miniaturization, as advertised early on by a number of researchers in the field. The small length scale well matched to the condensate size and proximity of a solid surface have opened up and driven further research possibilities. The first and perhaps most immediate example is the investigation of fundamental surface-induced forces, such as the van der Waals and Casimir–Polder forces. This field has progressed and expanded considerably due to the close and stimulating interaction between atom chip experimentalists and theorists. Furthermore, the repertoire of fields and interactions used on atom chips

Atom Chips. Edited by Jakob Reichel and Vladan Vuletić
Copyright © 2011 WILEY-VCH Verlag GmbH & Co. KGaA, Weinheim
ISBN: 978-3-527-40755-2

has grown to include radiofrequency and microwave potentials, resonant and far-detuned optical fields in miniature optical devices, as well as surface interactions with micro mechanical structures. In each case, the small-scale, near-field situation of the atom chip has been exploited in ingenious ways to create new and rich physical situations that go beyond the possibilities of macroscopic experiments. Examples include coupling of a BEC to an oscillating mechanical cantilever, cavity quantum electrodynamics experiments with BECs, and some of the most beautiful condensate interferometry experiments performed so far.

The combination of these features makes atom chips an interesting platform for quantum information and quantum simulation experiments. This has also motivated the development of the newest family of atom chips, surface-electrode-based ion traps, which present both similarities and interesting differences compared to their neutral-atom counterparts.

A third area has emerged where atom chips are used as a means to construct the most compact and robust ultra-cold atom devices. The very recent demonstration of BEC in microgravity was enabled by an atom chip. Trapped-atom clocks on atom chips are being explored as promising secondary frequency standards. The idea of "integrated atom optics" on atom chips as a means to build atom interferometers emerged with the first atom chip experiments, but is certainly still in its infancy today. Last but not least, experiments with BECs in cryogenic environments also benefit from the small size and robustness of atom chips.

This book represents a collective effort by the community of atom chip researchers to outline the state of their knowledge as of 2009/2010. Each chapter starts with a thorough introduction before exposing the state of the art on a specific topic. Additionally, there are introductory chapters describing the particularities of designing magnetic potentials and producing BECs on atom chips, as well as on atom chip fabrication. The latter is discussed in a tutorial style and sufficient detail to enable a researcher with minimal micro-fabrication knowledge to start fabricating atom chips. In this way, we hope that the book will be valuable for students and researchers who are entering the field of atom chips or are active in one of the neighboring fields, but also for anyone desiring to get an overview of this beautiful and active area of contemporary quantum physics.

Paris and Cambridge, June 2010 *Jakob Reichel and Vladan Vuletić*

List of Contributors

Jason M. Amini
Signature Technology Laboratory
Georgia Technology Research Institute
400 Tenth Street
Atlanta, GA 30318
USA

Seth Aubin
Department of Physics
College of William and Mary
Williamsburg, VA 23185
USA

Alma B. Bardon
Department of Physics
University of Toronto
60 St. George Street
Toronto, Ontario M5S 1A7
Canada

Isabelle Bouchoule
Laboratoire Charles Fabry de l'Institut
d'Optique
Campus Polytechnique RD 128
91127 Palaiseau Cedex
France

Joe Britton
Time and Frequency Division
National Institute of Standards and
Technology
325 Broadway
Boulder, CO 80305
USA

Michel Brune
Laboratoire Kastler Brossel de l'E.N.S.
24, rue Lhomond
75231 Paris Cedex 05
France

Robert Bücker
Atominstitut der Österreichischen
Universitäten
TU Wien
Stadionallee 2
1020 Wien
Austria

Tommaso Calarco
Institut für
Quanteninformationsverarbeitung
Universität Ulm
Albert-Einstein-Allee 11
89069 Ulm
Germany

Andreas Emmert
Laboratoire Kastler Brossel de l'E.N.S.
24, rue Lhomond
75231 Paris Cedex 05
France

Marcius H. T. Extavour
Ontario Power Generation
700 University Avenue
Toronto, Ontario
Canada, M5G 1X6

Ron Folman
Atom Chip Group
Ben-Gurion University
Be'er Sheva, 84105
Israel

József Fortágh
Physikalisches Institut
Universität Tübingen
Auf der Morgenstelle 14
72076 Tübingen
Germany

A. Günther
Physikalisches Institut
Universität Tübingen
Auf der Morgenstelle 14
72076 Tübingen
Germany

Peter Hannaford
Centre for Atom Optics & Ultrafast Spectroscopy
Faculty of Engineering and Industrial Science
Swinburne University of Technology
Serpells Lane,
Mail H38, PO Box 218
Hawthorn, Victoria 3122
Australia

Serge Haroche
Laboratoire Kastler Brossel de l'E.N.S.
24, rue Lhomond
75231 Paris Cedex 05
France

Carsten Henkel
Institut für Physik und Astronomie
Campus Golm
Universität Potsdam
14476 Potsdam
Germany

Edward A. Hinds
Centre for Cold Matter
Blackett Laboratory
Imperial College London
Prince Consort Road
London SW7 2AZ
United Kingdom

Thomas E. Judd
Physikalisches Institut
Universität Tübingen
Auf der Morgenstelle 14
72076 Tübingen
Germany

Lindsay J. LeBlanc
Department of Physics
University of Toronto
60 St. George Street
Toronto, Ontario M5S 1A7
Canada

Dietrich Leibfried
Time and Frequency Division
National Institute of Standards and Technology
325 Broadway
Boulder, CO 80305
USA

Ian D. Leroux
MIT Department of Physics
77 Massachusetts Avenue
Cambridge, MA 02139
USA

Adrian Lupaşcu
Institute for Quantum Computing
University of Waterloo
200 University Ave. W.
Waterloo, ON N2L 3G1
Canada

Stephanie Manz
Atominstitut der Österreichischen
Universitäten
TU Wien
Stadionallee 2
1020 Wien
Austria

Jason McKeever
Entanglement Technologies
Palo Alto, CA 94306
USA

Stefan Myrskog
Morgan Solar Inc.
30 Ordnance St.
Toronto, Ontario M6K 1A2
Canada

Antonio Negretti
Institut für
Quanteninformationsverarbeitung
Universität Ulm
Albert-Einstein-Allee 11
89069 Ulm
Germany

Gilles Nogues
Institut Neel
25 avenue des Martyrs
bâtiment T, BP 166
38042 Grenoble cedex 9

Jean-Michel Raimond
Laboratoire Kastler Brossel de l'E.N.S.
24, rue Lhomond
75231 Paris Cedex 05
France

Jakob Reichel
Laboratoire Kastler Brossel de l'E.N.S.
24, rue Lhomond
75231 Paris Cedex 05
France

Stefan Scheel
Blackett Laboratory
Imperial College London
Prince Consort Road
London SW7 2AZ
United Kingdom

Monika H. Schleier-Smith
MIT Department of Physics
77 Massachusetts Avenue
Cambridge, MA 02139
USA

Jörg Schmiedmayer
Atominstitut der Oesterreichischen
Universitäten
TU Wien
Stadionallee 2
1020 Vienna
Austria

Thorsten Schumm
Atominstitut der Österreichischen
Universitäten
TU Wien
Stadionallee 2
1020 Wien
Austria

Andrei Sidorov
Centre for Atom Optics & Ultrafast Spectroscopy
Faculty of Engineering and Industrial Science
Swinburne University of Technology
Serpells Lane,
Mail H38, PO Box 218
Hawthorn, Victoria 3122
Australia

David A. Smith
Atominstitut der Österreichischen Universitäten
TU Wien
Stadionallee 2
1020 Wien
Austria

Joseph H. Thywissen
Department of Physics
University of Toronto
60 St. George Street
Toronto, Ontario M5S 1A7
Canada

Philipp Treutlein
University of Basel
Department of Physics
Klingelbergstrasse 82
CH-4056 Basel
Switzerland

N. J. (Klaasjan) van Druten
Van der Waals-Zeeman Instituut
Universiteit van Amsterdam
Valckenierstraat 65-67
1018 XE Amsterdam
Netherlands

Vladan Vuletić
MIT Department of Physics
77 Massachusetts Avenue
Cambridge, MA 02139
USA

Christopher I. Westbrook
Laboratoire Charles Fabry de l'Institut d'Optique
Campus Polytechnique RD 128
91127 Palaiseau Cedex
France

David J. Wineland
Time and Frequency Division
National Institute of Standards and Technology
325 Broadway
Boulder, CO 80305
USA

C. Zimmermann
Physikalisches Institut
Universität Tübingen
Auf der Morgenstelle 14
72076 Tübingen
Germany

Part One Fundamentals

1
From Magnetic Mirrors to Atom Chips[1]

Andrei Sidorov and Peter Hannaford

1.1
Introduction

Following the advent of laser cooling and trapping techniques in the 1980s, a new exciting area of research, 'atom chips', has emerged in which sophisticated micron-scale structures on planar substrates are produced utilizing the latest technological developments in lithography and nanofabrication. These complex microstructures produce tiny magnetic field configurations which can trap, cool, and manipulate ensembles of ultra-cold atoms in the vicinity of a surface. Scaling down the dimensions of atom trapping geometry offers extended possibilities for the production and control of Bose–Einstein condensates (BECs). Enormous progress on the generation of BECs and quantum degenerate Fermi gases, on-chip matter–wave interferometers, and integrated detectors has been made in the last few years.

In the second section of this article we trace the historical evolution of this new field, from the first surface-based atom optical elements – magnetic mirrors – to the present-day micro-fabricated structures on a substrate – atom chips. In Section 1.3 we present the basic principles of magnetic mirrors for cold atoms and describe different types of magnetic mirror. Section 1.4 describes the production of a BEC on a permanent magnetic film atom chip; the application of this atom chip to probe the topology of magnetic fields using RF spectroscopy and to study the adiabatic splitting of a BEC in a double well for sensing asymmetric potentials; and investigations of the spatially dependent relative phase evolution of a two-component BEC. Finally, in Section 1.5 we describe a permanent magnetic lattice on an atom chip for trapping and manipulating multiple arrays of ultra-cold atoms and quantum degenerate gases.

---

[1] We dedicate this article to the memory of our friend and colleague Geoffrey I. Opat.

Atom Chips. Edited by Jakob Reichel and Vladan Vuletić
Copyright © 2011 WILEY-VCH Verlag GmbH & Co. KGaA, Weinheim
ISBN: 978-3-527-40755-2

1.2
Historical Background

In 1983 Opat [1] proposed the idea of using periodic arrays of electric fields on a planar substrate to reflect beams of polar molecules in a matter–wave interferometer. Almost a decade later the Melbourne group [2] reported the reflection of a beam of chloromethane (CH_3Cl) molecules from the exponentially decaying electric field above a periodic array of electrodes of alternating polarity.[2] Opat and coworkers [3] then extended these ideas to periodic arrays of magnetic fields to create a surface-based mirror for reflecting beams of laser-cooled atoms (Figure 1.1a). As slowly moving atoms in positive (or 'low-field seeking') magnetic states approach the periodic magnetic structure they experience an exponentially increasing magnetic field, with decay length $a/2\pi$ (where a is the period), and for sufficiently large magnetic fields the atoms are repelled by the structure. A number of schemes [3] were proposed for producing the periodic magnetic structures, including the use of arrays of magnets of alternating polarity, periodic magnetic fields 'recorded' on ferromagnetic substrates as in sound recording, and planar arrays of parallel wires alternately carrying electric current in opposite directions. Methods for fabricating the surface-based microstructures were suggested including lithographical techniques used in the electronics industry.

The first magnetic mirror for cold atoms was realized in 1995 by Roach et al. [4] by recording sinusoidal signals onto a magnetic audio-tape and observing the retroflection of laser-cooled rubidium atoms from the recorded structure. Soon after, Sidorov et al. [5] in Melbourne reported the retroreflection of cold cesium atoms from a 2-mm-period planar array of Nd-Fe-B magnets of alternating polarity. In 1997 the Sussex group demonstrated the focusing and multiple reflection of cold atoms bouncing on a curved magnetic mirror made from a concave-shaped floppy disk or video tape [6, 7].

One of the major challenges was to be able to scale the magnetic structures down to micron-scale periods, in order to produce a hard mirror with very short decay length and minimal finite-size effects. This problem was basically solved when it was shown [8] that a grooved periodic magnetic structure produces a magnetic field distribution that is essentially the same as that above an array of magnets of alternating polarity. The Melbourne group subsequently demonstrated the specular reflection of cold atoms from a 1-µm-period magnetic mirror constructed from micro-fabricated grooved structures coated with perpendicularly magnetized film [9, 10]. The scaling down of the period to 1 µm represented a significant advance in the miniaturization of surface-based atom optical elements.

In 1999 the Melbourne [11] and Harvard/Orsay/Gaithersburg [12] groups reported a magnetic mirror for cold atoms constructed from a planar array of current-carrying conductors lithographically patterned on a silicon wafer or a sapphire sub-

---

2) The beam of slowly moving polar molecules was produced in a gravity-assisted molecular beam line experiment (GAMBLE).

strate. A feature of this type of magnetic mirror is that the magnetic field may be readily varied, switched or modulated by varying the current in the conductors. Such magnetic mirrors were the first micro-fabricated surface-based optical elements for cold atoms and represented a significant step towards the development of more sophisticated atom chips based on micro-fabricated current-carrying conductors on a substrate [13–15].

In an earlier paper, Weinstein and Libbrecht [16] at CalTech had proposed various planar current-carrying wire geometries for constructing microscopic electromagnetic traps for cold atoms including the use of superconducting wire structures. In 1998 Vuletić et al. [17] and Fortágh et al. [18] in Munich demonstrated 3D microtraps for cold atoms based on a combination of electromagnets and permanent magnets or current-carrying conductors, and Drndic et al. [19] at Harvard reported the fabrication of micro-electromagnetic traps with geometries proposed in [16]. In 1999 the Munich group [20] reported the use of surface magnetic microtraps based on a 'U'-shaped wire quadrupole microtrap and a 'Z'-shaped wire Ioffe–Pritchard (IP) microtrap, with non-zero potential minimum to eliminate spin-flip losses, for the trapping of cold atoms on a substrate. Soon after, other groups [21–24] demonstrated the guiding of laser-cooled atoms by current-carrying wires, and Davis [25] proposed the use of permanent magnetic structures as miniature waveguides to transport cold atoms on a substrate. In 2000 the Innsbruck group [26] reported the trapping and guiding of cold atoms using a micro-fabricated circuit on a substrate, which they called an 'atom chip'. In the following year the Munich group [27] demonstrated a magnetic conveyer belt for transporting and merging cold atoms on an atom chip.

A major breakthrough came at the International Conference on Laser Spectroscopy in Snowbird in 2001 when Hänsel et al. [28] from Munich and Ott et al. [29] from Tübingen simultaneously announced the realization of a Bose–Einstein condensate (BEC) of ^{87}Rb atoms in a current-carrying magnetic microtrap on an atom chip. The use of miniature magnetic microtraps allowed the scaling down of the electric currents required to produce a BEC and the scaling up of the trap confinement and elastic collision rate, thus greatly simplifying and speeding up the production of a BEC. In 2006 Aubin et al. ([30], Chapter 12) in Toronto realized a degenerate Fermi gas of ^{40}K atoms on an atom chip by sympathetic cooling with ultra-cold ^{87}Rb atoms.

One of the next big challenges was to see if it was possible to perform coherence or interference experiments on ultra-cold atoms trapped on an atom chip at distances close to the surface (Chapters 4 and 5). In 2004 Treutlein et al. [31] in Munich reported the coherent manipulation of two hyperfine states of ultra-cold ^{87}Rb atoms in a current-carrying magnetic microtrap, with coherence times exceeding 1 s. This opened the way for Ramsey interferometry and miniature atomic clocks on an atom chip ([32], Chapter 8). However, on-chip interferometry by spatially splitting the condensate proved to be more challenging owing to difficulties of phase preservation and control of the condensate. In 2005 the MIT/Harvard groups [33] dynamically split a condensate by deforming a single-well magnetic trap into a double-well potential, but non-adiabatic evolution in a quartic potential during the splitting process led to an unpredictable relative phase. A major breakthrough came when the Heidelberg/Vienna group demonstrated a phase-preserving splitting scheme based on RF-induced adiabatic potentials ([34], Chapter 7), which allowed accurate control over the splitting process and the observation of reproducible interference fringes with a deterministic phase. In the same year the Boulder/Harvard groups [35] achieved splitting, reflection, and recombining of condensate atoms in a Michelson interferometer using a standing-wave light field in a waveguide on a chip. In 2007 the MIT/Harvard groups [36] reported the observation of phase coherence between two separated BECs on an atom chip for times up to about 200 ms after splitting the condensate. In 2009 the Munich group [37] demonstrated the coherent manipulation of BECs in a state-dependent potential with microwave fields on an atom chip, allowing the on-chip generation of multi-particle entanglement and quantum-enhanced metrology with spin-squeezed states [38] (Chapter 8).

Very recently, Deutsch et al. [39] in Paris have reported coherence times as long as 58 s in a Ramsey interferometer experiment on ^{87}Rb atoms trapped on an atom chip. The long coherence times are interpreted in terms of a spin self-rephasing mechanism induced by an identical spin rotation effect that occurs during collisions in the forward direction between two identical particles [40]

In parallel developments Sinclair et al. [41] in London produced a BEC on a permanent-magnet atom chip based on periodically magnetized videotape and Hall et al. [42, 43] in Melbourne produced a BEC in a microtrap on a TbGdFeCo permanent magnetic film atom chip. BECs have since been produced in microtraps on a Co-Cr-Pt hard disk [44] and on a Fe-Pt magnetic foil atom chip [45]. In 2007 Whitlock et al. in Amsterdam [46, 47] constructed a 2D asymmetric magnetic lattice with periods of 22 and 36 µm in orthogonal directions on a Fe-Pt film atom chip and the Melbourne group [48, 49] constructed a 1D 10-µm-period magnetic lattice on a TbGdFeCo film atom chip. Both groups demonstrated the loading of ultra-cold atoms into multiple sites of the permanent magnetic lattice. Magnetic lattices can be readily scaled to have a very large number of lattice sites and could form the basis of storage registers for quantum information processing.

Superconducting wires offer the prospect of an extremely low noise environment for trapped ultra-cold atoms. In 2006 Nirrengarten et al. ([50] and Chapter 10) in Paris reported the trapping of ultra-cold atoms on a superconducting atom chip and the following year Mukai et al. [51] in Tokyo demonstrated the trapping of atoms on a persistent supercurrent atom chip. In 2008 the Paris group [52] realized a BEC on a superconducting atom chip and the Tübingen group [53] demonstrated the Meissner effect using ultra-cold atoms trapped by a superconducting wire on an atom chip. The development of superconducting atom chips also opens the way to new fundamental quantum physics experiments such as the coupling of a BEC to superconducting loops [54].

Comprehensive review articles on magnetic mirrors, microtraps and atom chips for ultra-cold atoms and quantum degenerate gases have been published by Hinds and Hughes [55], Folman et al. [13], Reichel [14], and Fortágh and Zimmermann [15].

1.3
Magnetic Mirrors for Cold Atoms

1.3.1
Basic Principles

We first consider an atom with magnetic dipole moment $\boldsymbol{\mu}$ moving in an inhomogeneous magnetic field $\boldsymbol{B}(x, y, z)$. If the rate of change of the direction of $\boldsymbol{B}(x, y, z)$ as seen by the moving atom is slow compared with the atom's Larmor frequency, then the orientation of $\boldsymbol{\mu}$ can adiabatically follow the direction of the magnetic field and the position-dependent interaction potential $U_{\text{int}}(x, y, z) = -\boldsymbol{\mu} \cdot \boldsymbol{B}(x, y, z)$ exerts a gradient or Stern–Gerlach force on the atom given by $\boldsymbol{F}_{\text{grad}} = \nabla(\boldsymbol{\mu} \cdot \boldsymbol{B}) = -m_F g_F \mu_B \nabla B(x, y, z)$, where m_F is the magnetic quantum number, g_F the Landé factor, and μ_B the Bohr magneton. Thus, atoms oriented in positive (or low-field seeking) magnetic states ($m_F g_F > 0$) experience a negative force and are repelled by an increasing magnetic field, while atoms in negative (or high-field seeking) magnetic states ($m_F g_F < 0$) are attracted by an increasing magnetic field.

We now consider the inhomogeneous magnetic field produced by a periodic array of magnets having alternating perpendicular magnetization $M(y, z) = +M_z$ and $-M_z$ in the y direction with period a (Figure 1.1a). For an *infinite* array of long, thick ($t \gg a/2\pi$) magnets of width $a/2$, the components and the magnitude of the magnetic field above the array are given by [5]

$$B_y(y, z) = B_0 e^{-kz} \left(\sin ky - 1/3 e^{-2kz} \sin 3ky + \ldots \right),$$

$$B_z(y, z) = B_0 e^{-kz} \left(\cos ky - 1/3 e^{-2kz} \cos 3ky + \ldots \right),$$

$$|B(y, z)| = B_0 e^{-kz} \left(1 - 1/3 e^{-2kz} \cos 2ky + \ldots \right), \quad (1.1)$$

where $k = 2\pi/a$, and $B_0 = 8M_z$ (Gaussian units) is a characteristic surface magnetic field. For distances $z \gg a/4\pi$, Eq. (1.1) reduce to

$$\left[B_y(y, z);\ B_z(y, z) \right] \approx \left[B_0 e^{-kz} \sin ky;\ B_0 e^{-kz} \cos ky \right], \quad |B| \approx B_0 e^{-kz}. \quad (1.2)$$

Equation (1.2) indicates that for distances $z \gg a/4\pi$ the magnitude of the magnetic field is no longer dependent on y and falls off exponentially with distance z, with decay length $k^{-1} = a/2\pi$, while the direction of the magnetic field for a given distance z rotates in the yz-plane with period a as y varies (Figure 1.1a). Thus, an atom in a weak-field seeking magnetic state moving adiabatically towards the surface will be repelled by the exponentially increasing magnetic field, and if its magnetic potential energy exceeds the kinetic energy with which it approaches the surface the atom will be reflected. For a pure single-harmonic potential, Eq. (1.2) holds for all distances z above the array. For an array of magnets of finite thickness t, B_0 in Eq. (1.2) is replaced by $B_s = (1 - e^{-kt}) B_0$ [10, 56]. The distance of

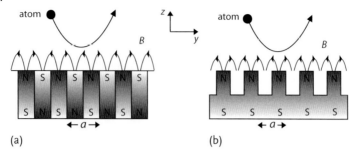

Figure 1.1 Schematic diagram of (a) a magnetic mirror consisting of perpendicularly magnetized elements of alternating polarity, (b) a grooved perpendicularly magnetized microstructure [10].

closest approach to the surface for an atom of mass M dropped from height h is $z_0 \approx (a/2\pi) \ln(m_F g_F \mu_B B_s / Mgh)$.

For small distances ($z < a/4\pi$) above the array, higher-order spatial harmonics in Eq. (1.1) contribute a corrugation, which have a relative amplitude given to good approximation by $C \approx 1/3(Mgh/m_F g_F \mu_B B_0)^2$ [56]. Such a corrugation leads to a diffusive component in the reflected atoms with an angular spread $\Delta\theta_{rms} \approx \sqrt{2}C$.

For a *finite* array of N magnets the magnetic field produced by the semi-infinite array of 'missing' magnets on the left (right) hand side of the mirror (Figure 1.1a) creates an effective magnetic field approximately the same as that of a magnet of width $a/4$ centered at $y = -Na/4$ ($+Na/4$) [5]. Thus, by placing two magnets, with width $a/4$ and opposite magnetizations, one at each end of the array, the magnetic field above a large area of a finite array will mimic the field above an infinite array.

Typical magnetic fields required to reflect cold atoms dropped from a height $h = 20$ mm range from 47 G for cesium to 2.5 G for lithium. The reflecting potential for a magnetic mirror with a characteristic surface field of $B_0 = 4.2$ kG (thick Nd-Fe-B magnets) is 24 µeV (280 mK).

A magnetic mirror may be turned into a spatial diffraction grating, or a diffractive beam splitter, for slowly moving atoms by applying a uniform bias field B_{1z} (or B_{1y}) [3, 57]. For distances $z \gg a/4\pi$ above the array, the components of the magnetic field are given by expressions similar to Eq. (1.2) but with the addition of B_{1z} to the right-hand side of the B_z component. The magnitude of the magnetic field is then given by

$$|B(y,z)| \approx \left(B_{1z}^2 + 2B_{1z} B_0 e^{-kz} \cos ky + B_0^2 e^{-2kz} \right)^{1/2}. \tag{1.3}$$

The second term in Eq. (1.3) is periodic in y, and may act as a spatial diffraction grating, while the third (mirror) term dies away rapidly compared with the grating term. This leads to the periodic corrugated pattern in Figure 1.11d.

A magnetic mirror may also be converted into a temporal diffraction grating, or an 'acousto-optic modulator', for cold atoms by vibrating the magnetic structure or applying an oscillating orthogonal magnetic field [58].

1.3.2
Experimental Realization of Magnetic Mirrors

1.3.2.1 Macroscopic Array of Rare-Earth Magnets of Alternating Polarity

To test the basic principles of the magnetic mirror, we first constructed a planar periodic array of 18 1-mm-wide Nd-Fe-B magnets of alternating polarity, with a 0.5-mm-wide magnet at each end to compensate for end-effects [5]. Hall probe measurements indicate that the B_y and B_z components above the array exhibit a sine and cosine dependence on y, while the magnitude of the magnetic field shows an exponential dependence on z, given by Eq. (1.2), with $a = 2.06$ mm and $B_0 = 4.2$ kG.

Time-of-flight (TOF) absorption measurements for laser-cooled cesium atoms optically pumped towards the $|F = 4, m_F = +4\rangle$ low field-seeking state and dropped from a height $h = 22$ mm onto the magnetic mirror exhibit two peaks (Figure 1.2). The first peak corresponds to atoms falling through the broad probe beam and the second peak, which was recorded with the probe beam turned on after the falling atoms had passed, corresponds to atoms reflected from the magnetic mirror. The relative intensities of the bounce and fall signals indicate a reflectivity close to 100 %. Similar measurements taken for atoms optically pumped towards the $|F = 4, m_F = -4\rangle$ high field-seeking state exhibit only weak reflection, confirming that the observed reflection is magnetic in origin. Studies of the spatial distribution of the incident and reflected atoms, by recording the transmission of a broad probe beam in a CCD array, indicate that the reflection is predominantly specular.

However, in order to produce a 'hard' magnetic mirror in which a cloud of atoms with a spread of velocities are all reflected from approximately the same plane, mag-

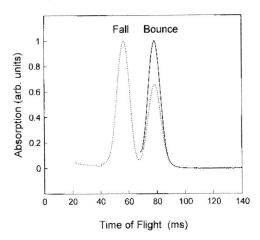

Figure 1.2 Time-of-flight absorption measurement showing the reflection of a beam of laser-cooled cesium atoms normally incident on a magnetic mirror constructed from a planar array of Nd-Fe-B magnets of alternating polarity. The probe beam was turned on before (dotted line) and after (solid line) reflection from the mirror [5].

netic structures with much smaller periods, preferably on the scale of a micron, are required.

1.3.2.2 Micro-Fabricated Grooved Magnetic Mirrors

The fabrication of an array of magnets of alternating polarity with periods down to 1 μm presents a considerable technical challenge. Model calculations (e.g., Figure 1.10c) show that a grooved perpendicularly magnetized structure (Figure 1.1b) produces a magnetic field distribution the same as that above an array of magnets of alternating polarity (Figure 1.1a), but with the characteristic surface field B_0 reduced by a factor of two [10]. Thus, the grooves behave as missing magnets of opposite polarity to that of the tiny magnets in the protrusions. Grooved magnetic structures may be readily fabricated with micron-scale periods and with excellent surface topology using micro-fabrication techniques such as electron-beam lithography, photolithography, or reactive ion etching.

We have found that by using perpendicularly magnetized films, such as $Co_{80}Cr_{20}$ or $Tb_6Gd_{10}Fe_{80}Co_4$, deposited on a grooved non-magnetic substrate, it is possible to fabricate high-quality magnetic structures with periods down to 1 μm [9, 10, 59]. Magnetic force microscopy (MFM) scans of the magnetic microstructures show that the variation of B_z with y is approximately sinusoidal, even at distances very close ($z \approx 0.05$ μm) to the surface. The sinusoidal shape at small distances is attributed to rounding of the top edges of the groove walls, which reduces the contribution of higher order spatial harmonics (Eq. (1.1)). The magnitude of the B_z component is found to decrease exponentially with distance z, with decay length $a/2\pi$. For grooved microstructures coated with ferromagnetic $Co_{80}Cr_{20}$ film [9, 10], the MFM scans show evidence of domain structure in the regions above the protrusions, with a size ~ 0.5 μm, while at distances greater than 0.5 μm the magnetic inhomogeneities have decayed away, consistent with a domain size of about 0.5 μm. Hysteresis loop measurements for the $Co_{80}Cr_{20}$ films exhibit a rhombohedral shape with a coercivity of about 1 kOe and a remanent magnetization of ~25 % of the saturation magnetization (typical of $Co_{80}Cr_{20}$), indicating that not all the domains remain oriented after the magnetizing process. The saturation magnetization for $Co_{80}Cr_{20}$ is ≈ 5 kG; so for grooved microstructures with $a = 1$ μm and $t = 0.2$ μm the characteristic surface field is estimated to be $B_s \approx 280$ G.

TOF absorption measurements for laser-cooled atoms dropped from a height $h = 18$ mm onto 15×15 mm^2 $Co_{80}Cr_{20}$ magnetic microstructures show strong reflection signals similar to Figure 1.2. For Cs $|F = 4, m_F = +4\rangle$ atoms incident on an $a = 1$ μm microstructure the distance of closest approach is $z_0 \approx 0.3$ μm. Measurements of the spatial distribution of the atom cloud at different times before and after reflection yield a value for the angular spread introduced by the reflection from the grooved mirror of $\Delta\theta_{\mathrm{rms}} = (16 \pm 20)$ mrad. To reduce the uncertainties further would require reducing the initial transverse velocity of the atoms, for example, by collimating the atom cloud or by reducing the temperature in the xy plane, or by extending the spatial distribution measurements out to much longer times after the first reflection.

The quality of the grooved magnetic microstructures could be improved, for example, by using *ferrimagnetic* $Tb_6Gd_{10}Fe_{80}Co_4$ films [59] instead of the $Co_{80}Cr_{20}$ film. MFM images taken for $Tb_6Gd_{10}Fe_{80}Co_4$ films show excellent magnetic homogeneity, with no evidence of domain structure. Such films have a magnetization $4\pi M_r \approx 2.8\,kG$, which for an $a = 1\,\mu m$, $t = 0.2\,\mu m$ magnetic microstructure gives a characteristic surface field $B_s \approx 640\,G$. Our calculations indicate that for the above parameters and using a small bias field to compensate end-effects and stray magnetic fields it should be possible to reduce corrugations so that the angular spread $\Delta\theta_{rms} < 1\,mrad$. This type of grooved magnetic microstructure appears to be the most promising approach to date for producing high-quality magnetic mirrors with micron-scale periods.

Although it is possible to fabricate grooved magnetic mirrors with periods smaller than 1 μm the incident atoms will then start to approach so close to the surface that van der Waals forces and the Casimir–Polder force become significant [13], (Chapters 4 and 5).

1.3.2.3 Micro-Fabricated Array of Current-Carrying Conductors

Magnetic mirrors based on periodic arrays of current-carrying conductors have a number of features. Each element in the series array carries the same current and hence produces the same magnetic flux, allowing the production of very flat mirrors, and the ability to control the current allows the magnetic field to be varied, switched, or modulated.

For an infinite array of *narrow* wire conductors alternately carrying equal currents I in opposite directions, the components and the magnitude of the magnetic field above the array are given by expressions similar to Eq. (1.1), but with a characteristic surface field $B'_0 = 8\pi I/a$ and a geometrical coefficient of unity instead of '1/3' in the second-order term [11]. For a finite array of wires, the magnetic field distribution above the array may approximate the infinite case by adding two compensating wires each carrying half the normal current to the ends of the array at half the normal spacing [11].

To demonstrate the principle of such a mirror we constructed an $a = 330\,\mu m$ planar array of 42 parallel gold conductors with two compensating wires on the ends (Figure 1.3) prepared on a silicon wafer by UV photolithography and electroplating techniques [11]. The gold conductors were 60 μm wide, 10 μm thick, and 8 mm long.

TOF absorption signals recorded for $|F = 4, m_F = +4\rangle$ cesium atoms dropped from a height $h = 20\,mm$ onto the current-carrying wire mirror when activated with 3 A, 12-ms current pulses to coincide with the arrival of the falling atoms indicate a reflectivity close to 100%. When the duration of the current pulse is reduced to much less than the spread of arrival times of the atoms, the TOF signal sharpens significantly due to 'velocity filtering' in the vertical direction. Studies of the dependence of the reflection signal on current through the wires show a reflection threshold at 0.45 A, followed by three steeply rising regions which are identified as representing the sequential reflection of atoms in the $m_F = +4$ (and $+3$), $+2$ and $+1$ magnetic states.

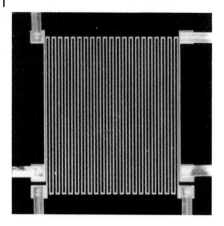

Figure 1.3 Photograph of the current-carrying wire mirror, consisting of a 330-μm-period array of 42 parallel gold wires and two compensating wires on the end [11].

Studies of the spatial distribution of the reflected atoms indicate that at the pulse current of 3 A, which corresponds to a characteristic surface field of $B'_0 = 230\,\text{G}$, and without activating the compensating end-wires the reflection is predominantly specular ($\Delta\theta_{\text{rms}} \approx 20\,\text{mrad}$), while at 1 A, corresponding to $B'_0 = 77\,\text{G}$, the reflection is much more diffuse ($\Delta\theta_{\text{rms}} \approx 120\,\text{mrad}$) as a result of atoms approaching close to the surface where corrugations due to higher order spatial harmonics become significant.

We estimate that it should be possible to reduce the period of the mirror by at least an order of magnitude, thereby producing a 'harder' mirror and also a flatter mirror by reducing corrugations due to end-effects. Model calculations for an $a = 30$-μm mirror with 10-μm square wires carrying 3 A and with no compensating end-wires indicate a relative corrugation $C \approx 0.003$ for $h = 20\,\text{mm}$, giving $\Delta\theta_{\text{rms}} \approx 4\,\text{mrad}$.

1.3.2.4 Magneto-Optical Recording of Magnetic Microstructures

Certain TbGdFeCo compositions, such as $\text{Tb}_{15}\text{Fe}_{79}\text{Co}_6$, exhibit a relatively low Curie temperature, allowing magnetic structures to be written with a focussed laser beam in the presence of an external magnetic field. Such magneto-optical recording offers the prospect of writing fine and also complex structures.

By mounting a pre-magnetized $\text{Tb}_{15}\text{Fe}_{79}\text{Co}_6$ film on an optical diffraction ruling engine and using a tightly focused diode laser beam in the presence of an external magnetic field oriented in the opposite direction to the film magnetization, magnetic microstructures have been written with linewidths down to about 1 μm and periods down to $a \approx 2\,\mu\text{m}$ [56]. The films are 0.18 μm thick and have a Curie temperature of about 220 °C, a room temperature coercivity of 1.1 kOe, a remanent (perpendicular) magnetization $4\pi M_r \approx 2.4\,\text{kG}$, and an estimated $B_0 \approx 100\text{--}200\,\text{G}$ (for $a = 2\,\mu\text{m}$). MFM images taken close to the surface of the ruled structure indicate that the magnetization of the ruled lines is quite uniform; however the unruled regions exhibit striped domain patterns due to some break-up of the pre-magnetized regions during the recording process [56]. At larger distances (≈ 0.5 μm) from the surface the magnetic field gradient due to the striped domain patterns is found to have decayed away and the MFM scans become approximately sinusoidal in the y direction, with an exponential fall off with distance z, with decay length $a/2\pi$.

TOF absorption measurements of laser-cooled cesium atoms dropped from a height $h = 25$ mm onto recorded magnetic structures exhibit strong reflection signals, and spatial distribution measurements of the reflected atoms indicate that the reflection is predominantly specular. It should be possible to improve the specularity by further optimization of the magneto-optical recording process and composition of the alloy.

1.4
The Magnetic Film Atom Chip

1.4.1
Background

Advances in photolithography and micro-fabrication techniques have led to the development of miniature surface-based current-carrying elements for the production of BECs and the precise manipulation of ultra-cold atoms, allowing the construction and integration of networks of microtraps, waveguides, and interferometers on the surface of a substrate [13–15, 55]. Miniaturization and scaling down the dimensions of atom traps allow the use of moderate electric currents in order to produce large magnetic field gradients and curvatures and hence a very tight confinement of atomic waves. These microstructures greatly increase elastic collision rates and allow condensates to be produced in just a few seconds of evaporative cooling.

Permanent magnetic materials may also be used to produce magnetic field configurations that can trap, guide, and precisely manipulate neutral atoms [10, 25, 60]. In a ferromagnetic material the magnetic domains generate magnetic fields which are equivalent to the fields produced by effective surface current loops flowing perpendicular to the magnetization vector around the domain edges. In a uniformly magnetized film with perpendicular anisotropy all the domains are aligned in the same direction and within the bulk the currents of neighboring domains cancel. At the same time a net effective current flows around the perimeter of the film with a magnitude given by the product of the magnetization and the film thickness ($I_{\text{eff}} = M_r \times t$). Using established micro-fabrication techniques such as photolithography, vapor deposition, or laser ablation, arbitrary shapes of magnetic film can be fabricated. As examples, a straight edge of a magnet in combination with a uniform external magnetic field will generate a 2D quadrupole potential [10] and a Z-shaped edge of a perpendicularly magnetized film will produce an IP magnetic trap [61]. The use of integrated materials (magnetic films, microwires, and electric field elements [62]) provides flexibility and extends the capability of integrated atom optics devices.

Various technical limitations can affect applications of current microcircuits on atom chips [13–15]. Large current densities are required to produce large magnetic field gradients and this can lead to excessive heat generation, microwires pealing off, and broken circuits. Current noise from power supplies increases heating rates and limits the condensate lifetime. Johnson thermal noise radiated by a micro-wire conductor leads to spin-flips which affect the BEC lifetime [63]. Permanent magnetic microstructures can replace current microcircuits on atom chips and overcome many of these problems.

Typical permanent magnetic films have a relatively large resistance and a thickness less than 1 µm which suppress radiated thermal magnetic noise. It is expected that permanent magnetic films will produce ultra-stable magnetic potentials and low heating rates. Magnetic films with perpendicular anisotropy, which are used in new-generation hard disk media, provide additional stability of the magnetic field and allow finer magnetic microstructures.

1.4.2
BEC on a Magnetic Film Atom Chip

Permanent magnetic microstructures can produce quadrupole magnetic potentials and with the use of external bias fields can allow adjustment of the position and tightness of the trap. However, the capture volume of the traps is small and the number of cold atoms trapped is not sufficient for efficient evaporative cooling. In order to improve the efficiency of atom capture we combine the permanent magnetic film with current-carrying wires using a two-layer structure (Figure 1.4) [43].

An H-shape plus two end-wire structure was machined in a 0.5-mm-thick silver plate and glued to a ceramic plate to form the bottom layer. The wires have a width of 1 mm and can carry a continuous current of 30 A with an associated temperature rise of no more than 40 °C. The H-shape structure can be arranged for both U and Z-shape currents for quadrupole and IP potentials, respectively.

The top layer of the chip is formed by two 300-µm-thick glass slides (40×23 mm^2) glued to the silver structure. High magnetic field gradients require a large effective surface current of the magnetic film and hence large values of the magnetization M_r and/or thickness t. We choose to use ferrimagnetic $Tb_6Gd_{10}Fe_{80}Co_4$ films which exhibit strong perpendicular magnetic anisotropy, an almost square hysteresis loop with a remanent magnetization $4\pi M_r = 2.8$ kG, a coercivity $4\pi H_c = 4$ kOe, and a Curie temperature $T_C \approx 300$ °C. The composition of the alloy is chosen to be a compromise between the desire to have large magnetization, high coercivity, and high Curie temperature. One of the slides is coated with a 0.9-µm-thick $Tb_6Gd_{10}Fe_{80}Co_4$ film (effective current 0.2 A). A long polished edge (Figure 1.4a) together with a bias field generates a quadrupole radial potential. Parallel currents in the two end-wires (separation 9 mm) provide axial confinement and conversion to a 3D IP trap. Both slides are finally coated with a 100-nm-thick gold film.

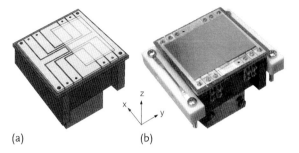

Figure 1.4 Schematic diagram (a) and photo (b) of the magnetic film atom chip [42].

We use a four-beam mirror magneto-optical trap (MMOT) [20] with a quadrupole magnetic field (10 G/cm) provided by two coils located outside the UHV chamber. Two of the laser beams are incident at 45° and are reflected by the gold coating of the chip. They are complemented by two counter-propagating horizontal beams. The volume of the trap is determined by the 35 mm diameter of the laser beams, and the center of the MMOT is located ∼ 5 mm below the chip surface. The trapping laser is detuned 18 MHz below the $F = 2 \rightarrow F = 3$ cycling transition in ^{87}Rb and has an intensity of 8 mW/cm^2. The repumping laser is combined with the four trapping beams and tuned to the $F = 1 \rightarrow F = 2$ transition.

The experimental cycle typically lasts 40 s and begins with the application of 6.5 A current through a dispenser (Figure 1.4b) producing ^{87}Rb atoms. About 5×10^8 atoms are collected in the MMOT in 10 s. We turn off the current through the dispenser and allow another 25 s for the UHV vacuum to recover (to typically 2×10^{-11} Torr) before transferring the trapped atoms to the chip-based potentials. The current through the external quadrupole coils is ramped down, the current through the U-wire is ramped up to 8 A and the bias magnetic field raised to 7.3 G. This procedure loads the rubidium atoms into the U-wire MMOT centered at 1.6 mm from the chip. A 2-ms far-detuned (−56 MHz) cooling stage allows the temperature of the atoms to be reduced from 140 to 40 µK. The trapping light and the U-wire current are switched off and a 200-ms σ^+-polarized pulse tuned to the transition $F = 2 \rightarrow F = 2$ optically pumps the atoms into the weak field-seeking state $|F = 2, m_F = 2\rangle$ in the presence of a bias field. A total of 4×10^7 atoms are loaded into the mode-matched IP trap when a current of 21.5 A is applied to the Z-shaped wire and the bias magnetic field is simultaneously increased to 13 G. The magnetic trap is further compressed by ramping the Z-wire current up to 31 A and the bias field up to 40 G. The compressed trap is located 550 µm from the film surface; the radial and axial trap frequencies are 530 and 18 Hz, respectively, and the estimated elastic collision rate is 160/s.

Evaporative cooling to the BEC transition proceeds in two stages. Firstly, RF current is applied to the two end-wires (Figure 1.4a) and its frequency is logarithmically ramped down in 9 s to 1.15 MHz. The number of trapped atoms decreases to 2×10^6 while the temperature is reduced to around 3 µK. The atoms are then loaded into the magnetic film trap by removing the current in the Z-wire and increasing the bias field to 47 G. Radial confinement of the trap is provided by the edge of

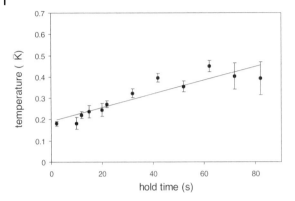

Figure 1.5 Measurements of the heating rate in a magnetic film microtrap located 210 μm from the surface [64].

the permanent magnetic field of the film and axial confinement is provided by two end-currents generated by an ultra-stable power supply. The frequency of a second RF field is ramped for 2 s from 1.5 down to 0.73 MHz. At the end of the evaporative cooling a BEC of typically 10^5 atoms is formed 100 μm from the surface. In order to minimize the effect of the permanent magnetic field from the film in a TOF expansion of the cloud we move the atoms to 1 mm from the film before releasing them.

The heating rate of an ultra-cold atomic cloud captured in the magnetic film trap 210 μm from the film (radial frequency 120 Hz) was measured by TOF temperature measurements after various hold times (Figure 1.5). The observed heating rate of 3.2 nK/s [64] is remarkably low, and is to be compared with the heating rate of atoms in the Z-wire trap of 270 nK/s. The low heating rate is a promising feature of the magnetic film atom chip.

1.4.3
Spatially Resolved RF Spectroscopy to Probe Magnetic Film Topology

Studies with atom chips based on current-carrying conductors have revealed that ultra-cold atomic clouds can exhibit fragmentation along the axial direction, which is related to the roughness of the wire surface and edges and to internal granular structure of the conductor [65]. This roughness leads to small deviations in the current path and correspondingly to a spatially varying axial magnetic field component which corrugates the trapping potential.

We have studied quantitatively the fragmentation of ultra-cold ^{87}Rb atoms on the magnetic film atom chip using a combination of spatially resolved imaging and RF spectroscopy of the atoms [66]. An atomic cloud was evaporatively cooled in the Z-wire IP trap down to 5 μK, loaded into the magnetic film trap at four different distances $z_0 = 57, 67, 87$, and 115 μm from the film and allowed to expand axially 5 mm by decreasing the end-currents to 0.5 A. The narrow energy distribution of the ultra-cold trapped atoms is greatly affected by small variations of the trapping

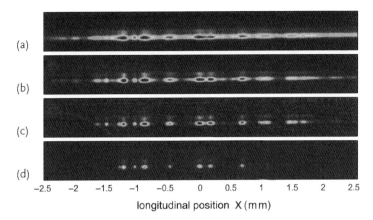

Figure 1.6 Absorption images of the atomic density in the elongated magnetic film trap located 67 μm below the film at different values of the cut-off frequency ν_f: (a) 1238 kHz, (b) 890 kHz, (c) 766 kHz, (d) 695 kHz [66].

potential, and the spatial distribution of the atoms reveals small corrugations in the potential [67]. A RF field is applied in a direction perpendicular to the quantization magnetic field and its frequency is swept from 2 MHz down to a cut-off frequency ν_f ranging from 1.4 to 0.5 MHz. The atoms in the high energy tail of the distribution are resonant with the RF field and are outcoupled to untrapped states leaving a truncated Boltzmann distribution. The bias field is turned off after the sweep, allowing the atoms to accelerate away from the film. The longitudinal distribution is unperturbed during the 1 ms expansion and is optically imaged with a spatial resolution of 5 μm. Variations in the axial extension of the cloud become noticeable when ν_f is reduced below 1.3 MHz (Figure 1.6a). Reducing the cut-off frequency further results in well-separated clumps of atoms which accumulate in the lowest potential wells (Figure 1.6d).

For a quantitative analysis of the experimental data we use an expression for the atomic density of a truncated Boltzmann distribution [66, 68] integrated over coordinates y and z

$$n(x, \beta) = n_0(x) \left[\mathrm{erf}\left(\sqrt{\beta}\right) - 2\sqrt{\beta/\pi}\, e^{-\beta}(1 + 2\beta/3) \right], \quad (1.4)$$

where the truncation parameter β is a function of the coordinate x, the cut-off frequency ν_f, the temperature T, and the value of the magnetic field $|B_x(x) + B_0|$ at the bottom of the trap:

$$\beta(x, \nu_f) = \left[m_F h \nu_f - m_F g_F \mu_B |B_x(x) + B_0| \right] / k_B T . \quad (1.5)$$

For each distance z_0 from the film a set of RF spectroscopy measurements is taken and the local value of the magnetic field at the coordinate x is extracted. The rms amplitude increases with distance z_0 (measured in μm) according to the approximate power-law (Figure 1.7)

$$B_{x,\mathrm{rms}} = 1.32 \times 10^4 \times z_0^{-1.85} \quad [\mu\mathrm{T}]. \quad (1.6)$$

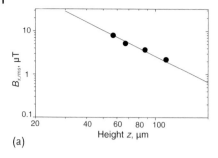 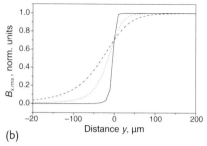

Figure 1.7 (a) Dependence of the magnetic-field roughness $B_{x,\mathrm{rms}}$ on height z. The solid line was obtained using Eq. (1.5). (b) Dependence of the normalized field roughness predicted by the model (Eq. (1.7)) on the transverse coordinate y for heights 10 μm (solid), 50 μm (dotted), and 100 μm (dashed line) [66].

Stylus profilometer measurements indicate that the rms roughness of the polished edge of the slide is around 10 nm over a 400-μm scale. The associated deviations of the effective current path cannot explain (i) the large rms values of the corrugated axial magnetic field and (ii) the observed $z_0^{-1.85}$ dependence. When the cloud is positioned $z_0 = 50$ μm below the magnetic surface substantial fragmentation is observed. However, when it is positioned 50 μm below the non-magnetic gold surface the density distribution exhibits a rather smooth profile. These observations indicate that the random component of the magnetic field originates from the bulk of the magnetic film as a result of inhomogeneous magnetization and not from variations of the film edge.

We have developed a model [66] to study how the variations in the film magnetization generate a random magnetic field. For white-noise fluctuations in the magnetization the rms amplitude of the magnetic noise is

$$B_{x,\mathrm{rms}} = \sqrt{\frac{3}{\pi} \frac{\mu_0 t d_0 \Delta M}{16 z^2}} \sqrt{1 + \frac{15}{8}\alpha - \frac{5}{4}\alpha^3 + \frac{3}{8}\alpha^5}, \qquad (1.7)$$

where $\alpha = y/\sqrt{y^2 + z^2}$, t is the film thickness, d_0 is the characteristic feature size of the domains, and ΔM is the rms value of the magnetization variations. Below the edge of the film ($y = 0$) the model predicts that the random magnetic field decays with a z_0^{-2} dependence which is consistent with the observed $z_0^{-1.85 \pm 0.3}$ dependence. This is slower than the $z_0^{-2.5}$ decay of a corrugated field below current-carrying microwires [69, 70]. The model predicts an even faster decay of the random component for $y < 0$ (Figure 1.7b). Using the experimental rms values and a characteristic domain size $d_0 \approx 5$ μm we estimate the level of inhomogeneity $\Delta M/M_s \simeq 0.3$, where M_s is the saturation magnetization of the TbGdFeCo film. If we assume the inhomogeneity originates from reversal of a small number of domains and assume a binomial distribution of the reversal then the mean magnetization is about $0.9 M_s$.

The observed magnetic inhomogeneity in the TbGdFeCo films is attributed to some deterioration during vacuum bake-out (140 °C for 4 days) of the magnetic film atom chip. It was found that such deterioration can be avoided by maintaining the temperature below 100 °C.

1.4.4
Adiabatic Splitting of a BEC for Asymmetric Potential Sensing

The idea of using adiabatic splitting of a BEC for sensing asymmetric potentials is based on the assumption that during an adiabatic splitting process the fractional number of atoms in each well of a double-well potential is proportional to the potential difference between the bottoms of the wells as long as the chemical potential is larger than the barrier height [71]. The situation is similar to two coupled pools of superfluids which have the same level of fluid and the deeper pool will contain more fluid than the shallower pool.

Studies of the magnetic field topology in Section 1.4.3 revealed a double-well structure near the center of the film ($x \sim 0$, Figure 1.6c,d) with very similar characteristics for both wells. One of the features of the random nature of the corrugated magnetic field is that the Fourier components decay exponentially with height and the short-wavelength components decay much faster than the long-wavelength components. As a result the barrier between the two wells decreases substantially when the atomic cloud is moved away from the film and the double-well structure merges into a single well.

Using various values of the bias field an ultra-cold cloud is positioned at different heights, and the 1D spatial distribution of the atoms is probed by optical imaging. The cloud temperature is measured by TOF expansion. The spatial dependence of the trapping potential is determined from the Boltzmann distribution law $V(x) = -k_B T \ln(n(x))$ and the double-well parameters (separation and barrier height) are extracted by fitting the measured potential to a model potential [64]. For a bias field of 2.6 G the coalescence point is located 175 μm from the film. Increased values of the bias field move the trap closer to the surface and split it into a double well in the longitudinal (x) direction. The measurements show that the barrier height can be adjusted from sub-kHz up to 100 kHz and the well separation can be up to 180 μm.

Accurate knowledge of the double-well potential was used for precise control of the splitting of the BEC. The splitting process and the associated distribution of atoms between the two wells are extremely sensitive to the presence of any asymmetry component Δ between the bottoms of the two wells [71]. In such a case the number of atoms in the deeper well is greater than in the shallower well and is determined by the trap frequencies of both wells (by sheer accident these parameters defined by the corrugated potential are almost identical) as well as external magnetic fields or gravity gradients. An asymmetric potential can be either compensated or intentionally introduced by applying a current imbalance between the end-wires on the chip (Figure 1.4a). A calibration of the applied potential asymmetry was carried out using RF spectroscopy [71]. Two BECs separated by 170 μm are initially produced in a symmetric double-well potential with a barrier height of several kHz. A large current imbalance of 1 A is introduced in the end-wires for a duration of 500 ms, thereby tilting the wells. A 1-ms RF pulse of variable frequency

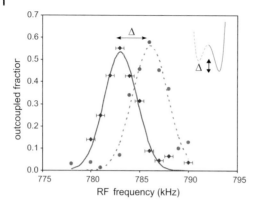

Figure 1.8 Outcoupling of atoms from two condensates in an asymmetric double well as a function of RF frequency [71].

ν_f is applied to resonantly outcouple those atoms from either well that are resonant with the RF field ($h\nu_f = \mu_B g_F B$). Optical imaging of the remaining atoms provides a measure of the number of outcoupled atoms as a function of ν_F (Figure 1.8). The resulting spectra yield an applied magnetic potential gradient of 32 Hz/μm for the current imbalance of 1 A.

A small condensate of 46 000 atoms is produced in a single well 175 μm from the film. A calibrated asymmetric potential is then applied via a current imbalance in the end-wires that is variable between ±120 mA. The bias magnetic field is slowly increased (over 2 s) between 2.5 and 2.7 G, splitting the condensate axially by 70 μm in the double well located 155 μm from the film. At this position the barrier height is about 25 % of the chemical potential. At the end of the adiabatic splitting the bias field is quickly increased (over 10 ms) to ramp the barrier above the chemical potential and to preserve the fractional distribution of the atoms. Absorption imaging determines the fractional number of atoms for various values of the applied calibrated asymmetry (Figure 1.9). The shot-to-shot noise level of 2.8 % pro-

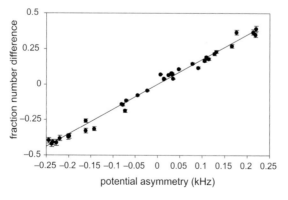

Figure 1.9 Measurements of the fractional number difference (circles) in the two wells as a function of the applied potential asymmetry. Solid line represents a linear fit [71].

vides a single-shot sensitivity of 17 Hz for the potential well energy imbalance for a 70 μm separation. This method of measuring the fractional distribution of atoms is suitable for the construction of sensors of gravitational field gradients. The demonstrated single-shot sensitivity to analogous changes in gravity is $\delta g/g = 2 \times 10^{-4}$ for a 70 μm separation [71] which corresponds to an 8-nm height difference between the two wells. At present the demonstrated sensitivity to potential gradients is well below the sensitivity of free-space atom interferometers [72] but surpasses the sensitivity of on-chip BEC interferometers which are currently limited by relatively short decoherence times.

1.4.5
Spatially Inhomogeneous Phase Evolution of a Two-Component BEC

Accurate knowledge of the phase of matter waves is an important factor in studying BEC coherence, and in potential future applications of BEC interferometers. We have studied the relative phase evolution of an elongated two-component BEC composed of the $|F = 1, m_F = -1\rangle$ and $|F = 2, m_F = 1\rangle$ hyperfine ground states of ^{87}Rb atoms [73], magnetically trapped on an atom chip [43]. These two states are particularly attractive for trapped-atom interferometry because the first-order Zeeman shift between them is canceled at a magnetic field of 3.23 G. To improve the relative phase detection we developed an imaging technique to simultaneously measure the column density of each hyperfine state during TOF expansion [73]. Our dual-state detection produces sub-percent uncertainty in the measured relative population while preserving the spatial mode of each state.

We prepare a condensate in a superposition of the two hyperfine states and interrogate the relative phase evolution using two-photon microwave-radiofrequency Ramsey interferometry. The inter- and intra-state scattering lengths have slightly different values and, as a result, the prepared superposition is in a non-equilibrium state [74], leading to undamped collective oscillations and complex spatial modes of each component. We focus [73] on the dynamics of the relative phase along the direction of weak (axial) confinement. The relative phase begins (after the first π/2 pulse) spatially uniform, then becomes inhomogeneous along the axial coordinate, varying by 1.9π across the condensate after 90 ms of evolution.

The second π/2 pulse converts spatial variations of the relative phase into spatial variations of the atomic density of each state. The time dependence of the relative phase is accompanied by interference in space and, as a result, spatial fringes appear in the absorption images (Figure 1.10) and change with the evolution time. The spatial dependence of the relative phase leads to inhomogeneous dephasing of the condensate wavefunction along the axial direction, and we observe a relatively fast decay of the Ramsey interference fringes (decay time ∼70 ms). We emphasize that, under our conditions, inhomogeneity of the relative phase of the condensate wavefunction is the dominant mechanism for the loss of the Ramsey fringes, rather than decoherence due to coupling to the environment or quantum phase diffusion.

Our observations of the spatial evolution of the relative phase and the inhomogeneous dephasing of the condensate wavefunction are well described by a mean-

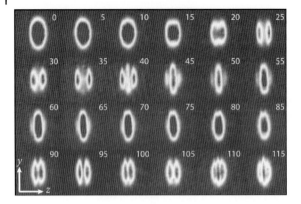

Figure 1.10 Single shot absorption images of the condensate in the state $|F=2, m_F=+1\rangle$ taken after 20 ms of free fall and expansion with varying Ramsey interferometry time (in ms). Application of the second $\pi/2$ pulse locally converts the relative phase into atom number density and leads to the appearance of spatial fringes along the z (axial) coordinate [73].

field formalism and supported by numerical solutions [73] of the coupled Gross–Pitaevskii equations for both components with decay terms corresponding to inter- and intra-state many-body loss mechanisms.

1.4.6
BEC on Other Permanent-Magnet Atom Chips

A number of permanent-magnet atom chips constructed from different types of magnetic media have been used for the production of BECs. The group at Imperial College, London [41] have used 3.5-µm-thick videotape with a sinusoidal pattern of in-plane magnetization. The pattern has a period of 106 µm and provides a surface magnetic field of 110 G, and in combination with a bias field of 1 G forms an array of waveguides 80 µm from the surface. ^{87}Rb atoms are initially trapped in the MMOT, loaded into an H-shape IP magnetic trap, further cooled via RF evaporation, and then loaded into the videotape microtrap with trap frequencies of 450 Hz (radial) and 15 Hz (axial) at a temperature of 10 µK. Further evaporative cooling in the permanent-magnet microtrap produces a condensate with about 8×10^4 rubidium atoms which is detected via anisotropic TOF expansion in free-flight or in the appearance of a bimodal distribution in a 1D waveguide expansion. Studies of axial center-of-mass motion and a breathing mode revealed decaying oscillations with a damping time of 180 ms which was associated with a slight roughness in the magnetic potential. The material of the videotape exhibits insulating properties and should lead to a strong suppression of spin-flips due to thermal Johnson noise. Measurements of the thermal cloud lifetime demonstrated long lifetimes ranging from 15 to 35 s at distances of 15 and 40 µm below the surface.

The group at MIT [44] have used a prototype hard-disk medium consisting of a 23-nm-thick Co-Cr-Pt-based oxide layer with perpendicular anisotropy and a 200-

nm-thick magnetically soft underlayer with a total magnetization of 470 emu/cm^3. A pattern with alternating magnetization of period 100 μm was written on the magnetic substrate. Ultra-cold atoms were loaded into a single microtrap in which axial confinement was accomplished by a Z-shape wire located below the disk. RF evaporation led to the production of 50 000 condensed atoms in the permanent magnetic microtrap with trap frequencies of 390 Hz (radial) and 9 Hz (axial). Optical imaging revealed a bi-modal distribution with a condensate fraction >80 %. A lifetime of ~30 s was reported for a height 40 μm above the surface. Tight radial confinement with a frequency of 16 kHz was demonstrated with a smaller period (10 μm) of the written structure. During 1D expansion of the BEC in a waveguide a break-up of the atomic cloud was detected at distances closer than 40 μm. This break-up is associated with perturbations of the axial magnetic confinement due to variations in the magnetic moments of domains and edge-effects during writing of the periodic magnetic structure.

The group in Amsterdam [45] have constructed a permanent-magnet atom chip consisting of in-plane magnetized 40-μm-thick Fe-Pt foil with a machined F-shaped self-biased IP trap with frequencies of 11 kHz (axial) and 30 Hz (radial) at 180 μm from the surface. The permanent magnetic trap was loaded by transporting atoms from a quadrupole magnetic trap by a combination of external magnetic fields. RF spectroscopy was employed to analyze the cloud temperature and the atomic distribution in the trap and to observe the onset of Bose condensation *in situ* without expansion from the trap.

1.5
Permanent Magnetic Lattice on a Magnetic Film Atom Chip

1.5.1
Background

Optical lattices have been used extensively to manipulate and control periodic arrays of ultra-cold atoms and quantum degenerate gases [75], for example, in studies of low-dimensional quantum gases [76, 77], quantum tunneling experiments such as the superfluid to Mott insulator quantum phase transition [78] and experiments on BECs in disordered potentials such as Anderson localization [79, 80]. Optical lattices also have potential application in quantum information processing since they can provide storage registers for qubits based on neutral atoms [81, 82].

An alternative approach for producing periodic lattices for ultra-cold atoms is based on periodic arrays of permanent magnetic microstructures on a magnetic film atom chip [49, 83, 84] using the magnetic micro-structure technology developed for magnetic mirrors [9, 10]. Simple 1D magnetic lattice structures consisting of arrays of 2D traps have been proposed [55, 83] and constructed using current-carrying conductors [85] and video tape [86], though loading of atoms into multiple lattice sites was not demonstrated. 2D lattices of magnetic microtraps with non-zero potential minima have been proposed based on crossed arrays of current-carrying conductors plus bias fields [87, 88] and on crossed arrays of perpendicularly magnetized grooved films plus bias fields [83]. Other geometries for constructing 2D magnetic lattices have also been proposed, including a single layer of periodic arrays of square-shaped magnets with three different thicknesses [83, 84] and periodic arrays of patterned asymmetric Z-shaped structures [46].

Permanent magnetic lattices have some distinctive characteristics. They have highly stable, reproducible potentials, low technical noise, and extremely low heating rates. They can have large barrier heights and large trap curvature, leading to high trap frequencies. They can be constructed with a wide range of periods, down to about 1 μm, and can have tailored complex potentials. They can be readily scaled to have a large number of lattice sites, for example, a 1-μm-period $10 \times 10\,\text{mm}^2$ 2D lattice has 10^8 sites, which could be very useful as a storage register for quantum information processing. They are static and not dynamic; however they can be used in conjunction with current-carrying wire traps to load atoms into the lattice. The atoms need to be prepared in low-field seeking states in order to be trapped, allowing RF evaporative cooling *in situ* and the use of RF spectroscopy. Finally, permanent magnetic lattices are suitable for integration onto an atom chip and for devices. Thus, magnetic lattices may be considered to be complementary to optical lattices, in much the same way as magnetic traps and optical dipole traps are complementary.

1.5.2
Basic Principles

1.5.2.1 One-Dimensional Magnetic Lattice

We first consider the simple case of a 1D magnetic lattice, similar to the periodic grooved magnetic mirror described in Section 1.3 but with the addition of bias fields B_{1x} and B_{1y} (Figure 1.11). For an infinite 1D magnetic array with period a, film thickness t and perpendicular magnetization, the components of the magnetic field for distances $z \gg a/4\pi$ above the array are given by [83]

$$[B_x; B_y; B_z] = \left[B_{1x}; B_{0y}e^{-kz}\sin ky + B_{1y}; B_{0y}e^{-kz}\cos ky\right], \quad (1.8)$$

where $k = 2\pi/a$, $B_{0y} = B_0(1 - e^{-kt})e^{kt}$ and $B_0 = 4M_z$ (Gaussian units).

The potential minimum (trap bottom), trap height, barrier heights, and trap frequencies (for a harmonic potential) are given, respectively, by [83]

$$B_{\min} = |B_{1x}|, \quad z_{\min} = \frac{a}{2\pi}\ln\left(\frac{B_{0y}}{B_{1y}}\right), \quad \Delta B^y = \left(B_{1x}^2 + 4B_{1y}^2\right)^{1/2} - |B_{1x}|,$$

$$\Delta B^z = \left(B_{1x}^2 + B_{1y}^2\right)^{1/2} - |B_{1x}|, \quad \omega_y = \omega_z = \frac{2\pi}{a}\left(\frac{m_F g_F \mu_B}{M|B_{1x}|}\right)^{1/2}|B_{1y}|.$$

$$(1.9)$$

All of these quantities may be controlled by the bias fields B_{1x} and B_{1y}.

Figure 1.11c–e shows numerically calculated contour plots of the magnetic fields for a finite 1D magnetic lattice and the parameters given in the caption. With no

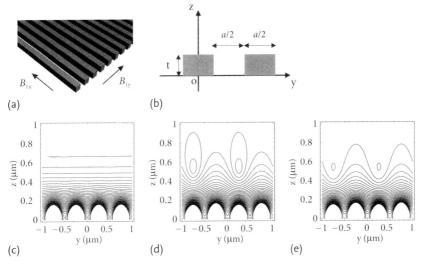

Figure 1.11 (a, b) 1D periodic array of parallel magnets with perpendicular magnetization. (c–e) Numerically calculated contour plots of the magnetic field in the central region in the y0z plane with (c) no bias fields, (d) $B_{1y} = -15\,\text{G}$, and (e) $B_{1x} = -20\,\text{G}$, $B_{1y} = -15\,\text{G}$. 1001 magnets, $a = 1\,\mu\text{m}$, $t = 0.05\,\mu\text{m}$, $l_x = 1000.5\,\mu\text{m}$, and $4\pi M_z = 3.8\,\text{kG}$ [83].

bias fields (Figure 1.11c), the magnitude of the magnetic field falls off exponentially with distance z, representing the simple case of a magnetic mirror [9, 10] as described in Section 1.3.1. With the addition of $B_{1y} = -15\,\text{G}$ (Figure 1.11d), the magnetic field develops 2D magnetic traps with zero potential minima; this configuration can give rise to spin-flips and is not suitable as a lattice for trapping atoms. With the addition of $B_{1x} = -20\,\text{G}$ and $B_{1y} = -15\,\text{G}$ (Figure 1.11e), the magnetic field has 2D magnetic traps with non-zero potential minima, where the trap bottom is given by $|B_{1x}|$.

1.5.2.2 Two-Dimensional Magnetic Lattice

We now consider a 2D magnetic lattice produced by two crossed periodic arrays of parallel magnets, thicknesses t_1 and t_2, separated by a small distance $s\ (< a/2\pi)$, and with bias fields B_{1x}, B_{1y} (Figure 1.12a) [83]. For an infinite *symmetrical* 2D magnetic lattice, the trap minimum, trap height, barrier heights, and trap frequencies (for a harmonic potential) are given, respectively, by

$$B_{\min} = c_1|B_{1x}|, \quad z_{\min} = \frac{a}{2\pi}\ln\left(\frac{c_2 B_{0x}}{B_{1x}}\right), \quad \Delta B^x = \Delta B^y = c_4|B_{1x}|,$$

$$\Delta B^z = c_5|B_{1x}|, \quad \omega_x = \omega_y = \frac{\omega_z}{\sqrt{2}} = \frac{2\pi}{a}\left(\frac{m_F g_F \mu_B c_3}{M|B_{1x}|}\right)^{1/2}|B_{1x}|,$$

(1.10)

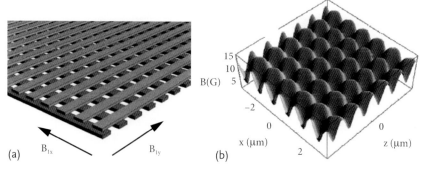

Figure 1.12 (a) 2D periodic array consisting of two crossed layers of parallel magnets with perpendicular magnetization. (b) 3D plot of the magnetic field. 2 × 1001 magnets, $a = 1\,\mu m$, $t_1 = 0.322\,\mu m$, $t_2 = 0.083\,\mu m$, $l_x = l_y = 1000.5\,\mu m$, $4\pi M_z = 3.8\,kG$, $B_{1x} = -4.08\,G$, $B_{1y} = -6.05\,G$, $B_{1z} = -0.69\,G$, $\nu_x = \nu_y = 232\,kHz$, $\nu_z = 329\,kHz$ [83].

with the constraint $B_{1y} = c_0 B_{1x}$ for a symmetrical lattice. In Eq. (1.10), $B_{0x} = B_0(1 - e^{-kt_2})e^{k(s+t_1+t_2)}$ and the c_i's are dimensionless constants involving only the geometrical parameters a, s, t_1 and t_2 of the magnetic arrays, given in [83]. All of the quantities in Eq. (1.10) may be controlled by the bias field B_{1x}. Figure 1.12b shows the numerically calculated magnetic field for a finite 2D magnetic lattice and the parameters given in the caption.

1.5.2.3 Permanent 1D Magnet Lattice for Ultra-Cold Atoms

We have constructed a 1D magnetic lattice with period $a = 10\,\mu m$ and dimensions $10 \times 10\,mm^2$ of perpendicularly magnetized $Tb_6Gd_{10}Fe_{80}Co_4$ magneto-optical film deposited on a grooved silicon wafer which was microfabricated by reactive ion etching [49]. The magnetic microstructure is coated with a reflecting gold film for use in a MMOT and mounted on an atom chip, similar to that described in Section 1.4. Beneath the atom chip is a silver plate which was machined to provide a U-wire for a quadrupole trap and 5-mm-long and 30-mm-long Z-wires for IP traps, which are required for loading the permanent magnetic lattice. The atom chip is mounted in the UHV chamber with the permanent magnetic microstructure facing down.

About 10^8 ^{87}Rb atoms are cooled and trapped in the MMOT and transferred to a compressed MOT produced by the U-current plus a bias field $B_{1x} = 7\,G$. The atoms are further cooled through polarization gradient cooling and optically pumped to the $|F = 2, m_F = +2\rangle$ low-field seeking state. The 5 mm-long Z-wire trap is then turned on, the wire current slowly ramped up to 35 A, and the bias field is simultaneously ramped up to $B_{1x} = 40\,G$. In this way about 5×10^7 atoms at a temperature of about 200 μK are transferred to the Z-wire trap. After waiting for the trapped atoms to thermalize, the atoms are RF evaporatively cooled to about 0.4 μK to create a BEC in the Z-wire trap about 250 μm below the chip surface where the magnetic field from the permanent magnetic microstructure is negligible. The y-bias field is then ramped up to $B_{1y} = 30\,G$ with $B_{1x} = 15\,G$ (trap bottom) to create a 1D magnetic lattice of 2D traps which overlaps the atom cloud. The barrier height

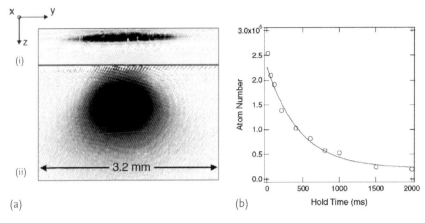

Figure 1.13 (a) Absorption image of atoms released from the magnetic lattice after (i) 4 ms and (ii) 16 ms time of flight. The broken structure in (i) is due to non-uniform reflection of the imaging beam from the grooved structure; (b) Number of atoms remaining in the magnetic lattice as a function of hold time. The solid line represents an exponential fit to the data points with a decay time of 0.45 s [49].

is typically 1.2 mK. About 45 % of the atoms are then transferred adiabatically from the Z-wire trap into the magnetic lattice.

After loading about 5×10^5 ^{87}Rb atoms at 15 µK from the 5-mm-long Z wire trap into the 10×10 mm^2 magnetic lattice, absorption images taken after 14 ms TOF indicate that atoms are trapped in about 150 of the 1000 lattice sites at an estimated 4 µm from the chip surface (Figure 1.13a). It should be possible to trap in most of the 1000 lattice sites by loading from the 30-mm-long Z wire trap. By releasing the atoms from the magnetic lattice (by switching off B_{1y}), the lifetime of the trapped atoms is found to be 0.45 s (Figure 1.13b). With our current pixel image resolution of 9 µm it is not possible to resolve atoms in the individual lattice sites of the $a = 10$ µm magnetic lattice.

Using parametric heating, the radial trap frequencies were found to range up to $\nu_{\text{rad}} = 90$ kHz. A small amount of axial confinement, $\nu_{\text{axial}} \approx 1$ Hz, is provided by the magnetic field from the surrounding magnetic film. The temperature of the trapped atoms in the magnetic lattice was determined by RF spectroscopy to be greater than 150 µK, which is much higher than the temperature, ~6 µK, determined from TOF expansion, due to adiabatic compression in the tightly confining magnetic microtraps. We note that the magnetic lattice microtraps are currently highly elongated ($\nu_{\text{rad}}/\nu_{\text{axial}} \approx 5 \times 10^4$), so that the trapped clouds resemble a 1D Bose gas, in which thermalization and hence evaporative cooling is extremely inefficient [77]. In a recent series of experiments [89], we have implemented controllable axial confinement by means of a Z-wire current and have optically pumped the $F = 2$, $m_F = +2$ atoms into the $F = 1$, $m_F = -1$ weak-field seeking state in order to reduce three-body recombination losses. With an aspect ratio $\nu_{\text{rad}}/\nu_{\text{axial}} \approx 40$, we have been able to load $\approx 5 \times 10^5$ atoms into 10 µm-period magnetic lattice and, after evaporative cooling, to reach temperatures of about 2 µK with a trap lifetime of ~ 10 s. This should allow us to reach quantum degeneracy in the individual magnetic lattice sites and to generate multiply arrays of condensates.

In order to perform quantum tunneling experiments in the magnetic lattice it will be necessary to reduce the lattice period down to $a \approx 2-4$ µm.

1.5.2.4 Other Permanent Magnetic Lattices

The group in Amsterdam [47] have recently realized a 2D array of microtraps above multiple Z-shaped patterns microfabricated in a 300-nm-thick Fe-Pt film ($M_r = 670$ kA/m). The magnetic lattice of ultra-cold ^{87}Rb $F = 2$, $m_F = 1$ atoms (periods 22 µm and 36 µm in orthogonal directions) was produced 10 µm above the surface with very tight confinement (radial and axial frequency 13.4 kHz). Atoms from a macroscopic Z-wire trap were loaded into more than 500 traps and individual traps were optically resolved. Three-body recombination losses were used to characterize the number and temperature of the atoms in the individual traps and show the onset of BEC in 70 sites. However, the average condensate fraction is less than 0.1 which is attributed to the very tight confinement. Transport of the atoms over several lattice sites in one direction and back was demonstrated. A focused laser beam was used to empty one site and move it over several periods. These studies are promising for building a scalable quantum information processor based on neutral-atom permanent-magnet chips.

1.6
Summary and Conclusions

After almost two decades since the initial ideas of surface-based magnetic optical elements for ultra-cold atoms were proposed, the use of atom chips for producing and manipulating clouds of ultra-cold atoms and quantum degenerate gases on a substrate has evolved into a very active new area of research. Atom chips allow the construction and integration of miniature networks of magnetic microtraps, waveguides, and interferometers into a single device. The combination of permanent magnetic materials with dynamic current-carrying conductors to produce complex magnetic field configurations offers a promising way of constructing stable low-noise atom chips with micron-scale elements. Such devices have a wealth of potential future applications as trapped-atom interferometers, sensors of potential gradients, chip-based atomic clocks, microsensors of magnetic field topology and storage registers for quantum information processing.

Acknowledgments

The authors thank the many colleagues who have contributed to the work reported in this article, including Alexander Akulshin, Russell Anderson, Tim Davis, Saeed Ghanbari, David Gough, Brenton Hall, Smitha Jose, Tien Kieu, Leszek Krzemien,

David Lau, Russell McLean, Geoffrey Opat, Timothy Roach, Wayne Rowlands, Falk Scharnberg, Mandip Singh, Chris Ticknor, Michael Volk, James Wang, and Shannon Whitlock. The work is supported by the Australian Research Council.

References

1 Opat, G.I. (1983) In *Proc. 3rd Marcel Grossmann Meeting of General Relativity*, Part B (ed. Hu Ning), Beijing, Science press, Amsterdam, North Holland, p 1491.
2 Wark, S. and Opat, G.I. (1992) *J. Phys. B* **25**, 4229.
3 Opat, G.I., Wark, S.J., and Cimmino, A. (1992) *Appl. Phys. B* **54**, 396.
4 Roach, T.M., Abele, H., Boshier, M.G., Grossman, H.L., Zetie, K.P., and Hinds, E.A. (1995) *Phys. Rev. Lett.* **75**, 629.
5 Sidorov, A.I., McLean, R.J., Rowlands, W.J., Lau, D.C., Murphy, J.E., Walkiewicz, M., Opat, G.I., and Hannaford, P. (1996) *Quant. Semiclass. Opt.* **8**, 713.
6 Hughes, I.G., Barton, P.A., Roach, T.M., Boshier, M.G., and Hinds, E.A. (1997) *J. Phys. B* **30**, 647.
7 Saba, C.V., Barton, P.A., Boshier, M.G., Hughes, I.G., Rosenbusch, P., Sauer, B.E., and Hinds, E.A. (1999) *Phys. Rev. Lett.* **82**, 468.
8 Sidorov, A.I., Lau, D.C., Opat, G.I., McLean, R.J., Rowlands, W.J., and Hannaford, P. (1998) *Laser Physics* **8**, 642.
9 Sidorov, A.I., McLean, R.J., Sexton, B.A., Gough, D.S., Davis, T.J., Akulshin, A., Opat, G.I., and Hannaford, P. (2001) *Compt. Rendus* **2**(4), 565.
10 Sidorov, A.I., McLean, R.J., Scharnberg, F., Gough, D.S., Davis, T.J., Sexton, B.J., Opat, G.I., and Hannaford, P. (2002) *Acta Phys. Polon. B* **33**, 2137.
11 Lau, D.C., Sidorov, A.I., Opat, G.I., McLean, R.J., Rowlands, W.J., and Hannaford, P. (1999) *Eur. J. Phys. D* **5**, 193.
12 Drndić, M., Zabow, G., Lee, C.S., Thywissen, J.H., Johnson, K.S., Prentiss, M., Westervelt, R.M., Featonby, P.D., Savalli, V., Cognet, L., Helmerson, K., Westbrook, N., Westbrook, C.I., Phillips, W.D., and Aspect, A. (1999) *Phys. Rev. A* **60**, 4012.
13 Folman, R., Krüger, P., Schmiedmayer, J., Denschlag, J., and Henkel, C. (2002) *Adv. At., Mol. Opt. Phys.* **48**, 263.
14 Reichel, J. (2002) *Appl. Phys. B* **75**, 469.
15 Fortágh, J. and Zimmermann, C. (2007) *Rev. Mod. Phys.* **79**, 235.
16 Weinstein, J.D. and Libbrecht, K.G. (1995) *Phys. Rev. A* **52**, 4004.
17 Vuletić, V., Fischer, T., Praeger, M., Hänsch, T.W., and Zimmermann, C. (1998) *Phys. Rev. Lett.* **80**, 1634.
18 Fortágh, J., Grossmann, A., Zimmermann, C. and Hänsch, T.W. (1998) *Phys. Rev. Lett.* **81**, 5310.
19 Drndić, M., Johnson, K.S., Thywissen, J.H., Prentiss, M., and Westervelt, R.M. (1998) *Appl. Phys. Lett.* **72**, 2906.
20 Reichel, J., Hänsel, W., and Hänsch, T.W. (1999) *Phys. Rev. Lett.* **83**, 3398.
21 Denschlag, J., Cassettari, D., and Schmiedmayer, J. (1999) *Phys. Rev. Lett.* **82**, 2014.
22 Müller, D., Anderson, D.Z., Grow, R.J., Schwindt, P.D.D., and Cornell, E.A. (1999) *Phys. Rev. Lett.* **83**, 5194.
23 Dekker, N.H., Lee, C.S., Lorent, V., Thywissen, J.H., Smith, S.P., Drndić, M., Westervelt, R.M., and Prentiss, M. (2000) *Phys. Rev. Lett.* **84**, 1124.
24 Key, M., Hughes, I.G., Rooijakkers, W., Sauer, B.E., Hinds, E.A., Richardson, D.J., and Kazansky, P.G. (2000) *Phys. Rev. Lett.* **84**, 1371.
25 Davis, T.J. (1999) *J. Phys. B: Quant. Semicl. Opt.* **1**, 408.
26 Folman, R., Krüger, P., Cassettari, D., Hessmo, B., Maier, T., and Schmiedmayer, J. (2000) *Phys. Rev. Lett.* **84**, 4749.
27 Hänsel, W., Reichel, J., Hommelhoff, P., and Hänsch, T.W. (2001) *Phys. Rev. Lett.* **86**, 608.
28 Hänsel, W., Hommelhoff, P., Hänsch, T.W., and Reichel, J. (2001) *Nature* **413**, 498.

29 Ott, H., Fortágh, J., Schlotterbeck, G., Grossmann, A., and Zimmermann, C. (2001) *Phys. Rev. Lett.* **87**, 230401.
30 Aubin, S., Myrskog, S., Extavour, M.H.T., LeBlanc, L.J., McKay, D., Stummer, A., and Thywissen, J.H. (2006) *Nature Phys.* **2**, 384.
31 Treutlein, P., Hommelhoff, P., Steinmetz, T., Hänsch, T.W., and Reichel, J. (2004) *Phys. Rev. Lett.* **92**, 203005.
32 Knappe, S., Shah, V., Schwindt, P.D.D., Hollberg, L., Kitching, J., Liew, L.-A., and Moreland, J. (2004) *Appl. Phys. Lett.* **85**, 1460.
33 Shin, Y., Sanner, C., Jo, G.-B., Pasquini, T.A., Saba, M., Ketterle, W., Pritchard, D.E., Vengalattore, M. and Prentiss, M. (2005) *Phys. Rev. A* **72**, 021604.
34 Schumm, T., Hofferberth, S., Andersson, L.M., Wildermuth, S., Groth, S., Bar-Joseph, I., Schmiedmayer, J., and Krüger, P. (2005) *Nature Phys.* **1**, 57.
35 Wang, Y.-J., Anderson, D.Z., Bright, V.M., Cornell, E.A., Diot, Q., Kishimoto, T., Prentiss, M., Saravanan, R.A., Segal, S.R., and Saijun, W. (2005) *Phys. Rev. Lett.* **94**, 090405.
36 Jo, G.-B., Shin, Y., Will, S., Pasquini, T.A., Saba, M., Ketterle, W., Pritchard, D.E., Vengalattore, M., and Prentiss, M. (2007) *Phys. Rev. Lett.* **98**, 030407.
37 Böhi, P., Riedel, M.F., Hoffrogge, J., Reichel, J., Hänsch, T.W., and Treutlein, P. (2009) *Nature Phys.* **5**, 592.
38 Riedel, M.F., Böhl, P., Li, Y., Hänsch, T.W., Sinatra, A., and Treutlin, P. (2010) *Nature* **464**, 1170.
39 Deutsch, C. Ramirez-Martinez, F., Lacroûte, C., Reinhard, F., Schneider, T., Fuchs, J.N., Piéchon, F., Laloë, F., Reichel, J., and Rosenbusch, P. (2010) *Phys. Rev. Lett.* **105**, 020401.
40 Lhuiller, C. and Laloë, F. (1982) *J. Phys. (Paris)* **43**, 197 and 225.
41 Sinclair, C.D.J., Curtis, E.A., Llorente Garcia, I., Retter, J.A., Hall, B.V., Eriksson, S., Sauer, B.E., and Hinds, E.A. (2005) *Phys. Rev. A* **72**, 031603.
42 Hall, B.V., Whitlock, S., Scharnberg, F., Hannaford, P., and Sidorov, A. (2005) *Laser Spectroscopy 17*, (eds E.A. Hinds, A. Ferguson, and E. Riis), World Scientific, Singapore, p. 275.
43 Hall, B.V., Whitlock, S., Scharnberg, F., Hannaford, P., and Sidorov, A. (2006) *J. Phys. B* **39**, 27.
44 Boyd, M., Streed, E.W., Medley, P., Campbell, G.K., Mun, J., Ketterle, W., and Pritchard, D.E. (2007) *Phys. Rev. A* **76**, 043624.
45 Fernholz, T., Gerritsma, R., Whitlock, S., Barb, I., and Spreeuw, R.J.C. (2008) *Phys. Rev. A* **77**, 033409.
46 Gerritsma, R., Whitlock, S., Fernholz, T., Schlatter, H., Luigjes, J.A., Thiele, J.-U., Goedkoop, J.B., and Spreeuw, R.J.C. (2007) *Phys. Rev. A* **76**, 033408.
47 Whitlock, S., Fernholz, T., Gerritsma, R., Schlatter, H., Luigjes, J.A., Thiele, J.-U., Goedkoop, J.B., and Spreeuw, R.J.C. (2009) *New J. Phys.* **11**, 023021.
48 Singh, M., Whitlock, S., Anderson, R., Ghanbari, S., Hall, B.V., Volk, M., Akulshin, A., McLean, R., Sidorov, A., and Hannaford, P. (2008) *Laser Spectroscopy 18* (eds J. Bergquist, L. Hollberg, and M. Kasevich), New York, World Scientific, p. 228.
49 Singh, M., Volk, M., Akulshin, A., Sidorov, A., McLean, R., and Hannaford, P. (2008) *J. Phys. B* **41**, 065301.
50 Nirrengarten, T., Qarry, A., Roux, C., Emmert, A., Nogues, G., Brune, M., Raimond, J.-M., and Haroche, S. (2006) *Phys. Rev. Lett.* **97**, 200405.
51 Mukai, T., Hufnagel, C., Kasper, A., Meno, T., Tsukada, A., Semba, K., and Shimizu, F. (2007) *Phys. Rev. Lett.* **98**, 260407.
52 Roux, C., Emmert, A., Lupascu, A., Nirrengarten, T., Nogues, G., Brune, M., Raimond, J.-M., and Haroche, S. (2008) *Eur. Phys. Lett.* **81**, 56004.
53 Cano, D., Kasch, B., Hattermann, H., Kleiner, R., Zimmermann, C., Koelle, D., and Fortágh, J. (2008) *Phys. Rev. Lett.* **101**, 183006.
54 Singh, M. (2009) *Opt. Express* **17**, 2600.
55 Hinds, E.A. and Hughes, I.G. (1999) *J. Phys. D* **32**, R119.
56 Lau, D.C., McLean, R.J., Sidorov, A.I., Gough, D.S., Koperski, J., Rowlands, W.J., Sexton, B.A., Opat, G.I., and Hannaford, P. (1999) *J. Opt. B* **1**, 371.
57 Davis, T.J. (2001) *Eur. J. Phys. D* **14**, 111.

58 Opat, G.I., Nic Chormaic, S., Cantwell, B.P., and Richmond, J.A. (2004) *J. Opt. B* **1**, 415 (1999).

59 Wang, J.Y., Whitlock, S., Scharnberg, F., Gough, D.S., Sidorov, A.I., McLean, R.J., and Hannaford, P. (2005) *J. Phys. D* **38**, 4015.

60 Hinds, E.A., Boshier, M.G., and Hughes, I.G. (1998) *Phys. Rev. Lett.* **80**, 645.

61 Eriksson, S., Ramirez-Martinez, F., Curtis, E.A., Sauer, B.E., Nutter, P.W., Hill, E.W., and Hinds, E.A. (2004) *Appl. Phys. B* **79**, 811.

62 Krüger, P., Luo, X., Klein, M.W., Brugger, K., Haase, A., Wildermuth, S., Groth, S., Bar-Joseph, I., Folman, R., and Schmiedmayer, J. (2003) *Phys. Rev. Lett.* **91**, 233201.

63 Henkel, C., Pötting, S., and Wilkins, M. (1999) *Appl. Phys. B* **69**, 379.

64 Whitlock, S. (2007) *Bose–Einstein condensates on a magnetic film atom chip* PhD thesis, Swinburne University of Technology, p. 92.

65 Fortágh, J., Ott, H., Kraft, S., Günther, A., and Zimmermann, C. (2002) *Phys. Rev. A* **66**, 041604.

66 Whitlock, S., Hall, B.V., Roach, T., Anderson, R., Volk, M., Hannaford, P., and Sidorov, A. (2007) *Phys. Rev. A* **75**, 043602.

67 Wildermuth, S., Hofferberth, S., Lesanovsky, I., Haller, E., Andersson, L.M., Groth, S., Bar-Joseph, I., Krüger, P., and Schmiedmayer, J. (2005) *Nature* **435**, 440.

68 Luiten, O.J., Reynolds, M.W., and Walraven, J.T.M. (1996) *Phys. Rev. A* **53**, 381.

69 Wang, D.W., Lukin, M.D., and Demler, E. (2004) *Phys. Rev. Lett.* **92**, 076802.

70 Estéve, J., Aussibal, C., Schumm, T., Figl, C., Mailly, D., Bouchoule, I., Westbrook, C.I., and Aspect, A. (2004) *Phys. Rev. A* **70**, 043629.

71 Hall, B.V., Whitlock, S., Anderson, R., Hannaford, P., and Sidorov, A. (2007) *Phys. Rev. Lett.* **98**, 030402.

72 McGuirk, J.M., Foster, G.T., Fixler, J.B., Snaddon, M.J., and Kasevich, M.A. (2002) *Phys. Rev. A* **65**, 033608.

73 Anderson, R.P., Ticknor, C., Sidorov, A.I., and Hall, B.V. (2009) *Phys. Rev. A* **80**, 023603.

74 Mertes, K.M., Merrill, J.W., Carretero-Gonzáles, R., Frantzeskakis, D.J., Kevrekidis, P.G., and Hall, D.S. (2007) *Phys. Rev. Lett.* **99**, 190402.

75 For example, Bloch, I. (2004) *Physics World* **98**(4), 25.

76 Tolra Laburthe, B., O'Hara, K.M., Huckans, J.H., Phillips, W.D., Rolston, S.L., and Porto, J.V. (2004) *Phys. Rev. Lett.* **92**, 19040.

77 Kinoshita, T., Wenger, T., and Weiss, D.S. (2006) *Nature* **440**, 900.

78 Greiner, M., Mandel, O., Esslinger, T., Hänsch, T., and Bloch, I. (2002) *Nature* **415**, 39.

79 Billy, J., Josse, V., Zuo, Z., Bernard, A., Hambrecht, B., Lugan, P., Clement, D., Sanchez-Palencia, L., Bouyer, P., and Aspect, A. (2008) *Nature* **453**, 891.

80 Roati, G., D'Errico, C., Fallani, L., Fattori, M., Fort, C., Zaccanti, M., Modugno, G., Modugno, M., and Inguscio, M. (2008) *Nature* **453**, 895.

81 Calarco, T., Hinds, E.A., Jaksch, D., Schmiedmayer, J., Cirac, J.I., and Zoller, P. (2000) *Phys. Rev. A* **61**, 022304.

82 Schrader, D., Dotsenko, I., Khudaverdyan, M., Miroshnychenko, Y., Rauschenbeutal, A., and Meschede, D. (2004) *Phys. Rev. Lett.* **93**, 150501.

83 Ghanbari, S., Kieu, T.D., Sidorov, A., and Hannaford, P. (2006) *J. Phys. B* **39**, 847.

84 Ghanbari, S., Kieu, T.D., and Hannaford, P. (2007) *J. Phys. B* **40**, 1283.

85 Günther, A., Kemmler, M., Kraft, S., Vale, C.J., Zimmermann, C., and Fortágh, J. (2005) *Phys. Rev. A* **71**, 063619.

86 Sinclair, D.J., Retter, J.A., Curtis, E.A., Hall, B.V., Llorente Garcia, I., Eriksson, S., Sauer, B.E., and Hinds, E.A. (2005) *Eur. Phys. J.* **35**, 105.

87 Yin, J., Gao, W., Hu, J., and Wang, Y. (2002) *Opt. Commun.* **206**, 99.

88 Grabowski, A. and Pfau, T. (2003) *Eur. Phys. J. D* **22**, 347.

89 Jose, S., Krzemien, L., Withlock, S., Singh, M., Sidorov, A., McLean, R., and Hannaford, P. *In preparation.*

2
Trapping and Manipulating Atoms on Chips
Jakob Reichel

2.1
Introduction

Atom chips have enabled original new research in a variety of fields, ranging from atom–surface interactions to quantum information. If one tries to classify atom chip experiments by their main reason to use chip technology rather than a more conventional technique, several distinct motivations appear. Each of them corresponds to a specific feature of atom chips.

In experiments studying surface forces (Chapters 4 and 5), properties of superconducting surfaces (Chapter 10), and atom–nanodevice coupling [1], this feature is the built-in *proximity of a surface* (which may be engineered to contain other nanodevices), combined with the *exquisite control over atomic position* enabled by micro-fabricated wire structures on the same surface. The latter advantage is also exploited in other contexts, such as positioning of atoms in a single antinode of an optical standing wave [2].

In studies of 1D quantum gases (Chapter 11), bosonic Josephson junctions and atomic beam splitters (Chapter 7), but also in quantum information processing (Chapter 9), it is the possibility to design *strongly confining, complex, multi-parameter potentials* which makes atom chips quite unique. Although today's atom chips are of course quite crude compared to integrated circuits in microelectronics, they do share key features such as miniaturization, integration, and the possibility to replicate many times a given functional block. Recent efforts to develop micro-fabricated ion traps or "ion chips" (Chapter 13) also fit into this context.

Finally, atom chips enable *simplified, robust and compact setups*, an advantage that can benefit most quantum gas experiments. This also combines favorably with the other advantages, as for example in the Fermi gas experiments of Chapter 12. Initially, this advantage was offset by the effort required to obtain a suitable chip and vacuum cell. This barrier is now being removed: the fabrication techniques are increasingly well described (Chapter 3 gives the most comprehensive review to date), and a complete chip-based vacuum cell for Bose–Einstein condensate (BEC) generation has become commercially available [3]. Robust and compact setups are

---

Atom Chips. Edited by Jakob Reichel and Vladan Vuletić
Copyright © 2011 WILEY-VCH Verlag GmbH & Co. KGaA, Weinheim
ISBN: 978-3-527-40755-2

especially important for field applications, making it natural to consider atom chips for compact atomic clocks (Chapter 8) and atom interferometers (Chapters 6 and 7) for navigation and geodesy. It is not a coincidence that an atom chip was used in the very recent, first demonstration of BEC in microgravity [4].

The purpose of this chapter is to introduce baseline knowledge that is used in most if not all atom chip experiments. Excellent general introductions to laser cooling and Bose–Einstein condensation exist (see for example [5–8]), so that these topics need not be discussed here. The emphasis in this chapter is on the design of magnetic traps from planar current distributions and on experimental aspects that are specific to atom chip experiments.

2.2
Overview of Trapping Techniques

Chips can be used to trap and manipulate atoms in many ways, using various interactions. Let us briefly review the types of potentials that have been demonstrated experimentally so far:

Magnetodynamic potentials are used in the vast majority of atom chips today: atoms are confined in the magnetic field of static or slowly varying currents flowing through micro-fabricated wires.

Permanent magnetic potentials can be generated by writing micro-scale patterns into a ferromagnetic film. These patterns can be designed to form traps, and particularly two-dimensional lattices. Although these fields cannot be turned off or modulated in real time, tunability can be achieved with external bias fields. BEC has been achieved in these traps, and they have been used to create two-dimensional arrays of traps. Permanent magnetic potentials are described in more detail in Chapter 1.

Electrostatic potentials exploit the static electric polarizability of neutral atoms. The small size of chip-based electrodes makes it possible to achieve sizable forces with reasonable voltage; a proof-of-principle experiment has been performed [9]. Neutral atoms are high-field seekers, however, so that a purely electrostatic trap is ruled out by Earnshaw's theorem.

Radiofrequency (RF) and microwave (MW) potentials couple to Zeeman or hyperfine transitions and can be used in several different ways to create or modify traps. RF potentials have established themselves as a way to create doublewell potentials for atom interferometry. They are discussed in Chapter 7. Chipbased microwave potentials are a new and very promising tool to create statedependent potentials for quantum information and spin squeezing. More details can be found in Chapter 9.

Optical fields are also amenable to integration via the use of integrated optics. It seems natural to consider integrated optical chips for atom trapping via the dipole force. So far however, there have been only a few attempts in this direction. Atoms have been trapped in an optical lattice generated by a micro-lens array [10], but the far-field approach of that experiment (the array was imaged into the vacuum chamber) lacked some of the features typically associated with chip-based traps. More recently, an array of small-scale magneto-optical traps (MOTs) has been realized by etching pyramid-shaped mirrors directly into an Si substrate [11]. 780-nm optical waveguides for detection of Rb atoms on chips are under development [12]. More information can be found in Chapter 3.

Time-averaged potentials are a method that can be used with all the physical coupling mechanisms above to create potentials with new properties. The atoms are exposed to a time-dependent potential. At each instant, the "time-orbiting potential" (TOP) provides confinement at a trap frequency ω_t, while being displaced periodically at a frequency $\omega_{TOP} \gg \omega_t$. The result is an effective potential which is the time average of the varying potential. The concept originally emerged to prevent Majorana losses in quadrupole traps [13]. On atom chips, it has been used successfully to reduce potential roughness [14] and may give access to "exotic" trap geometries like ring traps.

In this chapter, we mostly focus on magnetodynamic traps. Many types of magnetic potentials are straightforward to produce on chips. These include linear ("quadrupole") and quadratic ("Ioffe–Pritchard" or "IP") trapping potentials in two and three dimensions, as well as many other useful static and time-dependent potentials. The most widespread of them are discussed below. For some other desirable potentials, suitable planar current patterns are not easily found, and some are ruled out by Maxwell's equations. Examples of difficult potentials are strongly oblate traps ("pancakes") and ring-shaped traps with non-zero minima.

We will start with a brief description of the most elementary magnetic trapping potentials – quadrupoles and IP traps. Then we will proceed by analyzing a construction set of elementary wire layouts – such as the cross and the "H" – which enables the reader to understand and get a feeling for most of the layouts found on today's atom chips. For precise quantitative analysis and to check complex multi-wire patterns, numerical simulations must be used as a complement.

2.3
Magnetic Traps for Neutral Atoms

2.3.1
Magnetic Interaction

The interaction energy between a neutral atom and a magnetic field is much weaker than the atom's thermal energy at room temperature, even in the many-Tesla fields of superconducting magnets. Therefore, atoms have to be cooled before they can be magnetically trapped. (It is worth noting that neutrons were magnetically trapped before neutral atoms, despite their thousand-times smaller magnetic moment [15].) This discussion mainly follows [6].

Magnetic forces are comparatively strong for atoms with an unpaired electron, such as alkalis, resulting in magnetic moments μ_m of the order of[3] the Bohr magneton μ_B. The interaction of a magnetic dipole with an external magnetic field is given by

$$V(r) = -\boldsymbol{\mu}_m \cdot \boldsymbol{B}(r) = -\mu_m B(r) \cos\theta \ . \tag{2.1}$$

Classically, the angle θ between the magnetic moment and the magnetic field is constant due to the rapid precession of $\boldsymbol{\mu}_m$ about the magnetic field axis. Quantum-mechanically, the energy levels of a particle with angular momentum F and g factor g in a magnetic field are $E(m_F) = g\mu_B m_F B$, where m_F is the quantum number of the component of F along the direction of $\boldsymbol{B}$. The classical term $\cos\theta$ is now replaced by m_F/F; the classical picture of constant θ is equivalent to the system remaining in a state with fixed m_F.

Depending on the sign of gm_F, the particle experiences a magnetic force either towards minima of the field ($gm_F > 0$, *weak-field seeking state*) or towards maxima ($gm_F < 0$, *strong-field seeking state*). As Maxwell's equations do not allow a maximum of the magnetic field in free space [18, 19], only weak-field seekers can be trapped. Such states are usually prepared by optical pumping. It is worth noting that the trapped state, being a low-field seeker, is not the state of lowest energy in presence of the trapping field. This implies the possibility of exothermic two-atom collisions and corresponding loss channels for the trapped atoms.

2.3.2
Stability against Spin-Flip Losses

Because magnetic traps only confine weak-field seeking states, atoms will be lost from the trap if they make a transition to a strong-field seeking state. Such transitions can be induced by the motion in the trap, because the atom sees a field which is changing in amplitude and direction. The trap is stable only if the atom's magnetic moment adiabatically follows the direction of $\boldsymbol{B}$. This requires that the rate of change of the field's direction θ (in the reference frame of the moving atom) must be slower than the Larmor frequency:

$$\frac{d\theta}{dt} < \frac{\mu_m |\boldsymbol{B}|}{\hbar} = \omega_L. \tag{2.2}$$

The upper bound for $d\theta/dt$ in a magnetic trap is the trapping frequency. This adiabaticity condition is violated in regions of very small magnetic field, creating regions of trap loss due to "Majorana transitions" [20] into untrapped states.

---

3) Atoms with larger moments exist: bosonic chromium ($6\mu_B$) has been Bose-condensed [16], dysprosium ($10\mu_B$) has been trapped in a MOT [17].

2.3.3
Quadrupole Traps

Static magnetic traps can be subdivided into two classes: those in which the minimum is a zero crossing of the magnetic field, and those which have a minimum around a finite field value [21]. In the first case, the potential near the minimum can usually be approximated by a linear function, characterized by its gradient: $\boldsymbol{B} = B'_x x \boldsymbol{e}_x + B'_y y \boldsymbol{e}_y + B'_z z \boldsymbol{e}_z$, where Maxwell's equations require $B'_x + B'_y + B'_z = 0$. (We have chosen the coordinate axes to coincide with the axes of the quadrupole.) In macroscopic traps, this configuration is usually realized with two coils in "anti-Helmholtz" configuration [6]. The trapping potential, proportional to the field modulus, provides linear confinement. In the special case $B'_x = 0$, a two-dimensional quadrupole is obtained, which necessarily has $B'_y = -B'_z$.

Because $B = 0$ in the trap center, loss due to Majorana transitions is inevitable in quadrupole traps, but the loss rate depends on temperature. For atoms moving at a velocity v, the effective size of the "hole" in the trap is $\sqrt{2\hbar v/\pi\mu_m B'}$, which is about 1 µm for $\mu_m = \mu_B$, $v = 1$ m/s and $B' = 1000$ G/cm [13]. As long as the hole is small compared to the cloud diameter, the trapping time can be long (even longer than a minute), so that quadrupole traps can work very well if the atoms are not too cold.

2.3.4
Ioffe–Pritchard Traps

The lowest-order (and therefore tightest) trap which can have a non-zero field in the minimum is a harmonic trap. Indeed, in a quadrupole trap, the effect of a bias field is merely to displace the zero crossing. A harmonic trap with a bias field $B_{0\parallel}$ along x has an axial field of $B_x = B_{0\parallel} + B''x^2/2$. The leading term of the transverse field component B_z is linear, $B_z = B'z$. Applying Maxwell's equations, and assuming axial symmetry, leads to the following field configuration [6]:

$$\boldsymbol{B} = B_0 \begin{pmatrix} 1 \\ 0 \\ 0 \end{pmatrix} + B' \begin{pmatrix} 0 \\ -y \\ z \end{pmatrix} + \frac{B''}{2} \begin{pmatrix} x^2 - \frac{1}{2}(y^2 + z^2) \\ -xy \\ -xz \end{pmatrix}. \quad (2.3)$$

In an elementary macroscopic implementation, this field is generated by a 2D quadrupole consisting of four current bars along x and a pair of coils around x, spaced symmetrically with respect to $x = 0$ by a distance larger than the Helmholtz condition. On the axis, B_x thus has a quadratic minimum, providing confinement in this direction. This type of trap was first suggested and demonstrated for atom trapping by Pritchard [22, 23], and is similar to the Ioffe configuration discussed several decades earlier for plasma confinement [24]. It has become customary to refer to any trap of this field configuration as an Ioffe–Pritchard (IP) trap.

Around the origin, for $x \ll \sqrt{B_0/B'}$ and $y, z \ll B_0/B'$, the field modulus to second order is given by

$$|\boldsymbol{B}| \approx B_0 + \frac{B''}{2}x^2 + \frac{1}{2}\left(\frac{B'^2}{B_0} - \frac{B''}{2}\right)(y^2 + z^2) . \tag{2.4}$$

This field gives rise to a harmonic potential with trapping frequencies

$$\omega_x = \sqrt{\frac{\mu_m}{m}B''} \quad \text{and} \quad \omega_\perp = \sqrt{\frac{\mu_m}{m}\left(\frac{B'^2}{B_0} - \frac{B''}{2}\right)} . \tag{2.5}$$

These equations show that the aspect ratio $\omega_x/\omega_\perp$ of an ideal IP trap can be tuned from prolate (cigar shaped) to isotropic and on to oblate (pancake shaped) by adjusting B'' with respect to B'^2/B_0 (see [25] for an experimental demonstration).

If the atomic movement has larger amplitude, the linear term must be taken into account for the radial motion. The field modulus can then be written

$$|\boldsymbol{B}| \approx \sqrt{\left(B_0 + \frac{B''}{2}x^2\right)^2 + \left(B'^2 - B_0\frac{B''}{2}\right)(y^2 + z^2)} . \tag{2.6}$$

Equation (2.5) suggests that $\omega_\perp$ can be increased indefinitely by lowering B_0. Two restrictions must be taken into account. First, the transverse potential is quadratic only up to a transverse position $\rho = \sqrt{x^2 + y^2}$ given by $B(\rho) \lesssim B_0$. Beyond this point, the potential becomes linear and the trap frequency loses its meaning. Second, spin-flip losses (Section 2.3.2) must be avoided, which dictates a minimum B_0 to maintain a sufficient Larmor frequency. The spin-flip loss rate in an IP trap has been calculated for $F = 1$ in [26]. For $\omega_x \ll \omega_\perp$, it is

$$\gamma_M = 4\pi \omega_\perp \exp(-2\omega_L/\omega_\perp) . \tag{2.7}$$

In the experiment, B_0 is typically adjusted such that $\omega_L \geq 7\omega_\perp$ in the trap center to make spin-flip loss negligible ($\gamma_M \leq 10^{-5}\omega_\perp$).

2.3.5
Some General Properties of Magnetic Traps

Trap Depth A trap of finite depth containing a thermal atomic ensemble (temperature T) has losses due to atoms "boiling out of the trap". As a rule of thumb, the trap depth should be large compared to the mean atomic energy. Neglecting gravity, this leads to the condition

$$V_{max} = |\mu_m B_{max}| > \eta k_B T \tag{2.8}$$

with $\eta = 5 \dots 7$ in order to make this loss term negligible. More details on trap depth, volume, and maximum trapped atom number can be found in Chapter 12, Sections 12.4.3 to 12.4.5.

Compensation of Gravity For the restoring force $F \propto B'$ along the vertical axis, the minimum requirement is to compensate gravity: $B'_z \geq mg/\mu_m$. Taking the $|F = 2, m = 2\rangle$ state of ^{87}Rb as an example, the minimum field gradient is 15 G/cm.

Oscillation Frequency and Ground-State Size in a Harmonic Trap The oscillation frequency along the ith eigenaxis of a harmonic potential V is given by

$$\omega_i = \sqrt{\frac{1}{m}\frac{d^2 V}{dx_i^2}} = \sqrt{\frac{\mu_m}{m}\frac{d^2 B}{dx_i^2}}. \tag{2.9}$$

For ^{87}Rb, $|F = 2, m = 2\rangle$, the frequency $\nu_i = \omega_i/2\pi$ is conveniently calculated as

$$\nu_i = 12.7\,\text{Hz} \times \sqrt{\frac{d^2 B}{dx_i^2} \bigg/ \frac{\text{T}}{\text{m}^2}}.$$

The ground-state extension ($1/e$ radius of $|\Psi|^2$) is given by

$$\delta x_i = \sqrt{\frac{\hbar}{m\omega_i}} = \left(\frac{m\mu_m}{\hbar^2}\frac{d^2 B}{dx_i^2}\right)^{-\frac{1}{4}}. \tag{2.10}$$

For our example state, an oscillation frequency of $\nu = 1\,\text{kHz}$ corresponds to a ground-state extension $\delta x = 340\,\text{nm}$. To reduce δx by a factor α, B'' must increase by α^4 and, therefore, very small ground-state sizes become exceedingly difficult to achieve with macroscopic traps. With atom chips, it is possible to realize sizes much smaller than an optical wavelength, as will be shown below.

2.4
The Design of Wire Patterns for Magnetic Potentials

2.4.1
Conductor Elements and Multipoles

A multitude of static and time-dependent magnetic potentials can be created with planar arrangements of current-carrying wires. In the context of atom traps, planar and "pseudoplanar" wire layouts for quadrupole and IP traps were first proposed in [27], but were not used in experiments at that time, although they have been fabricated [28]. Most of the designs in use today derive from one of two basic schemes. The first is to superpose the field of a wire with a homogeneous field, as was done in the first chip-based atom trap [29]. The second is to use two or more parallel wires, as in the first atom guide on a chip [30]. We will start with a quick discussion of two-dimensional traps using either of these two schemes and then move on to analyze the elementary three-dimensional traps created by a conductor cross, an "H"- and a "U"-shaped wire. (As mentioned above, an alternative approach would be to start from the lowest-order multipole fields and analyze how these relate to elementary trapping potentials and to the fields produced by the various wire layouts, as done in [31].) Armed with the understanding of these elementary building blocks, we will proceed to discuss some more complex structures providing the function of an atomic "conveyer belt".

2.4.2
Wire Guide

An infinitely thin wire carrying a current I_1 creates a magnetic field of magnitude, gradient, and curvature

$$B(z) = \frac{\mu_0}{2\pi} \frac{I_1}{z}, \quad B'(z) = -\frac{\mu_0}{2\pi} \frac{I_1}{z^2}, \quad B''(z) = \frac{\mu_0}{\pi} \frac{I_1}{z^3}, \quad (2.11)$$

where z is the distance measured perpendicular to the wire axis. It can thus provide a strong gradient and curvature when z is small, but does not by itself constitute an atom trap, which requires a minimum of the field. One way to create a minimum, while maintaining the full field gradient of the wire, is to superpose a homogeneous external field $\tilde{B}$ perpendicular to the wire axis (Figure 2.1). In this section, we will take the wire axis to be e_x, and the external field along e_y. The wire field and the external field $\tilde{B} = \tilde{B}_y e_y$ cancel on a line parallel to the wire at a distance

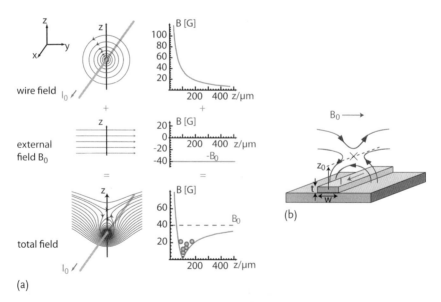

Figure 2.1 (a) Creating a two-dimensional trap with a wire and an external field. Top: wire field, center: external field, bottom: resulting total field. The left column shows magnetic field lines, the right column gives the magnitude of the field at $y = 0$ for a wire current $I_1 = 2$ A. (b) Waveguide for neutral atoms formed at a distance z_0 from a lithographically fabricated wire on a chip.

z_0 from the wire axis given by

$$z_0 = \frac{\mu_0}{2\pi} \frac{I_1}{\tilde{B}_y}. \tag{2.12}$$

In the vicinity of this line, the total field is well approximated by a two-dimensional quadrupole with gradient

$$B'(z_0) = -\frac{2\pi}{\mu_0} \frac{\tilde{B}_y^2}{I_1}. \tag{2.13}$$

To create a tight trap with a given current, it is obviously advantageous to reduce z_0, that is to increase $\tilde{B}_y$. This also makes the trap deeper, as the trap depth is simply given by $\tilde{B}_y$. Of course, when z_0 becomes comparable to the wire's lateral dimensions, the formulas above are no longer a good approximation. Analytical formulas for finite-sized wires exist, see Section 2.4.5 below.

Note that the homogeneous field $\tilde{B}_y$ can also be generated by several parallel wires or a wire of larger width, leading to a multi-wire guide which does not require coils. See [32] for a discussion of several such guides. Multi-wire guides also form the basis of atomic beam splitters discussed in Chapter 7.

The quadrupole guide is easily transformed into an IP guide (i.e., non-zero field in the minimum and quadratic variation around this minimum) by adding a homogeneous field component $\tilde{B}_x e_x$. As the field of the quadrupole guide has no component along the wire axis e_x, the total field modulus is the quadratic sum of the two fields. The field in the minimum is $\tilde{B}_x e_x$.

The thin-wire equations are useful for quick estimates, and are conveniently expressed in the following form:

$$B(z) = 2\,\text{G} \times \frac{I/\text{A}}{z/\text{mm}} \tag{2.14}$$

$$B'(z) = -20\,\text{G cm}^{-1} \times \frac{I/\text{A}}{(z/\text{mm})^2} \tag{2.15}$$

$$z_0 = 2\,\text{mm} \times \frac{I/\text{A}}{\tilde{B}_y/\text{G}} \tag{2.16}$$

$$B'(z_0) = -5\,\text{G cm}^{-1} \frac{(\tilde{B}_y/\text{G})^2}{I/\text{A}}. \tag{2.17}$$

2.4.3
Conductor Cross ("Dimple" Trap)

A very versatile three-dimensional IP trap can be created by adding a second, perpendicular wire with current I_2 to the IP guide of the preceding section (Figure 2.2) [33]. Here we take the axis of this wire to be e_y, and we call its field B_2. Its role is to provide the axial confinement. If I_2 is sufficiently small, the position of the minimum in the transverse yz plane is still approximately given by Eq. (2.12), and the axial confinement can be calculated by evaluating the field modulus on the axis

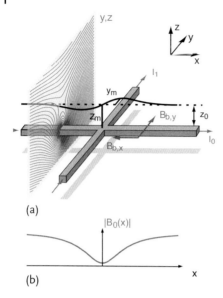

(a)

(b)

Figure 2.2 Trap formed above a conductor intersection ("dimple" trap). For $I_1 \gg I_2$, the trap can be understood as a superposition of the 2D quadrupole above the I_1 wire and the field of the I_2 wire, which is predominantly along e_x in the region of the trap minimum. (a) conductor configuration and transverse field; (b) longitudianal potential.

($y = 0, z = z_0$) of the unperturbed trap, which gives $B_0(x) \approx |\tilde{B}_x + B_{2,x}(x, z = z_0)|$ because the field of I_2 is predominantly along e_x in the region of interest. If the polarity of I_2 is chosen such that $\tilde{B}_x$ and $B_{2,x}(x, z = z_0)$ are of opposite sign, a quadratic field minimum forms above the wire intersection, as shown in Figure 2.2. Thus, the wire cross provides a simple and practically very useful IP trap. In terms of the IP equation (2.3), and still in the approximation of small I_2, the trap is characterized by a transverse gradient

$$B' = \frac{\mu_0}{2\pi} \frac{I_1}{z_0^2} \tag{2.18}$$

that depends on I_1 and $\tilde{B}_y$ (the components of the quadrupole guide), by a curvature along x

$$B'' = \left.\frac{\partial^2 B_{2,x}}{\partial x^2}\right|_{x=0, z=z_0} = \frac{\mu_0}{\pi} \frac{I_2}{z_0^3} \tag{2.19}$$

that is determined by I_2 and z_0, and a field in the center

$$B_0 = \left|\tilde{B}_x + \frac{\mu_0}{2\pi} \frac{I_2}{z_0}\right|, \tag{2.20}$$

given by the partial compensation of the I_2 field and $\tilde{B}_x$ in the trap center. The trap frequencies are well approximated by

$$\omega_x = \sqrt{\frac{\mu_m}{m} B''} \quad \text{and} \quad \omega_\perp = \sqrt{\frac{\mu_m}{m} \frac{B'^2}{B_0}}. \tag{2.21}$$

In the regime of this small-I_2 approximation, the trap is always cigar shaped and oriented roughly along e_x. For larger I_2, the axis rotates and also bends towards the chip at its ends. A ring-shaped quadrupole forms for $I_2 = I_1$. There is also a parameter range where slightly pancake-shaped traps can be formed. An in-depth discussion of the potentials generated by a wire cross beyond the small-I_2 approximation can be found in [31, 34].

2.4.4
"H", "Z", and "U" Traps

If we change the sign of I_2 in the configuration above, the potential along e_x at $z = z_0$ now has a maximum instead of a minimum above the intersection, while the transverse confinement is maintained. By adding one more wire parallel to I_2 and carrying the same current, one obtains two such maxima, between which atoms can be trapped (Figure 2.3). This is the conductor "H", another very general and useful configuration, of which the popular "Z" and "U" traps are approximations, as discussed in the next paragraph. The "H" can generate quadrupole and IP traps depending on the relative orientation (parallel or antiparallel) of the two I_2 currents. To simplify the analysis, let us first set $\tilde{B}_x = 0$. To first approximation then, as above, the transverse minimum remains on the straight line $z = z_0$ given by the guide, and the field on this line is now the sum of the two currents I_2. If these two currents have the same sign, the x components of their fields at $x = 0, y = 0, z = z_0$ add up, producing a non-zero field in the minimum. The resulting total potential is an IP trap. Conversely, when they have opposite sign, the two fields exactly cancel, and a quadrupole trap is obtained.

For $I_1 = I_2$, the IP potential of the "H" can be approximated by a single "Z"-shaped conductor and the quadrupole by a conductor "U", as in Figure 2.4. A pos-

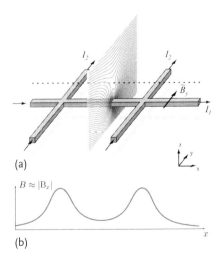

Figure 2.3 Trap formed above a conductor "H". (a) conductor configuration and transverse field; (b) longitudinal potential. (Figure adapted from [31].)

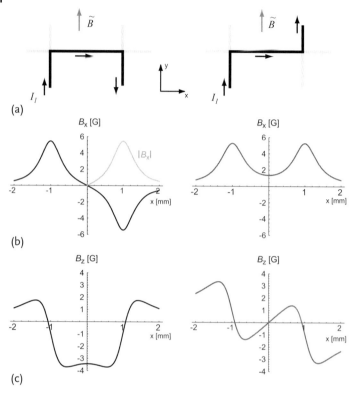

Figure 2.4 Quadrupole and IP traps generated by a wire "U" and "H". (a) current configurations; (b) and (c) field components B_x and B_z on the trap axis. Fields are calculated for a distance of 2 mm between the e_y wires, $I_1 = 2\,\text{A}$ and $\tilde{B}_y = 10\,\text{G}$, and a finite wire width of 50 μm. (Figure adapted from [31].)

itive side effect of using these simplified shapes is to reduce the position noise induced by current fluctuations, as some common-mode variations cancel out. The price to pay is a somewhat less symmetric shape of the potentials,[4] and of course less freedom on the aspect ratio due to the fixed $I_2 = I_1$.

Like the cross, the "H" is capable of producing more complex potentials, most notably double-well potentials (cf. Chapter 7).

2.4.5
Finite Wire Dimensions

For a cylindrical wire, the field outside the conductor is identical to that of an infinitely thin wire centered on the cylinder axis. Micro-fabricated wires, however, typically have a rectangular cross-section and an aspect ratio smaller than one (width greater than height). The field of a long conductor of this kind is well approximated

---

4) Note however that even the potential of the "H" does not retain the full symmetry of the conductor shape.

by that of an infinitely long wire of zero height, but non-zero width w, for which an analytical formula exists [33]:

$$B(z) = \frac{\mu_0}{\pi} \frac{I}{w} \text{arccot} \frac{2z}{w} = \frac{\mu_0}{\pi} \frac{I}{w} \left(\frac{\pi}{2} - \arctan \frac{2z}{w} \right). \tag{2.22}$$

Here, z is the distance from the wire surface. For $z \lesssim w$, the formula simplifies to

$$B(z) \approx \frac{\mu_0}{\pi} \frac{I}{w} \left(\frac{\pi}{2} - \frac{2z}{w} \right).$$

The surface field is

$$B_s = \frac{\mu_0}{2} \frac{I}{w}. \tag{2.23}$$

This field constitutes an upper limit to the trap depth that can be achieved. It equals the field created by the same current through an infinitely thin wire at $z = w/2$. The gradient of the field Eq. (2.22) is

$$B'(z) = -\frac{\mu_0}{2\pi} \frac{I}{z^2 + (w/2)^2}. \tag{2.24}$$

Instead of the $1/r^2$ dependence of the thin wire gradient, the broad wire gradient is of Lorentzian form. The gradient at the surface is

$$B'_s = -\frac{2\mu_0}{\pi} \frac{I}{w^2}, \tag{2.25}$$

which imposes a limit to the achievable trap frequency.

Figure 2.5 shows the result of a measurement in which Eq. (2.22) was experimentally verified for a conductor with $w = 300$ μm and ~ 10 μm height. The deviation from the thin-wire dependence becomes notable for $z \lesssim w$. Analytical formulas also exist for wires with rectangular cross-section [35].

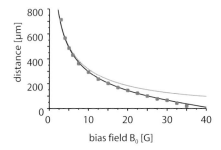

Figure 2.5 Measured distance z between the center of the trapped atom cloud and the surface as a function of the bias field $\tilde{B}$. The trapping field is generated with a "U"-shaped quadrupole wire as in Figure 2.4. The wire width is 300 μm, its current is kept constant at $I = 2$ A. Circles are experimental values obtained from absorption images, the solid line is the calculated position of the potential minimum obtained from Eq. (2.22) for $w = 300$ μm and $I = 2$ A. For comparison, the gray line gives the theoretical position for an infinitely thin wire.

2.4.6
Maximum Confinement

Compression plays a crucial role in BEC experiments because it increases the elastic collision rate $\gamma_e = n\sigma\bar{v}$, where n is the density, σ the elastic scattering cross-section, and $\bar{v}$ the mean velocity. γ_e sets the timescale for evaporative cooling and must be significantly larger than the rate of "bad", inelastic collisions. Adiabatic compression increases both n and $\bar{v}$, and even a sudden compression can still increase γ_e [6]. An upper limit is set by the inelastic three-body collision rate, which increases with n^2 and typically becomes the dominant loss channel for densities in the 10^{15} cm^{-3} range. In typical macroscopic traps, such densities occur only at the end of the evaporative cooling phase. The relatively weak compression in these traps is responsible for the long evaporation time, which in turn entails the requirement for a 10^{-11}–mbar UHV. In this subsection, we will quantify the confinement that can be reached with wires on chips.

2.4.6.1 Field Gradient

The preceding section 2.4.5 has shown that the maximum magnetic-field gradient B'_{max} created by a current-carrying straight wire of transverse dimension w is:

$$|B'_{max}| \approx \frac{\mu_0}{2} j, \tag{2.26}$$

with only a slight dependence on the wire's cross-sectional geometry (circular, rectangular, ...). For thin wires (up to a few µm), maximum current densities of $j \sim 10^{11}$ can be reached at room temperature (see Chapter 3). This leads to $|B'_{max}| \sim 6 \times 10^6$ G/cm. For a wire width $w = 1$ µm, this value would be reached with a total current $I = 0.1$ A.

2.4.6.2 Field Curvature and Trap Frequency

Because of the addition of a bias field (cf. Section 2.3.2), stable traps typically have a quadratic potential near the trap center. Taking into account the results of the previous section, what is the maximum curvature and trap frequency in a chip trap? The transverse frequency $\omega_\perp$ in a guide or a "dimple" trap is given by Eq. (2.21) and depends not only on B'_{max}, but also on the bias field B_0. We require $B_0 = \alpha \omega_\perp$ with typically $\alpha \sim 10^{-6}$ Gs to make spin-flip loss negligible (Section 2.3.4). This leads to

$$\omega_{max} = a_j j^{\frac{2}{3}} \tag{2.27}$$

with

$$a_j \sim \left(\frac{\mu_m}{\alpha m}\right)^{\frac{1}{3}} \mu_0^{\frac{2}{3}}.$$

For our example state and $j = 10^{10}$ A/m^2, the result is $\omega_{max} \sim 2\pi \times 1$ MHz.

In summary, for lithographic wires with a width of a few microns and below, maximum field gradients can reach about 10^7 G/cm and lead to trap frequencies up to a MHz. This is orders of magnitude above the values of conventional magnetic traps, explaining why chip traps can achieve fast evaporation, on the order of a second for reaching BEC. In fact, the limit is imposed by three-body loss rather than trap compression.

The very high trap frequencies we just calculated can be employed with single trapped atoms, as with trapped ions, to enter the Lamb–Dicke regime and potentially to resolve motional sidebands (see Chapter 13). The Lamb–Dicke regime is defined by $\eta \ll 1$, where $\eta = \sqrt{\omega_r/\omega_t}$ is the Lamb–Dicke factor, ω_t is the trap angular frequency, and $\omega_r = \hbar k_L^2/(2m)$ is the recoil angular frequency for an atom of mass m and a laser wave vector k_L. For optical transitions, $\omega_r/(2\pi)$ is typically a few kHz, so $\eta \gg 1$ can indeed be reached, whereas a trapping frequency larger than the linewidth of the cooling transition (i.e., the resolved sideband regime) will be hard or impossible to attain.

2.4.7
Combining Elements: Arrays, Conveyors and Others

More complex potentials can be constructed from the elements described above. A good example is the configuration of Figure 1.12 in Chapter 1 (p. 26). This is an array of wire crosses, each producing a potential well as described in 2.4.3 – the structure generates a two-dimensional lattice. As long as the trap-surface distance z_0 is smaller than the wire spacing, the trap frequencies for each well can be calculated to good approximation from the formulas in Section 2.4.3. Such an array is interesting because the currents can all be different, enabling row-column addressing and a wealth of modulation possibilities. To the author's knowledge, this particular structure has not been used in experiments yet. One-dimensional lattices have been realized and are described in Chapter 6.

Another interesting problem is the design of atomic "conveyors" – structures that transport an atomic cloud in a direction parallel to the chip surface. The first such conveyor was demonstrated in [36]. The meandering modulation conductors used in that conveyor (Figure 2.6a) can be considered as an approximation of the structure shown in Figure 2.6b (which requires two levels of metalization, while the meanders need only one). A second-generation conveyor that is based on the scheme of Figure 2.6b is demonstrated and described in detail in [37]. As shown in Figure 2.6c, it uses 5 independent transport currents per period instead of only two. Moreover, the single longitudinal (x-axis) wire is replaced by three parallel wires, enabling this conveyor to operate without external bias fields. These improvements enable significantly smoother transport than the former design, so that this conveyor could be used to transport atoms over distances up to 24 cm [37].

Other examples of more complex potentials are the various double-well potentials discussed in Chapter 7 or the coplanar waveguides described in Chapter 9.

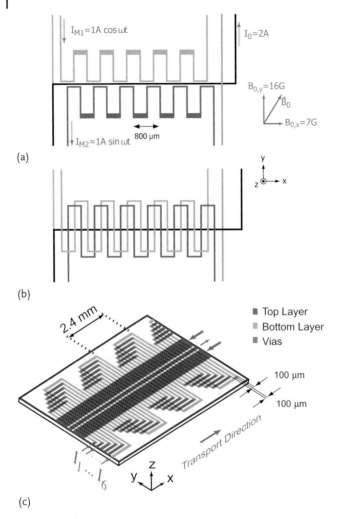

Figure 2.6 Magnetic conveyors. (a) The wire layout used in [36]. Currents and bias fields are indicated. This structure needs only one metalization layer. It can be seen as an approximation of the structure shown in (b), which needs two metalization layers and interconnects (vias) between them. (c) "Long-distance conveyor" (LDC), which has been used to transport a cloud of cold atoms over a distance of 24 cm.

2.5
Real Wires: Roughness and Maximum Current

2.5.1
Effect of Wire Roughness

The edges and the surface of wires on an atom chip inevitably have some roughness due to imperfections in the fabrication process. This leads to deviations of the current flow from a straight path, resulting in "corrugations" of the magnetic potentials they create. Under special conditions – specifically, for very elongated and cold atomic clouds close to the surface – these corrugations have been observed to break up the atomic cloud into several fragments (see [38] and references therein). At a distance d from the wire, only wire roughness on a length scale $s \geq d$ leads to potential corrugations with significant amplitude [38]. Features with $s \ll d$ average out at the position of the atoms. If the atomic ensemble is tightly trapped in all three dimensions, such that its size $R \ll d$, the corrugations manifest themselves only as a constant offset of the field at the trap center, which is easily compensated by adjusting the wire current. If, on the other hand, $R \gg d$, as in waveguiding experiments, the atoms will see the corrugated potential and break up into fragments of size $\geq d$. This shows that wire roughness can be a serious issue for waveguiding experiments on an atom chip. For most other experiments, it is of minor importance.

The amplitude and spectrum of the corrugations depends strongly on the fabrication method, including details such as the particular gold plating solution used in electroplating. The wires on which the corrugated potentials were first demonstrated turned out to be particularly rough (see for example Figure 2 in [39]). Much lower roughness has since been demonstrated, using various standard methods: in [39], gold evaporation and lift-off reduced edge roughness by more than two orders of magnitude compared to the electroplated wire result of the same group; [40] shows that electroplating can also produce excellent results (16 nm rms surface roughness in this case), similar results were obtained in [34]. Furthermore, remaining corrugations can be strongly reduced by a temporal averaging effect when modulating the wire current, as demonstrated in [14].

2.5.2
Heat Transport and Maximum Current

2.5.2.1 Wire–Substrate Interface

The maximum current that a chip wire can sustain is an important parameter, determining quantities such as the strongest confinement and the maximum trap–surface distance. Maximum-current problems also occur in microelectronics. So far, electromigration and other effects leading to aging on the timescale of years have not been a major concern for atom chip researchers, who have focused on short-time performance on the typical timescale of their experiments: what is the maximum current that can be applied for a few seconds, if the heating phase is followed by a cool-down phase of similar duration? In this situation, damage done by overheating is the major limiting factor.

We will start by considering a simple model given in [41] (Figure 2.7). The ohmic heat per unit length deposited in the wire is evacuated towards the substrate through an interface of thermal conductance K (in W/(m² K)). Here we suppose that the substrate remains at the initial temperature T_0. (We will come back to the case of large total power, where this condition is no longer fulfilled.) We also assume that the temperature is constant across the wire profile, which is true if the

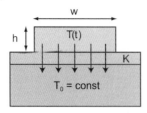

Figure 2.7 Heat transport model: Temperature is assumed to be homogeneous inside the wire of width w and height h. The ohmic heat is evacuated through a layer of thermal conductance K into a reservoir which is held at constant temperature T_0.

wire is not too thick. The temperature rise dT in the wire per time interval dt is then given by

$$Cwh\,dT = \frac{\rho(T)}{wh}(jwh)^2 dt - Kw(T - T_0)dt , \qquad (2.28)$$

where C is the volume heat capacity of the wire, w and h its width and height, and j the current density. Note that w factors out, as would be expected. Temperature variation of the conductivity is taken into account, $\rho(T) = \rho_0(1 + \alpha(T - T_0))$, which turns out to be essential. If the wire is initially at T_0 and a constant current is switched on at $t = 0$, the temperature evolution is

$$T(t) = T_0 + \frac{j^2 \rho_0 h}{K - \alpha j^2 \rho_0 h}\left(1 - e^{-t/\tau}\right) \quad \text{with} \quad \tau = \frac{hC}{K - \alpha \rho_0 j^2 h} . \qquad (2.29)$$

If j is too high, the increased heat transport due to the rising temperature difference is overwhelmed by a still stronger increase in the dissipated power due to the temperature dependence of ρ. In this case, $\tau < 0$, the temperature rises exponentially, and the wire is destroyed. This happens at a critical current density

$$j_c = \sqrt{K\alpha\rho_0 h} . \qquad (2.30)$$

Below this value, the wire temperature exponentially approaches an equilibrium value with a time constant τ.

The parameters of this model are well known a priori, except for K. Ideally, $K = k/h_i$, where h_i is the thickness of the interface layer and k its thermal conductivity (in W/(m K)). In reality, K also depends on parameters such as the deposition technique used to grow the wire, the presence of thin oxide layers and so on. In [41], K was deduced from the temperature rise (more precisely, from the measured increase in resistance). The authors found $K = 6.5, 3.5, 2.6, 2.3 \times 10^6$ W/(m² K) for evaporated gold wires on the following substrates, respectively: Si with 20 nm SiO$_2$, sapphire, Si with 500 nm SiO$_2$, and GaAs. Resistivity data for some relevant materials are listed in Table 3.1 on page 64. Inserting into Eq. (2.30), the model predicts a destruction threshold in the low 10^{11} A/m² range for micron-sized gold conductors at room temperature. This is indeed very close to measured values [28, 41]. For thicker wires, both calculated and measured maximum current densities are lower. This is simply because for a given j, a given section of the wire–substrate interface must carry away an increasing thermal power when h increases. The model also predicts the equilibrium temperature of the wire, which is less than 100 K above T_0

for $j_c/2$, again in good agreement with measurements. Finally, the time constant τ is in the micro-second range or below – generally too fast to be measured.

Nevertheless, one should not blindly trust the calculation to establish the maximum current for a given wire. If, for example, some irregularity in fabrication has led to a local constriction or bad adhesion to the substrate, the wire will locally heat and fail at this spot, like a fuse. A good way to test wires is to measure the rise in resistance. [41] suggests a resistance increase by 50 % as a safe maximum. The percentage may also be lower depending on the fabrication spread.

2.5.2.2 Heat Evacuation through the Substrate

When the total dissipated power becomes large – for larger cross-sections and/or total wire lengths – substrate heating can no longer be neglected. In the resistance-based temperature measurements, this shows up as a slow increase after the first, sudden step. To limit this heating, substrate materials with high thermal conductivity such as AlN and Si should be used, and should be attached to some form of heatsink. Cells with an "atom chip wall", as described in Section 2.7.2 below, provide an elegant solution, as the back side of the chip is outside the vacuum. A model taking into account the properties of the substrate and its coupling to a reservoir has been developed and experimentally confirmed in [42]. Taking the limiting case where there is no insulating layer between the wire and the substrate, the wire remains close to the temperature T_0 of the substrate underneath. T_0 however is no longer constant, but increases over time at a speed determined by the thermal contact between the substrate and the reservoir. The timescale of this increase is much slower than for the wire–substrate interface.

2.6
Loading Techniques

Neutral-atom traps are shallow, and laser cooling is a prerequisite to reaching the low kinetic energies that allows them to be trapped. The standard laser-cooling techniques are the magneto-optical trap and optical molasses, both requiring three mutually orthogonal pairs of laser beams. Unless a transparent substrate is used, this beam arrangement cannot be used close to a chip surface, and so chip traps require specially adapted laser cooling configurations. In the following sections, we discuss the two experimental configurations that are in widespread use today, and briefly discuss the optimum choice of parameters for transferring the cloud of laser-cooled atoms into the magnetic trap.

2.6.1
Mirror-MOT

One solution is to apply a mirror coating to the chip surface and to generate two of the six beams by reflection on this mirror, while the fourth and sixth beam are parallel to its surface [29] (Figure 2.8). There is a catch: the circular polarizations of

2 Trapping and Manipulating Atoms on Chips

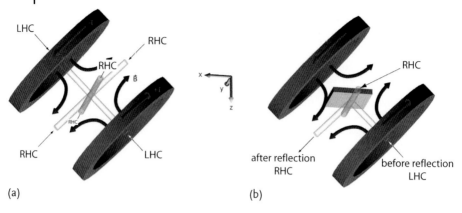

Figure 2.8 Standard MOT (a) and mirror-MOT (b). When the mirror (reflective chip substrate) is rotated by 45° with respect to the quadrupole axis, the polarization of the reflected beams is as required. (RHC: right-hand circular, LHC: left-hand circular polarization.)

the reflected beams must be the ones required by the MOT, which means that they must respect the symmetry imposed by the MOT's magnetic quadrupole field. It turns out that there is exactly one configuration that fulfils this requirement, shown in Figure 2.8b. The magnetic quadrupole axis is tilted by 45° with respect to the chip surface.

In some implementations, the mirror-MOT is charged from a cold atomic beam generated by a 2D+ MOT [43] in an attached cell with higher alkali vapor pressure [44]. This variant enables high atom numbers and fast loading with a simple experimental protocol, at the price of a somewhat increased complexity of the vacuum and laser setup.

2.6.2
Magnetic Elevator

Alternatively, atoms can be cooled in a standard MOT that is located far from the chip surface, so that the chip does not obstruct the MOT beams. In that case, a moveable trap must be employed to transport the atoms toward the chip after the MOT phase. Such a scheme was successfully employed in [45]. Here, the transport was magnetic and was achieved with a combination of two pairs of coils, which transported the atomic sample over a distance of 5 cm.

2.6.3
"Mode Matching"

After loading and precooling in the MOT, the atoms are compressed in some form of compressed or "dark" MOT, further cooled by a few milliseconds of polarization gradient cooling in the absence of magnetic fields (optical molasses), and finally optically pumped into a single Zeeman state (see for example [6]). In order to maintain the phase-space density thus achieved and start evaporation with a high collision rate, it is important to adapt the initial magnetic trap to the shape of the cold atomic cloud. For the case of a gaussian-shaped cloud transferred to a harmonic trap, an analytical formula for the required trap frequency as a function of the cloud's rms radius and temperature can be found [6]: the required trap frequency tends to be very low. But the trap must also be deep enough to contain the cloud – "spillover" must be avoided. Keeping in mind that the trap depth can only be as large as the difference between the external field – typically $\tilde{B}_y$ – and the field in the trap center B_0, Eq. (2.13) shows that low trap frequency and large depth simultaneously can only be achieved by maximizing the wire current.

2.7 Vacuum Cells

Chapter 3 gives a detailed description of chip fabrication. Here we discuss the assembly steps from the finished chip to the complete vacuum apparatus. These are the requirements:

- Maintain an acceptable vacuum. Traditional BEC experiments require a base pressure in the lower 10^{-11} mbar region. Chip experiments perform well with an order of magnitude higher pressure, which is due to the faster evaporation in the strongly compressed chip traps. This is still an ultra-high vacuum (UHV), so that materials must be carefully chosen and assembled.
- Good optical access to the chip area.
- Heat evacuation from the chip to outside the vacuum.
- Multi-wire electrical connections to the chip, and vacuum feedthroughs for these wires. In current designs, up to 30 DC connections are required, carrying up to 3 A each. The tendency is obviously rising. Some applications also involve radiofrequency and/or microwave connections.

Several designs exist that fulfill these requirements. All of them involve an ion pump and either a titanium sublimator or a non-evaporable getter (NEG). They differ in chip mounting and electrical feedthrough techniques, which in turn leads to significant differences in total size, weight, and complexity of the apparatus.

2.7.1 Traditional Cell

A first approach is to mount the chip completely inside the vacuum cell, using established standard techniques for electrical feedthrough and cell assembly (Figure 2.9a). As in traditional BEC apparatus, there are two options to achieve the optical access: standard viewports mounted on the flanges of a stainless-steel chamber, or a custom-made glass cell as shown in the figure. The glass cell is more compact and gives better optical access, but there are no established techniques to apply antireflection (AR) coatings to the inner surfaces. Furthermore, attaching the glass cell

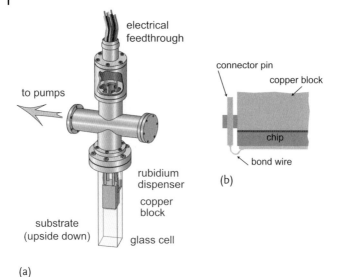

Figure 2.9 Traditional cell design. (a) Partial view of the vacuum cell (pumps not shown). (b) Detail showing chip connectorization.

to the rest of the apparatus (which is typically made from standard stainless-steel components of the "CF" type) requires special techniques such as Helicoflex or indium seals, which also tend to limit the bakeout temperature. Viewports can be ordered with AR coatings on both sides, can be baked out at 200 °C or even above, and this solution tends to be less expensive than a custom-made glass cell with its optically contacted surfaces. However, apart from the reduced optical access, the viewport solution leads to a larger cell, which in turn requires larger and higher-power magnetic coils for generating the homogeneous external fields.

The chip mount in such a design is typically a copper block attached to a CF flange. One issue is heat evacuation through this flange, stainless steel having a particularly bad heat conductivity (cf. Table 2.1 below). This can be solved by replacing the copper gasket that seals this flange by a thick copper ring. The chip mount is then attached to this ring, rather than to the flange. The ring is accessible from the outside, so that it can be water cooled, or a heat sink can be attached to it. (Note that the sealing surfaces of this ring must be reworked on the lathe each time the seal is opened.)

Apart from the larger size, the main technical difficulty in the traditional cell design is the chip connectorization and electrical feedthrough, especially when the pin count is high. Figure 2.9b shows one solution, applied in the first chip experiments in Munich [29, 46]: a row of gold-plated connector pins is attached to the chip holder. Wire bonding is used to connect the chip wires to the pins. (When using standard Au bonding wires with 25 µm diameter, five to ten bonds per pin should be used for maximum currents of a few Amperes.) Kapton-insulated copper wires with crimp plugs (available from several UHV parts manufacturers) connect the pins to a commercial electrical feedthrough. Other labs have developed similar techniques where the wire bonds are sometimes replaced by a spring-loaded connector system.

2.7.2
Compact Cell with Atom Chip Wall

In this more recent design, first described in [47], the chip forms one of the walls of the vacuum cell. This eliminates the need for a separate electrical feedthrough, greatly facilitates heat evacuation, enables easy connectorization with standard printed circuit board (PCB) connectors, and significantly reduces the size of the vacuum apparatus. Figure 2.10a shows one realization. Adequate choice of materials makes it possible to use standard spectroscopy cells, available as a catalog part from several manufacturers. This ensures good optical access as with the custom-made glass cells described above, but at a fraction of the cost. Because of the small volume of the cell, the pumping speed of the ion pump can now be much lower, further reducing the size and the weight of the system, as well as reducing the stray magnetic fields of the pump magnets. The complete system can be held in one hand.

Two options exist to join the chip and the cell (Figure 2.10b and c). UHV compatible two-component epoxy adhesives can be used between any two materials having similar thermal expansion coefficients (CTEs, see Table 2.1 and Table 3.1 on page 64). In particular, Pyrex is reasonably well matched to both AlN and Si substrates. Both combinations have been used with success in Munich and Paris, maintaining base pressures in the low 10^{-10} mbar range even after several years of use. Experience has shown that it is important to employ a special cure schedule: after assembly, the glue must be allowed to pre-cure at room temperature for much longer than its pot lifetime before being cured at high temperature. Reproducible

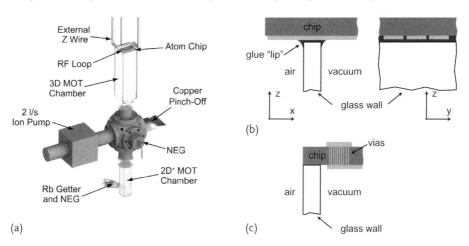

Figure 2.10 Compact cell with atom chip wall. (a) View of the complete vacuum cell (from [44], used with permission). (b) Detail of a glued chip-cell joint. (c) Detail of an anodically bonded chip-cell joint.

Table 2.1 Properties of some metals, ordered by resistivity. ρ resistivity; ρ' temperature coefficient of resistivity; k thermal conductivity; α linear thermal expansion coefficient; ρ_m mass density; T_m melting point.

Material	ρ $10^{-8}\,\Omega\mathrm{m}$	ρ' K^{-1}	k $\mathrm{W\,m^{-1}\,K^{-1}}$	α $10^{-6}\,\mathrm{K}^{-1}$	ρ_m $\mathrm{g\cdot cm^{-3}}$	T_m $^\circ\mathrm{C}$
Ag	1.63	0.0041	429	19.1	10.5	961.9
Cu	1.69	0.0043	401	17.0	8.96	1083
Au	2.20	0.0040	318	14.1	19.30	1064.4
Al	2.67	0.0045	237	23.5	2.70	660.4
Na	4.5	0.0055	128	71	0.97	97.8
In	8.8	0.0052	81.8	24.8	7.3	156.6
Fe	10.1	0.0065	80.4	12.1	7.87	1535
Pt	10.58	0.00392	71.6	9.0	21.45	1772
Rb	12.1	0.0048	58.2	9.0	1.53	38.9
stainl. steel	70–78	–	16.3	15–18	7.9	1400

results have been achieved with Epo-Tek 353ND left to pre-cure for at least 24 h before increasing the temperature. Remarkably, the fact that the CTE of the epoxy adhesives is a full order of magnitude higher than that of AlN, Si, or Pyrex (and still higher above the glue's glass transition temperature) does not appear to impact the reliability of the seal.

The glue seal easily fills the gaps between the chip wires, so that the latter can simply run straight across the glass interface (Figure 2.10b). One complication of the gluing method is the formation of a glue "lip" on the glass, which reduces optical access close to the chip surface. Once this lip has formed, it is difficult to remove without residue. Dispensing a homogenous and well-defined quantity of glue should avoid the problem, but requires virtuoso manual skills or special equipment.

In the special case of a Si-Pyrex seal, an industry-proven technique known as anodic bonding avoids glue altogether: Applying a high voltage across the interface while heating the pieces to more than 300 °C creates a stable, hermetically fused seal. The surfaces to be bonded must have low roughness and must be flat and parallel, which is increasingly difficult to achieve as the surface becomes larger. Furthermore, the chip wires may no longer run across the glass interface, so that hermetic vias must be employed as shown in Figure 2.10c. These requirements make anodic bonding more difficult to employ, but it enables higher bake-out temperatures and should allow still lower base pressures than glued seals. This technique is employed at JILA [44, 47] and in commercially available atom chip cells [3].

Additional information can be found in Chapter 3. Note that most of the techniques introduced here can also be used to simplify non-atom chip vacuum cells.

2.8
Conclusion and Outlook

In the first decade of their existence, atom chips have developed from a cool but exotic novelty into a serious option for many cold atom experiments. They have enabled the wide variety of results described in this book; yet, it seems reasonable to predict that these were only the first steps. In the beginning, atom chips were used almost exclusively by those who also developed them. This situation has changed. A small but growing number of laboratories now uses atom chips simply because it seems the right technology choice for their scientific goals: doing research *with* atom chips no longer requires a group to do research *on* atom chips. This is a healthy and encouraging development, to which several factors have contributed. Standard techniques have evolved, which have been reproduced in a large number of laboratories and are described in a host of articles. Importantly, it is becoming easier to obtain suitable chips. Atom chip vacuum cells have become commercially available. It also turned out that fairly modest micro-fabrication techniques are sufficient to produce standard atom chips. These techniques are increasingly well described in the literature, and are spreading in the community. This is occurring in a period where interaction between solid-state and atomic physicists is generally getting more intense, creating an interesting resonance.

On the way to a small and robust, "turnkey" cold atom experiment for lab and field applications, what remains to be done? The short answer is that systems coming close to this vision have already been demonstrated [4, 44]. However, their optical setups are still mounted mirror by mirror and screw by screw, in much the same way as in the first BEC experiments more than 15 years ago, even though they are sized down by an impressive factor. This way of assembling optical circuits is akin to electronics in the pre-integrated circuit age. At telecom wavelengths, much more advanced solutions exist and are available at low cost, but not at any suitable alkali wavelength. In this respect, the most important roadblock is no longer in the field of traps and vacuum apparatus, it is in the optical system.

Whether there is a future for the atom chip outside the research laboratory depends of course on the future of cold atom technologies in general. Today, the most well defined such technologies are frequency standards – including those in satellite navigation systems – and atom interferometry. Chip-based atomic clocks are under development (see Chapter 8) and compete with other novel secondary frequency standards. As far as chip-based atom interferometry is concerned, several novel beamsplitters have been demonstrated in proof-of-principle experiments and employed to study quantum gases (Chapter 7). Whether they can be useful for applications such as inertial navigation is an important open question. The answer depends on future developments in atom chip technology, but also and more fundamentally on the general properties of trapped-atom interferometers, which have been little explored so far.

Finally, research in quantum information has raised hopes for the emergence of "entanglement-based technologies" as a new class of technologies of which quantum computing might be only a first example. Moreover, already today, quantum information opens fascinating new perspectives on quantum physics, and these are developed in the interplay of theory and experiment. Recent experiments demonstrating squeezed-state generation (Chapters 8 and 9) and single-atom manipulation (Chapters 3 and 9) justify the hope that atom chips can make significant contributions to this field.

Acknowledgments

I am indebted to the former and present Ph.D. students and postdocs of the Paris and Munich atom chip groups. Their work has contributed to shaping the field, and their thesis manuscripts have had a lot of influence on this chapter.

References

1 Hunger, D., Camerer, S., Hänsch, T.W., König, D., Kotthaus, J.P., Reichel, J., and Treutlein, P. (2010) Resonant Coupling of a Bose–Einstein Condensate to a Micromechanical Oscillator. *Phys. Rev. Lett.* **104**, 143002.

2 Colombe, Y., Steinmetz, T., Dubois, G., Linke, F., Hunger, D., and Reichel, J. (2007) Strong atom-field coupling for Bose–Einstein condensates in an optical cavity on a chip. *Nature* **450**, 272.

3 www.coldquanta.com.

4 van Zoest, T., Gaaloul, N., Singh, Y., Ahlers, H., Herr, W., Seidel, S.T., Ertmer, W., Rasel, E., Eckart, M., Kajari, E., Arnold, S., Nandi, G., Schleich, W.P., Walser, R., Vogel, A., Sengstock, K., Bongs, K., Lewoczko-Adamczyk, W., Schiemangk, M., Schuldt, T., Peters, A., Konemann, T., Muntinga, H., Lammerzahl, C., Dittus, H., Steinmetz, T., Hansch, T.W., and Reichel, J. (2010) Bose–Einstein Condensation in Microgravity. *Science* **328**, 1540.

5 Metcalf, H.J. and van der Straten, P. (2003) Laser cooling and trapping of atoms. *J. Opt. Soc. Am. B* **20**, 887.

6 Ketterle, W., Durfee, D.S., and Stamper-Kurn, D.M. (1999) *Making, Probing and Understanding Bose–Einstein Condensates*, in *Proceedings of the International School of Physics "Enrico Fermi", Course CXL* (eds M. Inguscio, S. Stringari, and C. Wieman), IOS, Amsterdam, pp. 67–176.

7 Dalfovo, F., Giorgini, S., Pitaevskii, L.P., and Stringari, S. (1999) Theory of Bose–Einstein condensation in trapped gases. *Rev. Mod. Phys.* **71**, 463.

8 Leggett, A.J. (2001) Bose–Einstein condensation in the alkali gases: Some fundamental concepts. *Rev. Mod. Phys.* **73**, 307.

9 Krüger, P., Luo, X., Klein, M.W., Brugger, A., Haase, A., Wildermuth, S., Groth, S., Bar-Joseph, I., Folman, R., and Schmiedmayer, J. (2003) Trapping and Manipulating Neutral Atoms with Electrostatic Fields. *Phys. Rev. Lett.* **91**, 233201.

10 Dumke, R., Volk, M., Müther, T., Buchkremer, F.B.J., Birkl, G., and Ertmer, W. (2002) Micro-optical realization of arrays of selectively addressable dipole traps: A scalable configuration for quantum computation with atomic qubits. *Phys. Rev. Lett.* **89**, 097903.

11 Pollock, S., Cotter, J.P., Laliotis, A., and Hinds, E.A. (2009) Integrated magneto-optical traps on a chip using silicon pyramid structures. *Opt. Express* **17**, 14109.

12 Gleyzes, S., Amili, A.E., Cornelussen, R., Lalanne, P., Westbrook, C., Aspect, A., Estève, J., Moreau, G., Martinez, A., Lafosse, X., Ferlazzo, L., Harmand, J., Mailly, D., and Ramdane, A. (2009)

Towards a monolithic optical cavity for atom detection and manipulation. *Eur. Phys. J. D* **53**, 107.

13 Petrich, W., Anderson, M.A., Ensher, J.R., and Cornell, E.A. (1995) Stable, Tightly Confining Magnetic Trap for Evaporative Cooling of Neutral Atoms. *Phys. Rev. Lett.* **74**, 3352.

14 Trebbia, J.-B., Garrido Alzar, C.L., Cornelussen, R., Westbrook, C.I., and Bouchoule, I. (2007) Roughness Suppression via Rapid Current Modulation on an Atom Chip. *Phys. Rev. Lett.* **98**, 263201.

15 Kügler, K.-J., Paul, W., and Trinks, U. (1978) A magnetic storage ring for neutrons. *Phys. Lett.* **72B**, 422.

16 Griesmaier, A., Werner, J., Hensler, S., Stuhler, J., and Pfau, T. (2005) Bose–Einstein Condensation of Chromium. *Phys. Rev. Lett.* **94**, 160401.

17 Lu, M., Youn, S.H., and Lev, B.L. (2010) Trapping Ultracold Dysprosium: A Highly Magnetic Gas for Dipolar Physics. *Phys. Rev. Lett.* **104**, 063001.

18 Wing, W.H. (1984) On neutral particle trapping in quasistatic electromagnetic fields. *Prog. Quantum Electron.* **8**, 181.

19 Ketterle, W. and Pritchard, D.E. (1992) Trapping and Focusing Ground-State Atoms with Static Fields. *Appl. Phys. B* **54**, 403.

20 Majorana, E. (1932) Atomi orientati in campo magnetico variabile. *Nuovo Cimento* **9**, 43.

21 Bergeman, T., Erez, G., and Metcalf, H.J. (1987) Magnetostatic trapping fields for neutral atoms. *Phys. Rev. A* **35**, 1535.

22 Pritchard, D.E. (1983) Cooling Neutral Atoms in a Magnetic Trap for Precision Spectroscopy. *Phys. Rev. Lett.* **51**, 1336.

23 Bagnato, V.S., Lafyatis, G.P., Martin, A.G., Raab, E.L., Ahmad-Bitar, R.N., and Pritchard, D.E. (1987) Continuous stopping and trapping of neutral atoms. *Phys. Rev. Lett.* **58**, 2194.

24 Gott, Y.V., Ioffe, M.S., and Tel'kovskii, V.G. (1962) *Nuclear Fusion Supplement* **3**, 1045.

25 Moore, K.L., Purdy, T.P., Murch, K.W., Brown, K.R., Dani, K., Gupta, S., and Stamper-Kurn, D.M. (2006) Bose–Einstein condensation in a mm-scale Ioffe–Pritchard trap. *Appl. Phys. B* **82**, 533.

26 Gov, S., Shtrikman, S., and Thomas, H. (2000) Magnetic Trapping of Neutral Particles: Classical and Quantum-Mechanical Study of a Ioffe–Pritchard Type Trap. *J. Appl. Phys. D* **87**, 3989.

27 Weinstein, J.D. and Libbrecht, K.G. (1995) Microscopic magnetic traps for neutral atoms. *Phys. Rev. A* **52**, 4004.

28 Drndić, M., Johnson, K.S., Thywissen, J.H., Prentiss, M., and Westervelt, R.M. (1998) Micro-electromagnets for atom manipulation. *Appl. Phys. Lett.* **72**, 2906.

29 Reichel, J., Hänsel, W., and Hänsch, T.W. (1999) Atomic Micromanipulation with Magnetic Surface Traps. *Phys. Rev. Lett.* **83**, 3398.

30 Müller, D., Anderson, D.Z., Grow, R.J., Schwindt, P.D.D., and Cornell, E.A. (1999) Guiding Neutral Atoms Around Curves with Lithographically Patterned Current-Carrying Wires. *Phys. Rev. Lett.* **83**, 5194.

31 Hänsel, W. (2000) Magnetische Mikrofallen für Rubidiumatome. Ph.D. thesis, Ludwig-Maximilians-Universität München.

32 Thywissen, J.H., Olshanii, M., Zabow, G., Drndić, M., Johnson, K.S., Westervelt, R.M., and Prentiss, M. (1999) Microfabricated magnetic waveguides for neutral atoms. *Eur. Phys. J. D* **7**, 361.

33 Reichel, J., Hänsel, W., Hommelhoff, P., and Hänsch, T.W. (2001) Applications of Integrated Magnetic Microtraps. *Appl. Phys. B* **72**, 81.

34 Treutlein, P. (2008) Coherent manipulation of ultracold atoms on atom chips. Ph.D. thesis, Ludwig-Maximilians-Universität München.

35 Estève, J. (2004) Du miroir au guide d'onde atomique: effets de rugosité. Ph.D. thesis, Université Pierre et Marie Curie (Paris VI).

36 Hänsel, W., Reichel, J., Hommelhoff, P., and Hänsch, T.W. (2001) Magnetic Conveyer Belt for Transporting and Merging Trapped Atom Clouds. *Phys. Rev. Lett.* **86**, 608.

37 Long, R., Rom, T., Hänsel, W., Hänsch, T.W., and Reichel, J. (2005) Long distance magnetic conveyor for pre-

cise positioning of ultracold atoms. *Eur. Phys. J. D* **35**, 125.

38 Fortágh, J. and Zimmermann, C. (2007) Magnetic microtraps for ultracold atoms. *Rev. Mod. Phys.* **79**, 235.

39 Schumm, T., Estève, J., Figl, C., Trebbia, J.-B., Aussibal, C., Mailly, D., Bouchoule, I., Westbrook, C., and Aspect, A. (2005) Atom chips in the real world: the effetcs of wire corrugation. *Eur. Phys. J. D* **32**, 171.

40 Koukharenk, E., Moktadir, Z., Kraft, M., Abdelsalam, M.E., Bagnall, D.M., Vale, C., Jones, M.P.A., and Hinds, E.A. (2004) Microfabrication of gold wires for atom guides. *Sens. Actuators A* **115**, 600.

41 Groth, S., Krüger, P., Wildermuth, S., Folman, R., Fernholz, T., Schmiedmayer, J., Mahalu, D., and Bar-Joseph, I. (2004) Atom chips: Fabrication and thermal properties. *Appl. Phys. Lett.* **85**, 2980.

42 Armijo, J., Garrido Alzar, C.L., and Bouchoule, I. (2010) Thermal properties of AlN-based atom chips. *Eur. Phys. J. D* **56**, 33.

43 Dieckmann, K., Spreeuw, R.J.C., Weidemüller, M., and Walraven, J.T.M. (1998) Two-dimensional magneto-optical trap as a source of slow atoms. *Phys. Rev. A* **58**, 3891.

44 Farkas, D.M., Hudek, K.M., Salim, E.A., Segal, S.R., Squires, M.B., and Anderson, D.Z. (2009) A Compact, Transportable, Microchip-Based System for High Repetition Rate Production of Bose–Einstein Condensates. arxiv: 0912.0553.

45 Aubin, S., Extavour, M.H.T., Myrskog, S., LeBlanc, L.J., Estève, J., Singh, S., Scrutton, P., McKay, D., McKenzie, R., Leroux, I.D., Stummer, A., and Thywissen, J.H. (2005) Trapping Fermionic ^{40}K and Bosonic ^{87}Rb on a Chip. *J. Low Temp. Phys.* **140**, 377.

46 Hänsel, W., Hommelhoff, P., Hänsch, T.W., and Reichel, J. (2001) Bose–Einstein condensation on a microelectronic chip. *Nature* **413**, 498.

47 Du, S., Squires, M.B., Imai, Y., Czaia, L., A. Saravanan, R., Bright, V.M., Reichel, J., Hänsch, T.W., and Anderson, D.Z. (2004) Atom chip Bose–Einstein condensation in a portable vacuum cell. *Phys. Rev. A* **70**, 053606.

3
Atom Chip Fabrication

Ron Folman, Philipp Treutlein, and Jörg Schmiedmayer

3.1
Introduction

One of the key promises of atom chips is the building of a robust quantum laboratory by miniaturizing and integrating quantum optics and atomic physics tools on a single device, on a chip [1–3]. This vision follows the path taken previously by the micro-electronics and micro-optics fields. The advantages and strengths of the specific field, in our case quantum optics and atomic physics, are combined with the technological potential of microfabrication and (large scale) integration to build a robust platform for implementation of quantum operations. An important ingredient in developing such an integrated, micro-fabricated approach to manipulating atoms, molecules, or ions is the fabrication of the devices. The possibilities to combine vastly different technologies is thereby a key factor. This creates the technological basis for combining the best of the different quantum worlds of photons, atoms and solid-state in a single integrated quantum device.

This overview of atom chip fabrication is organized as follows: We first discuss the challenges to be faced when starting to conceive and fabricate chips for the (quantum) manipulation of atoms. We then describe the various ingredients and the corresponding fabrication methods. We focus not only on the currently most active and successful areas – current carrying wires and integrated photonics, but also look at more visionary approaches, examples being superconducting chips or the manipulation of atoms with real nanostructures such as carbon nanotubes.

Here we explicitly discuss the material engineering and fabrication of atom chips for the manipulation of neutral atoms, but the same concept of robustness and versatility through miniaturization and integration can also be applied to manipulate (polar) molecules, ions, or trapped electrons.

Atom Chips. Edited by Jakob Reichel and Vladan Vuletić
Copyright © 2011 WILEY-VCH Verlag GmbH & Co. KGaA, Weinheim
ISBN: 978-3-527-40755-2

3.2
Fabrication Challenges

In building atom-physics-based quantum devices on a chip, the challenge is not so much in micro- and nano-scale miniaturization, but rather in the integration of numerous technologies on the same device, and in the exceptional quality required from both the materials and the fabrication. Besides being UHV compatible, different requirements need to be addressed, depending on the specific application we have in mind.

- Current carrying wires are the workhorse of atom chips. Their key role is to provide magnetic fields for trapping and manipulation. These wires have to support high current densities, in some cases $> 10^7$ A/cm^2 [4, 5], many orders of magnitude larger than in a light bulb. Ultra-cold atoms are extremely sensitive to variations in the magnetic potential. These can be caused by non-uniform current flow in the wires. Current flow deviations of $\sim 10^{-6}$ rad can be seen in 1d BEC (Bose–Einstein condensate) experiments [6, 7]. A deviation of $> 10^{-4}$ rad makes a trap unusable for many experiments.
- Micro-structured magnetic films and permanent magnets provide a very attractive way to create strongly confining microtraps at very high integration density [8–12]. One of the key issues in fabrication thereby is again the homogeneity of the magnetic materials [13].
- Radio-frequency (RF) and microwave (MW) fields complement the static magnetic trapping potentials [14–20]. Coupling two different ground states creates dressed state potentials similar to optical dipole potentials. The wire structures and the substrates need to be chosen well to support precise microwaves.
- Electric fields offer a route to detailed structuring and manipulation [21]. At the micron scale the substrates and the fabricated structures have to withstand the large electric fields created at even moderate voltages.
- Multi-layer chips [20, 22, 23]: In an integrated atom chip, large structures (10 to 500 µm) for the initial trapping and cooling of atoms have to be combined with functional structures on the (sub) micrometer scale to achieve controlled manipulation at a length scale where tunnel coupling between sites becomes important. Often the versatile application of electric and magnetic potentials requires the crossing of wires on the chip without contact.
- In many applications the chip surface needs to be a high-quality mirror, be it for cooling and accumulating the atoms in a mirror-MOT (magneto-optical trap) or for imaging atoms close to the surface. Furthermore, reflection of light from the chip surface allows one to integrate optical dipole traps and optical lattices on the atom chip [24]. The latter puts very stringent requirements on the quality of the mirror surface.
- By integrating micro-optics and fiberoptics on the atom chip [25, 26] one can incorporate optical micromanipulation and efficient detection of atoms. The fabrication methods for the optics have to be compatible with the other processes. Integrating fully tunable optical microcavities pose an even larger challenge [27].

- Integration of superconducting elements on the chip will require compatibility with cryogenics.
- The atom chip is an ideal platform to integrate other solid-state quantum devices like Micro-Electro-Mechanical systems (MEMs) [28, 29], micro lenses and holes, miniature high-finesse cavities, current-carrying molecules or nanomagnets, crystalline materials, and so on with the atomic physics on the chip. Connecting atoms to superconducting qubits [30, 31] through circuit CQED (cavity quantum electrodynamics) [32, 33] will even require mK temperatures.
- Looking further into the future one can think of integrating more complex devices onto the chips like on-board light and particle sources, miniature vacuum volumes and pumps, and so on (see the last figure of this chapter (Figure 3.33)) to build an independent stand-alone device.

With all these diverse science possibilities each chip has an individual design, tailored to the physics and functions it is required to perform. What is needed is an interplay between physics, material engineering and fabrication, and compatibility of the methods.

3.3
The Substrate

The substrate onto which the atom chip is fabricated has to be compatible with the different fabrication techniques needed to implement the chip design and has to guarantee the robust operation of the device. For a comparison of different substrate materials see Table 3.1.

In general the most important physical properties for present-day atom chips stem from the fact that one of the key functions is to accumulate, cool and manipulate ultra-cold atoms using static or oscillating fields. The substrate of the atom chip has to support the wires, provide electrical insulation between them, and dissipate the heat generated in the wires. Therefore, a substrate with high thermal conductivity, a high heat capacity, and an insulating surface is needed. A second important requirement, needed for smooth magnetic potentials, is the ease of high-quality fabrication (few geometrical defects) and low wire surface roughness. This translates, for example, to very low surface roughness of the substrate.

A common substrate is a single crystal Si wafer in a (100) orientation with a thermal conductivity of $\sim 150\,W/(mK)$ at room temperature. Wafers are typically $\sim 500\,\mu m$ thick and polished with very small surface roughness, sufficient even for the smallest wires. They can be cleaved with a diamond scorer. Neither the resistivity of Si nor the 2 nm native oxide layer on the substrate are sufficiently insulating for controlling DC currents at the 10^{-5} level. Therefore, the wafer is covered by a SiO_2 insulation layer which provides a DC resistance of $> 40\,M\Omega$ between the wires. Since SiO_2 has a thermal conductivity of only $1.5\,W/(mK)$, the insulating layer should be as thin as possible [5, 34].

Table 3.1 Room-temperature properties of commonly used substrates for microfabrication. The properties given are typical values found in data sheets. The actual values can vary considerably for specific samples, depending on crystal orientation, purity, and fabrication procedure. The data was compiled from data sheets of the distributors, and a number of Internet resources like http://www.ioffe.ru, and [2].

Material	Therm. cond. $W\,m^{-1}\,K^{-1}$	Specific Heat $J\,Kg^{-1}\,K^{-1}$	ε_r	loss tang. 10 GHz 10^{-4}	Linear expans. $10^{-6}\,K^{-1}$	Density $g\,cm^{-3}$	Optical transm. nm
BeO	260–300	1000	6.7	30	8.4–9.0	2.86	
AlN	170–280	800	8.9	5	4.4–5.7	3.25	500–3000
BN	740	600	5–7	2–4	1.2–3	3.4	
InN	450	320	15		3–4	6.8	620–2500
GaN	1300	490	10		3–5.5	6.5	400–2500
Macor	1.5	790	5.9	50	13	2.52	
Sapphire	35–40	700	9–11	0.2–0.8	5.8	3.99	200–5500
Alumina	18–35	900	9.6	1	8.0	3.9	
SiN	10–43	680–800	10	10–20	3.3	2.4–3.3	
Fused quartz	1.46	670–740	3.8	1–4	0.54	2.2	180–2500
SiO$_2$	1.46	700	3.9	1–10	0.54	2.2	180–2500
BK7	1.11	858			8	2.51	400–1400
Pyrex	1.13	750	5.1	260	3.25	2.23	300–2500
Polyimide	0.1–0.35	1090	3.4	20–40	30–60	1.42	
Diamond	900–2000	470	5.7	2	0.8	3.52	400–300 000
GaAs	~55	330	10.9–12.9	<6	5.8	5.3	1500–14 000
Si	80–150	700	11.7	<10	4.7–7.6	2.34	1200–15 000
SiC	350–490	690	10.8	30	4.8	3.2	

AlN is a non-toxic polycrystalline ceramic with an excellent heat conductivity of typically 180 W/(mK) at room temperature. Compared with Si, AlN is less brittle. It can be laser machined and easily cleaved with a diamond scorer. A polished AlN surface (specified surface roughness < 40 nm) has a significant residual roughness, with isolated defects of micrometer size. This is a problem if structures < 10 μm are desired.

Both AlN and Si are common substrate materials for microwave circuits. AlN has a dielectric constant of $\varepsilon_r = 8.7$ and a loss tangent of $\tan\delta < 1\times 10^{-3}$ at 10 GHz. If only DC currents are used in the experiment, Si of any doping level can be employed as the substrate material. For microwave applications, however, it is very important to choose high-resistivity Si to avoid strong dielectric losses. A typical wafer has a resistivity of $\rho > 10^4\,\Omega\mathrm{cm}$, $\varepsilon_r = 11.9$, and $\tan\delta = 1\times 10^{-3}$ at 10 GHz.

Other substrates commonly used in atom chip fabrication are GaAs, sapphire, alumina, or SiO$_2$. Especially sapphire is interesting as a substrate for low-temperature microwave applications, where a loss tangent close to 10^{-6} has been observed. Sapphire is also interesting when using light, as it is transparent in the visible light range.

3.4 Lithography

In lithography, a pattern is transferred into a layer of resist on the chip surface. The patterned resist is subsequently used as a stencil for etching or depositing material. An extensive discussion of lithographic techniques can be found in [35]. In the following, we give a brief introduction to optical and electron-beam lithography, which are most commonly used for atom chip fabrication.

3.4.1 Optical Lithography

In optical- or photolithography, light is used for pattern transfer from a mask to a light-sensitive resist. The spatial resolution is, therefore, set by diffraction of light. Optical lithography is a parallel process which allows large chips to be structured in a short time, making it the dominant technique in the semiconductor industry. A typical photolithographic process involves several process steps as illustrated in Figure 3.1. In the following, we discuss this process in more detail.

Masks for photolithography – The photomask is a flat piece of glass (transparent to UV light) carrying a thin absorbing metal pattern (e.g., a 100 nm thick Cr layer) on one side. Depending on the resist system, the pattern on the mask is either a light field or dark field image of the structure to be fabricated. In research laboratories, the most common technique is contact lithography, where the metal pattern on the mask is in direct contact with the photoresist layer during exposure. A *mask aligner* is used to align mask and substrate with respect to each other and expose them to UV light (often, mercury lamps with spectral lines near 400 nm are used). The direct contact of mask and resist can result in defects. An alternative are projection masks, which are imaged by a high-resolution lens system onto the resist. Many commercially available masks are fabricated by laser lithography, that is by writing the pattern with a laser beam of $\simeq 0.7\ \mu m$ spot size into a thin

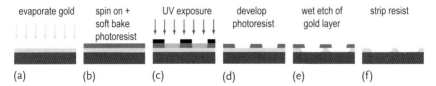

Figure 3.1 Process steps of optical lithography. (a) substrate preparation – (b) spinning resist and soft baking – (c) exposure and post-exposure treatment – (d) resist development – (e) transfer of the resist structure to the substrate – (f) resist stripping. The example uses a contact mask, a positive resist system, and pattern transfer to the metal layer by wet etching.

resist layer on the mask. Masks with smaller structures can be written by electron beam lithography (see below). For coarse structures (> 50 μm), the mask can be simply printed on a transparency. Laser lithography can also be used to avoid photomasks completely, by directly exposing the resist on each chip with the laser beam.

Substrate preparation – Lithography starts with a thorough cleaning of the substrate. Different (combinations of) wet or dry cleaning procedures are used, involving for example organic solvents in an ultrasonic bath, or a strong acid such as H_2SO_4, or an acid–oxidant combination (e.g., 'piranha etch'), or oxygen plasma cleaning. Subsequent to cleaning, some processes require deposition of homogeneous insulating and/or metal layers on the substrate. In the example of Figure 3.1, the substrate is covered with a thin Ti adhesion layer and a gold layer, which will later be patterned by wet etching.

Spinning resist and soft baking – A spin coater is used to apply a photoresist layer of a desired thickness (ranging from a few 100 nm to a few 100 μm, depending on the resist system and application). The photoresist is an organic polymer sensitive to UV radiation. After spin coating, the resist layer is soft baked (prebaked) to remove solvents and promote adhesion to the substrate.

Exposure and post-exposure treatment – The resist-coated substrate is aligned with respect to the photomask and subsequently illuminated with a well-defined intensity of UV light for a controlled duration (typically a few seconds). Finding the proper exposure dose is an important task during optimization of the process. The UV light induces a chemical reaction in the exposed areas of the photoresist, altering the solubility of the resist in a solvent. If a *positive resist* is used, the exposed areas are rendered soluble. In a *negative resist*, the exposed areas are rendered insoluble. After exposure, certain resists require further treatments. For example, an *image reversal resist* requires a post-exposure bake and subsequent flood exposure with UV light, thereby turning the originally positive resist into a negative one. The choice of the resist system for a given application depends on many parameters, such as the desired spatial resolution and resist thickness, the desired shape of the resist profile (positive vs. negative slope of the resist sidewalls, see Section 3.5), and the chemical stability in subsequent processing steps.

Resist development – During development, the resist is selectively dissolved, usually by submerging and agitating the substrate in a liquid developer solution. This transforms the latent image formed during exposure into a resist pattern that will serve as a mask in further deposition or etching steps. Careful control of the development time and developer temperature is required to reproducibly obtain the desired resist profile. A mild oxygen plasma treatment can be useful to remove unwanted resist left behind after development. Furthermore, some applications require postbaking of the developed resist to improve stability.

Transfer of the resist structure to the substrate – Various methods exist to transfer the resist structure to the substrate, and the most common ones for atom chip fabrication are discussed in Section 3.5 below. In the example of Figure 3.1, wires are defined by a wet etch of the metal layers with aqua regia.

Resist stripping – In the last step of the photolithographic process, the resist is removed (stripped). Similar to the procedures for substrate cleaning, a sequence of steps involving organic solvents, acids, acids+oxidants, or plasma cleaning is used. The additional requirement is that the chemicals and procedures employed should not attack the desired structures.

Spatial resolution – The spatial resolution of contact lithography is limited by the near-field diffraction of light at the structures in the mask and thus decreases if there is a gap between the mask and the substrate. Consider photoresist of thickness t exposed with light of wavelength λ through a mask with a pattern of equal lines and spaces of periodicity $2w$. The theoretical resolution, that is the minimum linewidth resolved, is given by $w_{min} = (3/2)\sqrt{\lambda(g + t/2)}$, where g is the gap between resist and mask [35]. To obtain high resolution, g has to be as small as possible. This means that edge beads and defects of the photoresist have to be avoided. The substrate has to be pressed against the mask. In a multi-layer process, the topology of the lower layers (if not planarized perfectly) will introduce gaps and thus decrease resolution. As an example, consider $\lambda = 400$ nm, $t = 1.5$ μm, and $g = 0$, which yields $w_{min} = 0.8$ μm. In practice, achieving such high resolution is possible, but challenging. Routinely, $w > 2-3$ μm is achieved.

Resist bleaching is a phenomenon that allows one, to some extent, to enhance resolution beyond w_{min}. During exposure, the resist becomes more transparent to the illuminating light. This allows one to define narrow features in thick resist, that is the fabrication of structures with high aspect ratio.

3.4.2
Electron-Beam Lithography

Electron-beam lithography allows one to fabricate significantly smaller structures than optical lithography. Moreover, the electron beam has a large depth of focus, facilitating patterning of substrates with uneven topography. E-beam lithography is a serial technique in which a narrow electron beam is scanned across a resist layer pixel-by-pixel. The resolution is limited by electron scattering in the resist and substrate to typically 10–100 nm, depending on resist thickness. The standard positive resist is polymethylmethacrylate (PMMA). Electron bombardment breaks the PMMA into fragments, the resulting shorter polymer chains are dissolved faster during development. Negative resists and multi-layer resist systems are available as well.

E-beam lithography is slow, and writing a complex pattern can easily take several hours, making the system prone to drifts and vibrations. Moreover, the size of the field that can be written without moving the substrate is typically only a few hundred μm, for larger areas stitching is necessary. Nevertheless, the superior resolution make e-beam lithography the method of choice in many applications. The basic steps involved in an electron-beam lithography process are similar to those discussed in the previous section. During exposure with the e-beam, the substrate is mounted in a vacuum chamber. No physical mask is needed, the pattern is defined in computer software.

Optical and electron-beam lithography can be combined [23]. The smallest and most critical structures in the chip center are defined by electron beam lithography, while optical lithography is used in a separate process step for uncritical 'large-scale' structures such as lead wires and contact pads at the edge of the chip.

3.5
Metallic Layers

Metallic wires, electrodes, or permanent magnets are the most commonly used structures on an atom chip. In the following we introduce the most important deposition and etching techniques used for the patterning of normal metals, discuss issues of roughness and homogeneity of the structures, and review the fabrication of structures out of special metals such as alloys, superconductors, semiconductors, and permanent magnets.

3.5.1
Deposition and Etching

Resist patterns created by lithography can be transferred to the substrate by *additive* (deposition) or *subtractive* (etching) techniques. The two most commonly used techniques for the fabrication of atom chip wires are additive techniques: electroplating and lift-off metallization.

3.5.1.1 Electroplating

Electroplating was among the first techniques used for atom chip fabrication [4, 36]. In this technique, the chip serves as the cathode of an electrolytic cell. Resist structures on the chip form a mold for the electroplated metal. Under an applied voltage, metal ions from the plating solution deposit on the areas not covered by resist. The amount of deposited material per unit time is controlled by the current flow. Electroplating is more time and material efficient than metal deposition through thermal evaporation or sputtering. It is, therefore, well suited also for structures with a thickness $\gg 1\,\mu$m. Furthermore, structures with high aspect ratio (wire height comparable to or higher than width) are easier to fabricate than with lift-off or etching techniques (see below). Electroplating is very versatile, it can also be used to fabricate vias and other non-planar structures [35]. A review of electroplating can be found in [37, 38], and its application to atom chips is discussed for example in [39–41].

Figure 3.2 illustrates a basic gold electroplating process (a detailed recipe can be found in [41]). Figure 3.2a: after substrate cleaning, a 2 Ti adhesion layer and a 50 nm Au seed layer are deposited by thermal evaporation. This layer serves as the cathode. Figure 3.2b: the chip is spin coated with a 6.5 μm thick layer of photoresist and subsequently patterned by optical lithography (Figure 3.2c,d). The resist structures should be slightly taller than the desired thickness of the wires. In our

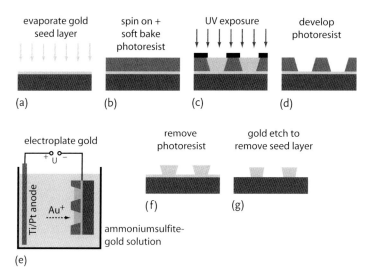

Figure 3.2 Electroplating. The process steps are explained in the text.

example, a positive tone resist with high pattern stability in acidic and alkaline plating baths is used.

In the following electroplating step (e), the chip and a Ti/Pt mesh anode are connected to a power supply and submerged in 1 liter of electroplating solution, whose temperature is carefully controlled in a water bath. In our example, a solution from Metakem is employed, which is based on ammoniumsulfite-gold(I) $[(NH_4)_3Au(SO_3)_2]$. It is commonly used for dentistry and jewelery and yields very smooth gold deposits of 99.99 % purity. To deposit gold, the desired plating current I is driven through the electrolytic cell. Stirring helps to avoid local depletion of the solution and thus ensures a more homogeneous gold layer. The thickness h of the gold layer deposited after a time t can be determined by Faraday's law: $h = a(ItM)/(nFS\varrho) = 1.1 \times 10^{-10}$ m^3/(A s) $\times (It)/S$, where M is the molar mass and ϱ the mass density of gold, F is Faraday's constant, $n = 1$ is the charge of the gold ions, and $a \approx 1$ the current efficiency for gold plating. The surface area S of the deposited gold film is given by the surface area of the wire layout plus the contact pads and connectors. In the parallel plate configuration, the current density is approximately constant over the exposed gold areas of the chip. (f) After the wires are electroplated to the desired thickness, the resist is stripped. In the final step (g), the Au seed layer and the Ti adhesion layer are removed by a wet etch with aqua regia. This isolates the electroplated wires from each other. The gold etch also attacks the wires, increasing surface roughness. It is, therefore, desirable to work with a seed layer as thin as possible. Figure 3.3a–d shows electroplated gold structures on Si chips fabricated with this process.

The quality of the electroplated wires crucially depends on the choice of the plating solution and on the process parameters. Different solutions (and also the same solution at different process parameters) deposit gold of very different grain

3 Atom Chip Fabrication

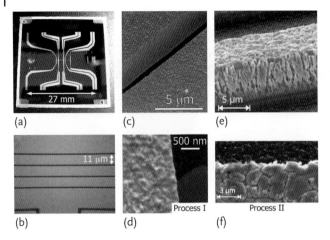

Figure 3.3 Electroplated gold wires fabricated with two different processes. (a–d) Wires of 5 μm thickness on a Si chip, fabricated with the process of [41]. (a) Entire chip with contact pads. (b) Close-up of (a). The gaps between wires are 3 μm wide. (c,d) SEM pictures of the wires. The grain size of the gold is 200 nm, the rms surface roughness is 15 nm. (e,f) For comparison, SEM pictures of wires fabricated with the process of [42] are shown (thickness 4.5 μm). Grains of micrometer size are visible, the rms roughness is 200 nm.

size [38]. Moreover, the complex bath chemistry can be disturbed by contaminants. Some processes result in deposits with micrometer grain size [42], resulting in inhomogeneous current flow in the wires and thus roughness of the generated potentials. Optimized electroplating processes, however, result in structures with very low roughness [39, 41]. In Figure 3.3c–f, SEM pictures of electroplated gold wires fabricated with the processes of [41, 42] are shown in comparison. The roughness of the wires of [41] is about one order of magnitude smaller than that of [42].

3.5.1.2 Evaporation and Lift-Off Metallization

Evaporation and lift-off metallization is a standard technique for fabricating metal structures of high quality [35]. In this technique, patterned resist acts as a mask for evaporated gold. After removal of the resist ("lift-off"), only the desired gold structures remain on the chip. Gold layers deposited by thermal or electron-beam evaporation often have smaller surface roughness than electroplated wires. This is important for generating smooth trapping potentials for waveguides or 1d traps [43] and for using the wire layer simultaneously as a mirror for a MOT [36, 44, 144, 146, 152]. Evaporation allows very precise control over the thickness of the deposited material. However, the thickness is limited by the slow deposition rate and substantial material consumption to a few μm at maximum. An alternative for thicker structures is sputtering of the metal, or electroplating (see previous section). Evaporation and lift-off was adapted to atom chip fabrication as described in [5, 23, 40, 41, 44, 45].

An example of a lift-off sequence is illustrated in Figure 3.4 (a detailed recipe can be found in [41]). Figure 3.4a: the substrate is spin-coated with a 1.6 μm thick layer

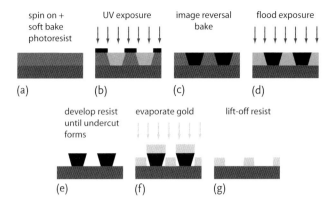

Figure 3.4 Lift-off sequence with an image reversal resist. Gold is deposited by e-beam evaporation. The process steps (a)–(g) are explained in the text.

of photoresist. An image reversal resist is used, which allows one to create resist sidewalls with a negative slope ("undercut"), which is crucial for lift-off metallization. Figure 3.4b: the resist is exposed for a few seconds with UV light through a mask which bears a negative image of the wires to be fabricated. Due to the small exposure dose, the resist is not fully exposed down to the substrate. This leads to the negative slope of the resist sidewalls after image reversal and development. Figure 3.4c: an image reversal bake on a hot plate at a carefully controlled temperature cross-links the resist in the exposed areas and thus renders it insoluble in the developer and insensitive to further exposure. Figure 3.4d: subsequently, the chip is flooded with UV light, which now renders the previously unexposed areas soluble. Figure 3.4e: these areas are removed during development. The development time is carefully adjusted to control the formation of the undercut. Residues of photoresist are removed by a short oxygen plasma cleaning step. Subsequent baking of the resist structures on a hot plate improves resist stability. Figure 3.4f: a 1 µm thick layer of gold is deposited on top of a 3 nm thick Ti adhesion layer by e-beam evaporation in a UHV evaporation chamber. The resist acts as a mask for the gold, while the undercut prevents gold in the resist trenches from sticking to gold on top of the resist. Figure 3.4g: lift-off is performed in a bath of hot acetone. This removes the resist and the gold on top of it and leaves behind the desired structures. If necessary, lift-off can be forced by agitation or by mild ultrasound.

For successful lift-off, the resist layer should be somewhat thicker than the thickness of the deposited gold layer. The resist undercut can be adjusted by changing exposure time, image-reversal temperature, and development time. A strong undercut can be seen in an optical microscope as a bright outline of the resist edges. However, for very small periodic structures, mechanical and thermal stability of the resist requires that development be stopped before the bright outline is visible. For process optimization, the undercut can be observed with an electron microscope, see Figure 3.5.

Figure 3.6 shows wires that were fabricated by evaporation and lift-off [5, 41]. The surface and edge roughness of the wires is small, resulting in very smooth mag-

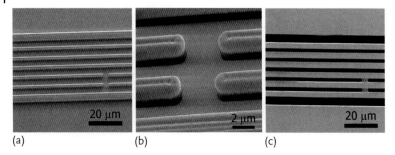

Figure 3.5 SEM images of gold wires before and after lift-off. (a) Resist structure with deposited gold on top. (b) Close-up of (a), the undercut of the resist sidewalls is visible. (c) Gold wires remaining after lift-off.

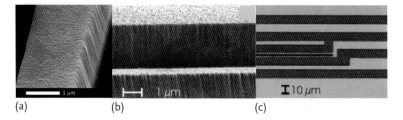

Figure 3.6 SEM images of micrometer-size atom chip wires fabricated by evaporation and lift-off. (a) Gold wire, 1 μm thick and 1.7 μm wide [41]. The grain size of the gold is ∼ 100 nm, the rms surface roughness is 3 nm. (b,c) Gold wires of [5] with 50–80 nm grain size.

netic potentials. Moreover, the gold wire layer can serve as a high-quality mirror if the gaps between the wires are sufficiently narrow.

3.5.1.3 Wet and Dry Etching

The simplest subtractive technique is wet etching, which is illustrated in Figure 3.1 above. After the resist has been patterned, the chip is submerged and agitated in a suitable etchant (aqua regia in the case of gold wires), which removes the metal layer in the areas not covered by the resist. Wet etching results in relatively rough wire edges. Moreover, the metal is removed isotropically, and wet etching is thus not suitable for fabricating structures with steep side walls and high aspect ratio. By contrast, some dry etching techniques such as reactive ion etching [35] are very anisotropic, and can thus be used to fabricate high aspect ratio structures with steep edges. Brute force ion etching (called ion-beam milling) may also be used if a thin layer of metal needs to be etched and the lift-off method is not feasible, for example, when high evaporation temperatures are used (e.g., if one needs large grain size), and the resist may only be put on after the evaporation process (this method has been used in [7]).

For wet and dry etching, it is essential to choose an etchant that effectively removes the material to be patterned, but leaves intact additional structures on the chip and the functionality of the etch mask. The etch mask does not have to be a resist structure, it can also be a suitable patterned metal or dielectric layer. Two

3.5 Metallic Layers

very useful papers with hundreds of etch rates for a large number of materials in different wet and dry etches are [46, 47]. They are also useful for choosing the right chemicals for cleaning and resist stripping tasks.

Besides etching, subtractive patterning techniques include focused ion-beam milling (see the next section), which can be used to structure arbitrary materials, and laser machining, which is useful for example for hole drilling. These techniques differ from etching in that they do not require a physical mask.

3.5.1.4 Designing Potentials by Postprocessing the Wires

In general the magnetic potentials on atom chips are formed by the subtraction of two (large) magnetic fields, the field of a current carrying wire and a (homogeneous) bias field (see Chapter 2). The value of the magnetic field at the potential minimum δB_{min} is determined by the angle between the field of the wire and the bias field. A small change of the current direction results in a significant change in the trapping potential. A deviation of the current flow by $\vartheta \sim 10^{-5}$ rad can lead to $\delta B_{min} \sim 1$ mG which corresponds to 67 nK for trapped Rb atoms.

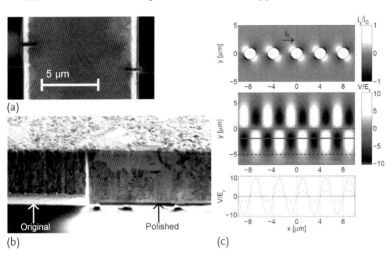

Figure 3.7 Examples of wires sculptured by FIB. (a) *Double well* created by making two 1 μm deep, 0.2 μm wide cuts in opposite edges of a trapping wire. (b) *Polishing* the wire edge as imaged by the FIB itself. The polished region (on the right) is smoother and reveals details of the structure of the gold wire. The contrast in the image is given by the ion reflection coefficient, varying with the relative orientation of the crystal axes of the gold grains. The picture shows a small region of 9.2 × 5.1μm² at the end of the polished section. (c) *Magnetic lattice* with periodicity $a = 4$ μm created by a string of holes with diameter $D = a/2$ in a wire with thickness $d = a/2$ and width $w = 10$ μm. Top: Current density j_y in the transverse direction. Center: The potential modulation V at a height of $z = 6$ μm, scaled to the characteristic energy $E_r = \hbar^2 k_a^2/2m_{Rb}$ (lattice vector $k_a = 2\pi/a$). Bottom: The potential along the lines drawn in the central image. The maximum modulation with a peak-to-peak depth $V_0 \simeq 20E_r$, is obtained at $y = 2$ μm (continuous curve) while $V_0 \simeq 6E_r$ for $y = 5$ μm (dashed curve).

One can implement slight changes in the current path deliberately by sculpturing the bulk of a lithographically patterned plane conductor [48] and thereby fine tune or even design local features in the trapping potential. The contributions of a current flow pattern with wave vector k to the magnetic field are exponentially damped with height z as e^{-kz} when receding from the wire. Therefore, in order to achieve a modulation of a fraction η_0 of the maximum achievable field at a minimum structure size $\lambda_{min} = 2\pi/k_{max}$ one has to stay at distances of $z < -\log \eta_0 / k_{max}$ (for details see [48, 49, 51]).

This postprocessing of the chip structures can be done with a Focused Ion Beam (FIB) technique [50], which allows one to create modifications with high precision (< 20 nm) and large aspect ratios (height/width > 30). The Heidelberg/Vienna group experimentally demonstrated the power of this technique by sculpturing a 10 μm wide and 2.5 μm thick gold conductor on a Si substrate [48, 51] (Figure 3.7).

3.5.2
Effects of Roughness and Homogeneity of the Fabricated Structures

A small change of the current direction results in a significant change in the trapping potential. Irregular current flow will result in uneven potentials with significant roughness.

After the first creation of a BEC on an atom chip a number of groups observed large roughness in the magnetic potentials holding the atoms. In many cases even a 1 μK *thermal* cold atom cloud was fragmented into many components when approaching distances below 100 μm from the wire [42, 52–55]. This potential variation originates from inhomogeneous magnetic-field components ΔB in the direction *parallel* to the current carrying wire [56] and can be attributed to variations in the current flow direction in the chip wires. Common to all but one of these experiments was, that they used electroplated wires to hold the trapping currents (the experiment by the Sussex/Imperial College group used a drawn copper wire [54]). In an experiment by the Orsay group, the large potential roughness observed was clearly identified as coming from the rough edges of the fabricated wires, and the model of Wang et al. [57] provided a full quantitative explanation [42, 55]. In many cases the observed modulations were so strong that they prevented the creation of a continuous elongated BEC.

At about the same time in 2003 it became clear that this is not the general case. Such strong potential roughness was not found in the experiments at Heidelberg [43] using atom chips with gold wires evaporated on a Si substrate and structured by the lift-off technique [5]. Nevertheless, a very small potential roughness remains which is also magnetic in origin. Measurements of the potential landscape over the whole width of a 100 μm wide wire, by scanning the position of the condensate across the wire using the BEC magnetic field microscope [6, 58], show no significant increase near the wire edges. A detailed analysis clearly showed that the edge roughness model cannot explain the observed potential variations [43]. Recent detailed experiments on different gold wires [7] revieled that the remaining potential roughness has to come from variations of the local properties of the conductance in the gold layer.

Even more surprising are the findings of Aigner *et al.* [7] where the current flow in three different wires with different grain size and wire thickness was probed. The measurements showed that the thinnest wire with the largest grains had the smoothest current flow. A detailed analysis of the measured potential landscape and a comparison to the surface roughness of the gold layer clearly show that the local properties of the metal, and not imperfections in the wire boundary are the key to the current flow variations in the wires with the smoothest potentials [60].

It is interesting to compare the potential roughness observed on different chips and in different experiments. In most applications of atom chips an important parameter is how large the potential roughness ΔB is in comparison to the energy scale of the transverse confinement $\omega_\perp$. A stringent requirement for 1d experiments is $\hbar\omega_\perp \gg \mu \Delta B$, where $\mu = \mu_B g_F m_F$ is the magnetic moment of the trapped atomic state. The magnetic confinement $\omega_\perp \propto d^2 B/dr^2$ is determined by

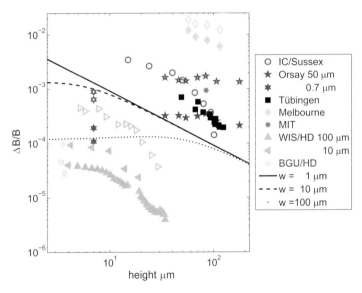

Figure 3.8 Comparison of different atom chip potential roughness measurements. Data is taken from figures in the published literature and Ph.D. theses and displayed according to the experimental group. Filled symbols denote rms values, and data displayed as open symbols are peak-to-valley maximum height of the roughness. *Sussex*: data from a gold-coated copper wire [65]; *Orsay*: data from an electroplated wire [42, 55], and recent data from a five-wire setup on an evaporated gold chip; *Tübingen*: data from electroplated wires (PhD thesis, J. Fortagh); *Melbourne*: data from a permanent-magnet atom chip [13]; *MIT*: data from electroplated wires; *WIS/HD*: data from the analysis chips fabricated at the Weizmann Institute of Science: a 100 μm wire [43], 2003 data from various 10 μm wires [43, 59]. The open triangles give the peak-to-valley of the worst ever roughness observed in HD: a 10 μm wire (Ph.D. thesis, L. Della Pietra). *BGU/HD*: Recent data from three different wires fabricated at Ben-Gurion and analyzed in HD [7].

the magnetic-field gradient dB/dr and the longitudinal Ioffe field B_{Ioffe}. To achieve the same ratio $R_{\omega\perp} = \mu \Delta B/\hbar \omega_\perp$, the allowed relative magnetic field roughness for a trap built at height h above a flat wire of finite width w scales like:

$$\frac{\Delta B}{B} < R_{\omega\perp} \frac{2w}{(w^2 + 4h^2)\left(\frac{\pi}{2} - \arctan\left[\frac{2h}{w}\right]\right)} \sqrt{\frac{\mu_B g_F m_F}{M B_{\text{ip}}}}. \qquad (3.1)$$

Figure 3.8 shows a comparison of a large variety of measured potential roughness data from different laboratories.

3.5.3
Special Metals

3.5.3.1 Alloys

One of the high-priority goals in atom chip research is the increase of lifetime and coherence time for ultra-cold atoms trapped in magnetic potentials close to the surface. This is important for both scientific aims and technological applications. Progress towards this goal demands the control and reduction of magnetic noise produced by the metallic components of the atom chip. Randomly fluctuating magnetic fields are generated by thermal current noise in the conducting chip elements and reduce the number of trapped atoms (losses), increase their temperature (heating) and lead to a phase uncertainty in the atom's state (decoherence) – see, for example, [1] and references therein. Theoretical analysis of the magnetic noise generated by a normal metal [61–64] predicts a fast reduction of the lifetime τ with the decrease of the distance z_t between the trapped atom and the metal surface (trap height); this is in excellent agreement with lifetime measurements [65–67]. At a trap height less than 10–20 μm, thermal magnetic noise may exceed all other harmful influences on the atom cloud (technical noise due to current supply instability, residual gas collisions, stray magnetic fields) and could present the dominant limit for the lifetime.

As is well known by now following the work of Carsten Henkel (see the corresponding chapter in this book), cooling of normal metals such as gold or copper will not reduce the harmful effect of thermally induced noise on the atoms. Hindering processes such as spin-flip (reducing trap lifetime), heating, and decoherence will thus not be suppressed by cooling the atom chip. The reason for this rather counter-intuitive behavior has to do with the fact that the strength of the relevant magnetic noise is proportional to $R = T/\rho$ where T is the metal surface temperature and ρ its resistivity. As ρ is typically linearly dependent on T (due to phonon scattering) R does not become smaller with dropping temperatures and typically even becomes larger. As presented in the next section, superconductors are expected to have much smaller noise, but on the other hand present other challenges such as sensitivity to external magnetic fields and inhomogeneous current distribution. The question, therefore, arises if one may find a material where on the one hand cold temperatures do lower the level of R and on the other, a simple material is used.

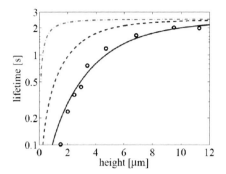

Figure 3.9 Comparison of trapping lifetimes of ^{87}Rb atoms above a copper wire on an atom chip [67] with a theoretical calculation (solid line) – taken from [68]. Predicted lifetimes are also shown for a similar wire made of an alloy of Ag with 5.5 % Au content, cooled down to $T = 77$ K (dashed) and 4.2 K (dash-dotted). Note that the calculation in [68] differs from the one made in [67]. Van der Waals forces are not taken into account. Let us also mention that the maximum noise reduction factor of 75 is not visible, due to the affect of the technical noise and/or background gas collisions limiting the lifetime in this experiment to a maximum of $\tau_{tech} = 2.5$ s [69].

Indeed alloys have such features. On the one hand they are a simple metal easily deposited (e.g., with sputter), with no sensitivity to external magnetic fields and with an homogeneous current distribution, and on the other their R drops with dropping temperatures. The reason for this behavior lies in the special dependency of ρ on T in alloys. At low temperatures, resistivity due to imperfections and impurities, if large enough, may become dominant over that dependent on phonon scattering. Alloys have impurities and hence the resistance becomes a constant at low temperatures. This is explained in detail in [68], which also presents the data for numerous metals. As an example of the benefits of such a material, we present in Figure 3.9 how the spin-flip rate behaves as a function of distance to the surface. The fact that such a simple choice of material may make a difference of up to two orders of magnitude in the lifetime, shows clearly the important role material engineering can play in atom chips.

3.5.3.2 Superconductors

In the last few years, the application of superconducting materials to atom chips has been widely discussed as a perspective to extend the lifetime of cold atoms [64, 70–73]. A recent theoretical estimate [72] of the magnetic noise caused by a superconductor in the Meissner state showed that the lifetime of atoms trapped above a superconducting layer would be, at least, six orders of magnitude longer than above a normal metal in the same geometry. The analysis presented in [73] predicts an atom lifetime of 5000 s at a trap height of 1 µm. For comparison, at the same height in a normal metal trap the lifetime is less than 0.1 s [67].

Details of experiments with superconductors may be found in Chapter 10. Here we give a brief account with emphasis on fabrication. Two first realizations of atom chips with superconducting elements have been reported in [74–76]. In both setups, the trapped atoms were ^{87}Rb. In the Paris experiment, the current-carrying wires (in "U" and "Z" shape) were made of niobium and operated at about 4.2 K. The obtained atom spin relaxation time τ_s was estimated as 115 s. This value is comparable to the best one achieved for atoms trapped near normal-metal wires [77]. In the second experiment [76], special efforts were undertaken to reduce the influence of technical noise. Utilizing a MgB$_2$ film, a "Z"-shaped wire was fabricated as a part of a closed superconducting loop and operated in the persistent current regime. This permits one to disconnect the current supply and get rid of its instability, that is, technical noise. To our knowledge, in both experiments the trap lifetimes were limited by processes other than the magnetic noise generated by the superconducting elements of the atom chip. Furthermore, to the best of our knowledge, to date no atoms were taken in these two experiments to a distance below a few tens of μm from the surface.

The Paris atom chip was made on a 65 mm × 30 mm silicon wafer (thickness 360 μm) with a 500 nm insulating oxidized layer. It was coated by a 900 nm thick layer of Nb by cathodic plasma sputtering. A "U"-wire (width 280 μm) is used for the on-chip mirror-MOT, and a "Z"-wire (width 40 μm) for the magnetic Ioffe–Pritchard trap. The cross-section of the "Z"-wire was 40 × 0.9 μm^2. The wires and contact pads are produced by standard optical lithography with a laser-printed mask followed by reactive ion etching. The resulting wire edge precision is about 5 μm. A gold layer of thickness 200 nm was deposited on the niobium wire to increase its reflection. The current through the "Z"-wire reached 1.5 A, close to the critical current value of 1.94 A. The critical current density obtained for the niobium wire was $J_c \approx 5 \times 10^6$ A/cm^2 at 4.2 K.

In the Japanese group, a magnesium diboride MgB$_2$ film of 1.6 μm thickness was grown by molecular-beam epitaxy (MBE) on a sapphire C substrate (10 × 10 × 0.5 mm^3). The cross-section of the "Z"-wire was 100 × 1.6 μm^2. The persistent superconducting current of $I = 2.4$ A in a "Z"-structure was significantly lower than the critical value of $I_c \approx 16$ A (the critical current density at 4.2 K is about $J_c \approx 10^7$ A/cm^2 and the transition temperature of MgB$_2$ is $T_c \approx 35$ K). The top of the MgB$_2$ thin film was coated with a thin gold layer to prevent radiation heating and for better reflection. The circuit pattern on the chip was produced by removing the unnecessary part of the MgB$_2$ thin film by ion milling. This group has also demonstrated trapping atoms on a Nb film [78].

Recently, two additional groups joined the experimental effort: The Singapore group recently built a chip with an atomic interferometer structure on the basis of a high-T_c supercoonducting film [79, 80]. The YBa$_2$Cu$_3$O$_{7-\delta}$ (YBCO) film was grown by epitaxy on a yttria- stabilized zirconia (YSZ) single crystalline substrate. The lattice constants of YBCO and YSZ are matched, allowing homogeneous growth of the superconducting material. The final thickness of the YBCO film was 600–800 nm. On top of the YBCO film a 200 nm thick layer of gold was deposited to protect the superconducting material. Structuring of the chip is performed by two different techniques. These are standard optical lithography followed by a wet-chemical etching as well as direct femto-second (fs) laser ablation. The standard lithography procedure has a resolution limit of about 1 μm for the structure size.

With the fs-laser ablation procedure the desired patterns on the chip are realized by locally removing the gold and YBCO layers with focused laser pulses (2 µJ, 130 fs, center wavelength 800 nm), resulting in insulating regions between the superconducting structures. In this laser-assisted structuring technique, it is crucial that the film is not heated. In case of heating, oxygen would be lost and the superconducting properties of the thin film would degrade. This requirement sets the demand for using fs-laser pulses and may at times set wet etching as the preferable method. In the final chip, the roughness of the surface was almost negligible with an rms value of the surface height of only 2.3 nm. The critical current density J_c at liquid nitrogen temperature was 2×10^6 A/cm^2. Recently, this group has managed to trap atoms in the field of a vortex [81].

In the Tübingen group [82], a cylindrical niobium wire of diameter 0.125 mm has been used. The wire was mechanically clamped to copper plates firmly attached to a helium cryostat. A distance of about 25–30 µm was achieved between the atoms and the super current (where one has to take into account the super current radius for each temperature).

Aside from low thermal magnetic noise, the application of superconductors in atom chips may be advantageous also due to high current densities without Joule heating, and practically zero electric fields across the superconducting elements. In addition, as noted, technical noise can be reduced by inducing a persistent current. Hence there is significant motivation to make available simple fabrication procedures for superconductors. Recently, two works have analyzed in detail the trapping parameters of cold atoms in magnetic traps made by type I and type II superconductors [83, 84].

3.5.3.3 Semiconductors

In the future, atom chips might be built like integrated circuits from a single semiconductor substrate. For example, the current used for trapping and manipulating atoms will flow in an epitaxially grown layer that is covered by an insulating layer on top. Such an approach has the advantage that the epitaxially grown material is more uniform and better controlled than the wires deposited onto the surface. In addition, such a structured semiconductor atom trap may be used to probe the current flow in the semiconductor, by using the cold atoms as a magnetic field microscope [6, 7, 58].

First steps to develop a semiconductor-based atom chip were carried out by the Heidelberg/Weizmann team [34]. The chips were fabricated from a GaAs wafer grown using molecular beam epitaxy (MBE) at the Weizmann Institute of Science. Such a chip contains a super lattice of 20 layers of GaAs and AlGaAs which are used as an etch stop layer for selective etching. On top of it, about one micron of Si-doped (n+) GaAs is grown as a conductive layer. The measured 3d charge carrier density is $N \sim 6.4 \times 10^{18}$, the mobility is $\mu = 1160$ cm^2/V s, and the 2d resistivity of this layer is 8.5 Ω.

The wire structures are then fabricated into the GaAs wafer by UV lithography and dry etched by reactive ion etching, which is stopped when the super lattice below the doped layer is reached. The etch results in straight walls. In the pat-

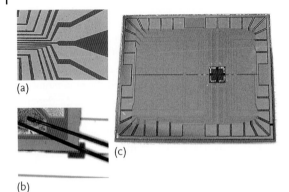

Figure 3.10 Example of an atom chip fabricated in a GaAs substrate. (a) Central region of the chip. The deep etched trenches defining where the doped layer is removed are shown in dark. (b) Wire bonds connected to the alloyed metal pads on the GaAs chip. (c) Overview of the GaAs chip mounted on a carrier chip.

terned semiconductor wires the charge carriers are located in the doped layer below the surface. They are contacted by alloying: Several metal films are evaporated in defined order and thickness (50 Å Ni, 400 Å Ge, 800 Å Au, 200 Å Ni, and 2000 Å Au) onto the regions, where the semiconductor is going to be contacted. The first four layers are to be alloyed into the surface and contact the semiconductor, the top gold layer supports enough metal to allow bonding. Alloying is accomplished by a well-defined thermal cycle heating the sample up to 430 °C [34].

Finally, the small GaAs chip is mounted on a standard gold layer carrier atom chip by gluing, and contacted by wire bonding. The atoms are to be trapped and cooled on the carrier atom chip, and then transported to the semiconductor chip. Pictures from a fabricated and mounted chip can be seen in Figure 3.10.

3.5.4
Permanent Magnets

Most atom chips rely on a pattern of current-carrying wires to generate the necessary magnetic fields. By contrast, it is also possible to use patterned films of permanently magnetized material for this purpose (see Chapter 1). The ideal magnetic material for use in atom chips has a large remanent magnetization and coercivity, regardless of the shape, is very homogeneous, has a high Curie temperature, is corrosion resistant, and UHV compatible. Several groups have developed fabrication processes for permanent-magnet atom chips, see for example [10, 11, 85–92] and references therein, and we discuss a few examples in the following.

Commercial magnetic storage media such as videotape or hard disks provide a convenient way to prepare small magnetic structures. In [85] an atom mirror was fabricated by etching of a common hard drive. The hard drive provides a large area of thin magnetic film whose remnant magnetic field and coercivity can be as large as 7 and 3 kG, respectively. The atom mirror is fabricated by etching 2 μm wide, 100 nm deep trenches into the film. Standard photolithography is used to create the etch mask. The cobalt alloy of the magnetic film is granular, which enhances the coercivity and allows one to magnetize the material in plane and parallel to the short axis of the magnetic strips.

The Sussex/IC group demonstrated that a pattern of magnetization recorded in commercial videotape can be used to generate an array of elongated magnetic microtraps [86]. Videotape can store patterns with feature sizes down to a few micrometers using simple commercial recording equipment adapted in the laboratory. It is designed to hold data reliably for long periods of time and has a high coercivity (1500 Oe), making the magnetization insensitive to the presence of magnetic bias fields. The tape used in [86] has a 3.5 μm thick magnetic layer containing iron-composite needles, 100 nm long with 10 nm radius, which are set in a glue and aligned in parallel. The film is supported by a polymer ribbon 11 μm thick. Remarkably, the tape is UHV compatible and is able to withstand baking at 120 °C. The same group has also investigated various other permanent magnetic materials for atom chips, such as magneto-optically patterned CoPt thin films [87].

The Amsterdam group has found that FePt is a magnetic material ideally suited for use in atom chips [10, 88–90]. Their first chip was made out of a 40 μm thick foil of FePt [10]. An "F"-like shape was cut out by spark erosion and glued onto a mirror. This produces a self-biasing Ioffe–Pritchard trap. The second generation atom chip uses structures lithographically written into a FePt film [89]. Figure 3.11a shows a two-dimensional FePt pattern with the magnetization oriented perpendicular to the film, which has been used to create a large two-dimensional array of magnetic microtraps [12]. The 300 nm thick FePt film is grown in the ordered face-centered-tetragonal (fct) phase, which is magnetically hard with a high uniaxial anisotropy

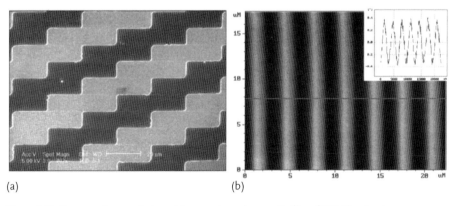

(a) (b)

Figure 3.11 Permanent-magnet atom chips. (a) SEM image of the patterned magnetic film of [89]. The light gray areas correspond to the FePt pattern and the darker regions are the Si substrate. The film can be used to create a two-dimensional magnetic lattice potential. (b) Magnetic force micrograph of the patterned magnetic film of [91]. The structure is a 150 nm thick $Gd_{10}Tb_6Fe_{80}Co_4$ film on a 140 nm thick Cr underlayer on a Si grating structure with a period of 3 μm. The grooves are represented by the light regions and the inset shows a cross-section of the MFM signal along the indicated line.

constant. The measured remanent magnetization is $M_r = 670$ kA/m, the remanent to saturation magnetization ratio $M_r/M_s = 0.93$, and the coercivity $H_c = 0.95$ T. After baking for 3 h at 150 °C in air, the magnetization had reduced by only 3 %, and the film was found to be stable against typical time-varying fields applied in the experiment. The film is patterned using optical lithography and argon-plasma etching. After patterning, the film is coated with a 100 nm reflective gold overlayer and magnetized in a 5 T magnetic field.

The Melbourne group uses atom chips based on perpendicularly magnetized multi-layered TbGdFeCo/Cr films [11, 91, 92]. These films have a large perpendicular magnetic anisotropy and are suitable for the production of periodically grooved, micron-scale structures. Figure 3.11b shows an example of a patterned film that can be used to produce a one-dimensional magnetic lattice. To pattern such magnetic films, a periodically grooved structure is first fabricated into a Si substrate by reactive ion etching. The TbGdFeCo film is then deposited on this grooved structure by magnetron sputtering. Because the magnetic anisotropy is found to deteriorate for a film thickness above 250 nm, a multi-layer structure is deposited, consisting of several layers of TbGdFeCo (160 nm) and Cr (100 nm) to achieve a total thickness of about 1 μm. A 10 nm Cr film and finally a 150 nm gold film are deposited on top of this structure in order to produce a mirror for the mirror-MOT. Measurements show that the remanent magnetization of the film is about 3 kG and the coercivity is about 6 kOe [92].

3.5.5
Metal Outlook

As the fabrication of magnetic-field-based atom chips evolves in the future, one would also need to look at additional forms of materials. Aside from the above-mentioned metals, alloys, superconductors, semiconductors and permanent magnets, it stands to reason that additional classes of new materials will be investigated in the near future. Here, as an example, we shall focus on nanowires, molecules and metallic crystals, which are already being studied in several groups.

Nanowires (Figure 3.12) constitute an interesting system. While the current may be sufficiently high to maintain magnetic traps, the Johnson noise may be sufficiently low to enable long spin-flip lifetimes at small atom–surface distances. Lifetimes of seconds may be reached for atom–surface distances of less than one micrometer [49]. Fabricating nanowires may only be done through direct e-beam writing, and even then, requires considerable attention to details such as edge roughness and contact resistivities. We note that due to surface scattering, the nanowire resistivity may be high, an issue which may be solved by using crystalline nanowires, namely molecules, which are our next topic.

Molecules may form interesting conductors for numerous reasons. First, they may be able to sustain extremely high current densities. Second, they may suppress hindering effects such as van der Waals and Casimir–Polder (CP) forces, corrugations due to imperfections causing electron scattering, and noise giving rise to spin-flips, heating and decoherence. Third, molecules have a relatively sharp ab-

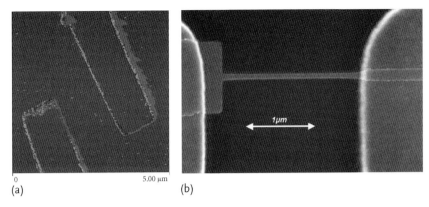

Figure 3.12 (a) An AFM scan of a single-wall carbon nanotube (2 μm long connected to Pd pads [95]). Around 40 μA were pushed through this CNT. (b) Nanowire: 50 nm wide and 20 nm thick [49] (fabrication of both samples done at the Ben-Gurion University fabrication facility).

sorption spectrum and hence may be put much closer to sensitive optical devices in comparison with normal metal wires. Finally, one may also look into a future in which molecules will self-assemble into circuits (forming molecular electronics), and similarly may be used to assemble current-carrying atom chip circuits.

A specific molecular scale structur that has recently attracted attention is the Carbon Nanotube (CNT). The CNT may hold current densities of up to 10^9 A/cm^2 (two orders of magnitude higher than normal metals), may have ballistic transport (hence negligible scattering causing corrugations), and is also expected to produce less noise and CP forces. Two articles describe the advantages of CNTs [93, 94]. To the best of our knowledge, to date, no atoms have been trapped in the magnetic field of a current pushed through a CNT but several groups, such as the Tübingen group, are pursuing this goal.

CNTs come in different shapes and forms: from single-wall CNTs to multi-wall CNTs, from semiconducting to metallic, and from suspended to substrate based. While a multi-wall CNT offers more current, it is also less crystalline in nature. While a suspended CNT offers less CP it probably also enables less current density due to the lack of a heat sink. It is unclear what the ultimate current limit of single-wall CNTs is but while 20 μA seem to be enough for trapping at a height of a few hundred nanometers (also taking into account the CP force) [95], experiments at the Ben-Gurion University group show that CNTs of a few μm length are able to carry up to 50 μA.

Typically a Chemical Vapor Deposition sample will contain many CNTs with random orientations, lengths, and degrees of straightness. A preliminary survey is taken by AFM and, if suitable CNTs are seen, grids of Ti alignment marks (e.g., 2 μm long, 1 μm wide, 40 nm thick, 10 μm spacing) are made by e-beam lithography (PMMA resist baked at 200 °C) and thermal evaporation over several small areas of the Si wafer. After lift-off with NMP at 80 °C these areas are then carefully scanned by AFM; co-ordinates for suitable CNTs are measured with respect to the Ti alignment marks.

The next step is to contact both ends of the CNT with parallel leads in order to form the "Z" shape used for Ioffe–Pritchard atom chip microtraps. Pd leads are usually thought to have the best contact resistance to CNTs and are made by using a second e-beam lithography step with PMMA (the same as before) and are typically 30–40 nm thick and 1 μm wide (see Figure 3.12).

In the future, it is expected that CNTs will be grown in a controlled way along predefined paths. One such method, includes growth of CNTs along atomic steps in the substrate [96, 97]. An alternative method for controlling the CNT position and orientation utilizes an imprint process [98]. In such a case where the CNT's position and direction is determined by the user, one may also etch the substrate before depositing the CNT, evaporate or electroplate a loading wire into the etched channel, deposit an insulating layer and obtain a smooth substrate by Chemical Mechanical Polishing (CMP) or other planarization methods described in this chapter, and then place the CNT on top – hence obtaining a truly monolithic CNT atom chip.

Before moving on, let us note that nanowires and molecules such as CNTs may be used also for other purposes. For example, ideas have emerged where plasmonic waves on such elements may be used to trap and couple light between nearby atoms. Another example, which was realized experimentally, includes utilizing CNTs for high efficiency atom counting. The experiment was conducted in the Tübingen group [99]. In Figure 3.13 we show the CNTs used to create a high electric field for the purpose of ionization and ion counting.

Another class of materials that has so far not been used in atom chips is crystalline metals. There may be several advantages in utilizing such materials. To start with, they may offer reduced electron scattering and thus produce potentials with

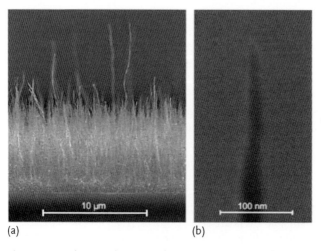

Figure 3.13 (a,b) Vertical CNTs used to create a high electric field for ionization of neutral atoms. The ions are then detected with single atom precision. A close-up image is shown in (b).

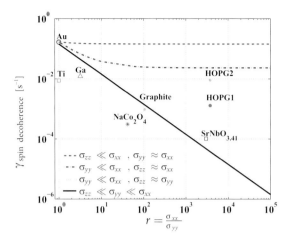

Figure 3.14 The spin decoherence rate as a function of the electrical anisotropicity of the material [100]. The different lines represent different types of anisotropic materials (see legend). Several material examples are also shown.

less corrugation. They may also be used in their electrically anisotropic form to significantly reduce decoherence even at room temperature (Figure 3.14) [100].

Fabrication of atom chips using these materials is not trivial. To start with, these materials may not be deposited through evaporation and need to be grown with specialized methods. Typically, one can buy layers of these materials after which they will need to be bonded to a substrate (as they usually come as thick leafs or attached to dissolvable substrates), thinned (without causing too much surface roughness), and patterned. Obviously, standard lift-off may not be used here as the photoresist may only be put on top of the material. Shadow masks followed by wet etching, ion-beam milling or plasma etching will need to be utilized. In addition to the above, one would also have to utilize normal evaporated metals to make electrical contacts as the above crystalline materials may not have suitable conductivity in all directions.

3.6
Additional Features

3.6.1
Planarization and Insulation

Dielectric layers for planarization and electrical insulation are essential for the fabrication of multi-layer chips. Besides the dielectric and planarization properties, good heat conductivity, ultra-high vacuum compatibility, and chemical stability in subsequent processing steps are important criteria for the choice of suitable materials.

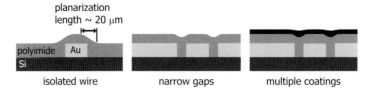

Figure 3.15 Planarization and insulation with polyimide. Narrow gaps are easier to planarize than broader, isolated features. Multiple coatings improve planarization [41].

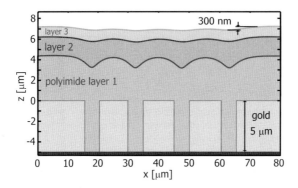

Figure 3.16 Planarization results [41]. Surface topography of three polyimide layers above small gaps in a lower gold layer, measured with an AFM. Note the different scaling of the two axes. On this chip, the polyimide layers were 4.3, 1.9, and 1.0 μm thick.

Spin-on polyimide has been used in the multi-layer atom chip designs of [23, 41]. It provides good planarization, electrical insulation, and is UHV compatible. Moreover, some polyimides can be directly structured by UV lithography. On the other hand, polyimide has a thermal conductivity which is about three orders of magnitude smaller than that of gold or silicon. Therefore, the polyimide layer has to be as thin as possible to allow for large current densities in the wires of the upper gold layer. To process polyimide, the chip is spin-coated with the polyamic ester. Subsequent curing of the polyimide is carried out at temperatures of typically 300–400 °C. Fully cured polyimide is resistant to solvents and most acids, but can be dry etched in an oxygen plasma [47].

As illustrated in Figure 3.15, a layer of polyimide does not globally planarize a chip surface with an uneven topography. It rather acts as a "low-pass filter" which locally smooths out the topography. Only features with a lateral extension smaller than the planarization length, which is ∼ 20 μm for the polyimide used in [41], can be fully flattened out [101]. The degree of planarization is higher for several thin coatings compared with a single thick coating of equal total thickness. Figure 3.16 shows AFM measurements of the polyimide surface topography above gold wire structures. Thin gaps (5 μm wide) between 5 μm thick wires are planarized to a step height of 300 nm with three layers of polyimide. If necessary, the thickness of the polyimide could be reduced by subsequent back-etching.

3.6.2
On-Chip Mirrors

It is often desirable to have a high-quality mirror as the uppermost layer on the atom chip. A mirror is required for the operation of a mirror-MOT [36]. Moreover, it enables imaging of the atoms very close to the chip surface, using an absorption imaging beam that is reflected from the surface [102]. The mirror can also be used for reflecting laser beams that generate optical dipole traps.

High-quality dielectric mirror coatings can be transferred to the chip using a replica technique [103, 104], which is illustrated in Figure 3.17. In this technique, a detachable mirror coating supplied on a transfer substrate is glued onto the desired chip area using vacuum-compatible epoxy glue. The transfer substrate is manually removed, leaving the coating on the atom chip. Various types of dielectric coatings on transfer substrates are commercially available. By cutting the transfer substrate to the desired size, it is possible to only partially cover the atom chip with the mirror [29, 105]. The transfer technique can also be used for homogeneous metal layers [103]. The advantage compared to evaporating the metal layer is that pretty good planarization of underlying structures can be achieved.

Covering the uppermost wire layer with a dielectric mirror or metal mirror on a dielectric spacer is not always desired. The mirror layer increases the minimum distance of the atoms to the wires, thereby reducing the attainable field gradients. If the uppermost wire layer contains microwave guiding structures, the mirror changes the waveguide impedance and introduces additional losses or shields the atoms from the microwave field. In these cases, a viable option is to use the uppermost wire layer itself as a mirror [20, 44]. This requires high-quality metal deposits and narrow gaps between the wires. Experiments have shown that a mirror-MOT is remarkably tolerant against the distortions of the light field caused by the gaps, even for gap sizes of the order of 10 μm. However, the gaps between wires cause clearly visible distortions on high-resolution absorption images taken with light reflected from the mirror.

Even if no dielectric mirror is desired, it can be advantageous to cover the uppermost metal layer on the chip by a few nanometers of a dielectric such as SiO_2. This

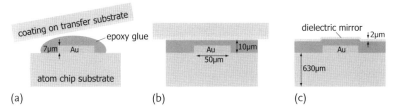

Figure 3.17 Fabricating a dielectric mirror on the atom chip using the transfer technique [103, 104]. (a) A small drop of UHV-compatible epoxy glue is dispensed on the atom chip. (b) The transfer substrate with the dielectric coating is sandwiched onto the chip. (c) After curing the epoxy, the transfer substrate is lifted off, leaving the dielectric mirror on the atom chip. On a structured chip, epoxy shrinking causes small modulations of the mirror surface.

prevents alkali atoms from contaminating the metal layer over time and degrading its reflectivity or even creating shortcuts between wires. A thin SiO_2 layer could also reduce the electric fields from alkali adsorbates on the surface.

3.6.3
Multi-Layer Chips

Many applications of atom chips in both fundamental science and technology require chips with complex multi-layer wire patterns. A multi-layer chip allows for much greater flexibility in trap design by avoiding wire crossings which would arise in a single layer. This is crucial for example for experiments on quantum information processing, which will eventually require large arrays of qubits in individually addressable microtraps on a single chip (see Chapter 9). Some proposals for quantum gates require the integration of microwave guiding structures on the chip in addition to the already complex DC wire structures [17, 20]. Moreover, it is often necessary to combine wires of very different size and thickness on the same chip, for example because the experimental sequence starts with an ensemble of atoms in a large-volume trap, while the actual experiment is performed with a small Bose–Einstein condensate in a micrometer-scale trap. In all these cases, multiple wire layers separated by dielectric insulation layers are beneficial. Here we discuss several examples of multi-layer atom chips which have been used in experiments. A general introduction to multi-layer microfabrication is given in [101].

The simplest way to produce a "multi-layer" chip is to fabricate two single-layer chips and glue them on top of each other. This greatly simplifies fabrication because the processes used for the two layers are completely independent and the chips can be independently discarded if something goes wrong. This technique has been used for example in the experiments of [22], where a carrier chip with a standard layout of larger wires for initial trapping and transporting of atom clouds carries several experiment chips with dedicated micro- and nanostructures for the main experimental task, see Figure 3.18a. The carrier chip used in [22] is itself a two-layer structure, with one wire layer on the front side and another one on the back side of the same substrate. For interconnecting the two layers, wire vias have to be fabricated through the 250 μm thick substrate. The vias are made by laser cutting of holes 400 μm in diameter and electroplating gold inside.

In [105], a two-layer chip is described in which the two wire layers are on the same side of the substrate, see Figure 3.18b. The chip is fabricated using thick-film hybrid technology. This fabrication process is mainly employed in high-power electronics and is based on screen-printing with metallic and dielectric printing pastes. The desired structure is lithographically transferred to a fine-gauge printing mesh. The printing paste is then squeezed through this mesh onto the substrate. Large-format substrates and thick conductor layers with high current capability are standard with this technology. However, the structures are grainy and the minimum feature size as well as the thickness of the layers is of the order of tens of micrometers and thus not as small as in direct lithographic techniques.

3.6 Additional Features | 89

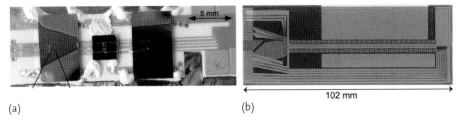

Figure 3.18 (a) Atom chip of [22]. Three different experiment chips with micrometer size structures are glued on top of a carrier chip with larger structures for positioning of the atoms. (b) Atom chip of [105]. This magnetic conveyor belt is a two-layer structure with vias, fabricated by thick-film hybrid technology.

While relatively simple to implement and useful in many cases, these techniques have two main drawbacks: there is a relatively large distance between the two wire layers and it is difficult to accurately align the structures on the different layers with respect to each other. To overcome these limitations, several groups have developed custom multi-layer chip fabrication processes where the metal layers are fabricated by optical or electron-beam lithography and separated by thin insulating layers of at most a few µm thickness [23, 34, 41]. This enables precisely aligned crossed wire configurations on a micrometer scale, greatly enhancing the flexibility in designing potentials for the atoms.

Figure 3.19 shows pictures of the multi-layer chip described in [23]. It is fabricated using a combination of optical and electron-beam lithography, gold deposition by evaporation, and lift-off. On this chip, larger wire structures are fabricated on top of smaller ones, facilitating planarization of the lower layer. The two wire layers are separated by a polyimide layer, which is structured by optical lithography to cover only regions where conducting structures will cross. The insulation layer is

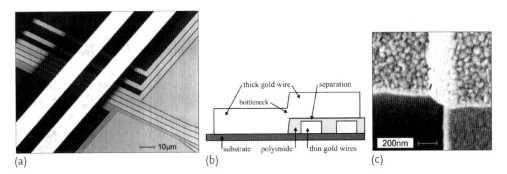

Figure 3.19 Multi-layer atom chip of [23]. (a) SEM micrograph of the central part of the chip. 10 µm wide wires with 1.4 µm high cross structures created by e-beam lithography. The smallest features are 300 nm gaps between 700 nm wide and 140 nm high wires. Electrical insulation of the two layers is provided by 500 nm thick polyimide pads, visible as a partially transparent layer. (b) Cross-section of chip (not to scale). The step in the upper gold wire causes a bottleneck of reduced wire cross-section. (c) SEM top view of the step. The wire runs from left to right in the upper half of the picture. In the lower right part the polyimide pad running from top to bottom is visible.

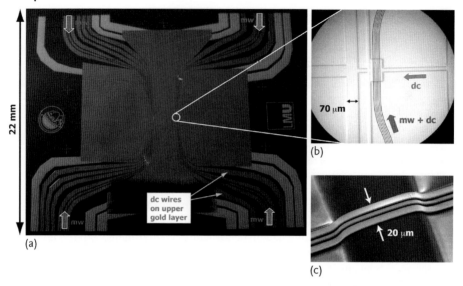

Figure 3.20 Multi-layer atom chips of [20, 41]. (a) Photograph of a whole chip. Two gold wire layers are separated by a thin polyimide layer (see main text). Wires on the lower layer are lighter because of the overlying polyimide. (b) Close-up of the central region of the chip. Parallel wires on the upper layer cross wires on the lower layer. (c) SEM micrograph of a different chip. Small gold wires (thickness 800 nm) forming a coplanar microwave guide cross a large transverse wire of 7 μm thickness. A single polyimide layer of only 4 μm thickness between the wires already provides sufficient smoothing of the step.

thinned in an ozonator to about 500 nm and cured. This layer thickness proves to be sufficient to insulate the two conducting planes while providing good heat transfer and keeping the step height the top layer wires have to surmount to a minimum (see Figure 3.19b,c).

Figure 3.20 shows pictures of multi-layer chips described in [20, 41]. The chips are fabricated using optical lithography. The chip shown in Figure 3.20a,b has a lower wire layer of 5 μm thick electroplated gold, which is covered by three layers of spin-on polyimide with a total thickness of 6 μm. On top of the polyimide, a 1 μm thick gold wire layer is fabricated by evaporation and lift-off. On this chip, the small wires are on top of the larger ones. The three polyimide layers provide very good planarization. The smallest structures on the upper wire layer are several parallel wires of 2.5 μm width, separated by 2 μm gaps. Some of the wires on the upper layer form integrated microwave guiding structures. Figure 3.20c shows a different chip, on which such a microwave guide crosses a thick transverse wire on the lower gold layer, demonstrating the smoothing of the step by the polyimide.

3.7
Current Densities and Tests

An important parameter in all atom chips is the electric performance of the fabricated structures. Atom chips require current densities in the excess of 10^6 A/cm^2 in continuous use. For a wire (height h and width w) to support such high current densities $j = I/wh$, the ohmic heat $P = RI^2$ created in the wires has to be efficiently transferred into the substrate and removed. The temperature rise and eventually the destruction limit of a wire depends not only on j and the length of the current pulse τ but also on the geometry and on material parameters like heat conductivity λ and heat capacity C_V. In a typical atom chip the wire is separated from the thermally conducting substrate by a thin electrical insulating layer which at the same time is a bad thermal conductor. This leads to two heat removal mechanisms on very different time scales [5].

The first process concerns the heat flow from the wire to the substrate through the insulating layer. In a 1d model, the timescale of this process is given by $\tau_{\text{fast}} = (C_V h)/(k - h j^2 \rho)$ where k is the thermal conductance through the insulating layer and ρ the temperature-dependent resistivity of the wire. For a typical Au chip wire $\tau_{\text{fast}} \sim 1\,\mu$s which is very short when compared to other timescales of the experiments and the temperature difference ΔT saturates to:

$$\Delta T(t) = \frac{h j^2 \rho (\Delta T = 0)}{k - h j^2 \rho} \left(1 - e^{-t/\tau_{\text{fast}}}\right). \tag{3.2}$$

The second process concerns the heat transport in the substrate and is much slower: in a 2d model with a point-like heat source on the surface of a half-space substrate the temperature increase is then given by

$$T_s(t) = \frac{h w \rho j^2}{2\pi\lambda} \Gamma\left(0, \frac{C_V w^2}{4\pi^2 \lambda t}\right) \approx \frac{\rho I j}{2\pi\lambda} \ln\left(\frac{4\pi^2 \lambda t}{C_V w^2}\right) \tag{3.3}$$

where the (small) temperature dependence of the resistivity is neglected.

These two models together accurately reproduce the results obtained in a two-dimensional numerical calculation as long as the substrate can be treated as a heat sink, which holds for short times (typically 100 ms–1 s). For longer times the heat transport out of the substrate has to be taken into account.

Atom chip wires are typically tested using a four-point measurement, pushing a constant current through the wires while recording the change of the respective voltage drop with time. For simple single-layer chips the latter is a good indicator of the change in resistivity caused by the Ohmic heating of the wires. In typical tests [5] a silicon substrate with a thin SiO$_2$ layer shows the lowest immediate increase of temperature, and the smallest long-term heating. Silicon with a thick SiO$_2$ layer does show stronger heating, which is even worse for sapphire and GaAs substrates. In accordance with the above model the highest current densities are tolerated by the thinnest wires with the smallest width. A 700 nm wide and 140 nm high wire carried currents of up to 60 mA over 10 s, corresponding to a current den-

3 Atom Chip Fabrication

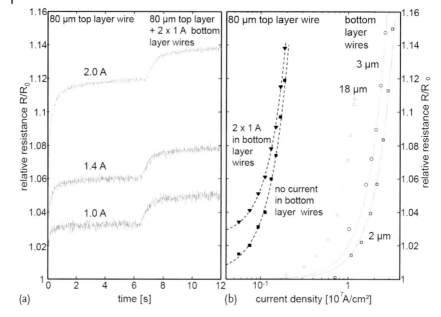

Figure 3.21 (a) Temperature evolution of a 80 μm top layer trapping wire for different applied currents. After ~ 6 s, an additional current of 1 A is sent through two 500 μm bottom layer confinement wires. (b) Temperature evolution for different current densities in various chip wires. Solid lines are theoretical predictions according to a simple dissipation model which applies to bottom layer wires in direct contact with the substrate [5]. Reduced heat dissipation reduces the current density for top layer wires, currents in the bottom layer wires lead to additional heating (dashed lines to guide the eye).

sity of 6×10^7 A/cm². Similar tests were recently carried out with a chip fabricated on an AlN substrate [106].

Tests of multi-layer atom chip structures (Figure 3.21) [23] show a similar behavior, but the combination of short sub-μm wires connected to large structures, and the simultaneous current through crossing wires does lead to delicate failure modes. The resistivity change is not a firm indicator of the temperature any more because different parts of the wire contribute differently to the changes. A simple rule of how much resistivity increase one can safely tolerate cannot be applied. Limits have to be found for each individual wire category.

Finally, atom chips also create electric fields. Very small structures naturally create large electric fields. This is due to sharp corners as well as small gaps. For example, a moderate voltage of 60 V translates to an applied electric field of over 200 kV/mm over 300 nm gaps between small wires. It has been shown that atom chips can indeed sustain very high electric fields without breakdown allowing one to create very steep potential surfaces and to implement versatile state-selective combinations of electric and magnetic potentials [21].

3.8
Photonics on Atom Chips

The interaction between atoms and light is an essential ingredient in preparing, detecting, and manipulating atoms. It is, therefore, natural to integrate micro-optics on the atom chip.

3.8.1
Fiber-Based Integrated Optics

A simple and very well-developed technology that allows one to integrate light with the atom chip is fiberoptics. It can be directly mounted on the chip surface. Components that don't need active alignment, such as cavities with mirrors implanted in the fiber, or tapered fibers for dipole traps or for delivering of excitation light, can be directly integrated on the chip. Here micro-fabricated SU8 holding structures are the method of choice [26]. This strategy will not work for structures that need to be actively aligned during operation; For example actuators are needed for high finess fiber cavities [116–120].

3.8.1.1 SU8 – Holding Structures

SU8 [107] is an epoxy-based negative photoresist which can be patterned using 365–436 nm UV light, and which is compatible with Ultra High Vacuum requirements. It has very high mechanical, chemical, and thermal stability. Its specific properties facilitate the production of > 100 μm thick structures with very smooth sidewalls with a sizable undercut. SU8 is widely used to fabricate diverse microcomponents ranging from optical planar waveguides with high thermal stability and controllable optical properties to mechanical parts such as microgears [108] for engineering applications, and micro-fluidic systems for chemistry [109].

To mount fibers on the chip, the Heidelberg/Vienna team fabricate two SU8 ridges which are separated by a distance a few hundreds of nanometer smaller than the fiber diameter, and have a height of typically 60 % of the fiber diameter. The layout of the used alignment structure with fibers is shown in Figure 3.22. This design includes funnels to simplify assembly. To avoid angular misalignment the total length of the alignment structure is quite long (6000 μm). The structure is divided into several subsegments to reduce stress induced by thermal expansion. The fibers can then be inserted from the top by applying a little pressure on the fiber. Inside the holding structure the fibers are held precisely by the undercut, and can be shifted longitudinally with high precision. To fix the fibers they are glued with UV-cured glue.

The quality of the SU8 alignment structures can be assessed by mounting a split fiber-based optical resonator, with mirrors inside the two mounted fibers. The finesse of the resonator then strongly depends on losses introduced in the gap by misalignment of the two fibers [110]. Starting with a fiber resonator with a finesse of $\mathcal{F} = 152.8 \pm 1.1$, the fiber is cut between the mirrors to introduce a gap and the new surfaces are polished. The two separate fiber pieces are then introduced into

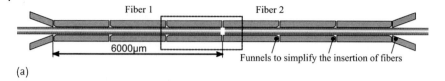
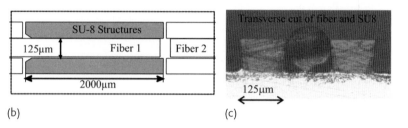

Figure 3.22 (a) Layout of the alignment structure and a magnified part (dotted rectangle) in (b). (c) Fiber in a SU8 structure mounted on a gold-coated atom chip. The atom chip and the SU8 structures have been cut with a wafer saw. The SU8 maintains structural integrity during the cutting.

the SU8 structures and mounted with a small (< 1 μm) gap size. The finesse of the split and mounted resonator was measured to be $\mathcal{F} = 132.0 \pm 1.3$. This corresponds to a coupling loss from fiber to fiber of $0.29 \pm 0.04\,\%$. Neglecting other additional losses, this corresponds to a transversal misalignment of < 100 nm or an angular misalignment of $< 5\times 10^{-3}$ rad. When changing the temperature of the substrate between 20 and 70 °C the finesse of the mounted fiber resonator shows no change.

3.8.1.2 Fiber-Based Fluorescence Detector

A simple fluorescence detector capable of high-fidelity detection of single atoms can be built using a single-mode tapered lensed fiber to deliver the excitation light to a very small spot and a multi-mode fiber mounted at right angle to collect the fluorescent light (Figure 3.23). To eliminate background light and to protect the photon counter, the light passes through an interference filter before being directed to the single-photon counting module (SPCM).

In the implementation by the Vienna group [112–114] the fibers are arranged at 45° to the guide and orthogonal to one another. The detector is fully integrated on the atom chip by mounting the two fibers on the chip surface in lithographically defined holders fabricated from SU8 resist guaranteeing very accurate and ultra-stable passive alignment [26]. The collection fiber with NA (Numerical Aperture) 0.275 collects 1.9 % of the fluorescence photons, and an average of 1.1 photons is counted for each atom. In the absence of atoms, less than 10^{-8} of the excitation light is scattered into the collection fiber and the total background is dominated by the SPCM dark count rate. With such a low background a single detected photon implies that an atom is present in the detection region. With an average of 1.1 photons/atom, the detector achieved single atom detection with 66 % efficiency, and a typical signal-to-noise ratio of 100. Most remarkable is that the stray light

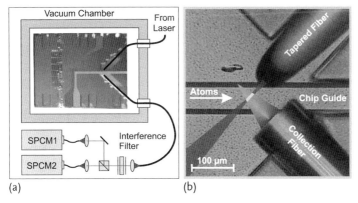

Figure 3.23 (a) Basic layout of the detector and the atom chip in the vacuum chamber. A magnetic guide (light gray) transports the atoms from the Z-shaped wire trap (dark gray) to the detector. The excitation light is delivered by a tapered fiber, and the fluorescence light is collected by a multi-mode fiber. Both are connected to the optics outside the vacuum chamber by a fiber feed-through [111]. The collected light then passes through a 3 nm wide (FWHM) interference filter centered at 780 nm before being directed to the single- photon counting module(s) (SPCM). For high efficiency atom detection a single SPCM is employed while correlation measurements require two SPCMs in a Hanbury Brown–Twiss like configuration. (b) A microscope image of the detection region on the chip. The multi-mode fiber collects light from a cone (light gray) determined by its NA. The overlap of this cone with the excitation light from the tapered fiber (dark gray) defines the detection region (white).

level is so low that atoms can propagate freely in the magnetic guide for more than a second with the detection light on, and be detected one-by-one when passing the detection region. The single atom detection was verified by measuring the photon correlation function, and values of $g^{(2)}(0) < 0.05$ were obtained [113].

The efficiency of the detector is limited by the NA of the collection fiber, while the SNR is limited by the background. Using a low noise photon counter with a dark count rate < 25 cps, a total background of 55 cps can be achieved. Using a commercially available fiber with NA = 0.53 will allow one to collect 4.5 photons from each atom, and a single atom detection efficiency of ≈ 95 % could be reached for an interaction time of 20 μs. With two detection fibers of NA = 0.53 and with an interaction time of 20 μs, a single atom detection efficiency beyond 99 % seems feasible. Such a system would reach 9 photons/atom. This will allow true atom counting by transient count rate analysis.

3.8.1.3 Fiber Cavities

The integration of high-finesse optical fiber cavities on atom chips has enabled spectacular experiments on cavity quantum electrodynamics with a Bose–Einstein condensate [27] as well as high-fidelity single atom detection and preparation [115], an essential ingredient of chip-based quantum information processing (see Chapter 9). In these experiments, the atom chip allows precise positioning of the atoms in the cavity field. Fiber cavities advantageously combine small mode volume with direct access to the high-intensity part of the intra-cavity field. Moreover, the light can be coupled directly in and out of the cavity through the fiber, avoiding the need for sensitive coupling optics. Here we discuss different methods to fabricate chip-based fiber cavities.

In [116–118], plano-concave microcavities are described, where the curved mirror is etched into a silicon wafer. The plane mirror is a flat, cleaved fiber tip with a high reflectivity coating applied to it, see Figure 3.24a. Arrays of concave mirrors are fabricated in silicon by wet-etching isotropically through circular apertures in a lithographic mask using a mixture of HF and HNO_3 in acetic acid. In order to investigate cavities of various lengths, the etching parameters are adjusted to produce wafers with a range of mirror radii between 30 and 250 µm. The etched surface typically has an r.m.s. roughness of 5 nm. If the concave mirror templates are coated with sputtered gold, a cavity finesse $\mathcal{F}$ of the order of a few hundred results. Alternatively, a dielectric coating designed for 99.99 % reflectivity can be used [116]. In this case, a cavity finesse of $\mathcal{F} = 6000$ can be achieved for a 15 µm long cavity, limited by the surface roughness of the curved mirror. A cavity of this kind but with $\mathcal{F} = 280$ was used for atom detection with less than one atom on average in the cavity mode [118].

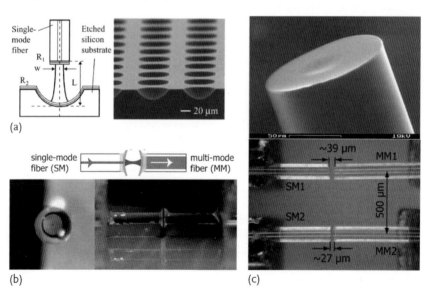

Figure 3.24 Optical fiber cavities on atom chips. (a) Plano-concave cavity of [116, 117] with a finesse of up to $\mathcal{F} = 6000$. Left: The plane mirror is formed by coating a fiber tip, the concave mirror is formed by coating an etched silicon wafer. Right: SEM micrograph of an array of mirrors on a silicon wafer. (b) Fiber Fabry–Pérot cavity with $\mathcal{F} = 1000$ described in [104, 119]. Two fibers functionalized with a concave dielectric mirror face each other. Left photograph: a single-mode fiber processed with a concave mirror. Right: mounted cavity of 27 µm length on an atom chip. (c) Fiber Fabry–Pérot cavity with $\mathcal{F} = 37\,000$ used in [27, 115]. Top: SEM image of a laser-machined fiber end face. Bottom: Two fiber cavities of different length mounted on the same atom chip.

A different type of fiber cavities was developed in [104, 119], see Figure 3.24b. The cavities are of the Fabry–Pérot type and are formed with miniature spherical mirrors positioned on the end of single- or multi-mode optical fibers. The concave mirrors are fabricated from a convex template and a lift-off step. For large radii of curvature, ≥ 500 µm, the template is a commercial ball lens whereas for smaller radii, 100–500 µm, a custom silica microlens with a surface roughness ~ 1 nm is used. It is coated in one run with a release layer and a dielectric Bragg mirror with nominal reflectivity of 99.7 %. A cleaved single mode fiber is then positioned immediately above the center of the coated lens by maximizing the back reflection of a laser beam coupled into the fiber. The fiber is then glued in place with a UV-curing epoxy, after which the application of a small force is sufficient to detach the mirror from the original substrate. The result is a fiber functionalized with a concave mirror, see Figure 3.24b. Two fiber tips facing each other form the cavity. The finesse is $\mathcal{F} = 1000$, currently limited by the quality of the multi-layer coatings. The detection of small ensembles of cold Rb atoms guided through such a cavity on an atom chip was demonstrated in [104].

In the experiments of [27, 115], an improved fiber Fabry–Pérot cavity with a much higher finesse of $\mathcal{F} = 37\,000$ is used. The cavity design is based on a new laser-machining process [120] where a single, focused CO_2 laser pulse creates a concave depression in the cleaved fiber surface, see Figure 3.24c. The machining is performed in a regime where thermal evaporation occurs, while melting is restricted to a thin surface region, avoiding global contraction into a convex shape. The surface is then coated with a dielectric mirror coating.

The observed $\mathcal{F} = 37\,000$ is limited by the sub-optimum coating. Ultimately, a finesse of $\mathcal{F} = 150\,000$ should be achievable based on the measured surface roughness of about 0.2 nm rms. Radii of curvature can be fabricated down to 50 µm and probably below. Because of the small mirror diameter (smaller than the fiber diameter, which is typically 125 µm), the mirrors can approach each other very closely (down to a few optical wavelengths) without touching. The result is a very small mode waist between 1 and 2 µm, and a small mode volume down to a few cubic wavelengths. These high-finesse fiber cavities have enabled atom-chip-based cavity quantum electrodynamics experiments in the strong coupling regime as well as high-fidelity single-atom detection and preparation [27, 115].

3.8.2
Microlens and Cylindrical Lens

While fibers certainly hold promise, they also possess numerous disadvantages. For example, they require alignment and gluing. Because of their radius of typically 60 µm, their interaction with atoms is usually at a similar height above the surface and this hinders the advantages of traps close to the surface (high gradients and low power consumption). To overcome this problem one would have to etch paths for the fibers within the substrate. However, the main disadvantage of fibers is in the scaling problem. Because of their size and the need for individual handling, it is hard to imagine hundreds and perhaps thousands of traps addressed individu-

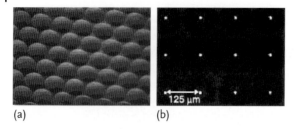

Figure 3.25 (a) An hexagonal array of spherical microlenses (taken from [122]); (b) the light intensity distribution at the focal plane.

ally by arrays of fibers. This has prompted an intensive effort aimed at integrating photonics into standard atom chips in order to achieve monolithic devices. The first interaction of cold atoms with substrate-based optical elements was achieved in 2002 in the Hannover group [121] (see for example Figure 3.25).

A review of these elements may be found in [122] (see for example Figure 3.25). Here circular and cylindrical lenses are formed with the goal of creating dipole force traps and guides. These guides may then be combined to create beam splitters and interferometers [123]. A review of how these lenses may be fabricated can be found in [124].

3.8.3
Microdisks and Microtoroids

The first substrate-based high-finesse devices to interact with cold atoms appeared in 2006 in two papers from the groups of Kimble and Vahala [125] and from the group of Mabuchi [126]. These experiments showed single atom sensitivity. To the best of our knowledge, these were also the first non-fiber photonic elements of any kind to be put into the vacuum at close vicinity to the atoms. Hence, perhaps they may deserve the title of the first photonic atom chips. They were based on microtoroids sustaining a whispering gallery mode of light [127]. These seminal experiments still used fibers (tapered) in order to couple the light into the disks and this required delicate positioning, gluing, and tuning. Furthermore, no trapping of atoms has taken place in these experiments and this will presumably be aspired to in the next generation setups.

Let us review in more detail the above experiments. In the experiments performed by the Mabuchi group a quality factor of $Q \approx 10^6$ (defined as the ratio of the frequency to the line width or loss rate) was achieved by etching a 9 μm SiN disk. Here work was done with commercially available Si wafers with a 250 nm layer of SiN deposited by Low Pressure Chemical Vapor Deposition (LPCVD). A highly circular mask was made by e-beam lithography and resist reflow and was transferred to the SiN by way of C_4F_8/SF_6 plasma dry etch. Potassium hydroxide wet etch selectively removed the underlying ⟨100⟩ Si substrate until the SiN microdisk was supported by a small diameter Si pillar. A final cleaning step to remove organic materials from the disk surface was performed using a $H_2SO_4 : H_2O_2$ wet

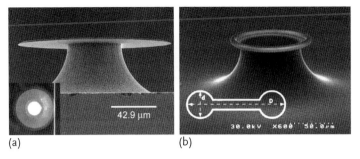

Figure 3.26 (a) A microdisk (taken from [129]). The disk itself is made of 2 μm thick oxide. The inset shows the fiber which couples the light into the microdisk. (b) An ultra-high-finesse toroid (taken from [130]).

etch. To fine tune the disk to the atomic resonance, a series of timed etches of a 20 : 1 diluted 49 % HF solution were employed. The resonance was shown to blue shift at a rate of 1.1 nm/min. Further fine tuning can be achieved by temperature variations and a dependence of 0.012 nm/K was measured.

Similar work was done in the group of Kimble [125]. Their microtoroid was developed by the group of Vahala, where quality factors exceeding $Q \approx 10^8$ were achieved by surface tension reflow following the melting of the surface with a CO_2 laser [128]. The laser, having typical intensities of 100 MW/cm^2, is centered on the disk, but because of the Si pillar, only the disk periphery is melted. Here, surface roughness below the nanometer scale needs to be achieved. For the purposes of interacting with atoms, quality factors exceeding 10^6 were sufficient and those were realized with disk diameters of about 100 μm without the need for laser melting [129]. First, silicon wafers with 2 μm of silicon oxide thermally grown (silica) were patterned by standard optical lithography and wet etch of the oxide using buffered HF. The silica disks were then isolated from the silicon substrate with an isotropic silicon dry etch using XeF_2 (3 Torr). Like before, the resonator is now suspended and being supported by a silicon pillar. See Figure 3.26 for examples.

3.8.4
Mounted and Fully Integrated Fabry–Pérots

Similar to fully integrated microdisks, monolithically fabricated Fabry–Pérots (FPs) offer robustness, accuracy and scalability. As the mode volume achieved by fiber based FPs is already extremely small, it is not expected that integrated FPs would be able to decrease the mode volume by more than say factor 10 at most. Hence, the advantage in realizing these elements is really in the fact that they enable scalability. Different from the case of the microdisk where the atom interacts with an evanescent field, here the atom interacts with a real light mode. This may enable numerous advantages, one being the distance from a material object. Attempts to realize such optical elements are being made by several groups. This includes the Orsay group of Aspect and Westbrook where a photonic waveguide is being etched

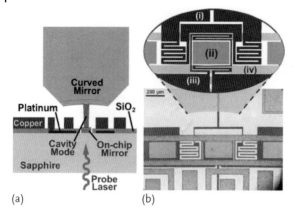

Figure 3.27 (a) Flat dielectric mirror on a sapphire substrate (taken from [132]). The mirror is made of a combination of alternating layers of silicon dioxide and tatalum pentoxide (see text). Around the mirror one may observe the platinum heaters for actuating the mirror. The second mirror is curved and mounted on a piezo stage. (b) Details of mirror region. Top: (i) thermometer wire, (ii) dielectric mirror pad, (iii) heater wire, (iv) waveguide wire.

in the middle and the mirrors are realized by way of Bragg gratings implanted in the waveguide. Taking into account the interface between the waveguide and the vacuum, this forms a "double" FP of sorts and was analyzed in detail in [131]. To the best of our knowledge, the only group so far to realize a non-fiber miniature FP mounted on the surface of an atom chip is the Berkeley group of Dan Stamper-Kurn. Though not monolithically fabricated (but rather assembled), this system may already be considered as integrated. Let us briefly review their work.

The first effort [132] was made by using a sapphire substrate both as the base for the atom chip wires and also for high-reflectivity planar mirrors (Figure 3.27). The vision is to place a curved mirror above the planar "mirror pad" forming a stable mode. Here, the flat dielectric mirrors which were used were made of alternating layers of SiO_2 and Ta_2O_5 (Research Electro-Optics, Boulder) totaling about 5 μm in thickness. With a 400 nm thermally evaporated aluminum protective layer in place, Reactive Ion Etching (RIE) was used to etch away unwanted mirror areas (100 sccm CF_4 and 10–20 sccm O_2 at 85 mTorr and RF power of 0.4 W/cm^2). About 5 μm of mirror edge were lost in this process, achieving finally a finesse of about $\mathcal{F} = 10^5$.

One valuable outcome of this work is the high level of control realized over the temperature of the atom chip surface, and the discovery that one could actuate a high-finesse mirror thermally to achieve a closed-loop control of the Fabry–Pérot cavity spacing with a 1 MHz bandwidth. This is a powerful new capability which would enable robust Cavity QED since the kHz-level acoustic noise can be now suppressed immensely.

A second "cheaper" and less integrated project now under way utilizes Deep Reactive Ion Etching (DRIE) to thin down a silicon wafer to just <100 μm thickness, including the wires atop that thinned substrate. A hole is then micromachined through this thinned wafer. Mirrors mounted on a separate piezo mounting are

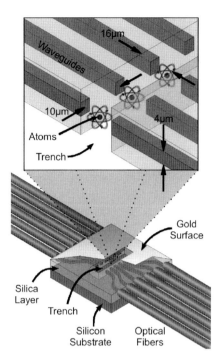

Figure 3.28 An array of monolithically fabricated atom-light junctions [133].

positioned on both sides of the chip so that a vertical cavity mode pierces through the horizontal chip surface. In this way, the mirrors are decoupled from the atom chip (useful for both thermal and vibrational isolation), and may be reused for many iterations of the atom chip. In addition one may use cheaper materials and fabrication for the chip itself.

Finally, let us note a recent result from the group of Ed Hinds at Imperial College [133] in which for the first time a fully integrated (monolithically fabricated) array of atom-light junctions has been operated (Figure 3.28). This may be considered the first real step towards a scalable system.

3.8.5
Planar Optics

Planar optics (i.e., optics with no curved surfaces) are well established, and although they are ideal for standard fabrication techniques, they have almost not been exploited in the field of atom chips. To the best of our knowledge only one publication exists describing such work [134]. In this work by J. Weiner and H.J. Lezec, a Fresnel (diffraction) lens has been etched by a Focused Ion Beam (FIB) (Figure 3.29). The overall size of the element is still quite large (about $200 \times 200\ \mu m^2$) but the method may be considerably miniaturized. For example, it has also been suggested to trap and measure atoms via a long focal length plasmon lens [135],

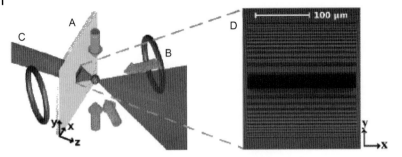

Figure 3.29 The cold atom setup and a zoom-in on the FIB etched 1d Fresnel diffraction lens (taken from [134]).

made so far also by FIB. Here light passes the substrate through a 250 nm hole. Such holes may be produced by the beam of a FIB or electron-beam-induced etching (utilizing gas injectors inside an e-beam machine), as these beams have a very small diameter of several nanometers. To the best of our knowledge, while a small divergence beam has been produced by a plasmon lens, a focus has not. Another idea for an extremely small planar optical element is to utilize the fast growing field of sub-wavelength optics. Here, a transparent material may be etched with high aspect ratios to produce a phase mask which gives rise to optical control. As the resolution required in these elements is below that of optical lithography, and as e-beam lithography is time and price consuming, such elements will most likely be produced by nano-imprint methods where a hard mask possessing the negative image is brought into contact with a soft material. The mask can, for example, be made of sapphire. Currently, DRIE protocols exist for sapphire, for example with Nickel masks allowing for an etch selectiveness of 1:7.

3.8.6
Photonics Outlook

It is clear that the eventual goal of the photonics effort should be to use light on the surface of the atom chip to trap, manipulate, and measure atoms (down to a single atom) in a way that allows integration with other atom chip techniques such as magnetic traps, and utilizing configurations that require standard monolithic fabrication techniques in order to achieve robust and accurate scalability. This probably means solid-state light wave guides which will transfer light from fibers glued at the edge of the chip to monolithic optical elements in the center of the chip, as in the recent achievement by the Hinds group (Figure 3.28). Waveguides with low scattering, that is, small surface roughness, are a challenge in their own right. For example, annealing a SiO waveguide for 20 h at 1000 °C may reduce its surface roughness from several nanometer rms to less than one. The eventual goal would of course be to have no fibers at all, namely to have the light sources and the photodetectors integrated on the chip itself so that only electronics will need to interface with the chip. While the above is the general vision, it is unclear what the final "winning" monolithic optical element would be. Considering the variety of demands including also the eventual price tag, it is likely that for every specific application another type of element would perform best.

Concerning the integration of microdisks, the first issue is that of simultaneous trapping allowing for long interrogation times. Standard magnetic trapping poses the challenge of light-induced spin-flips to un-trappable states. In addition, a metal with a wide absorption spectrum put in the vicinity of a high-finesse mode may significantly reduce its finesse. Magnetic traps based on molecules (e.g., CNT) with very sharp absorption peaks may be one way to solve this problem. Another way may be through simultaneous optical trapping and sensing. Here blue and red modes create a potential minimum next to the disk. Counter propagating modes can create arrays of traps. A fabrication tolerance analysis of such a scenario was carried out in [136]. The second issue requiring development is that of tunability. So far tunability has only been done by thermal control. In [127], other tunability mechanisms have been analyzed. These include fabricating the disk from piezo material where a voltage would change the disk diameter or utilizing an Electro-Optical (EO) material where a voltage would induce a change of index. Both these options require extensive materials and fabrication development. For example, standard EO materials are crystalline in nature and may not be evaporated onto the wafer (e.g., $LiNbO_3$). Etching these materials with a high surface quality is also very challenging. A possible solution may be to use polymers with EO features that may be deposited through evaporation. A third issue is that of light coupling. While disk fabrication without the existence of a nearby waveguide enables one to utilize a variety of methods to achieve high finesse (in which case a tapered fiber is later brought to the vicinity of the disk), monolithic fabrication of both disk and waveguide demand either a single-step fabrication in which methods are used for the disk that do not harm the waveguide or significantly affect the gap between the two, or alternatively a double-step fabrication which involves accurate alignment and reduces the yield. As parameters like the shape of the waveguide, its surface roughness, and the exact dimension of the gap to the microdisk may significantly affect the light coupling and the disk finesse, the joint fabrication of these two elements is a demanding challenge.

Concerning the integration of FPs: Two issues seem to be the most challenging problems on the way to the development of such a device. The first is 3d fabrication and the second tunability. For a stable mode to exist at least one of the two mirrors must have a 3d profile or alternatively both mirrors must have a 2d profile where the overall structure is 3d. A well-established fabrication protocol for 3d structures actually does not exist. There are attempts to produce 3d structures by way of controlled electron penetration and scattering within an e-beam resist. Similarly, there are attempts to utilize laser light where the focus exposes 3d structures in a photoresist. However, these are crude methods without good control over parameters like curvature and the surface roughness achieved so far is not good enough. Even if these problems were to be solved, there still remains the issue of how to evaporate a mirror (be it metallic or dielectric) onto these highly curved surfaces. The second issue involves tunability. Different from the case of the microdisk, here tunability would involve a real movement of the mirrors (unless some of the cavity mode is in a material, as in the above-described "double" FP, in which case the index of refraction may be changed as in the microdisks, or unless some thermal mechanism like that developed by the Dan Stamper-Kurn group may be employed). Such a movement is possible through MEMs or piezo stages, but then the interface to the waveguides bringing and taking the light becomes complex.

Aside from the above-noted elements realized so far, there are also numerous other ideas in the early stage of development, and below we note three as an example. The first is photonic crystals or band gap materials. In [137], a feasibility study was done concerning atom detection via the interaction with a light mode confined in the wafer by a 2d array of holes. Here the features of the light mode are critical as there needs to be a significant component of light outside the wafer. In a second example, it has been suggested to use the evanescent field of waveguides for trapping and interacting with atoms. Two color schemes (red detuned and blue detuned) create a trapping minimum while counter propagating beams create standing waves and, therefore, arrays of traps. The affect of surface roughness on the feasibility of such an idea was analyzed in [136]. These ideas would require considerable fabrication development and nanometer-scale control over parameters such as edge and surface roughness, as well as high aspect ratio etching. In a third example, light is confined to propagation in nanostructures, and may be used for strong interaction with atoms and for efficient transport of light (see for example the work of M. Lukin, P. Hemmer, P. Zoller et al. [138, 139]).

Last but not least, photonics will also include considerable work on the substrate itself. Using transparent substrates such as sapphire, one would be able to introduce planar optics such as the Fresnel lens described previously or even sub-wavelength optics. 3d fabrication may also be utilized such as in the pyramid- or bowl-shaped substrate etching done by the Hinds group (and collaborators from Southampton) for the creation of an on-chip integrated Magneto-Optical Trap [140] or optical cavity [116, 117]. Transparent substrates may also enable evanescent blue detuned light to form stable near-surface traps with charged electrodes [141].

3.9
Chip Dicing, Mounting, and Bonding

After the microstructures on an atom chip have been fabricated, the chip has to be cut to the desired size, mounted on a heat sink, integrated into a UHV vacuum chamber, and connected to the outside world. Figure 3.30 shows examples of such complete atom chip systems.

In Figure 3.30a, the atom chip of [20, 41] is shown. The main "experiment chip" consists in this case of two layers of gold wires on a Si substrate (cf. Figure 3.20). It is cut to size with a diamond blade and subsequently glued onto a "carrier chip" with larger gold wires on an AlN substrate. The carrier chip serves as a heat sink and electrical vacuum feed through. The epoxy glue is UHV-compatible, heat conductive, but electrically insulating, and care is taken that the gap between the chips

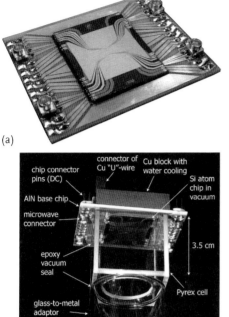

Figure 3.30 Atom chip mounting, bonding, and contacting. (a) Multi-layer microwave atom chip of [20, 41]. The two-layer experiment chip is glued and wire bonded to a single-layer carrier chip. Microwave and DC connectors are soldered onto the front and back side of the carrier chip. (b) Atom chip shown in (a) attached to a glass cell vacuum chamber. UHV compatible epoxy seals the vacuum. (c) Atom chip system of [142]. The chip is attached to a vacuum flange with electrical feedthroughs. Copper structures behind the chip are used as a heat sink and to create large-volume traps.

is completely filled with epoxy to avoid virtual leaks in the vacuum system. The wires on the experiment chip are wire bonded to the carrier chip, using up to 15 gold bond wires per chip wire. The carrier chip substrate has arrays of holes on each side, which were laser-cut before the wires were fabricated. The holes allow one to solder wires on the front side of the chip to pins of a connector fed through from the back side, using a 70 In/30 Pb reflow solder paste. A connector on the back side has the advantage that it does not inhibit optical access to the chip.

The atom chip is attached to a copper block, which is kept at a constant temperature by water cooling. Millimeter-scale machined copper structures are integrated into this block. Such U- or Z-shaped metal structures are a useful tool to create large-volume quadrupole or Ioffe–Pritchard type magnetic fields in the initial trapping stages of the experiment [44, 142].

Different techniques exist to integrate atom chips into the UHV vacuum chambers required for ultra-cold atom experiments. One approach is to mount the atom chip assembly onto a vacuum flange with electrical feedthroughs, see Figure 3.30c [44, 142, 143]. In the example of Figure 3.30b, a different technique is used, in which the carrier chip is directly connected to a glass cell vacuum chamber. This technique, which was first described in [144], eliminates the need for separate electrical vacuum feedthroughs, as the carrier chip wires can be directly connected from outside the vacuum chamber. To seal the vacuum, either UHV-compatible epoxy glue [41] or anodic bonding techniques [145] can be used. This technique allows one to further miniaturize ultra-cold atom experiments [145], and complete systems consisting of an atom chip attached to a miniaturized glass cell vacuum chamber are commercially available.

3.10
Further Integration and Portability

The atom chip was born in three laboratories around 1999 (Munich, Innsbruck, and Boston). The vision of its makers and those who followed was of a robust and accurate, miniature and portable system that could replace the large cold-atom systems. Such an achievement would not only enable subtle experiments in fundamental physics but could also enable technological applications such as clocks, sensors, and quantum information processing.

For this to happen numerous technologies must be further developed and integrated:

- Substrates: Although substrates (wafers) may seem a trivial matter, they are an extremely important issue. For example, when using standard silicon wafers, one reduces the possibility to transmit visible light through the wafer. One also adds in such a case the requirement for an insulation layer which usually conducts heat badly, thus disconnecting between the current carrying wires and their heat sink. Sapphire, for example, is transparent in the visible and does not require an insulating layer while being a good heat conductor. However, it is very hard (although possible) to etch sapphire, and adhesion to metallic layers may be more problematic. Sapphire is a good example for other considerations, as it may be used when RF with low losses is required (e.g., in ion chips). Transparent wafers also enable easy back-side alignment for back-side fabrication. It is very likely that any substrate used in the future will require robust deep etch processes. Vias for through-wafer electrical contacts (see following), loading wires (for magnetic loading) and loading slits (e.g., in ion chips where the particle source is likely to be on the back side of the chip), and eventually miniature vacuum cells within the wafers (closed by anodic bonding for example) – all these would require deep and at times high aspect ratio etching.
- Metallic and insulating layers: Assuming electromagnetic, electric, and magnetic fields will continue to play a major role in atom chips, metallic and insulation layers will continue to need improved material engineering and fabrication protocols. Numerous methods may be combined. For example, loading wires which are shut down during subtle quantum operations and which are far away from the atoms, may be made of thick layers (e.g., 10–100 µm) by way of electroplat-

ing on the back side of a thin wafer (e.g., 100 µm) or in deep etched grooves on the front side of a thick wafer. Finer wires made by evaporation with thicknesses up to 2 µm will be put closer to atoms for traps and guides in which quantum operations take place. Here, for complex trap and guide geometries, numerous layers may be evaporated with planarized insulating layers in between.

- "Exotic" materials: These pose a fabrication challenge in terms of compatibility of materials and fabrication protocols. For example, as described above, crystalline materials (e.g., electrically anisotropic materials) may prove to be very advantageous in fighting potential corrugations and spin-flip, heating and dephasing due to noise. Such materials may not be evaporated and their patterning may not be done by normal means of lift-off (of photoresist). Here shadow-mask techniques need to be employed for plasma etching or ion-beam milling. Finally, integration with nanowires, molecules such as CNTs and other nanodevices such as mechanical oscillators will require the integration of even more materials and fabrication techniques as described above. For example, while single-wall CNTs usually grow in a Chemical Vapor Deposition (CVD) oven, delicate electrical contacts may be done by Electron-Beam-Induced Deposition (EBID), and contact wires by thermal evaporation which may be either done by an e-gun (in the case of palladium) or by thermal evaporation (in the case of gold).

- Photonics: The integration of photonics means that all light sources and photodetectors would eventually be on the chip so that no light connections via fibers are required. These light sources would deliver the light to the different optical systems through single- or multi-mode dielectric waveguides having cross-sections of a few micrometers, or alternatively through nanostructures (plasmonics), as noted in the photonics outlook section. The optical elements themselves have been discussed above and may include waveguides, micro-curved lens, planar lens, and FP or micro-disk cavities. Transparent substrates may also enable evanescent blue detuned light to form stable near-surface traps with charged electrodes. Atom laser cooling would also need to be miniaturized and here a single-beam MOT would perhaps be utilized. Such miniature pyramid/cone MOTs have been developed in several groups [140, 146, 147]. For effective atom cooling, manipulation and measurement, frequency lock schemes will also need to be integrated onto the chip. Indeed, recently several new miniaturized lock schemes have been suggested (e.g., see [148]).

- Electronics: To increase the signal over noise, and to reduce the amount of electrical connections to the chip, it is plausible that future chips will require some level of electronics to be integrated onto them. For example, fabrication of photodiodes and preamplifiers for the readout, or feedback electronics and high-voltage drivers for the laser lock. It stands to reason that such electronics will be fabricated on the back side of the wafer and connected to the atom optics side by means of electrical vias.

- Vacuum: The issue of miniature vacuum cells has been the focus of considerable effort and will undoubtedly continue to be so. Similar to the effort in the context of atom cells for room temperature atom optics devices made by the group of

John Kitching [149], it is possible that future miniature cells will be contained inside the substrate itself. Here techniques of wafer anodic bonding, as well as deep etch and MEMs, will play a major role. Blue LEDs may also be used to achieve a fast transition from the state of bad vacuum required for good MOT loading, to the state of good vacuum required for long trap and interrogation times [150].

As an example of the state-of-the-art of miniature vacuum systems and electric contacts we present recent work from the groups of Dana Anderson and Victor Bright (Figure 3.31) [151]. This paper describes a new method for fabricating through-wafer interconnects in atom-trapping chips used in ultra-high-vacuum atom optics cells for Bose–Einstein condensation experiments. A fabrication process was developed which uses copper electroplating to seal the vias. The advantages of using feedthrough atom-trapping chips are the simple micro-fabrication process and the reduction of the overall chip area bonded to the glass of the atom-trapping cell. The results demonstrate that more than 10 A of current can be conducted through the vias while the vacuum can be held at around 10^{-11} Torr pressure at room temperature. The yield rate of fabricated via interconnects in this process after anodic bonding (requires heating to 425 °C) is 97 %.

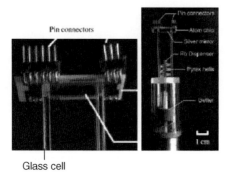

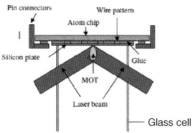

Figure 3.31 A new atom chip configuration in which the atom chip itself is one facet of the vacuum chamber and the electrical contacts are done through the substrate by way of vias (see text for details). Taken from [151].

Figure 3.32 Miniature vacuum system (taken from [151]).

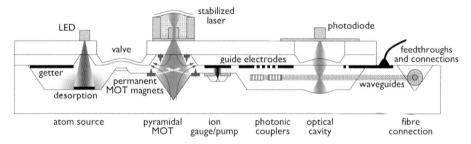

Figure 3.33 A futuristic vision of the atom chip (courtesy of Tim Freegarde). Here all the concepts of miniature vacuum, miniature light and particle sources, photonics and micro-magnetic traps, all come as an integrated device on top of one substrate.

The group has also pioneered work on ultra-low-volume vacuum systems as part of an effort to complement the miniaturization of atom optics which state-of-the-art fabrication enables. In Figure 3.32 we present their chamber.

We conclude this section with a beautiful illustration of a future atom chip made by Tim Freegarde of Southampton (Figure 3.33). Here one finds many of the features discussed above. Deeply etched substrate, miniature vacuum system including miniature particle sources, pumps and valves, single-beam MOT, on-board light sources and photodetectors, and so on.

3.11
Conclusion and Outlook

Fabrication of atom chips has come a long way from the simple gold layers of the first demonstration experiments. Ten years later we have numerous different technologies involved in building a chip, and new combinations are developed with each new experimental task to be implemented. This trend of utilizing the best material engineering has to offer, will continue.

The atom chip has proven to be a versatile platform in integrating different experimental technologies for quantum manipulation of atoms. Its success has led to similar developments for the manipulation of ions and molecules.

In the future, an additional force driving atom chip development will be the increasing demand for integrating different quantum systems into larger hybrid quantum systems, combining the best of the quantum worlds to create a versatile quantum technology. For example, integration with photonic quantum technology will allow atoms to be a versatile quantum memory in quantum repeater nodes for long-distance quantum communications. Integrating atomic systems with superconducting quantum circuits and the emerging field of circuit QED will allow one to transfer quantum information processed by MW photons into long-lived hyperfine states. Integration with nano-mechanical devices will provide a way to couple and read out mechanical vibrations.

For all these one will need to constantly combine and improve the different micro and nano technologies. These continuing developments will allow the atom chip to become a genuine platform for quantum technologies.

Acknowledgments

We would like to sincerely thank Ramon Szmuk for his assistance.

References

1 Folman, R., Krüger, P., Schmiedmayer, J., Denschlag, J., and Henkel, C. (2002) Microscopic atom optics: from wires to an atom chip. *Adv. At. Mol. Opt. Phys.* **48**, 263.

2 Reichel, J. (2002) Microchip traps and Bose–Einstein condensation. *Appl. Phys. B* **74**, 469.

3 Fortágh, J. and Zimmermann, C. (2007) Magnetic microtraps for ultracold atoms. *Rev. Mod. Phys.* **79**, 235.

4 Drndić, M., Johnson, K.S., Thywissen, J.H., Prentiss, M., and Westervelt, R.M. (1998) Micro-electromagnets for atom manipulation. *Appl. Phys. Lett.* **72**, 2906.

5 Groth, S., Krüger, P., Wildermuth, S., Folman, R., Fernholz, T., Mahalu, D., Bar-Joseph, I., and Schmiedmayer, J. (2004) Atom chips: Fabrication and thermal properties. *Appl. Phys. Lett.* **85**, 2980.

6 Wildermuth, S., Hofferberth, S., Lesanovsky, I., Haller, E., Andersson, L., Groth, S., Bar-Joseph, I., Krüger, P., and Schmiedmayer, J. (2005) Microscopic magnetic-field imaging. *Nature* **435**, 440.

7 Aigner, S., Pietra, L.D., Japha, Y., Entin-Wohlman, O., David, T., Salem, R., Folman, R., and Schmiedmayer, J. (2008) Long-range order in electronic transport through disordered metal films. *Science* **319**, 1226.

8 Davis, T.J. (1999) Atomic de Broglie waveguides and integrated atom optics using permanent magnets. *J. Opt. B* **77**, 408.

9 Vengalattore, M., Rooijakkers, W., and Prentiss, M. (2002) Ferromagnetic atom guide with in situ loading. *Phys. Rev. A* **66**, 053403.

10 Barb, I., Gerritsma, R., Xing, Y.T., Goedkoop, J.B., and Spreeuw, R.J.C. (2005) Creating Ioffe–Pritchard micro-traps from permanent magnetic film with in-plane magnetization. *Eur. Phys. J. D* **35**, 75.

11 Hall, B.V., Whitlock, S., Scharnberg, F., Hannaford, P., and Sidorov, A. (2006) A permanent magnetic film atom chip for Bose–Einstein condensation. *J. Phys. B* **39**, 27.

12 Whitlock, S., Gerritsma, R., Fernholz, T., and Spreeuw, R.J.C. (2009) Two-dimensional array of microtraps with atomic shift register on a chip. *New J. Phys.* **11**, 023021.

13 W. andB. V. Hall, S., Roach, T., Anderson, R., Volk, M., Hannaford, P., and Sidorov, A.I. (2007) Effect of magnetization inhomogeneity on magnetic microtraps for atoms. *Phys. Rev. A* **75**.

14 Zobay, O. and Garraway, B.M. (2001) Two-dimensional atom trapping in field-induced adiabatic potentials. *Phys. Rev. Lett.* **86**, 1195.

15 Lesanovsky, I., Schumm, T., Hofferberth, S., Andersson, L.M., Krüger, P., and Schmiedmayer, J. (2006) Adiabatic radio frequency potentials for the coherent manipulation of matter waves. *Phys. Rev. A* **73**, 033619.

16 Lesanovsky, I., Hofferberth, S., Schmiedmayer, J., and Schmelcher, P. (2006) Manipulation of ultracold atoms in dressed adiabatic RF-potentials. *Phys. Rev. A* **74**, 033619.

17 Treutlein, P., Hänsch, T.W., Reichel, J., Negretti, A., Cirone, M.A., and Calarco, T. (2006) Microwave potentials and optimal control for robust quantum gates on an atom chip. *Phys. Rev. A* **74**, 022312.

18 Courteille, P.W., Deh, B., Fortágh, J., Günther, A., Kraft, S., Marzok, C., Slama, S., and Zimmermann, C. (2006) Highly versatile atomic micro traps generated by multifrequency magnetic field modulation. *J. Phys. B* **39**, 1055.

19 Schumm, T., Hofferberth, S., Andersson, L., Wildermuth, S., Groth, S., Bar-Joseph, I., Schmiedmayer, J., and Krüger, P. (2005) Matter-wave interferometry in a double well on an atom chip. *Nature Phys.* **1**, 57.

20 Böhi, P., Riedel, M.F., Hoffrogge, J., Reichel, J., Hänsch, T.W., and Treutlein, P. (2009) Coherent manipulation of Bose–Einstein condensates with state-dependent microwave potentials on an atom chip. *Nature Phys.* **5**, 592.

21 Krüger, P., Luo, X., Klein, M.W., Brugger, K., Haase, A., Wildermuth, S., Groth, S., Bar-Joseph, I., Folman, R., and Schmiedmayer, J. (2003) Trapping and manipulating neutral atoms with electrostatic fields. *Phys. Rev. Lett.* **91**, 233201.

22 Günther, A., Kemmler, M., Kraft, S., Vale, C.J., Zimmermann, C., and Fortagh, J. (2005) Combined chips for atom-optics. *Phys. Rev. A* **71**, 063619.

23 Trinker, M., Groth, S., Haslinger, S., Manz, S., Betz, T., Bar-Joseph, I., Schumm, T., and Schmiedmayer, J. (2008) Multi-layer atom chips for versatile atom micro manipulation. *Appl. Phys. Lett.* **92**, 254102.

24 Gallego, D., Hofferberth, S., Schumm, T., Krüger, P., and Schmiedmayer, J. (2009) Optical lattice on an atom chip. *Opt. Lett.* **34**, 3463.

25 Quinto-Su, P., Tscherneck, M., Holmes, M., and Bigelow, N. (2004) On-chip optical detection of laser cooled atoms. *Opt. Expr.* **12**, 5098.

26 Liu, X., Brenner, K.-H., Wilzbach, M., Schwarz, M., Fernholz, T., and Schmiedmayer, J. (2005) Fabrication of alignment structures for a fiber resonator by use of deep-ultraviolet lithography. *Appl. Opt.* **44**, 6857.

27 Colombe, Y., Steinmetz, T., Dubois, G., Linke, F., Hunger, D., and Reichel, J. (2007) Strong atom-field coupling for Bose–Einstein condensates in an optical cavity on a chip. *Nature* **450**, 272.

28 Treutlein, P., Hunger, D., Camerer, S., Hänsch, T.W., and Reichel, J. (2007) Bose–Einstein condensate coupled to a nanomechanical resonator on an atom chip. *Phys. Rev. Lett.* **99**, 140403.

29 Hunger, D., Camerer, S., Hänsch, T.W., König, D., Kotthaus, J.P., Reichel, J., and Treutlein, P. (2010) Resonant coupling of a Bose–Einstein condensate to a micromechanical oscillator. *Phys. Rev. Lett.* **104**, 143002.

30 Verdu, J., Zoubi, H., Koller, C., Majer, J., Ritsch, H., and Schmiedmayer, J. (2009) Strong magnetic coupling of an ultracold gas to a superconducting waveguide cavity. *eprint arXiv:0809.2552*.

31 Petrosyan, J., Bensky, G., Kurizki, G., Mazets, I., Majer, J., and Schmiedmayer, J. (2009) Reversible state transfer between superconducting qubits and atomic ensembles. *Phys. Rev. A* **97**, 040304.

32 Wallraff, A., Schuster, D.I., Blais, A., Frunzio, L., Huang, R.S., Majer, J., Kumar, S., Girvin, S.M., and Schoelkopf, R.J. (2004) Strong coupling of a single photon to a superconducting qubit using circuit quantum electrodynamics. *Nature* **431**, 162.

33 Schoelkopf, R. and Girvin, S. (2008) Wiring up quantum systems. *Nature* **451**, 664.

34 Groth, S. (2006) Development, fabrication, and characterisation of atom chips. Ph.D. thesis, University of Heidelberg.

35 Madou, M.J. (2002) *Fundamentals of Microfabrication: The Science of Miniaturization*, CRC Press, New York, 2nd edn.

36 Reichel, J., Hänsel, W., and Hänsch, T.W. (1999) Atomic Micromanipulation with Magnetic Surface Traps. *Phys. Rev. Lett.* **83**, 3398.

37 Ruythooren, W., Attenborough, K., Beerten, S., Merken, P., Fransaer, J., Beyne, E., Van Hoof, C., De Boeck, J., and Celis, J.P. (2000) Electrodeposition for the synthesis of microsystems. *J. Micromech. Microeng.* **10**, 101.

38 Schlesinger, M. and Paunovic, M. (eds) (2000) *Modern Electroplating*, Wiley & Sons, New York, 4th edn.

39 Koukharenko, E., Moktadir, Z., Kraft, M., Abdelsalam, M.E., Bagnall, D.M., Valeb, C., Jones, M. P. A., and Hinds, E.A. (2004) Microfabrication of gold wires for atom guides. *Sens. Act. A* **115**, 600.

40 Lev, B. (2003) Fabrication of micromagnetic traps for cold neutral atoms. *Quantum Inf. Comput.* **3**, 450.

41 Treutlein, P. (2008) *Coherent manipulation of ultracold atoms on atom chips*. Ph.D. thesis, Ludwig-Maximilians-Universität München and Max-Planck-Institut für Quantenoptik, Published as MPQ report 321.

42 Estève,, J., Aussibal, C., Schumm, T., Figl, C., Mailly, D., Bouchoule, I., Westbrook, C.I., and Aspect1, A. (2004) Role of wire imperfections in micromagnetic traps for atoms. *Phys. Rev. A* **70**, 043629.

43 Kruger, P., Andersson, L.M., Wildermuth, S., Hofferberth, S., Haller, E., Aigner, S., Groth, S., Bar-Joseph, I., and Schmiedmayer, J. (2007) Potential roughness near lithographically fabricated atom chips. *Phys. Rev. A* **76**, 063621 (see also *eprint arXiv:condmat/0504686 (v1)*).

44 Folman, R., Krüger, P., Cassettari, D., Hessmo, B., Maier, T., and Schmiedmayer, J. (2000) Controlling Cold Atoms using Nanofabricated Surfaces: Atom Chips. *Phys. Rev. Lett.* **84**, 4749.

45 van Es, J.P.J., Wicke, P., van Amerongen, H.A., Rétif, C., Whitlock, S., and van Druten, J.N. (2009) Box traps on an atom chip for one-dimensional quantum gases. *preprint arXiv:0911.5250*.

46 Williams, K. and Muller, R. (1996) Etch rates for micromachining processing. *J. Microelectromech. Syst.* **5**, 256.

47 Williams, K.R., Kishan, G., and Wasilik, M. (2003) Etch Rates for Micromachining Processing – Part II. *J. Microelectromech. Syst.* **12**, 761.

48 Pietra, L.D., Aigner, S., Hagen, C.V., Groth, S., Bar-Joseph, I., Lezec, H., and Schmiedmayer, J. (2007) Designing potentials by sculpturing wires. *Phys. Rev. A* **75**, 063604.

49 Salem, R., Japha, Y., Chabé, J., Hadad, B., Keil, M., Milton, K., and Folman, R. (2010) Nanowire atomchip traps for submicron atom–surface distances. *New J. Phys.* **12**, 023039.

50 Meingailis, J. (1987) *J. Vac. Sci. Technol. B* **5**, 469.

51 Pietra, L.D., Aigner, S., vom Hagen, C., Lezec, H.J., and Schmiedmayer, J. (2005) Cold atoms near surfaces: designing potentials by sculpturing wires, in *J. Phys. Conf. Ser.*, Institute of Physics Publishing, **19**, 30–33.

52 Fortágh, J., Ott, H., Kraft, S., Günther, A., and Zimmermann, C. (2002) Surface

effects in magnetic microtraps. *Phys. Rev. A* **66**, 041604R.

53 Leanhardt, A.E., Shin, Y., Chikkatur, A.P., Kielpinski, D., Ketterle, W., and Pritchard, D.E. (2003) Bose–Einstein condensates near a microfabricated surface. *Phys. Rev. Lett.* **90**, 100404.

54 Jones, M.P.A., Vale, C.J., Sahagun, D., Hall, B.V., Eberlein, C.C., Sauer, B.E., Furusawa, K., Richardson, D., and Hinds, E.A. (2004) Cold atoms probe the magnetic field near a wire. *J. Phys. B: At. Mol. Opt. Phys.* **37**, L15.

55 Schumm, T., Estève, J., Figl, C., Trebbia, J.-B., Aussibal, C., Nguyen, H., Mailly, D., Bouchoule, I., Westbrook, C.I., and Aspect, A. (2005) Atom chips in the real world: the effects of wire corrugation. *Eur. Phys. J. D* **32**, 171.

56 Kraft, S., Günther, A., Ott, H., Wharam, D., and Fortágh, Z.C.J. (2002) Anomalous longitudinal magnetic field near the surface of copper conductors. *J. Phys. B* **35**, L469.

57 Wang, D.-W., Lukin, M.D., and Demler, E. (2004) Disordered Bose–Einstein condensates in quasi-one-dimensional magnetic microtraps. *Phys. Rev. Lett.* **92**.

58 Wildermuth, S., Hofferberth, S., Lesanovsky, I., Groth, S., Bar-Joseph, I., Krüger, P., and Schmiedmayer, J. (2006) Sensing electric and magnetic fields with Bose–Einstein Condensates. *Appl. Phys. Lett.* **88**, 264103.

59 Krüger, P. (2004) *Coherent matter waves near surfaces*. Ph.D. thesis, University of Heidelberg.

60 Japha, Y., Entin-Wohlman, O., David, T., Salem, R., Aigner, S., Schmiedmayer, J., and Folman, R. (2008) Model for organized current patterns in disordered conductors. *Phys. Rev. B* **77**.

61 Varpula, T. and Poutanen, T. (1984) Magnetic field fluctuations arising from thermal motion of electric charge in conductors. *J. Appl. Phys.* **55**, 4015.

62 Henkel, C., Pötting, S., and Wilkens, M. (1999) Loss and heating of particles in small and noisy traps. *Appl. Phys. B* **69**, 379.

63 Sidles, J., Garbini, J., Dougherty, W., and Chao, S.H. (2003) The classical and quantum theory of thermal magnetic noise, with applications in spintronics and quantum microscopy. *Proc. IEEE* **91**, 799.

64 Scheel, S., Rekdal, P., Knight, P., and Hinds, E. (2005) Atomic spin decoherence near conducting and superconducting films. *Phys. Rev. A* **72**, 42901.

65 Jones, M.P.A., Vale, C.J., Sahagun, D., Hall, B.V., and Hinds, E.A. (2003) Spin coupling between cold atoms and the thermal fluctuations of a metal surface. *Phys. Rev. Lett.* **91**, 080401.

66 Harber, D., McGuirk, J., Obrecht, J., and Cornell, E. (2003) Thermally induced losses in ultra-cold atoms magnetically trapped near room-temperature surfaces. *J. Low Temp. Phys.* **133**, 229.

67 Lin, Y., Teper, I., Chin, C., and Vuletić, V. (2004) Impact of the Casimir–Polder potential and Johnson noise on Bose–Einstein condensate stability near surfaces. *Phys. Rev. Lett.* **92**, 050404.

68 Dikovsky, V., Japha, Y., Henkel, C., and Folman, R. (2005) Reduction of magnetic noise in atom chips by material optimization. *Eur. Phys. J. D* **35**, 87.

69 Lin, Y.-J. private communications.

70 Skagerstam, B., Hohenester, U., Eiguren, A., and Rekdal, P. (2006) Spin decoherence in superconducting atom chips. *Phys. Rev. Lett.* **97**, 70401.

71 Rekdal, P. and Skagerstam, B. (2007) Decay processes in the presence of thin superconducting films. *Phys. Rev. A* **75**, 22904.

72 Skagerstam, B. and Redkal, P. (2007) Photon emission near superconducting bodies. *Phys. Rev. A* **76**, 052901.

73 Hohenester, U., Eiguren, A., Scheel, S., and Hinds, E. (2007) Spin-flip lifetimes in superconducting atom chips: Bardeen–Cooper–Schrieffer versus Eliashberg theory. *Phys. Rev. A* **76**, 33618.

74 Nirrengarten, T., Qarry, A., Roux, C., Emmert, A., Nogues, G., Brune, M., Raimond, J.-M., and Haroche, S. (2006) Realization of a superconducting atom chip. *Phys. Rev. Lett.* **97**, 200405.

75 Roux, C., Emmert, A., Lupascu, A., Nirrengarten, T., Nogues, G., Brune, M., Raimond, J.M., and Haroche, S. (2008) BEC on a superconducting atom chip. *Eur. Phys. Lett.* **81**, 56004.

76 Mukai, T., Hufnagel, C., Kasper, A., Meno, T., Tsukada, A., Semba, K., and Shimizu, F. (2007) Persistent supercurrent atom chip. *Phys. Rev. Lett.* **98**, 260407.

77 Zhang, B., Henkel, C., Haller, E., Wildermuth, S., Hofferberth, S., Krüger, P., and Schmiedmayer, J. (2005) Relevance of sub-surface chip layers for the lifetime of magnetically trapped atoms. *Eur. Phys. J. D* **35**, 97.

78 Shimizu, F., Hufnagel, C., and Mukai, T. (2009) Stable neutral atom trap with a thin superconducting disc. *Phys. Rev. Lett.* **103**, 253002.

79 Müller, T., Wu, X., Mohan, A., Eyvazov, A., Wu, Y., and Dumke, R. (2008) Towards a guided atom interferometer based on a superconducting atom chip. *New J. Phys.* **10**, 073006.

80 Müller, T., Zhang, B., Fermani, R., Chan, K., Wang, Z., Zhang, C., Lim, M., and Dumke, R. (2009) Trapping of ultra-cold atoms with the magnetic field of vortices in a thin film superconducting microstructure. *Arxiv preprint arXiv:0910.2332*.

81 Müller, T., Zhang, B., Fermani, R., Chan, K., Wang, Z., Zhang, C., Lim, M., and Dumke, R. (2009) Trapping of ultra-cold atoms with the magnetic field of vortices in a thin film superconducting microstructure. *Arxiv preprint arXiv:0910.2332*.

82 Cano, D., Kasch, B., Hattermann, H., Kleiner, R., Zimmermann, C., Koelle, D., and Fortágh, J. (2008) Meissner effect in superconducting microtraps. *Phys. Rev. Lett.* **101**, 183006.

83 Cano, D., Kasch, B., Hattermann, H., Koelle, D., Kleiner, R., Zimmermann, C., and Fortágh, J. (2008) Impact of the Meissner effect on magnetic microtraps for neutral atoms near superconducting thin films. *Phys. Rev. A* **77**, 63408.

84 Dikovsky, V., Sokolovsky, V., Zhang, B., Henkel, C., and Folman, R. (2009) Superconducting atom chips: advantages and challenges. *Eur. Phys. J. D* **51**, 247.

85 Lev, B., Lassailly, Y., Lee, C., Scherer, A., and Mabuchi, H. (2003) Atom mirror etched from a hard drive. *Appl. Phys. Lett.* **83**.

86 Sinclair, C.D.J., Retter, J.A., Curtis, E.A., Hall, B.V., Garcia, I.L., Eriksson, S., Sauer, B.E., and Hinds, E.A. (2005) Cold atoms in videotape micro-traps. *Eur. Phys. J. D* **35**, 105.

87 Eriksson, S., Ramirez-Martinez, F., Curtis, E.A., Sauer, B.E., Nutter, P.W., Hill, E.W., and Hinds, E.A. (2004) Micron-sized atom traps made from magneto-optical thin films. *Appl. Phys. B* **79**, 811.

88 Xing, Y.T., Eljaouhari, A., Barb, I., Gerritsma, R., Spreeuw, R.J.C., and Goedkoop, J.B. (2004) Hard magnetic FePt films for atom chips. *Phys. stat. sol. (c)* **1**, 3702.

89 Gerritsma, R., Whitlock, S., Fernholz, T., Schlatter, H., Luigjes, J.A., Thiele, J.U., Goedkoop, J.B., and Spreeuw, R.J.C. (2007) Lattice of microtraps for ultracold atoms based on patterned magnetic films. *Phys. Rev. A* **76**, 33408.

90 Xing, Y.T., Barb, I., Gerritsma, R., Spreeuw, R. J. C., Luigjes, H., Xiao, Q.F., Retif, C., and Goedkoop, J.B. (2007) Fabrication of magnetic atom chips based on FePt. *J. Magn. Magn. Mater.* **313**, 192.

91 Wang, J.Y., Whitlock, S., Scharnberg, F., Gough, D.S., Sidorov, A.I., McLean, R.J., and Hannaford, P. (2005) Perpendicularly magnetized, grooved GdTbFeCo microstructures. *J. Phys. D* **38**, 4015.

92 Singh, M., Volk, M., Akulshin, A., Sidorov, A., McLean, R., and Hannaford, P. (2008) One-dimensional lattice of permanent magnetic microtraps for ultracold atoms on an atom chip. *J. Phys. B* **41**, 065301.

93 Peano, V., Thorwart, M., Kasper, A., and Egger, R. (2005) Nanoscale atomic waveguides with suspended carbon nanotubes. *Appl. Phys. B* **81**, 1075.

94 Fermani, R., Scheel, S., and Knight, P. (2007) Trapping cold atoms near carbon nanotubes: Thermal spin flips and Casimir–Polder potential. *Phys. Rev. A* **75**, 62905.

95 Petrov, P., Machluf, S., Younis, S., Macaluso, R., David, T., Hadad, B., Japha, Y., Keil, M., Joselevich, E., and Folman, R. (2009) Trapping cold atoms using surface-grown carbon nanotubes. *Phys. Rev. A* **79**, 43403.

96 Ismach, A., Segev, L., Wachtel, E., and Joselevich, E. (2004) Atomic-step-templated formation of single wall car-

bon nanotube patterns. *Angew. Chem. Int. Ed.* **43**, 6140.

97 Ismach, A., Kantorovich, D., and Joselevich, E. (2005) Carbon nanotube graphoepitaxy: Highly oriented growth by faceted nanosteps. *J. Am. Chem. Soc.* **127**, 11554.

98 Huang, X. M. H., Caldwell, R., Huang, L., Jun, S.C., Huang, M., Sfeir, M.Y., O'Brien, S.P., and Hone, J. (2005) Controlled placement of individual carbon nanotubes. *Nano Lett.* **5**, 1515.

99 Grüner, B., Jag, M., Stibor, A., Visanescu, G., Häffner, M., Kern, D., Günther, A., and Fortágh, J. (2009) Integrated atom detector based on field ionization near carbon nanotubes. *gas].*

100 David, T., Japha, Y., Dikovsky, V., Salem, R., Henkel, C., and Folman, R. (2008) Magnetic interactions of cold atoms with anisotropic conductors. *Eur. Phys. J. D* **48**, 321.

101 Wilson, S.R. and Tracy, C.J. (eds) (1993) Handbook of Multilevel Metallization for Integrated Circuits: Materials, Technology, and Applications. Noyes Publications, Westwood, New Jersey, USA.

102 Schneider, S., Kasper, A., Hagen, C.V., Bartenstein, M., Engeser, B., Schumm, T., Bar-Joseph, I., Folman, R., Feenstra, L., and Schmiedmayer, J. (2003) Bose–Einstein condensation in a simple microtrap. *Phys. Rev. A* **67**, 023612.

103 Reichel, J., Hänsel, W., Hommelhoff, P., and Hänsch, T.W. (2001) Applications of integrated magnetic microtraps. *Appl. Phys. B* **72**, 81.

104 Steinmetz, T., Balocchi, A., Colombe, Y., Hunger, D., Hänsch, T.W., Warburton, R.J., and Reichel, J. (2006) Stable fiber-based Fabry–Pérot cavity. *Appl. Phys. Lett.* **89**, 111110.

105 Long, R., Rom, T., Hänsel, W., Hänsch, T.W., and Reichel, J. (2005) Long distance magnetic conveyor for precise positioning of ultracold atoms. *Eur. Phys. J. D* **35**, 125.

106 Armijo, J., Garrido, A.C., and Bouchoule, I. (2010) Thermal properties of AlN-based atom chips. *Eur. Phys. J. D* **56**, 33.

107 del Campo, A. and Greiner, C. (2007) SU-8: a photoresist for high-aspect-ratio and 3D submicron lithography. *J. Micromech. Microeng.* **17**, R81.

108 Seidemann, V., Rabe, J., Feldmann, M., and Büttgenbach, S. (2002) SU8-micromechanical structures with in situ fabricated movable parts. *Microsyst. Tech.* **8**, 348.

109 Duffy, D., Schueller, J.M.O., and Whitesides, G. (1998) Rapid prototyping of microfluidic systems in poly(dimethylsiloxane). *Anal. Chem.* **70**, 4974.

110 Wilzbach, M., Haase, A., Schwarz, M., Heine, D., Wicker, K., Liu, X., Brenner, K.-H., Groth, S., Fernholz, T., Hessmo, B., and Schmiedmayer, J. (2006) Detecting neutral atoms on an atom chip. *Fortschr. Phys.* **54**, 746.

111 Abraham, E.R.I. and Cornell, E.A. (1998) Teflon feedthrough for coupling optical fibers into ultrahigh vacuum systems. *Appl. Opt.* **37**, 1762.

112 Wilzbach, M., Heine, D., Groth, S., Liu, X., Hessmo, B., and Schmiedmayer, J. (2009) A simple integrated single-atom detector. *Opt. Lett.* **34**, 259.

113 Heine, D., Wilzbach, M., Hessmo, B., and Schmiedmayer, J. (2009) Single-atom detector: photon and atom statistics. *Phys. Rev. A* **97**, 021804R.

114 Heine, D., Rohringer, W., Fischer, D., Wilzbach, M., Raub, T., Loziczky, S., Liu, X., Groth, S., Hessmo, B., and Schmiedmayer, J. (2010) A single atom detector integrated on an atom chip: Fabrication, characterization and application. *New J. Phys.* **12**, 095005, eprint arXiv:1002.1573.

115 Gehr, R., Volz, J., Dubois, G., Steinmetz, T., Colombe, Y., Lev, B.L., Long, R., Estève,, J. and Reichel, J. (2010) Cavity-based single atom preparation and high-fidelity hyperfine state readout. To be published in *Phys. Rev. Lett.*, preprint arXiv:1002.4424.

116 Trupke, M., Hinds, E.A., Eriksson, S., Curtis, E.A., Moktadir, Z., Kukharenka, E., and Kraft, M. (2005) Microfabricated high-finesse optical cavity with open access and small volume. *Appl. Phys. Lett.* **87**, 211106.

117 Eriksson, S., Trupke, M., Powell, H.F., Sahagun, D., Sinclair, C.D.J., Curtis, E.A., Sauer, B.E., Hinds, E.A., Moktadir,

Z., and Gollasch, C.O. (2005) Integrated optical components on atom chips. *Eur. Phys. J. D* **35**, 135.

118 Trupke, M., Goldwin, J., Darquié, B., Dutier, G., Eriksson, S., Ashmore, J.P., and Hinds, E.A. (2007) Atom detection and photon production in a scalable, open, optical microcavity. *Phys. Rev. Lett.* **99**, 63601.

119 Steinmetz, T. (2008) Resonator-Quantenelektrodynamik auf einem Mikrofallenchip. Ph.D. thesis, Ludwig-Maximilians-Universität München and Max-Planck-Institut für Quantenoptik.

120 Hunger, D., Steinmetz, T., Colombe, Y., Deutsch, C., Hänsch, T.W., and Reichel, J. (2010) Fiber Fabry–Perot cavity with high finesse. *New J. Phys.* **12**, 065038.

121 Dumke, R., Volk, M., Müther, T., Buchkremer, F., Birkl, G., and Ertmer, W. (2002) Micro-optical realization of arrays of selectively addressable dipole traps: A scalable configuration for quantum computation with atomic qubits. *Phys. Rev. Lett.* **89**, 97903.

122 Birkl, G., Buchkremer, F., Dumke, R., and Ertmer, W. (2001) Atom optics with microfabricated optical elements. *Opt. Commun.* **191**, 67.

123 Dumke, R., Müther, T., Volk, M., Ertmer, W., and Birkl, G. (2002) Interferometer-type structures for guided atoms. *Phys. Rev. Lett.* **89**, 220402.

124 Borrelli, N.F. (1999) Microoptics Technology: Fabrication and applications of lens arrays and devices. *CRC*.

125 Aoki, T., Dayan, B., Wilcut, E., Bowen, W., Parkins, A., Kippenberg, T., Vahala, K., and Kimble, H. (2006) Observation of strong coupling between one atom and a monolithic microresonator. *Nature* **443**, 671.

126 Barclay, P.E., Srinivasan, K., Painter, O., Lev, B., and Mabuchi, H. (2006) Integration of fiber-coupled high-Q SiNx microdisks with atom chips. *Appl. Phys. Lett.* **89**.

127 Rosenblit, M., Horak, P., Helsby, S., and Folman, R. (2004) Single-atom detection using whispering-gallery modes of microdisk resonators. *Phys. Rev. A* **70**, 53808.

128 Armani, D., Kippenberg, T., Spillane, S., and Vahala, K. (2003) Ultra-high-Q toroid microcavity on a chip. *Nature* **421**, 925.

129 Kippenberg, T., Spillane, S., Armani, D., and Vahala, K. (2003) Fabrication and coupling to planar high-Q silica disk microcavities. *Appl. Phys. Lett.* **83**, 797.

130 Spillane, S., Kippenberg, T., Vahala, K., Goh, K., Wilcut, E., and Kimble, H. (2005) Ultrahigh-Q toroidal microresonators for cavity quantum electrodynamics. *Phys. Rev. A* **71**, 13817.

131 Horak, P., Klappauf, B.G., Haase, A., Folman, R., Schmiedmayer, J., Domokos, P., and Hinds, E. (2003) Possibility of single-atom detection on a chip. *Phys. Rev. A* **67**, 43806.

132 Purdy, T. and Stamper-Kurn, D. (2008) Integrating cavity quantum electrodynamics and ultracold-atom chips with on-chip dielectric mirrors and temperature stabilization. *Appl. Phys. B* **90**, 401.

133 Kohnen, M., Succo, M., Petrov, P., Nyman, R., Trupke, M., and Hinds, E. (2009) An integrated atom-photon junction. *Nature Photonics in print*, Arxiv preprint arXiv:0912.4460.

134 Gay, G., Alloschery, O., V. de Lesegno, B., O'Dwyer, C., Weiner, J., and Lezec, H. (2006) *Optical Response of Nanostructured Surfaces: Experimental Investigation of the Composite Diffracted Evanescent Wave Model.*

135 Lezec, H., Degiron, A., Devaux, E., Linke, R., Martin-Moreno, L., Garcia-Vidal, F., and Ebbesen, T. (2002) Beaming light from a subwavelength aperture. *Science* **297**, 820.

136 Rosenblit, M., Japha, Y., Horak, P., and Folman, R. (2006) Simultaneous optical trapping and detection of atoms by microdisk resonators. *Phys. Rev. A* **73**, 63805.

137 Lev, B., Srinivasan, K., Barclay, P., Painter, O., and Mabuchi, H. (2004) Feasibility of detecting single atoms using photonic bandgap cavities. *Nanotechnology* **15**, S556.

138 Akimov, A., Mukherjee, A., Yu, C., Chang, D., Zibrov, A., Hemmer, P., Park, H., and Lukin, M. (2007) Gen-

eration of single optical plasmons in metallic nanowires coupled to quantum dots. *Nature* **450**, 402.

139 Chang, D., Thompson, J., Park, H., Vuletić, V., Zibrov, A., Zoller, P., and Lukin, M. (2009) Trapping and manipulation of isolated atoms using nanoscale plasmonic structures. *Phys. Rev. Lett.* **103**, 123004.

140 Trupke, M., Ramirez-Martinez, F., Curtis, E., Ashmore, J., Eriksson, S., Hinds, E., Moktadir, Z., Gollasch, C., Kraft, M., Prakash, G.V., *et al.* (2006) Pyramidal micromirrors for microsystems and atom chips. *Appl. Phys. Lett.* **88**, 071116.

141 Shevchenko, A., Lindvall, T., Tittonen, I., and Kaivola, M. (2004) Microscopic electro-optical atom trap on an evanescent-wave mirror. *Eur. Phys. J. D* **28**, 273.

142 Wildermuth, S., Krüger, P., Becker, C., Brajdic, M., Haupt, S., Kasper, A., Folman, R., and Schmiedmayer, J. (2004) Optimized magneto-optical trap for experiments with ultracold atoms near surfaces. *Phys. Rev. A* **69**, 030901.

143 Hänsel, W., Hommelhoff, P., Hänsch, T.W., and Reichel, J. (2001) Bose–Einstein condensation on a microelectronic chip. *Nature* **413**, 498.

144 Du, S., Squires, M.B., Imai, Y., Czaia, L., Saravanan, R., Brigh, V., Reichel, J., Hänsch, T.W., and Anderson, D.Z. (2004) Atom-chip Bose–Einstein condensation in a portable vacuum cell. *Phys. Rev. A* **70**, 053606.

145 Farkas, D.M., Hudek, K.M., Salim, E.A., Segal, S.R., Squires, M.B., and Anderson, D.Z. (2009) A compact, transportable, microchip-based system for high repetition rate production of Bose–Einstein condensates. *preprint arXiv:0912.0553*.

146 Lee, K., Kim, J., Noh, H., and Jhe, W. (1996) Single-beam atom trap in a pyramidal and conical hollow mirror. *Opt. Lett.* **21**, 1177.

147 Kim, J., Lee, K., Noh, H., Jhe, W., and Ohtsu, M. (1997) Atom trap in an axicon mirror. *Opt. Lett.* **22**, 117.

148 Groswasser, D., Waxman, A., Givon, M., Aviv, G., Japha, Y., Keil, M., and Folman, R. (2009) Retroreflecting polarization spectroscopy enabling miniaturization. *Rev. Sci. Instrum.* **80**, 093103.

149 Liew, L.A., Knappe, S., Moreland, J., Robinson, H., Hollberg, L., and Kitching, J. (2004) Microfabricated alkali atom vapor cells. *Appl. Phys. Lett.* **84**, 2694.

150 Aubin, S., Extavour, M., Myrskog, S., LeBlanc, L., Esteve, J., Singh, S., Scrutton, P., McKay, D., McKenzie, R., Leroux, I., *et al.* (2005) Trapping fermionic ^{40}K and bosonic ^{87}Rb on a chip. *J. Low Temp. Phys.* **140**, 377.

151 Chuang, H.C.R., Anderson, D.Z., and Bright, V.M. (2008) The fabrication of through-wafer interconnects in silicon substrates. *J. Micromech. Microeng.* **18**, 045003.

152 Schneble, D., Gauck, H., Hartl, M., Pfau, T., and Mlynek, J. (1999) In Optical Bose-Einstein Condensation in Atomic Gases, Proceedings of the International School of Physics "Enrico Fermi", Inguscio, M., Stringari, S., and Wieman, C., *IOS Press*, Amsterdam, p. 469.

Part Two Ultracold Atoms near a Surface

4
Atoms at Micrometer Distances from a Macroscopic Body
Stefan Scheel and E.A. Hinds

4.1
Introduction

Miniaturization is a key idea behind the development of atom chips [1–5]. The micrometer-sized structures that can be formed on a chip – either by patterns of permanent magnetization or by micro-fabricated wires – allow very strong field gradients to be produced with rather modest magnetizations or currents. This makes it easy to form tightly trapped clouds of cold atoms, which are required both for rapid evaporation to make Bose–Einstein condensates [6–8] and for subsequent rapid manipulation. As well as producing strong trapping forces, miniaturization also opens the possibility of controlled movement of atoms on the micrometer length scale. This is important, for example, in realizing matter–wave interferometry [9] using the center-of-mass de Broglie waves on a chip, or for controlled tunneling between traps [10], or for other more elaborate quantum circuits based on the flow and interaction of neutral atoms [11].

The atoms being manipulated by the chip have to be held some distance away from it, which is normally similar to the characteristic transverse length scale of the structures on the surface. This connection between perpendicular and transverse length scales derives from the Laplace equation, which the magnetostatic potential ϕ must satisfy [1]. For example, the Fourier component of ϕ with transverse variation e^{ikx} decays away from the surface as e^{-kz}. If there is also some periodicity in y, the decay with z is even faster. Thus, the high-frequency Fourier components generated by structures on the chip are exponentially attenuated at distances larger than their characteristic transverse length scales. The result is this: if one wants to obtain the full benefit of micrometer-sized structures on the surface of the chip, it is necessary to hold the atoms at micrometer-scale distances from the surface. Because of this, it is natural to hold atoms at a distance in the range 1–100 μm away from the surface of a chip.

As soon as this became possible in the laboratory, experiments revealed that the coupling between the atom and the surface at these short distances can produce significant effects. At the heart of this coupling are the fluctuations of the electromagnetic field, whose strength can be greatly enhanced close to the surface in comparison with the field of free space. For example, radio-frequency magnetic noise can drive spin-flip transitions in the atoms, causing them to be lost from a magnetic trap. At audio frequencies, the gradient of this fluctuating field can excite motion of the atoms in the trap thereby heating the cloud. As well as producing dissipative effects, the atom–field coupling gives rise to an average force – the Casimir–Polder force – which can perturb the trap potential very strongly at sub-micrometer distances from the surface. In the conventional surface physics literature, forces of this kind are known as "long-range" interactions because they involve no overlap of the atomic wavefunction with that of the surface. This name may seem paradoxical in the cold-atom physics community, where millimeters or centimeters are often a more natural length scale, however we will continue to describe these interactions as long range. A more important length scale is set by the wavelength of the electromagnetic field responsible for a given effect. For example, the Casimir–Polder force results mainly from electric field fluctuations in the optical domain, for which the distance range 1–100 μm lies in the far field. By contrast, the loss due to spin-flips is a near-field effect since radio-frequency wavelengths are enormous in comparison with the atom–surface distance.

In this chapter, we consider that the atom and the surface are both electrically neutral and we assume that they are weakly coupled to each other so that their responses are linear. This is a good approximation in the case of these long-range interactions. We also consider that each atom interacts with the surface without regard to any other atoms in the cloud, which is a good approximation for the atom densities explored to date experimentally. With these approximations in mind, we set up in Section 4.2 a quantum theory for the electromagnetic fields in which the mode functions of the elementary electromagnetic quantum excitations include not only the usual far-field behavior of photons, but also the full near-field behavior due to the associated excitation of surface modes. In Section 4.3 we discuss how fluctuations of the material-assisted field drive spin-flips in a paramagnetic atom held by a magnetic trap near the surface. The same fluctuating interaction also heats the trapped atom cloud. This leads us to consider surfaces at low temperature and, in Section 4.3.2, to discuss superconductors and the prospect of extremely slow spin relaxation rates. In Section 4.3.5 we turn to the heating of polar molecules near a surface, which proceeds through the *electric* dipole coupling and is an important effect when considering the possibility of molecules trapped on a chip. While a particular Fourier component of the fluctuating field can drive a resonant transition, the broad-band mean-square field induces an energy level shift – the Casimir–Polder shift – which we discuss in Section 4.4. We consider Casimir–Polder interactions between various combinations of electric and magnetic bodies and in several geometries.

4.2
Principles of QED in Dielectrics

The theory of quantum electrodynamics (QED) in dielectric media provides a full and convenient framework for describing long-range atom–surface interactions. The basic idea is to incorporate the dielectric materials associated with the surface into the quantum theory of light using statistical principles to obtain a self-consistent description of the quantized light field in the presence of the macroscopic matter. This matter-assisted field will then be allowed to perturb an atom placed outside the surface.

We start with a brief outline of QED in the presence of magneto-dielectric media, as developed in [12–16]. We assume that the arrangement of magneto-dielectric (or metallic) bodies can be described by a complex dielectric permittivity $\varepsilon(\mathbf{r}, \omega) = \varepsilon_R(\mathbf{r}, \omega) + i\varepsilon_I(\mathbf{r}, \omega)$ and a complex magnetic permeability $\mu(\mathbf{r}, \omega) = \mu_R(\mathbf{r}, \omega) + i\mu_I(\mathbf{r}, \omega)$ whose real and imaginary parts, as a consequence of causality, have to satisfy the Kramers–Kronig relations

$$\varepsilon_R(\mathbf{r}, \omega) - 1 = \frac{1}{\pi} \mathcal{P} \int_{-\infty}^{\infty} d\omega' \frac{\varepsilon_I(\mathbf{r}, \omega')}{\omega' - \omega},$$

$$\varepsilon_I(\mathbf{r}, \omega) = -\frac{1}{\pi} \mathcal{P} \int_{-\infty}^{\infty} d\omega' \frac{\varepsilon_R(\mathbf{r}, \omega')-1}{\omega' - \omega}, \qquad (4.1)$$

and similarly for $\mu(\mathbf{r}, \omega)$. Here $\mathcal{P}$ indicates the principal value. Further, as a function of a complex frequency argument, the permittivity and the permeability satisfy the relations

$$\varepsilon^*(\mathbf{r}, \omega) = \varepsilon(\mathbf{r}, -\omega^*), \quad \mu^*(\mathbf{r}, \omega) = \mu(\mathbf{r}, -\omega^*). \qquad (4.2)$$

They are analytic in the upper half-plane, without poles or zeros, and they converge to unity for sufficiently large frequencies. In most cases it is not necessary to include magnetic responses as the relevant susceptibilities are almost always close to their vacuum values. However, with the theory of Casimir–Polder forces in mind (see Section 4.4), it is instructive to keep the description general enough to include magnetic effects.

In the Schrödinger picture, the operator of the electric field strength can be decomposed into [14–16]

$$\hat{E}(\mathbf{r}) = \int_0^\infty d\omega\, \hat{E}(\mathbf{r}, \omega) + \text{h.c.}, \qquad (4.3)$$

$$\hat{E}(\mathbf{r}, \omega) = \sum_{\lambda=e,m} \int d^3s\, G_\lambda(\mathbf{r}, \mathbf{s}, \omega) \cdot \hat{f}_\lambda(\mathbf{s}, \omega) \qquad (4.4)$$

with the definitions

$$G_e(r, r', \omega) = i\frac{\omega^2}{c^2}\sqrt{\frac{\hbar}{\pi\varepsilon_0}\varepsilon_I(r, \omega)}\, G(r, r', \omega)\,, \qquad (4.5)$$

$$G_m(r, r', \omega) = -i\frac{\omega}{c}\sqrt{\frac{\hbar}{\pi\varepsilon_0}\frac{\mu_I(r, \omega)}{|\mu(r, \omega)|^2}}\left[G(r, r', \omega) \times \overleftarrow{\nabla}'\right]\,, \qquad (4.6)$$

where the dyadic Green function $G(r, s, \omega)$ is the fundamental solution to the Helmholtz equation

$$\nabla \times \frac{1}{\mu(r, \omega)}\nabla \times G(r, s, \omega) - \frac{\omega^2}{c^2}\varepsilon(r, \omega) G(r, s, \omega) = \delta(r - s)\,. \qquad (4.7)$$

We will come shortly to the meaning of $\hat{f}_\lambda(s, \omega)$, but first we discuss some properties of G.

For materials that are reciprocal, that is those without optical activity, the Onsager–Lorentz reciprocity theorem demands that the transpose G^T satisfies

$$G^T(r, s, \omega) = G(s, r, \omega)\,. \qquad (4.8)$$

One consequence of this is that the transmission coefficient from either side of a dielectric material is the same up to a phase, that is there are no one-way mirrors. Further, the dyadic Green function inherits its analytic properties from the dielectric permittivity and the magnetic permeability, hence

$$G^*(r, s, \omega) = G(r, s, -\omega^*)\,. \qquad (4.9)$$

Most importantly, it fulfils an integral relation

$$\sum_{\lambda=e,m}\int d^3 s\, G_\lambda(r, s, \omega) \cdot G_\lambda^+(r', s, \omega) = \frac{\hbar}{\pi\varepsilon_0}\frac{\omega^2}{c^2}\,\text{Im}\, G(r, r', \omega) \qquad (4.10)$$

which is a consequence of the linear fluctuation-dissipation theorem and can be quite easily derived from the Helmholtz equation (4.7) [14–16]. It should be remarked that there are two separate levels of linear response involved in this theory – the response of the polarization (and magnetization) to an external electromagnetic field and the response of the electric field to an external perturbation, Eq. (4.4). The integral relation Eq. (4.10) is a consequence of the second type of linear response. By the linearity of electromagnetism, G can be decomposed into a bulk contribution $G^{(0)}(r, s, \omega)$ and a scattering part $G^{(S)}(r, s, \omega)$ that is responsible for satisfying the boundary conditions at the interfaces between dielectric bodies.

The dynamical variables $\hat{f}_\lambda(r, \omega)$ in Eq. (4.4) denote two independent sets of bosonic vector fields with the commutation rules

$$\left[\hat{f}_\lambda(r, \omega), \hat{f}_{\lambda'}^\dagger(r', \omega')\right] = \delta_{\lambda\lambda'}\delta(r - r')\delta(\omega - \omega')\,. \qquad (4.11)$$

They describe collective excitations of the electromagnetic field and the absorbing dielectric matter, and not free-space photonic excitations alone. We see that Eq. (4.4) generalizes the usual free-space mode expansion

$$\hat{E}(r) = i \sum_{\mu} \omega_\mu A_\mu(r) \hat{a}_\mu + \text{h.c.} \tag{4.12}$$

to allow for the presence of background materials. Here the mode functions $A(r)$ are replaced by the Green tensor times the square root of the imaginary part of the permittivity. The consistency of this quantization scheme in dielectrics can be proved by noting that the fundamental QED equal-time commutation relation between the operators of the electric field strength and the magnetic induction is [14–16]

$$\left[\hat{E}(r), \hat{B}(r')\right] = \frac{i\hbar}{\varepsilon_0} \nabla \times \delta(r - r'), \tag{4.13}$$

as in quantum electrodynamics in free space.

The vacuum fluctuations can be expressed using Eqs. (4.4), (4.10) and (4.11) as

$$\langle 0|\hat{E}(r,\omega) \otimes \hat{E}^\dagger(r',\omega')|0\rangle = \frac{\hbar\omega^2}{\pi\varepsilon_0 c^2} \operatorname{Im} G(r, r', \omega) \delta(\omega - \omega'), \tag{4.14}$$

$$\langle 0|\hat{B}(r,\omega) \otimes \hat{B}^\dagger(r',\omega')|0\rangle = \frac{\hbar\mu_0}{\pi} \operatorname{Im}\left[\nabla \times G(r, r', \omega) \times \overleftarrow{\nabla}'\right] \delta(\omega - \omega'). \tag{4.15}$$

The thermal expectation value at any mode frequency ω is just the vacuum value, given above, multiplied by a correction to account for the thermal occupation number of the mode:

$$\langle \hat{E}(r,\omega) \otimes \hat{E}^\dagger(r',\omega')\rangle_T = \left[\bar{n}_{\text{th}}(\omega) + 1\right] \langle 0|\hat{E}(r,\omega) \otimes \hat{E}^\dagger(r',\omega')|0\rangle, \tag{4.16}$$

$$\langle \hat{E}^\dagger(r,\omega) \otimes \hat{E}(r',\omega')\rangle_T = \bar{n}_{\text{th}}(\omega) \langle 0|\hat{E}(r,\omega) \otimes \hat{E}^\dagger(r',\omega')|0\rangle, \tag{4.17}$$

$$\langle \hat{B}(r,\omega) \otimes \hat{B}^\dagger(r',\omega')\rangle_T = \left[\bar{n}_{\text{th}}(\omega) + 1\right] \langle 0|\hat{B}(r,\omega) \otimes \hat{B}^\dagger(r',\omega')|0\rangle, \tag{4.18}$$

$$\langle \hat{B}^\dagger(r,\omega) \otimes \hat{B}(r',\omega')\rangle_T = \bar{n}_{\text{th}}(\omega) \langle 0|\hat{B}(r,\omega) \otimes \hat{B}^\dagger(r',\omega')|0\rangle. \tag{4.19}$$

Here the mean occupation number $\bar{n}_{\text{th}}(\omega) = \left[e^{\hbar\omega/(k_B T)} - 1\right]^{-1}$ is given by the Bose–Einstein distribution.

Finally, the time evolution of the electromagnetic fields is governed by the bilinear Hamiltonian

$$\hat{H}_F = \sum_{\lambda=e,m} \int d^3r \int_0^\infty d\omega \hbar\omega \, \hat{f}_\lambda^\dagger(r,\omega) \cdot \hat{f}_\lambda(r,\omega). \tag{4.20}$$

This bears a close resemblance to the Hamiltonian $\hat{H} = \sum_\mu \hbar\omega_\mu \hat{a}_\mu^\dagger \hat{a}_\mu$ of free-space QED in which the $\hat{a}$ are the photonic amplitude operators. The similarity stems from the fact that for all linear response theories one can find a microscopic model Hamiltonian that is bilinear in the coupling between the free electromagnetic field and the material degrees of freedom [17]. The dynamical variables $\hat{f}_\lambda(\mathbf{r}, \omega)$ then follow from a Bogoliubov transformation of the electromagnetic and matter operators. This in effect yields a quasi-free field theory as can be seen from the operator expansion (4.4) and the Hamiltonian (4.20).

4.3
Relaxation Rates near a Surface

4.3.1
Spin Flips near a Dielectric or Metallic Surface

The desire to manipulate neutral atoms in tight traps and with micrometer precision has led to the development of magnetic traps on atom chips, produced either by current-carrying wires or by permanently magnetized structures on the chip [1–5]. The magnetic field is designed to have a local minimum at $B = B_0$, where the atoms are trapped. An atom in angular momentum eigenstate $|F, m_F\rangle$ has (weak field) Zeeman energy $g_F \mu_B m_F B$, where g_F is the g-factor, μ_B is the Bohr magneton, and m_F is the magnetic quantum number. For atoms in a state with $g_F m_F > 0$, the minimum in the magnetic field forms a trap. By contrast, atoms having $g_F m_F = 0$ experience no Zeeman interaction potential, while those with $g_F m_F < 0$ find themselves on a potential maximum and are expelled from the region. Thus the trapped state $|F, m_F\rangle$ is a metastable one, below which there lie untrapped states accessible by spin flip transitions. A true ground-state trap could be formed if there were instead a local maximum of the static magnetic field, but this is not possible in free space [18]. For atoms close to the center of the trap, the energy released by a single spin flip, $|\Delta m_F| = 1$, is $|g_F| \mu_B B_0$. Since B_0 is typically of the order of 100 µT, the resonant frequency for a spin flip is of the order of 1 MHz.

Spin-flip transitions can be driven by technical noise, for example, coherent RF signals unintentionally coupled to the current in the chip wires at a frequency close to the transition frequency. Those residual RF currents would result in magnetic fields that decay as $1/d$ for a thin wire, where d is the atom–surface distance, and even more slowly for thicker wires. The power scales as the square of the magnetic field, and hence a spin transition rate proportional to $1/d^2$ or less is expected [19]. To estimate the associated technical requirements, let us suppose that an atom stays for 1 s over a current-carrying wire. The Rabi frequency that generates the spin flips is, therefore, 0.5 Hz. Since the Bohr magneton is approximately 10^6 Hz/G, the RF magnetic field must be smaller than 1 µG which, for an atom being 1 mm away from the current-carrying wire, amounts to current fluctuations of less than 1 µA [20]. This requires unusually high DC current stability (relative stability better than 10^{-6}) which is challenging but certainly achievable.

However, even when such experimental imperfections are eradicated, thermal magnetic noise emanating from the material of the chip itself can significantly reduce the lifetime of the trapped atoms [21]. It is thus of some importance to understand the properties of those more fundamental fluctuations and to consider how they may be tailored to one's needs. In the following, we use the noise power of the field fluctuations, as derived in Section 4.2, to calculate the atomic spin relaxation rates, and we compare our results with experimental observations.

We write the Hamiltonian of the coupled atom–field system as

$$\hat{H} = \hat{H}_F + \hat{H}_A + \hat{H}_{\text{int}} \tag{4.21}$$

with $\hat{H}_F$ being the Hamiltonian (4.20) of the medium-assisted electromagnetic field and with $\hat{H}_A = \sum_i \hbar \omega_i |i\rangle\langle i|$ being the free atomic Hamiltonian in terms of the energy eigenstates $|i\rangle$ of the atom. For the present purpose, we take the interaction between the two as the magnetic dipole coupling $\hat{H}_{\text{int}} = -\hat{\boldsymbol{\mu}} \cdot \hat{\boldsymbol{B}}(\boldsymbol{r}_A)$. Here the operator for the medium-assisted magnetic induction is

$$\hat{\boldsymbol{B}}(\boldsymbol{r}) = \int_0^\infty d\omega \, \hat{\boldsymbol{B}}(\boldsymbol{r},\omega) + \text{h.c.}, \tag{4.22}$$

$$\hat{\boldsymbol{B}}(\boldsymbol{r},\omega) = \frac{1}{i\omega} \sum_{\lambda=e,m} \int d^3s \, [\nabla \times \boldsymbol{G}_\lambda(\boldsymbol{r},\boldsymbol{s},\omega)] \cdot \hat{\boldsymbol{f}}_\lambda(\boldsymbol{s},\omega), \tag{4.23}$$

which follows directly from Eq. (4.4) and the application of Faraday's law. Heisenberg's equations of motion can be solved, after making the Markov approximation, to obtain a transition rate for the electron spin. For this purpose, we take an atom in its electronic ground state with $L = 0$, and we neglect the nuclear spin because of the tiny electron-to-proton mass ratio m_e/m_p so that the magnetic moment vector is proportional to the electronic spin operator $\hat{\boldsymbol{S}}$. The transition rate between two electron spin states $|i\rangle$ and $|f\rangle$ is then [22]

$$\Gamma = \frac{2\mu_0(\mu_B g_S)^2}{\hbar} \langle f|\hat{\boldsymbol{S}}|i\rangle \cdot \text{Im}\left[\nabla \times \boldsymbol{G}(\boldsymbol{r}_A,\boldsymbol{r}_A,\omega_A) \times \overleftarrow{\nabla}\right] \cdot \langle i|\hat{\boldsymbol{S}}|f\rangle (\bar{n}_{\text{th}} + 1), \tag{4.24}$$

where $\bar{n}_{\text{th}} = [e^{\hbar\omega/(k_B T)} - 1]^{-1}$ is, as before, the mean thermal photon number at the ambient temperature T. This formula can alternatively be derived directly from Fermi's Golden Rule, taking the spectral density of the magnetic field noise given in Eq. (4.18). The Green function $\boldsymbol{G}(\boldsymbol{r}_A,\boldsymbol{r}_A,\omega_A)$ contains all relevant information about the electrical and geometrical properties of the dielectric or metallic bodies in the vicinity of the atom.

The observation of this effect was first reported in [23], with the experimental arrangement shown schematically in Figure 4.1a. Cold ^{87}Rb atoms were magnetically trapped in their $|F=2, m_F=2\rangle$ ground state close to the surface of a 185-μm-radius copper wire, coated with 55 μm of Al and covered in a 10-μm ceramic sheath. The experiment measured the lifetime for the loss of atoms from the trap as a function of the distance between the cloud and the surface of the wire in the range 27–85 μm. Because of electrical power dissipated in the wire, its temperature

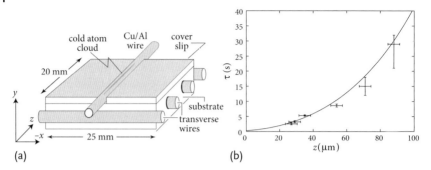

Figure 4.1 (a) Schematic of experimental apparatus to test the rate of thermally induced spin-flips near a metallic wire (not to scale); (b) atom-loss lifetime τ as a function of atom–surface distance z. Experimental data taken from [23], theoretical curve taken from [22].

$T = 380$ K was well above that of the room-temperature gold substrate. Figure 4.1b shows these measured times, plotted versus the atom–surface distance. The line plotted on the same graph gives the result of a subsequent calculation based on Eq. (4.24), using the known dyadic Green function (not reproduced here) for cylindrical multilayers [24]. Further details of the calculation are given in [22]. There is excellent agreement between the theory and the experiment, which confirmed the dependence on transition frequency as well as distance. Harber et al. [25] tested the spin-flip rates of magnetically trapped atoms near plane copper and titanium surfaces, providing further confirmation of the effect.

In more modern atom chips, the geometry is typically planar, with a thick insulating substrate supporting a conducting or magnetic layer of thickness h and resistivity ρ that provides the microscopic magnetic field patterns for the chip. When combined with the vacuum above the chip, this forms a three-layer planar structure, for which there are three relevant length scales. These are the thickness h of the active layer, the skin depth $\delta = \sqrt{2\rho/(\mu_0 \omega_A)}$ of this layer at the frequency ω_A of the spin-flip transition, and the atom–surface distance d. We take it that the skin depth of the substrate is much longer than that of the magnetic layer, as it will be for an insulating or semiconducting substrate. The spin-flip transition wavelength $2\pi c/\omega_A$ is also long. These lead to three relevant extreme cases, that can be summarized by the following expression [21, 26]:

$$\frac{\tau}{\tau_0} = \left(\frac{8}{3}\right)^2 \frac{1}{\bar{n}_{\mathrm{th}}+1} \left(\frac{\omega_A}{c}\right)^3 \begin{cases} \frac{d^4}{3\delta}, & \delta \ll d, h, \\ \frac{\delta^2 d}{2}, & \delta, h \gg d, \\ \frac{\delta^2 d^2}{2h}, & \delta \gg d \gg h. \end{cases} \quad (4.25)$$

Here, τ_0 is the trapping lifetime in free space which, for a transition frequency of 400 kHz, amounts to 3×10^{25} s [27].

Figure 4.2 illustrates how these regimes are related, taking as an example the case of an atom placed 50 μm from the surface, with a transition frequency of $2\pi \times 560$ kHz. The solid line, computed by inserting the relevant Green function into Eq. (4.24), is for a thick film – the case of large h. When the skin depth (resistivity)

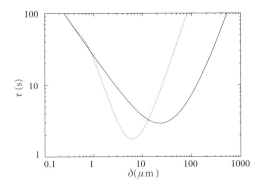

Figure 4.2 Trapping lifetime as a function of skin depth for a thin substrate layer ($h = 1\,\mu$m, dashed line) and a half-space (solid line). The atom–surface distance was chosen as $d = 50\,\mu$m, the transition frequency is $\omega_A = 2\pi 560$ kHz. Figure taken from [26].

is small enough, the spin-flip lifetime τ scales as $1/\delta$ and increasing δ causes a drop in the lifetime as given by the first line of Eq. (4.25). When $\delta \simeq d$, the lifetime reaches a minimum, after which it increases again, eventually approaching the δ^2 form in the second line of Eq. (4.25). The dashed line, for a thin film $h = 1\,\mu$m, exhibits a similar evolution with increasing δ. Starting from the same $1/\delta$-dependence, the spin-flip lifetime decreases to a minimum before rising to another δ^2 asymptote, this time with the extra factor of $d/h = 50$, as given by the third line of Eq. (4.25). We can understand the behavior in the large δ (insulator) limit by recalling that the fluctuation-dissipation theorem requires the noise power to be proportional to the imaginary part of the permittivity and hence to $1/\delta^2$. As for the dependence on h and d, we see that the $1/d$ spin-flip rate for the thick film is just the integral of the h/d^2 partial rates due to the constituent thin films. Thus, it appears that the noise fields produced by sources at different depths within the film add incoherently in this limit [28]. This is not so in the short skin-depth limit, where $\tau \propto d^4/\delta$. Naturally, the thickness of the film does not appear in this formula since it is only the top region of thickness $\sim \delta$ that contributes to the field outside the surface.

For the typical case illustrated in Figure 4.2 (transition frequency $\omega_A = 2\pi \times 560$ kHz), the skin depths for Ag, Cu, Au, and Al are in the range of 85–113 μm. These are remarkably close to the thick-film, minimum-lifetime condition $\delta \sim d$. With a thinner, 1 μm film, as illustrated by the dotted line in Figure 4.2, the minimum moves to shorter skin depth, leading to a substantially longer lifetime because these materials lie on the "insulator" side of the minimum. However, something more is required if one wants to achieve spin-flip lifetimes of many seconds for atoms within a few micrometers of the surface. Three possibilities emerge. First, one can use a material of lower conductivity, which is normally the case for atom chips based on patterns of permanent magnetization (see Chapter 1). Sinclair et al. [8] have used a piece of videotape to form a permanent-magnet atom chip with which they demonstrated this lengthening of the lifetime. Second, if the chip continues to be made of conductors such as gold or copper, then it should either be very thin and/or it should be cooled in a cryostat to reduce the thermal noise power, which is proportional to $k_B T$. Third, the film can be superconducting, in which case the superconducting screening of the magnetic field, assisted by the small value of $k_B T$ at low temperature, leads to a strong suppression of the field noise. The next section considers the superconducting case in a little more detail.

4.3.2
Spin Flips near a Superconductor

The theoretical results presented in Section 4.3.1 indicate that atoms placed a few micrometers away from a superconducting surface can enjoy very low spin-flip rates. However, that theory applies to normal conductors and has to be extended in order to treat the case of a superconductor correctly [26, 29, 30]. Knowing that the skin depth will be small, we focus our attention on the first line of Eq. (4.25). For a bulk superconductor with conductivity $\sigma(\omega) = \sigma'(\omega) + i\sigma''(\omega)$ and $\sigma''(\omega) \ll \sigma'(\omega)$, this expression for the spin flip lifetime can be written as [30]

$$\frac{\tau}{\tau_0} = \left(\frac{8}{3}\right)^2 \frac{(\omega\mu_0)^{1/2} d^4}{\bar{n}_{\text{th}} + 1} \left(\frac{\omega}{c}\right)^3 \frac{\sigma''(\omega)^{3/2}}{\sigma'(\omega)}. \tag{4.26}$$

The task is then to determine the complex conductivity. The simplest approach is the London two-fluid model [31], which assumes that there are both normal and superconducting charge carriers. Purely normal carriers obey Ohm's law $j_n = \sigma_n E$ and have a real conductivity in the Drude model of $\epsilon_0 \omega_p^2/\gamma$, where ω_p is the plasma frequency and γ is the elastic scattering rate. The superconducting component satisfies the London relation $\Lambda \frac{\partial j_s}{\partial t} = E$. The fraction of normally conducting electrons follows the Gorter–Casimir expression $n_n(T)/n_0 = (T/T_c)^4$ [32]. Combining these facts, one obtains the conductivity

$$\sigma(\omega) = \varepsilon_0 \omega_p^2 \left\{ \frac{1}{\gamma} \left(\frac{T}{T_c}\right)^4 + \frac{i}{\omega} \left[1 - \left(\frac{T}{T_c}\right)^4\right] \right\}, \tag{4.27}$$

which is shown in the graphs of Figure 4.3a,b by the dashed lines. For comparison, the solid lines show the very different result of the BCS theory, while the symbols plot results obtained using the more sophisticated Eliashberg theory, described in [30]. One sees that although the two-fluid model does not fully capture the rich dynamics of superconductors (as it neglects coherence effects and dissipation) it does give a relatively accurate and intuitive picture of the complex conductivity. Graphs 4.3c,d plot the spin-flip lifetimes derived from Eq. (4.26) using the various calculated conductivities. At a temperature of 4 K, it is clear that the spin-flip lifetime due to these thermal fluctuations can exceed 10^4 s even for atom–surface distances as small as 1 μm. This much-increased lifetime is due in part to the hundredfold reduction in temperature from 300 to 4 K, but there is also a large factor resulting from the very short penetration length of the magnetic field in the superconductor.

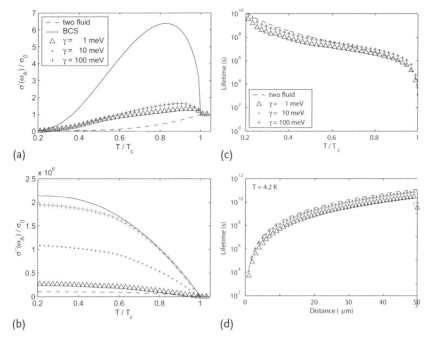

Figure 4.3 Temperature dependence of (a) real part $\sigma'(\omega_A)$ and (b) imaginary part $\sigma''(\omega_A)$ of the optical conductivity, normalized to the normal-state conductivity $\sigma_0 = \varepsilon_0 \omega_p^2 \gamma$. The atomic spin-flip frequency is $\omega_A = 2\pi 500\,\text{kHz}$, and $T_c = 9.2\,\text{K}$ is the superconductor transition temperature. The different lines correspond to the results for the two-fluid model (dashed line), using $\delta_0 = 16\,\mu\text{m}$ and $\lambda_L = 35\,\text{nm}$, BCS theory (solid line), and Eliashberg theory symbols for three elastic impurity scattering rates γ. (c),(d): Spin-flip lifetime τ_A as a function of (a) temperature (for 10 μm atom–surface distance) and (b) distance (at 4.2 K). Figures taken from [30].

First experimental results have now appeared for atoms trapped near superconducting surfaces [33–35] (see Chapter 10). In these experiments, the trapping lifetimes were significantly shorter than we calculate here because of atom loss due to technical noise. Cano *et al.* have pointed out [36], and observed [35], an interesting side effect of the Meissner effect on a trap formed by a wire and a bias field. When the wire is superconducting, the uniform field lines of the bias field are diverted around the outside of the wire, causing the bias to be more intense than normal close to the wire and making the trap lie closer to the wire than it would otherwise. The amount of flux excluded by the wire varies with its temperature and, therefore, so does the position of the trap.

From the viewpoint of atomic physics, the magnetic noise leading to loss of atoms from a trap may seem to be a nuisance. However, the same effect can also be seen as a new way to probe the physics of the surface through its field noise. For example, Figure 4.4 shows the power spectrum of the flux noise due to vortex pairs in a thin superconductor for a variety of temperatures close to the Kosterlitz–Thouless–Berezinski transition temperature. The figure is taken from [37], where

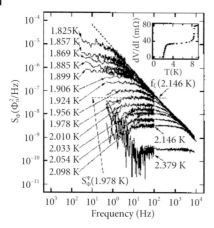

Figure 4.4 Spectral density of magnetic flux noise versus frequency for 15 temperatures above T_{KTB}. Dashed lines have slope −1 and 0. Inset shows dV/dI versus T. Figure taken from [37].

this noise is measured using a SQUID loop roughly 1 mm in diameter. On the basis of this data, Scheel et al. [38] have calculated that cold atoms can be a sensitive probe of vortex flux noise. One type of measurement would be to map out the magnetic noise by observing the spin-flip rate as a function of position. Another would be to use interferometry to investigate transverse coherence of the noise. We discuss transverse relaxation in the next section, but our comments there are brief since Chapter 7 is dedicated to interferometry.

4.3.3
Transverse Spin Relaxation

The first demonstration of transverse atomic spin coherence on a chip was described in [39]. A 600-nK thermal cloud of ^{87}Rb atoms was trapped magnetically in the $|F = 1, m_F = -1\rangle$ state at an adjustable distance from the chip surface. Using a 2-photon "$\pi/2$" pulse of RF-plus-microwave field, these atoms were placed in an equal coherent superposition of their original state and the $|F = 2, m_F = +1\rangle$ state, which differs in energy by the 6.8 GHz hyperfine splitting. The evolution of relative phase between the two states thus constitutes a 6.8 GHz clock, which can be read out by the standard Ramsey method of applying a second $\pi/2$ pulse. When the frequency sum of the two applied fields is not exactly equal to the atomic frequency, the beat note between the two is seen as an oscillation in the final state population according to the time interval between the pulses. This Ramsey oscillation is plotted in Figure 4.5. The experiment showed a very slow damping of the Ramsey fringes, with a decay time of approximately 3 s. In fact, the long-term stability of the clock was not limited by decoherence, but by the local reference oscillator, demonstrating the suitability of atom chips for practical quantum instruments. An important feature of this experiment was the use of a 3.23-G bias field at the center of the trap. This gave the two states identical g-factors, making the clock frequen-

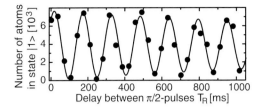

Figure 4.5 Ramsey oscillations between the $|F=1, m_F=-1\rangle$ and $|F=2, m_F=+1\rangle$ states of ^{87}Rb, measured with atoms held 9 µm from the surface a chip. An exponentially damped fit to the fringes yields a coherence lifetime of 2.8 ± 1.6 s. Figure taken from [39].

cy insensitive to small-amplitude fluctuations, such as those due to the thermal noise.

This experiment also suggests that one might make a tuneable magnetic-field noise meter by arranging for the g-factors of the two states to be different, either through a different choice of DC field or by using a different pair of Zeeman sublevels. Then, the clock coherence would accumulate phase noise in proportion to the integrated magnetic field noise and the fluctuations of the Ramsey fringes could be used as a very sensitive probe of the magnetic field fluctuations at the site of the atom cloud. In the context of vortex pairs, mentioned in the last section, it would also be of interest to probe the correlations in the noise between two different positions on the chip in order to obtain information about the size, density, and correlations of the vortices. For this, it has been proposed [38] that one might split an atom cloud into a coherent superposition of two separated clouds. To the extent that the magnetic noise is the same on both halves of the cloud, it does not affect their relative phase. However, if the noise is uncorrelated over the splitting distance the phase difference becomes random. Thus, after recombining the two halves, the phase noise of the interference pattern provides information about the spatial correlation function of the magnetic noise.

4.3.4
Heating

In Sections 4.3.1 and 4.3.2, we have mainly concentrated on the magnetic dipole interaction as a mechanism for driving spin flips that de-excite the atoms and lead to atom loss. The excitation energy of the atom is dissipated in the surface and the atom is ejected from the trap by the static magnetic field gradients. However, there can be more than one trapped level, for example, the $m_F = 2$ and $m_F = 1$ states in the $F = 2$ ground state of ^{87}Rb are both trapped. In that case, a decay from $m_F = 2$ to $m_F = 1$ can be followed by a transition back to $m_F = 2$, also due to the magnetic field fluctuations. The analysis of this heating rate follows closely the formulae that we have already presented, the only difference being that $\bar{n}_{th}$ replaces $\bar{n}_{th} + 1$ in Eqs. (4.25) and (4.26) because there is no spontaneous excitation. In principle then, the spin system is able to come into thermal equilibrium with the surface when the populations of the two states are in the ratio $\bar{n}_{th}/(\bar{n}_{th} + 1)$. In practice, the loss of atoms from the cloud through transitions to lower, un-trapped states prevents thermal equilibrium from being reached.

It is worth noting that the spring constant of the trap for an $m_F = 1$ atom is half of that for an $m_F = 2$ atom. Atoms prepared in the $m_F = 2$ state and subsequently stimulated down to $m_F = 1$, will, therefore, oscillate with typically $\sqrt{2}$ times their initial amplitude of motion. On being re-excited to $m_F = 2$, the atom is typically further from the center of the trap than it was at the outset, and therefore the center-of-mass motion is heated. This coupling provides one mechanism for the trap vibration to come into equilibrium with the spin system. There is also direct heating of the trap vibrations as a result of forces exerted on the atoms by the fluctuating magnetic field gradients. This last mechanism of heating has been considered in detail in Section 3.2 of [21], where it is evaluated and found to be very weak.

4.3.5
Electric Dipole Coupling of Molecules to a Surface

Cold atoms close to a hot surface are heated magnetically, as we have discussed above. This is the dominant effect for atoms, despite the fact that the electric dipole coupling is normally by far the stronger of the two. The reason is that electric dipole transitions have to change the parity of the atom and must, therefore, involve electronic transitions. These typically occur in the visible at frequencies of the order of 500 THz, whereas the noise spectrum of the field cuts off above $k_B T/h \simeq 6$ THz at room temperature. Thus, it is the mismatch between the thermal spectrum of the surface and the electronic excitation spectrum of the atom that protects it from electric dipole coupling to the noise in the surface field. The same does not apply to polar molecules, whose rotational and vibrational transition dipoles are driven by the electric field at low frequencies where the surface field noise is strong [40]. Since there is currently great interest in the possibility of trapping molecules close to surfaces, we develop the theoretical framework used above to treat the case of electric dipole coupling.

The internal dynamics of the molecule is described by the equations of motion

$$\dot{\sigma}_{nn}(t) = -\Gamma_n \sigma_{nn}(t) + \sum_k \Gamma_{kn} \sigma_{kk}(t) , \qquad (4.28)$$

$$\dot{\sigma}_{mn}(t) = \left[-i\tilde{\omega}_{mn} - \frac{1}{2}(\Gamma_n + \Gamma_m)\right] \sigma_{mn}(t) , \quad m \neq n . \qquad (4.29)$$

The total relaxation rate Γ_n of a given atomic level n is the sum of all individual inter-atomic transition rates Γ_{nk}, $\Gamma_n = \sum_k \Gamma_{nk}$, with

$$\begin{aligned}\Gamma_{nk} = {} & \frac{2\mu_0}{\hbar} \tilde{\omega}_{nk}^2 d_{nk} \cdot \mathrm{Im}\, G\left(r_A, r_A, |\tilde{\omega}_{nk}|\right) \cdot d_{kn} \\ & \times \{\Theta(\tilde{\omega}_{nk})[\bar{n}_{\mathrm{th}}(\tilde{\omega}_{nk}) + 1] + \Theta(\tilde{\omega}_{kn})\bar{n}_{\mathrm{th}}(\tilde{\omega}_{kn})\} . \end{aligned} \qquad (4.30)$$

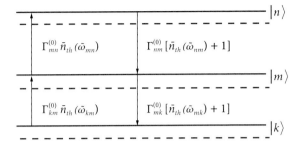

Figure 4.6 Three molecular energy levels $|k\rangle$, $|m\rangle$, and $|n\rangle$ connected by upward and downward electric dipole transitions at the rates shown. Here $\Gamma_{ij}^{(0)}$ represents the zero-temperature parts of the transition rate. The dashed lines indicate that each bare atomic energy level is shifted by the amount $\delta\omega_i$.

These transition rates follow directly from application of Eq. (4.14) together with the thermal corrections $\bar{n}_{\text{th}}(\tilde{\omega}_{nk}) + 1$ and $\bar{n}_{\text{th}}(\tilde{\omega}_{kn})$ for downward and upward transitions, respectively as illustrated in Figure 4.6. This same interaction with the medium-assisted electromagnetic field also shifts the atomic transition frequencies ω_{mn} to

$$\tilde{\omega}_{mn} = \omega_{mn} + \delta\omega_m - \delta\omega_n, \tag{4.31}$$

where the level shift $\delta\omega_n = \sum_k \delta\omega_{nk}$ of a given atomic level n is made up of contributions

$$\delta\omega_{nk} = \frac{\mu_0}{\pi\hbar} \mathcal{P} \int_0^\infty d\omega\, \omega^2 \left\{ \mathbf{d}_{nk} \cdot \text{Im}\,\mathbf{G}^{(S)}(\mathbf{r}_A, \mathbf{r}_A, \omega) \right.$$
$$\cdot \mathbf{d}_{kn} \left[\frac{\bar{n}_{\text{th}}(\omega) + 1}{\tilde{\omega}_{nk} - \omega} + \frac{\bar{n}_{\text{th}}(\omega)}{\tilde{\omega}_{nk} + \omega} \right]$$
$$\left. + \frac{\omega|\mathbf{d}_{nk}|^2}{6\pi c} \left[\frac{\bar{n}_{\text{th}}(\omega)}{\tilde{\omega}_{nk} - \omega} + \frac{\bar{n}_{\text{th}}(\omega)}{\tilde{\omega}_{nk} + \omega} \right] \right\}. \tag{4.32}$$

Here we have split the Green function into $\mathbf{G}^{(0)} + \mathbf{G}^{(S)}$, as discussed in Section 4.2. The first term in the integral is due to the scattering part $\mathbf{G}^{(S)}$ and involves both the thermal fluctuations and the vacuum fluctuations. The second term is the free-space level shift, due to thermal photons alone. In this term, we omit the shift caused by the free-space vacuum fluctuations, known as the Lamb shift, because that is already present in the measured energy levels. In most cases of interest, the level shifts are too small to be significant and it is appropriate to set $\tilde{\omega}_{mn} = \omega_{mn}$. In that case, the heating rate for a ground-state molecule can be written as

$$\Gamma_0 = \sum_k \frac{2\mu_0}{\hbar} \omega_{k0}^2 \bar{n}_{\text{th}}(\omega_{k0}) \mathbf{d}_{0k} \cdot \text{Im}\,\mathbf{G}(\mathbf{r}_A, \mathbf{r}_A, \omega_{k0}) \cdot \mathbf{d}_{k0}. \tag{4.33}$$

The advantage of this approximation is that the temperature dependence resides solely in the mean thermal photon numbers $\bar{n}_{\text{th}}(\omega_{k0})$ and all information about the geometric arrangement of macroscopic bodies is in the Green function $G(\mathbf{r}_A, \mathbf{r}_A, \omega_{k0})$. We have already tacitly made the same approximation in Section 4.3.1 where we discussed the magnetic interactions.

Let us consider the case of a molecule close to a wide, thick slab of permittivity $\varepsilon(\omega)$ and permeability $\mu(\omega)$. The scattering part of the Green function at spatial coincidence, $\mathbf{G}^{(S)}(\mathbf{r}_A, \mathbf{r}_A, \omega)$, is given by [16, 41]

$$\mathbf{G}^{(S)}(\mathbf{r}, \mathbf{r}, \omega) = \frac{i}{8\pi} \int_0^\infty dq \, \frac{q}{\beta_0} e^{2i\beta_0 d}$$
$$\times \left[\left(r_s - \frac{\beta_0^2 c^2}{\omega^2} r_p \right) (\mathbf{e}_x \mathbf{e}_x + \mathbf{e}_y \mathbf{e}_y) + 2\frac{\beta_0^2 c^2}{\omega^2} r_p \mathbf{e}_z \mathbf{e}_z \right], \quad (4.34)$$

where

$$r_s = \frac{\mu(\omega)\beta_0 - \beta_1}{\mu(\omega)\beta_0 + \beta_1}, \quad r_p = \frac{\varepsilon(\omega)\beta_0 - \beta_1}{\varepsilon(\omega)\beta_0 + \beta_1} \quad (4.35)$$

are the Fresnel reflection coefficients for s- and p-polarized waves and

$$\beta_0 = \sqrt{\frac{\omega^2}{c^2} - q^2}, \quad \beta_1 = \sqrt{\frac{\omega^2}{c^2}\varepsilon(\omega)\mu(\omega) - q^2} \quad (4.36)$$

are the z-components of the wave vector $\mathbf{k} = (\mathbf{q}, \beta)$ in free space (β_0) and inside the body (β_1), respectively. To leading order in $\omega d/c$ the Green tensor is well approximated by

$$\mathbf{G}^{(S)}(\mathbf{r}, \mathbf{r}, \omega) = \frac{c^2}{32\pi\omega^2 d^3} \frac{\varepsilon(\omega) - 1}{\varepsilon(\omega) + 1} (\mathbf{e}_x \mathbf{e}_x + \mathbf{e}_y \mathbf{e}_y + 2\mathbf{e}_z \mathbf{e}_z). \quad (4.37)$$

The dependence on r_s and thus on the permeability $\mu(\omega)$ has vanished in this limit which justifies our earlier claim that magnetic properties can frequently be disregarded. For most cases, the permeability is very close to unity anyway. Inserted into the ground-state heating rate (Eq. (4.33)), one finds that

$$\Gamma_0(d) = \Gamma_0^{(0)} \left[1 + \left(\frac{d_{nr}}{d} \right)^3 \right], \quad (4.38)$$

where

$$d_{nr} = \frac{c}{\omega_{k0}} \sqrt[3]{\frac{\text{Im}\varepsilon(\omega_{k0})}{2|\varepsilon(\omega_{k0}) + 1|^2}} \quad (4.39)$$

is a scale length associated with the non-retarded limit. At distances that are large compared with d_{nr}, the heating rate of the molecule is just that due to the free-space blackbody field $\Gamma_0^{(0)}$. However, when the molecule is closer to the surface than d_{nr}, there is a dramatic (cubic) increase in the heating. For a metal whose dielectric response can be described by a Drude permittivity

$$\varepsilon(\omega) = 1 - \frac{\omega_p^2}{\omega(\omega + i\gamma)} \quad (4.40)$$

with Plasma frequency ω_p and damping constant γ, the scale length can be approximated in the limit $\omega_{k0} \ll \gamma \ll \omega_p$ by

$$d_{nr} = c \sqrt[3]{\frac{\gamma}{2\omega_p^2 \omega_{k0}^2}} . \quad (4.41)$$

The ratio ω_p^2/γ for ordinary metals is of the order of 10^{18} rad/s [40, 42]. Thus, for rotational transitions with frequencies of 10–100 GHz, the corresponding scale length d_{nr} is a few tens of micrometers. For vibrational transitions at a few THz, this distance reduces to a micrometer or less. This means that molecules trapped a few micrometers away from a surface are liable to have their rotational heating very strongly amplified by the proximity of the surface, whereas the vibrational heating will be approximately that in free space at the same temperature. When Buhmann et al. [40] examined numerical solutions of Eq. (4.33), they found that Eq. (4.38) did not describe the intermediate region $d \simeq d_{nr}$ particularly well. Empirically, the total rate in this range is much better approximated (normally to within 1 %) by

$$\Gamma_0(d) = \Gamma_0^{(0)} \left[1 + \left(\frac{d_0}{d}\right)^2 + \left(\frac{d_{nr}}{d}\right)^3 \right], \quad (4.42)$$

where the new characteristic distance d_0 is given by

$$d_0 \simeq \frac{3c}{4} \sqrt[4]{\frac{\gamma}{2\omega_p^2 \omega_{k0}^3}} . \quad (4.43)$$

We emphasize that Eq. (4.42) provides a useful empirical description of this intermediate regime, but we do not consider that the d^{-2} term corresponds to any new physical effect. Figure 4.7, taken from [40], shows the distance z_c at which the heating of the molecule has twice the free-space value, both for rotational and for vibrational heating. A wide variety of diatomic polar molecules and surface materials are considered. The solid lines indicate the slope $\omega^{-3/4}$ corresponding to the empirical d_0 of Eq. (4.43).

It seems relevant here to comment on the motional heating of ions in a trap – also due to the fluctuation of the electric field – for which similar scaling laws with distance have been reported in the literature [43, 44]. Because the oscillation frequencies of the ion are low, one has $d \ll d_{nr}$, and the noise field that heats the motion is the strongly enhanced near-field of the surface. Since the trap electrodes are not at all well approximated by a plane slab, one needs to consider how the geometry affects the Green function and hence the heating rate. As a rule of thumb, the heating rate in the non-retarded limit increases with the dimensionality of the body. For example, the d^{-3}-dependence of the heating rate for an infinite half-space becomes a d^{-4}-dependence for a thin plate or a wire. The reason for this behavior can be found in the interpretation of the imaginary part of the Green tensor (or rather $\omega^2 \text{Im} G$, see Eq. (4.14)) as the local density of states (LDOS). The LDOS comprises all local eigenmodes of the system under consideration, and thus determines the strength of the electromagnetic field fluctuations felt by the ion, atom, or

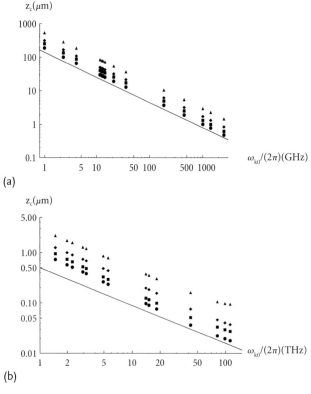

Figure 4.7 Distance z_c at which the heating of the molecule has twice the free-space value vs. frequency of the molecular transition. (a) Rotational heating. (b) Vibrational heating. Surface materials are gold (circles), iron (squares), platinum (diamonds), and ITO (triangles). Frequencies of the plotted data points correspond (left to right) to RbCs, KRb, NaRb, LiCs, BaF, LiRb, YbF, CaF, NaCs, KCs, LiH, NH, OD(a), and OH(a). Solid lines indicate the slope corresponding to the $\omega^{-3/4}$ frequency dependence given by the empirical Eq. (4.42).

molecule. This dependence on the dimensionality of the electrodes might explain the odd scaling law $d^{-3.8\pm0.6}$ obtained in [43] for electrodes that are intermediate between a plane and a thin wire.

4.4
Casimir–Polder Forces

Up to now, we have mainly focussed our attention on dissipative effects due to transitions driven within the atom or molecule by the resonant Fourier components of the electric or magnetic field noise. In this section we turn our attention to the shift of energy levels, to which all frequencies contribute. This was briefly introduced in Eq. (4.32), but then we immediately neglected it as a small effect. This approxi-

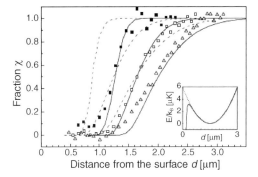

Figure 4.8 Remaining atom fraction χ in a trap at distance d from the dielectric surface for a condensate (solid squares), and for thermal clouds at 2.1 µK (open squares) and 4.6 µK (triangles). The solid (dashed) lines are calculated with (without) Casimir–Polder potential for the condensate and the thermal clouds (left to right). The inset shows the trapping potentials for $C_4 = 8.2 \times 10^{-56}$ J m^4 (solid line) and $C_4 = 0$ (dotted line). Figure taken from [45].

mation was appropriate in the context of studying dissipative effects because the spectral density of the noise does not change appreciably over the frequency shifts in question. However, the forces arising from the gradient of these level shifts, known as dispersion forces, can have a significant effect on the motion of an atom or molecule as we now discuss. In the case of an atom near a magneto-dielectric body, the dispersion force is known as the Casimir–Polder force, and is dominated by the optical part of the noise spectrum, for which the atom experiences the far field since $d \gg \lambda/(2\pi)$. In this limit, the Casimir–Polder force produces a strong attraction to the surface at distances below 1 µm. For example, the insert in Figure 4.8, taken from [45], shows the effect of the Casimir–Polder force on the trapping potential for Rb atoms held by a magnetic trap near a Cu surface. On the side closest to the surface, the trap potential is bent over and the depth of the trap is reduced to only 3 µK when the center of the trap is 1.5 µm away from the surface. The data in Figure 4.8 show how atoms are lost from the trap as it approaches the surface because of the Casimir–Polder force. This effect imposes a significant limit on efforts to miniaturize atom chip traps [46]. It is for this reason that we have not considered atom–surface distances below 1 µm in the preceding sections of this article. It has been noted [47–49] that the attractive Casimir–Polder force may be balanced against the repulsive optical dipole force of an evanescent light field at the surface. This makes it possible to form a trap a little closer to the surface, offering the hope of moving towards slightly sub-micrometer structures in the future.

The Casimir–Polder potential can be calculated from Eq. (4.32). We set $\bar{n}_{\text{th}}$ to zero since there is no appreciable thermal occupation of the visible parts of the spectrum. Re-writing the frequency as $\omega = -iu$, the off-resonant contributions to the shift of the lth atomic level are [41, 50, 51]

$$U_l^{\text{off}}(\mathbf{r}_A) = \frac{\mu_0}{\pi} \sum_k \int_0^\infty du \frac{\omega_{kl} u^2}{\omega_{kl}^2 + u^2} \mathbf{d}_{lk} \cdot \mathbf{G}^{(S)}(\mathbf{r}_A, \mathbf{r}_A, iu) \cdot \mathbf{d}_{kl}. \quad (4.44)$$

If the atom is excited, there can also be resonant transitions, corresponding to real photon emission, which contribute a further shift

$$U_l^{res}(\mathbf{r}_A) = -\mu_0 \sum_k \Theta(\omega_{lk})\omega_{lk}^2 \mathbf{d}_{lk} \cdot \text{Re}\, \mathbf{G}^{(S)}(\mathbf{r}_A, \mathbf{r}_A, \omega_{lk}) \cdot \mathbf{d}_{kl}. \qquad (4.45)$$

These shifts arise from the residues at the poles. For a ground-state atom ($l = 0$) there are no resonant terms, and the ground-state Casimir–Polder potential can be written as

$$U_0(\mathbf{r}_A) = -\frac{\hbar\mu_0}{2\pi} \int_0^\infty d\omega\, \omega^2 \text{Im}\,\text{Tr}\left[\boldsymbol{\alpha}_0(\omega) \cdot \mathbf{G}^{(S)}(\mathbf{r}_A, \mathbf{r}_A, \omega)\right] \qquad (4.46)$$

where

$$\boldsymbol{\alpha}_0(\omega) = \alpha_0(\omega)\mathbf{1} = \lim_{\epsilon \to 0} \frac{2}{3\hbar} \sum_k \frac{\omega_{k0}}{\omega_{k0}^2 - \omega^2 - i\omega\epsilon} |\mathbf{d}_{k0}|^2 \mathbf{1} \qquad (4.47)$$

is the ground-state polarizability of a spherically symmetric atom [$\mathbf{1}$: unit dyad]. Equation (4.46) allows for the physical interpretation of the ground-state Casimir–Polder potential as being due to correlations of the fluctuating electromagnetic field ($\text{Im}\,\mathbf{G}$) with the corresponding induced electric dipole of the atomic system ($\text{Re}\,\boldsymbol{\alpha}$) plus the correlations of the fluctuating electric dipole moment ($\text{Im}\,\boldsymbol{\alpha}$) with its induced electric field ($\text{Re}\,\mathbf{G}$) [52]. Meschede et al. [53] discuss exactly how much of the energy derives from the fluctuations of the field and how much from the fluctuations of the dipole. They show that if the separation between field and dipole fluctuations is to be physically meaningful, the shift must come in equal amounts from the two sources.

For an atom at a distance d away from a half-space with dielectric permittivity $\varepsilon(\omega)$ and magnetic permeability $\mu(\omega)$, one can distinguish between two extreme cases [54]. The long-distance (retarded) approximation is valid if the atom–surface distance d is larger than the largest atomic transition wavelength as well as the largest medium resonance wavelength. In this case,

$$U_0(d) = -\frac{3\hbar c \alpha_0(0)}{64\pi^2 \varepsilon_0 d^4} \int_1^\infty dv \left[\left(\frac{2}{v^2} - \frac{1}{v^4}\right) \frac{\varepsilon(0)v - \sqrt{\varepsilon(0)\mu(0) - 1 + v^2}}{\varepsilon(0)v - \sqrt{\varepsilon(0)\mu(0) - 1 + v^2}} \right.$$
$$\left. - \frac{1}{v^4} \frac{\mu(0)v - \sqrt{\varepsilon(0)\mu(0) - 1 + v^2}}{\mu(0)v - \sqrt{\varepsilon(0)\mu(0) - 1 + v^2}} \right]. \qquad (4.48)$$

In the short-distance (non-retarded) limit, one finds that

$$U_0(d) = -\frac{C_3}{d^3} + \frac{C_1}{d} \qquad (4.49)$$

with non-negative coefficients

$$C_3 = \frac{\hbar}{16\pi^2\varepsilon_0}\int_0^\infty du\,\alpha_0(iu)\frac{\varepsilon(iu)-1}{\varepsilon(iu)+1}, \quad (4.50)$$

$$C_1 = \frac{\mu_0\hbar}{16\pi^2}\int_0^\infty du\,u^2\alpha_0(iu)$$
$$\times\left[\frac{\varepsilon(iu)-1}{\varepsilon(iu)+1}+\frac{\mu(iu)-1}{\mu(iu)+1}+\frac{2\varepsilon(iu)[\varepsilon(iu)\mu(iu)-1]}{[\varepsilon(iu)+1]^2}\right]. \quad (4.51)$$

It should be noted that the sign of the Casimir–Polder potential in the long-distance limit, Eq. (4.48), is not fixed. For purely dielectric materials ($\mu = 0$) the potential is attractive, whereas for purely (or at least strongly) magnetic materials it switches to a repulsive potential. This is most easily seen in the limit of weak (magneto-) dielectric properties in which Eq. (4.48) reduces to [54]

$$U_0(d) \approx -\frac{\hbar c\alpha_0(0)}{640\pi^2\varepsilon_0 d^4}\left[23\chi_e(0) - 7\chi_m(0)\right], \quad (4.52)$$

where $\chi_e(0) = \varepsilon(0) - 1 \ll 1$ and $\chi_m(0) = \mu(0) - 1 \ll 1$ are the static electric and magnetic susceptibilities, respectively. Similarly, in the short-distance limit, Eq. (4.49), any material that shows no dielectric response at all ($\varepsilon = 1$) necessarily induces repulsive Casimir–Polder forces. Hence, it may be possible to tune the long-range surface potentials to one's needs by choosing appropriate materials or structuring the surface to enhance or suppress particular electromagnetic response properties. Possible methods include photonic band-gap structures as well as metamaterials [55]. Of course, the shorter range behavior is not susceptible to these methods since an atom close enough to the surface sees the local material rather than the average over the artificial structure.

The scaling laws for the quantum electrodynamic forces between electric or magnetic atoms and thin or thick plates are summarized in Table 4.1 [41]. As a rule of thumb, the force between an atom and a d-dimensional body increases as z^d relative to the corresponding atom–atom force. However, because of the nonadditivity of the dispersion forces, it may not be possible to derive exact expressions involving macroscopic bodies from those involving microscopic bodies since this would require one to know the form of all the n-atom contributions. Hence, knowledge of the Casimir–Polder force between an atom and a macroscopic body does not imply a correct extrapolation to the Casimir force between two macroscopic bodies.

The first quantitative measurements of the Casimir–Polder force were performed on a ground-state ^{23}Na beam passing through a micrometer-sized gold cavity [56]. The Casimir–Polder force deflected a significant fraction of the atomic beam towards the cavity mirrors. The intensity of the beam transmitted through the cavity as a function of the cavity width L, $I(L)$ is shown in Figure 4.9 normalized to $I(L = 6\,\mu m)$ [the opacity is defined there as the ratio $I(L = 6\,\mu m)/I(L)$]. As the relevant $3S$–$3P$ transition in ^{23}Na is at 589 nm, the long-distance approximation,

Table 4.1 Asymptotic power laws for the forces between (a) two atoms (van der Waals force), (b) an atom and a small sphere, (c) an atom and a thin ring, (d) an atom and a thin plate, (e) an atom and a half-space (Casimir–Polder forces) and (f) for the force per unit area between two half-spaces (Casimir force). In the table heading, p stands for a polarizable object and m for a magnetizable one. The signs + and − denote repulsive and attractive forces, respectively.

Distance → Polarizability →		Retarded p ↔ p	Retarded p ↔ m	Nonretarded p ↔ p	Nonretarded p ↔ m
(a)	r	$-\dfrac{1}{r^8}$	$+\dfrac{1}{r^8}$	$-\dfrac{1}{r^7}$	$+\dfrac{1}{r^5}$
(b)	r_A	$-\dfrac{1}{r_A^8}$	$+\dfrac{1}{r_A^8}$	$-\dfrac{1}{r_A^7}$	$+\dfrac{1}{r_A^5}$
(c)	ρ_A	$-\dfrac{1}{\rho_A^8}$	$+\dfrac{1}{\rho_A^8}$	$-\dfrac{1}{\rho_A^7}$	$+\dfrac{1}{\rho_A^5}$
(d)	z_A	$-\dfrac{1}{z_A^6}$	$+\dfrac{1}{z_A^6}$	$-\dfrac{1}{z_A^5}$	$+\dfrac{1}{z_A^3}$
(e)	z_A	$-\dfrac{1}{z_A^5}$	$+\dfrac{1}{z_A^5}$	$-\dfrac{1}{z_A^4}$	$+\dfrac{1}{z_A^2}$
(f)	z	$-\dfrac{1}{z^4}$	$+\dfrac{1}{z^4}$	$-\dfrac{1}{z^3}$	$+\dfrac{1}{z}$

Eq. (4.48), applies, and the $1/L^4$-scaling is observed (dashed line in Figure 4.9a) whereas the short-distance $1/L^3$-law could be conclusively ruled out (black solid line in Figure 4.9a).

More recently, dipole oscillations in magnetically trapped ^{87}Rb Bose–Einstein condensates have been used to measure Casimir–Polder forces near a dielectric surface [57]. The gradient of the Casimir–Polder potential force modifies the spring constant of the trap to produce a fractional frequency shift γ_z given by

$$\gamma_z = \frac{\omega_z' - \omega_z}{\omega_z} \approx \frac{1}{2} \frac{\partial_z^2 U_0(z)}{\omega_z^2 m_{Rb}} \quad (4.53)$$

with ω_z' and ω_z being the shifted and unshifted trap frequencies in the direction perpendicular to the surface, respectively. Figure 4.9b shows measurements made by this method using Rb atoms near a fused silica surface ($\varepsilon(0) = 3.83$) [57]. The dashed, solid, and dotted lines are theoretical curves for $T = 0$, 300, and 600 K,

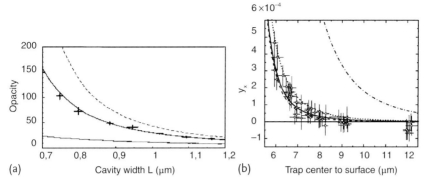

Figure 4.9 (a) Dashed curve: long-distance limit of the Casimir–Polder interaction, black curve: short-distance limit, thin curve: no interaction. Figure taken from [56]; (b) Normalized frequency shift of the center-of-mass oscillation frequency of a Bose–Einstein condensate near a dielectric surface. Figure taken from [57].

respectively. The dash-dotted line represents an extrapolation to the short-distance potential (Eq. (4.49)) which, given the dominant dipole transition wavelength in ^{87}Rb of 780 nm is not applicable at the experimental distance values.

Although the Casimir–Polder potential is attractive, it changes so rapidly close to a surface that it can reflect atoms through the mechanism of quantum reflection [58]. The quantum reflection of a de Broglie wavepacket by a rapidly decreasing potential is exactly analogous to the reflection of light at a dielectric interface where the refractive index decreases. In both cases, significant reflection is observed if the local wavenumber normal to the surface,

$$k_\perp(d) = \frac{1}{\hbar}\sqrt{m^2 v_\perp^2 - 2m\, U(d)} \qquad (4.54)$$

in the case of the atom, changes by more than $k_\perp$ over a distance $1/k_\perp$.

Shimizu [59] was able to demonstrate this using laser-cooled metastable Ne atoms in the state $1s_3\left[(2s)^5 3p : {}^1P_0\right]$. These were dropped from a small (100 μm diameter) magneto-optical trap to fall through a collimating aperture and then onto a silicon {1,0,0} surface or onto a BK7 glass surface. The reflected atoms were detected using a micro-channel plate to amplify the electrons released by de-excitation of the metastable atoms. Figure 4.10 shows the measured reflectivity versus the velocity at which the atoms approached the Si surface. At the lowest normal velocity of approximately 1 mm/s, the reflectivity exceeded 30 %. The solid line shows the calculated reflectivity, based on quantum reflection from the Casimir–Polder potential. The two are in excellent agreement. One nice feature of the experiment is that atoms reflecting from the repulsive inner wall of the potential, immediately outside the surface, are not detected because they are de-excited from the metastable state by their interaction with the surface and are, therefore, invisible to the detector.

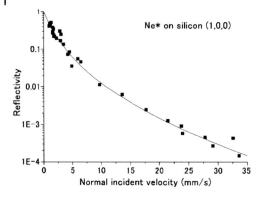

Figure 4.10 Atom reflectivity vs. the normal incident velocity on the Si(1,0,0) surface. The solid curve is the reflectivity calculated using the Casimir–Polder potential. Figure taken from [59].

4.5
Closing Remarks

Over the last ten years, there has been very rapid progress in miniaturizing the structures that form magnetic, electric and optical fields and in integrating these structures to make atom chips. Throughout this development, numerous technical obstacles have been overcome. For example, the need for well-behaved magnetic fields has led to methods of fabricating wires and magnetic thin films so that they have more homogenous interiors and smoother, straighter edges [60–63]. Also, the need to deliver light to specific parts of the chip has stimulated ideas for integrating optical devices into the wafer [64–67]. In order to take maximum advantage of these technical advances, it is desirable to bring ultra-cold atoms close to the surface of the chip, where quantum fluctuations start to affect one's ability to trap the atoms and to manipulate them coherently. The dissipative effects of resonant noise can be reduced by several strategies discussed above, including the use of low temperatures and the choice of insulating or superconducting materials. However, the dispersion forces are more difficult to control because they are broad-band phenomena and generate very strong power-law forces close to a surface. Even so, these forces may be used to advantage at distances down to a few hundred nanometers (e.g., [49]). Understanding the origin of these effects has helped us to recognize the fundamental limits in chip designs and shows that the atom chip provides an exciting new toolbox for atom manipulation very close to surfaces. As this field develops further, the Casimir–Polder potential, perhaps shaped by microstructures on the surface or by using novel artificial metamaterials is likely to play a significant role.

Acknowledgments

We acknowledge fruitful discussions with S.Y. Buhmann, U. Hohenester, and M.R. Tarbutt. This work was funded in parts by the UK Engineering and Physical Sciences Research Council, the European Commission (Atomchips, FastNet, Conquest, and SCALA programmes), and the Royal Society.

References

1 Hinds, E.A. and Hughes, I.A. (1999) *J. Phys. D* **32**, R119.
2 Folman, R., Krüger, P., Schmiedmayer, J., Denschlag, J., and Henkel, C. (2002) *Adv. At. Mol. Opt. Phys.* **48**, 263.
3 Reichel, J. (2002) *Appl. Phys. B* **75**, 469.
4 Henkel, C., Schmiedmayer, J., and Westbrook, C. (2006) *Eur. Phys. J. D* **35**, 1, and following papers.
5 Fortágh, J. and Zimmermann, C. (2007) *Rev. Mod. Phys.* **79**, 235.
6 Ott, H., Fortágh, J., Schlotterbeck, G., Grossmann, A., and Zimmermann, C. (2001) *Phys. Rev. Lett.* **87**, 230401.
7 Hänsel, W., Hommelhoff, P., Hänsch, T.W., and Reichel, J. (2001) *Nature* (London) **413**, 498.
8 Sinclair, C.D.J., Curtis, E.A., Llorente Garcia, I., Retter, J.A., Hall, B.V., Eriksson, S., Sauer, B.E., and Hinds, E.A. (2005) *Phys. Rev. A* **72**, 031603(R).
9 Berman, P.R. (ed.) (1997) *Atom Interferometry*, Academic Press.
10 Rab, M., Cole, J.H., Parker, N.G., Greentree, A.D., Hollenberg, L.C.L., and Martin, A.M. (2008) *Phys. Rev. A* **77**, 061602(R).
11 Seaman, B.T., Krämer, M., Anderson, D.Z., and Holland, M.J. (2007) *Phys. Rev. A* **75**, 023615.
12 Scheel, S., Knöll, L., and Welsch, D.-G. (1998) *Phys. Rev. A* **58**, 700.
13 Knöll, L., Scheel, S., and Welsch, D.-G. (2001) in *Coherence and Statistics in Photons and Atoms* (ed. J. Peřina), Wiley, New York, pp. 1–63.
14 Dung, H.T., Buhmann, S.Y., Knöll, L., Welsch, D.-G., Scheel, S., and Kästel, J. (2003) *Phys. Rev. A* **68**, 043816.
15 Raabe, C., Scheel, S., and Welsch, D.-G. (2007) *Phys. Rev. A* **75**, 053813.
16 Scheel, S. and Buhmann, S.Y. (2008) *Acta Phys. Slov.* **58**, 675.
17 Suttorp, L.G. and Wubs, M. (2004) *Phys. Rev. A* **70**, 013816.
18 Earnshaw, S. (1842) *Trans. Camb. Philos. Soc.* **7**, 97.
19 Leanhardt, A.E., Shin, Y., Chikkatur, A.P., Kielpinski, D., Ketterle, W., and Pritchard, D.E. (2003) *Phys. Rev. Lett.* **90**, 100404.
20 Scheel, S., Pachos, J.K., Hinds, E.A., and Knight, P.L. (2006) in *Quantum Coherence: From Quarks to Solids* (eds W. Pötz, J. Fabian, and U. Hohenester), *Lect. Notes Phys.* **689**, Springer, Heidelberg.
21 Henkel, C., Pötting, S., and Wilkens, M. (1999) *Appl. Phys. B: Lasers Opt.* **B69**, 379.
22 Rekdal, P.K., Scheel, S., Knight, P.L., and Hinds, E.A. (2004) *Phys. Rev. A* **70**, 013811.
23 Jones, M.P.A., Vale, C J., Sahagun, D., Hall, B.V., and Hinds, E.A. (2003) *Phys. Rev. Lett.* **91**, 080401.
24 Li, L.-W., Leong, M.-S., Yeo, T.-S., and Kooi, P.-S. (2001) *J. Electromagn. Waves Appl.* **14**, 961.
25 Harber, D.M., McGuirk, J.M., Obrecht, J.M., and Cornell, E.A. (2003) *J. Low Temp. Phys.* **133**, 229.
26 Scheel, S., Rekdal, P.K., Knight, P.L., and Hinds, E.A. (2005) *Phys. Rev. A* **72**, 042910.
27 Purcell, E.M. (1946) *Phys. Rev.* **69**, 681.
28 Varpula, T. and Poutanen, T. (1984) *J. Appl. Phys.* **55**, 4015.
29 Skagerstam, B.-S.K., Hohenester, U., Eiguren, A., and Rekdal, P.K. (2006) *Phys. Rev. Lett.* **97**, 070401.
30 Hohenester, U., Eiguren, A., Scheel, S., and Hinds, E.A. (2007) *Phys. Rev. A* **76**, 033618.
31 London, F. and London, H. (1935) *Proc. R. Soc. Lond., Ser. A* **149**, 71.
32 Gorter, C.S. and Casimir, H.B.G. (1934) *Z. Phys.* **35**, 963.
33 Nirrengarten, T., Qarry, A., Roux, C., Emmert, A., Nogues, G., Brune, M., Raimond, J.-M., and Haroche, S. (2006) *Phys. Rev. Lett.* **97**, 200405.

34 Mukai, T., Hufnagel, C., Kasper, A., Meno, T., Tsukada, A., Semba, K., and Shimizu, F. (2007) *Phys. Rev. Lett.* **98**, 260407.

35 Cano, D., Kasch, B., Hattermann, H., Kleiner, R., Zimmermann, C., Koelle, D., and Fortágh, J. (2008) *Phys. Rev. Lett.* **101**, 183006.

36 Cano, D., Kasch, B., Hattermann, H., Koelle, D., Kleiner, R., Zimmermann, C., and Fortágh, J. (2008) *Phys. Rev. A* **77**, 063408.

37 Shaw, T.J., Ferrari, M.J., Sohn, L.L., Lee, D.-H., Tinkham, M., and Clarke, J. (1996) *Phys. Rev. Lett.* **76**, 2551.

38 Scheel, S., Fermani, R., and Hinds, E.A. (2007) *Phys. Rev. A* **75**, 064901.

39 Treutlein, P., Hommelhoff, P., Steinmetz, T., Hänsch, T.W., and Reichel, J. (2004) *Phys. Rev. Lett.* **92**, 203005.

40 Buhmann, S.Y., Tarbutt, M.R., Scheel, S., and Hinds, E.A. (2008) *Phys. Rev. A* **78**, 052901.

41 Buhmann, S.Y. and Welsch, D.-G. (2006) *Prog. Quantum Electron.* **31**, 51.

42 Palik, E.D. (ed.) (1991) *Handbook of Optical Constants of Solids II*, Academic Press, New York.

43 Turchette, Q.A., Kielpinski, D., King, B.E., Leibfried, D., Meekhof, D.M., Myatt, C.J., Rowe, M.A., Sackett, C.A., Wood, C.S., Itano, W.M., Monroe, C., and Wineland, D.J. (2000) *Phys. Rev. A* **61**, 063418.

44 Labaziewicz, J., Ge, Y., Antohi, P., Leibrandt, D., Brown, K.R., and Chuang, I.L. (2008) *Phys. Rev. Lett.* **100**, 013001.

45 Lin, Y.-J., Teper, I., Chin, C., and Vuletić, V. (2004) *Phys. Rev. Lett.* **92**, 050404.

46 Fermani, R., Scheel, S., and Knight, P.L. (2007) *Phys. Rev. A* **75**, 062905.

47 Landragin, A., Courtois, J.-Y., Labeyrie, G., Vansteenkiste, N., Westbrook, C.I., and Aspect, A. (1996) *Phys. Rev. Lett.* **77**, 1464.

48 Long, R., Steinmetz, T., Hommelhoff, P., Hänsel, W., Hänsch, T.W., and Reichel, J. (2003) *Phil. Trans. R. Soc. Lond. A* **361**, 1375.

49 Bender, H., Courteille, P., Zimmermann, C., and Slama, S. (2009) *Appl. Phys. B: Lasers Opt.* **96**, 275.

50 Wylie, J.M. and Sipe, J.E. (1984) *Phys. Rev. A* **30**, 1185; (1985) *ibid.* **32**, 2030.

51 Buhmann, S.Y., Knöll, L., Welsch, D.-G., and Dung, H.T. (2004) *Phys. Rev. A* **70**, 052117.

52 Henkel, C., Joulain, K., Mulet, J.-P., and Greffet, J.-J. (2002) *J. Opt. A: Pure Appl. Opt.* **4**, 109.

53 Meschede, D., Jhe, W. and Hinds, E.A. (1990) *Phys. Rev. A* **41**, 1587.

54 Buhmann, S.Y., Welsch, D.-G., and Kampf, T. (2005) *Phys. Rev. A* **72**, 032112.

55 Smith, D.R., Padilla, W.J., Vier, D.C., Nemat-Nasser, S.C., and Schultz, S. (2000) *Phys. Rev. Lett.* **84**, 4184.

56 Sukenik, C.I., Boshier, M.G., Cho, D., Sandoghdar, V., and Hinds, E.A. (1993) *Phys. Rev. Lett.* **70**, 560.

57 Harber, D.M., Obrecht, J.M., McGuirk, J.M., and Cornell, E.A. (2005) *Phys. Rev. Lett.* **72**, 033610.

58 Friedrich, H., Jacobi, G., and Meister, C.G. (2002) *Phys. Rev. A* **65**, 032902.

59 Shimizu, F. (2001) *Phys. Rev. Lett.* **86**, 987.

60 Koukharenko, E., Moktadir, Z., Kraft, M., Abdelsalam, M.E., Bagnall, D.M., Vale, C., Jones, M.P.A., and Hinds, E.A. (2004) *Sens. Actuators A* **115**, 600.

61 Schumm, T., Esteve, J., Figl, C., Trebbia, J.-B., Aussibal, C., Nguyen, H., Mailly, D., Bouchoule, I., Westbrook, C.I., and Aspect, A. (2005) *Eur. Phys. J. D* **32**, 171.

62 Whitlock, S., Hall, B.V., Roach, T., Anderson, R., Volk, M., Hannaford, P., and Sidorov, A.I. (2007) *Phys. Rev. A* **75**, 043602.

63 Moktadir, Z., Darquié, B., Kraft, M., and Hinds, E.A. (2007) *J. Mod. Opt.* **54**, 2149.

64 Eriksson, S., Trupke, M., Powell, H.F., Sahagun, D., Sinclair, C.D.J., Curtis, E.A., Sauer, B.E., Hinds, E.A., Moktadir, Z., Gollasch, C.O., and Kraft, M. (2005) *Eur. Phys. J. D* **35**, 135.

65 Trupke, M., Ramirez-Martinez, F., Curtis, E.A., Ashmore, J.P., Eriksson, S., Hinds, E.A., Moktadir, Z., Gollasch, C., Kraft, M. Prakash, G.V., and Baumberg, J.J. (2006) *Appl. Phys. Lett.* **88**, 071116.

66 Colombe, Y., Steinmetz, T., Dubois, G., Linke, F., Hunger, D., and Reichel, J. (2007) *Nature* (London) **450**, 272.

67 Trupke, M., Goldwin, J., Darquie, B., Dutier, G., Eriksson, S., Ashmore, J., and Hinds, E.A. (2007) *Phys. Rev. Lett.* **99**, 063601.

5
Interaction of Atoms, Ions, and Molecules with Surfaces
Carsten Henkel

Introduction

It is quite remarkable that atoms (and molecules or ions) can actually be trapped near the solid surface of an atom chip, at distances of a few micrometers. Given typical shallow trapping potentials and the low kinetic temperatures of the atoms, this is a situation that is manifestly not in thermal equilibrium: an ultra-cold cloud at some µK or below should simply evaporate out of its trap if it equilibrates with the chip components, no matter whether these are in a cryostat or at room temperature. The reason for the exquisite (meta)stability of atom chip traps is rooted in the quite slow exchange of heat between the trapped particles and their "hot" environment because the two are only weakly coupled. This is not only good news for a calculation of the relevant rates (of energy exchange and particle loss), on the one hand, because perturbation theory can be applied. It also turns out, on the other, that for typical atom chip settings the rates can be experimentally measured, in particular when metallic structures are involved. We review in this chapter the progress made within the last decade for calculating and detecting neutral atom–surface interactions in atom chips. For completeness, we also discuss results obtained for charged particles in front of a solid surface, which are of relevance for ion chips (see Chapter 13).

5.1
Qualitative Overview

Neutral atoms interact with a nearby surface via their electric and magnetic dipole moments. Conversely, the surface radiates electromagnetic fields over a broad frequency spectrum. We shall see that this radiation does not have the simple and universal properties of blackbody radiation; its characteristic frequency and spatial scales are rather linked to the surface material. A main focus in this chapter will be on metallic surfaces, as these are often used for producing magnetic traps.

Atom Chips. Edited by Jakob Reichel and Vladan Vuletić
Copyright © 2011 WILEY-VCH Verlag GmbH & Co. KGaA, Weinheim
ISBN: 978-3-527-40755-2

5.1.1
Electromagnetic Dipole Moments

A neutral atom does not have a permanent *electric dipole moment* d if it is in an electronic state of well-defined parity. Molecules can have a non-zero d if their rotation axes are aligned. Transition dipole moments occur in atoms as matrix elements $\langle e|d|g\rangle$ of the electric dipole operator between different states (of opposite parity). The product ea_0 [electron charge times Bohr radius] can be taken as a typical order of magnitude. To define a characteristic energy scale, we first calculate

$$\frac{(ea_0)^2}{4\pi\varepsilon_0(1\,\mu\text{m})^3} \approx (k_B)40\,\text{nK} \approx (2\pi\hbar)800\,\text{Hz}\,. \tag{5.1}$$

This means that two electric dipoles, placed a few micrometers apart, experience an interaction energy of a few nanoKelvin. This estimate illustrates the significance of (long range) interactions between ultra-cold polarized molecules.

In neutral atoms, electric dipole moments are not permanent, but fluctuate spontaneously or are induced by external fields. The latter contribution, $d_{\text{ind}} = \alpha E$, is conveniently described by the (linear) polarizability

$$\alpha^{(g)}(\omega) = \sum_e \frac{(2\omega_{eg}/3\hbar)|\langle e|d|g\rangle|^2}{\omega_{eg}^2 - i\Gamma_{eg}\omega - \omega^2}\,, \tag{5.2}$$

where ω is the angular frequency of the external field and the sum is over excited states $|e\rangle$ at energies $\hbar\omega_{eg}$ above the state $|g\rangle$ under consideration. (This one does not need to be the ground state.) Γ_{eg} is the spontaneous decay rate for the process $|e\rangle \to |g\rangle$ (the "natural linewidth" of the level $|e\rangle$). Equation (5.2) assumes that the polarizability is isotropic (induced dipole always parallel to the field).

In typical order of magnitude, consider the static polarizability: $\alpha^{(g)}(0)/\varepsilon_0$ is a volume of order $\sim a_0^3$, therefore $\alpha^{(g)}(0) \sim 10^{-42}\,\text{F}\,\text{m}^2 \sim (ea_0)10^{-13}\,\text{m/V}$. Static fields of reasonable magnitude thus induce very small dipole moments on the "atomic scale" ea_0. Electronically excited states are much more polarizable because their electron clouds are larger in size. The polarizability shows a resonance at the Bohr frequencies $\omega \approx \omega_{eg}$, where the induced dipole is enhanced by a factor ω_{eg}/Γ_{eg}. For electronically excited states in hydrogen-like atoms, this ratio is $\sim 10^9$. A resonant field can thus induce an electric dipole moment of magnitude comparable to ea_0. This case is typically avoided in optical traps because the spontaneous emission rate would become too large for a resonant excitation. The induced, off-resonant dipole then scales like $1/(\omega_{eg} - \omega - i\Gamma_{eg}/2)$ at not too large detuning (i.e., $|\omega - \omega_{eg}| \ll \omega_{eg}$).

Finally, the polarizability also characterizes the spectral strength $S^d(\omega)$ of spontaneous fluctuations of the dipole operator. This quantity provides the Fourier spectrum of the dipole correlation function

$$\langle g|d_i(t)d_j(t')|g\rangle - \langle g|d_i(t)|g\rangle\langle g|d_j(t')|g\rangle = \int_{-\infty}^{+\infty} \frac{d\omega}{2\pi} e^{-i\omega(t-t')} S_{ij}^d(\omega)\,. \tag{5.3}$$

For an atom in the ground state (zero temperature, isotropic), it is proportional to the imaginary part of the polarizability (the absorption)

$$S_{ij}^d(\omega) = \begin{cases} 2\hbar \delta_{ij} \operatorname{Im} \alpha^{(g)}(\omega), & \omega > 0 \\ 0, & \omega < 0 \end{cases}. \quad (5.4)$$

The asymmetry of the spectrum in the variable ω is fundamentally related to the existence of a state of lowest energy. Integrating over all frequencies, we get the expected closure relation $\langle g | \boldsymbol{d}^2 | g \rangle = \sum_e |\langle e | \boldsymbol{d} | g \rangle|^2$. In hydrogen-like atoms, this sum gets strong contributions from the D1 and D2 lines, but transitions to other levels, the excitation of core electrons, and the continuum of non-bound states together contribute between 5 % (Li) to 45 % (Cs) [1].

The *magnetic dipole moment* $\boldsymbol{\mu}$ has matrix elements of typical magnitude $\mu_B = e\hbar/2m$, the Bohr magneton. In the electronic ground state of hydrogen-like atoms (orbital angular momentum $l = 0$), it is dominated by the electron spin $\boldsymbol{S}$, $\boldsymbol{\mu} = -\mu_B g \boldsymbol{S}/\hbar$ ($g \approx 2.00232$ the gyromagnetic ratio of the electron). A typical dipole–dipole interaction energy is negligibly small,

$$\frac{\mu_0 \mu_B^2}{4\pi (1\,\mu\text{m})^3} \sim 2\,\text{pK}, \quad (5.5)$$

because it involves the square of the Sommerfeld fine structure constant relative to the electric dipole interaction. For this reason, magnetic interactions are only relevant for particular atoms like chromium ($|\langle \boldsymbol{\mu} \rangle| \approx 6\mu_B$), despite the fact that atomic magnetic dipoles are permanent. Another qualitative difference to the electric dipole is the dominance of low frequencies: indeed, the relevant Bohr frequencies are either zero (degenerate magnetic sublevels, in optical traps, for example), of the order of the Larmor frequency eB/m in a magnetic field (Zeeman splitting, typically in the MHz range), or given by the hyperfine splitting due to the interaction with the nuclear spin (typically a fraction of a GHz to a few GHz, increasing with the atomic weight). This means that for magnetic fields that resonantly excite the corresponding transitions, thermally excited photons have to be taken into account. We show below that this is at the origin of the finite trap lifetime in atom chips based on metallic microstructures.

5.1.2
Electromagnetic Field Strengths

We distinguish between the "trapping fields" that provide the confinement potentials for the atom chip (discussed in Chapters 1 and 2), on the one hand, and fluctuating fields, on the other. The character of these fluctuations strongly depends on the relevant frequencies: in the visible range, for example, only quantum (or "vacuum") fluctuations are present ($\hbar\omega \gg k_B T$), while at frequencies relevant for magnetic dipole transitions, a very large thermal occupation prevails ($\hbar\omega \ll k_B T$). In addition, technical noise and fluctuations of the atom chip currents (shot noise) provide a relevant noise floor at even lower frequencies. In all cases, one has to take into account the dependence on the distance from the chip surface, and on the geometry and material parameters of the microstructures. There is no universal behavior as for blackbody radiation.

The basic quantity that characterizes the spectral strength of a fluctuating field is the cross-correlation spectrum. Let us consider as an example the electric field $E(r, t)$ and its spectrum S_{kl}^E. The latter is a tensor because of the different field components, and is defined by the Fourier expansion

$$\langle E_k(r, t) E_l(r', t') \rangle - \langle E_k(r, t) \rangle \langle E_l(r', t') \rangle = \int_{-\infty}^{+\infty} \frac{d\omega}{2\pi} e^{-i\omega(t-t')} S_{kl}^E(r, r', \omega). \tag{5.6}$$

By taking the sum over diagonal tensor components of S_{kl}^E, we spectrally resolve the electric energy density. In free space, this gives the famous Planck formula

$$\frac{\varepsilon_0}{2} \operatorname{tr} S^E(r, r, \omega) = \frac{\hbar\omega/4}{1 - e^{-\beta\omega}} \frac{2\omega^2}{\pi c^3}, \tag{5.7}$$

where $\beta = \hbar/k_B T$ is the inverse temperature. The operator ordering here is such that at positive frequencies, $1/(1 - e^{-\beta\omega})$ is the average thermal occupation number $\bar{n}(\omega)$ plus one quantum of vacuum fluctuation. At negative frequencies, $1/(1 - e^{-\beta\omega}) = -\bar{n}(-\omega)$. These factors are unity (respectively, zero) for typical electric-dipole transition frequencies (an exception are optically active vibrational transitions in molecules, see Chapter 4) and are both equal to $\sim k_B T/(\hbar\omega) \gg 1$ for typical magnetic dipole Bohr frequencies. The last factor in Eq. (5.7), $2\omega^2/(\pi c^3)$, is the classical mode density per unit volume and unit frequency, summed over the two photon polarizations. It arises from the wavevector integration over the dispersion relation,

$$\frac{2\omega^2}{\pi c^3} = 2 \int \frac{d^3 k}{(2\pi)^3} 2\pi \delta(\omega - c|k|). \tag{5.8}$$

The magnetic energy density in free space is also given by Eq. (5.7).

Near the chip surface, the Planck formula (5.7) does not apply because the former modifies the photon mode functions due to reflection and scattering. This is well known for transitions in the visible [2, 3]. At lower frequencies, another effect becomes much more important: additional modes appear that are related to the thermal excitations of the chip material, chiefly its electric charges and currents. This can lead to radiation spectra that are enhanced by several orders of magnitude. The spectra are no longer universally given and must be calculated from suitable models for the reflection properties of the surface. We consider here for simplicity a planar chip of infinite lateral extent and comment below on microstructures of finite lateral size.

The electromagnetic noise can be reduced or enhanced by the surface. Reduction occurs, for example, at a perfectly reflecting surface where the mode functions must have a node on the boundary: this boundary condition quenches the field noise up to distances of a fraction of the wavelength $\lambda = 2\pi c/\omega$. For normally conducting metals, however, this happens only for one polarization component, the so-called TM-case (electric field in the plane of incidence spanned by the k-vector and the surface normal, with a component perpendicular to the surface). The other polarization (TE) shows enhanced noise, in particular at low frequencies, because the thermal motion of charge carriers in the surface generates fluctuating fields that do not appear in the far field (i.e., at distances beyond λ).

We outline in Section 5.1.3 a convenient way to calculate the electromagnetic noise spectrum at thermal equilibrium, using Green function techniques. The results for the electric and magnetic energy spectra close to a metallic surface are [4, 5] (see Chapter 4, Eq. (4.25))

$$\frac{\varepsilon_0}{2} \mathrm{tr}\, S^E(\mathbf{r},\mathbf{r},\omega) \approx \frac{1}{1-\mathrm{e}^{-\beta\omega}} \frac{\hbar}{8\pi z^3} \mathrm{Im}\frac{\varepsilon(\omega)-1}{\varepsilon(\omega)+1}, \quad z \ll \min(\delta(\omega), \lambda)$$

$$\frac{1}{2\mu_0} \mathrm{tr}\, S^B(\mathbf{r},\mathbf{r},\omega) \approx \frac{1}{1-\mathrm{e}^{-\beta\omega}} \frac{\hbar}{4\pi\delta^2 z}, \quad z \ll \delta(\omega) \ll \lambda,$$

(5.9)

where $\varepsilon(\omega)$ is the dielectric function of the surface material. These spectra are measured at a distance z from the surface that is large compared to the atomic scale and small compared to the so-called skin depth $\delta(\omega) = c/(\omega \mathrm{Im}\sqrt{\varepsilon(\omega)})$. In the microwave range and below, $\delta(\omega) \approx (2/(\mu_0\sigma_0\omega))^{1/2}$ is the length scale set by the frequency ω and the electromagnetic diffusion coefficient $(\mu_0\sigma_0)^{-1}$ (σ_0 is the DC conductivity). The enhancement with respect to the far-field blackbody spectrum scales with $(\lambda/z)^3 \gg 1$ (electric field) and $\lambda^3/(\delta^2 z) \gg 1$ (magnetic field).

The power law dependences of Eq. (5.9) on distance are characteristic for a metal with a local response (no spatial dispersion, normal skin effect) and arise from the integration over k-vectors with projection $K \gg \omega/c$ on the surface. These "evanescent" waves give rise to quasistatic fields where the exponential factor becomes $\approx \exp(-2Kz)$. This dependence determines the characteristic k-vectors as $K \sim 1/(2z)$, unless the reflection coefficients at the surface provide another scale. This happens for example at distances larger than the skin depth, where the power laws cross over to $1/z^2$ (and $1/z^4$) for the electric (and magnetic) energy spectra [5].

These electromagnetic field spectra need to be integrated over all frequencies to get the dispersion interactions of the van der Waals–Casimir–Polder type for polarizable atoms. The resulting expressions are in some limiting cases very similar to the static response of the surface: perfect screening of electric fields and transparency for magnetic fields in the case of normal metals (not superconductors). We give typical orders of magnitude and other limiting cases in Section 5.2.

5.1.3
Digression: Surface Green Functions

This Section provides a few formulas for field spectra and Green tensors at a planar surface. The reader not interested in this technical material may jump to Section 5.2.

The noise spectrum at thermal equilibrium and the Green function are linked by the fluctuation–dissipation theorem [6, 7] that takes the form

$$S_{ij}^E(\mathbf{r}, \mathbf{r}', \omega) = \frac{2\hbar}{1 - e^{-\beta\omega}} \operatorname{Im} G_{ij}(\mathbf{r}, \mathbf{r}', \omega), \quad (5.10)$$

where $G_{ij}(\mathbf{r}, \mathbf{r}', \omega)$ is the so-called electric Green tensor (or dyadic). A similar relation holds for the magnetic noise spectrum and the magnetic Green tensor, see below. Equation (5.4) is a fluctuation–dissipation relation at $T = 0$. We recall that the Green tensor gives the electric field $E_i(\mathbf{r}) = G_{ij}(\mathbf{r}, \mathbf{r}', \omega) d_j$ at frequency ω radiated by an oscillating unit electric dipole located at position $\mathbf{r}'$. In this formulation, the noise spectrum leaves us with a classic electrodynamics problem: find the radiation of a source dipole near the chip surface. This radiation field can be split into two components: the "direct" radiation $\mathbf{r}' \to \mathbf{r}$ one would observe in free space and a "reflected part" due to the presence of the surface. The latter part depends linearly on surface reflection coefficients $r_{\text{TM,TE}}$, and in the calculation, one has to integrate over k-vectors parallel to the surface. For many applications, the reflected Green tensor evaluated at the source dipole, $G_{ij}^{(r)}(\mathbf{r}, \mathbf{r}, \omega)$, suffices: it is diagonal with elements $G_{xx}^{(r)}(z, \omega) = G_{yy}^{(r)}(z, \omega)$ and $G_{zz}^{(r)}(z, \omega)$ that only depend on the distance z to the surface [4, 7]

$$G_{xx}^{(r)}(z, \omega) = \frac{i}{8\pi\varepsilon_0} \int_0^\infty \frac{K\,dK}{k_z} e^{2ik_z z}$$
$$\times \left[\frac{\omega^2}{c^2} r_{\text{TE}}(K, \omega) + \left(K^2 - \frac{\omega^2}{c^2} \right) r_{\text{TM}}(K, \omega) \right]$$

$$G_{zz}^{(r)}(z, \omega) = \frac{i}{4\pi\varepsilon_0} \int_0^\infty \frac{K\,dK}{k_z} e^{2ik_z z} K^2 r_{\text{TM}}(K, \omega), \quad (5.11)$$

where the reflection coefficients are chosen such that they reduce to $r_{\text{TE}} \to (1-n)/(1+n)$ and $r_{\text{TM}} \to (n-1)/(n+1)$ at normal incidence ($K = 0$, $n = \sqrt{\varepsilon(\omega)}$). The wavevector $k_z = \sqrt{\omega^2/c^2 - K^2}$ with $\operatorname{Im} k_z \geq 0$ for $K \geq \omega/c$. Its limit $k_z \to iK$ for large K ensures the exponential cutoff with distance z for evanescent waves. (The Green function of Chapter 4, Eq. (4.34), multiplied with $\omega^2 \mu_0$, has the same meaning as our one.)

The magnetic Green tensor, $H_{ij}(\mathbf{r}, \mathbf{r}', \omega)$, gives the field $\mathbf{B}(\mathbf{r})$ radiated by a unit magnetic dipole at position $\mathbf{r}'$. It can be expressed in terms of a double curl (with respect to $\mathbf{r}$ and $\mathbf{r}'$) of the electric tensor [8] (cf. Chapter 4, Eq. (4.24)). In the Fourier expansion used here at a planar surface, this link can be simplified further because the TE and TM polarization modes are rotated one into the other by taking the curl. In this way, it suffices to multiply Eq. (5.11) by $\mu_0 \varepsilon_0 = 1/c^2$ and to make the exchange $r_{\text{TE}} \leftrightarrow r_{\text{TM}}$.

5.2
Interaction Potentials

The strongest surface interaction for neutral atoms is due to fluctuating electric dipole moments, with a characteristic frequency ω_{eg} typically in the visible or near-infrared (for alkali atoms). Two distance scales are relevant: the transition wavelength $\lambda_{eg} = 2\pi c/\omega_{eg}$ (a fraction of a micrometer) and the thermal wavelength $\lambda_T = c\beta = \hbar c/k_B T$ (a few micrometers). Experimental measurements with atom chips have been performed in the range of 1...10 micrometers: at short distances, the atom–surface interaction distorts the trapping potential and opens a loss channel towards the surface; at large distances, accurate measurements of the trap oscillation frequency give access to the curvature of the interaction potential. The magnetic dipole contribution is small and has not yet been measured.

5.2.1
Charges and Permanent Dipoles

Let us consider to begin with the simplest case, a particle with charge q at a distance z from the surface. It "feels" an electrostatic interaction with its image "below" the surface. The interaction potential is

$$V_q(z) = -\frac{1}{8\pi\varepsilon_0} \frac{\varepsilon(0) - 1}{\varepsilon(0) + 1} \frac{q^2}{2z}, \tag{5.12}$$

where $\varepsilon(0)$ is the static relative permittivity of the surface material and $2z$ the distance between the charge and its image. For a metal, $\varepsilon(0) \to \infty$ and the "electrostatic reflection coefficient" $(\varepsilon(0) - 1)/(\varepsilon(0) + 1) \to 1$.

To get a feeling for the orders of magnitude, take an "ion chip" (Chapter 13) with a typical distance of order $z = 100\,\mu m$ and an ion of $M = 100$ atomic mass units. The potential Eq. (5.12) can then destabilize a harmonic trap of frequency $\sim 100\,kHz$, this number scaling like $M^{-1/2} z^{-3/2}$. The effective confinement in a miniaturized ion trap must be stronger than this value.

A permanent electric dipole moment $\langle d \rangle$, that occurs in polar molecules, interacts in a similar way with an image dipole. The interaction energy can be found by differentiating Eq. (5.12) with respect to r and r_{im}, leading to

$$V_d(z) = -\frac{1}{8\pi\varepsilon_0} \frac{\varepsilon(0) - 1}{\varepsilon(0) + 1} \frac{\langle d \rangle^2 + \langle d_z \rangle^2}{(2z)^3}, \tag{5.13}$$

where d_z is the dipole projection perpendicular to the surface. The anisotropy with respect to the orientation of the dipole is characteristic for (quasi)static dipoles. At a distance $z = 1\,\mu m$, the interaction energy is of the order $40\,nK$, taking the usual estimate ea_0 for the dipole moment. At this level, the atom–surface interaction represents a significant perturbation of the trapping potential. As a rule of thumb, the attractive interaction prevents trapping at distances shorter than $\sim 100\,nm$.

5.2.2
Van der Waals Potential

An atom with a fluctuating electric dipole interacts with a surface according to the van der Waals–Casimir–Polder potential that is given at any distance by

$$V_{\text{vdW-CP}}(z) = -\frac{\hbar}{2\pi}\text{Im}\int_{-\infty}^{+\infty}\frac{d\omega}{1-e^{-\beta\omega}}\alpha_{ij}(\omega)G_{ji}^{(r)}(\mathbf{r},\mathbf{r},\omega), \qquad (5.14)$$

where $\alpha_{ij}(\omega)$ is the atomic polarizability Eq. (5.2) and $G_{ji}^{(r)}$ is that part of the electric Green tensor that is due to reflection from the surface. Strictly speaking, Eq. (5.14) applies to a thermalized atom, but can be used for a ground-state atom provided the relevant frequencies ω_{eg} are much larger than typical temperatures ($\beta\omega_{\text{eg}} \gg 1$). See Section 5.2.4 for a more general case. For the alkali atoms, the electric dipole fluctuations in the ground-state $|g\rangle$ are dominated by virtual transitions to the first excited state $|e\rangle$ (the D1 and D2 lines). Because their frequencies are well beyond the thermal range, the integration can be carried out over positive frequencies only, neglecting $e^{-\beta\omega}$ in the denominator. (This becomes relevant at large distance, see below and Section 4.3.5 in Chapter 4.)

The van der Waals–London regime applies at atom–surface distances much smaller than the corresponding wavelength, $z \ll \lambda_{\text{eg}}$. The potential is then dominated by the instantaneous (nonretarded) response of the surface to the electric dipole fluctuations and can be interpreted in terms of an image dipole. One finds an energy shift proportional to the squared electric-dipole matrix elements $|\langle g|d_i|e\rangle|^2$, with a power law $1/z^3$ and a frequency average of the surface "reflectivity" in the non-retarded limit, weighted by the atomic polarizability Eq. (5.2) [7]

$$V_{\text{vdW}}(z) = -\frac{1}{4\pi\varepsilon_0(2z)^3}\sum_e(|\langle g|d|e\rangle|^2 + |\langle g|d_z|e\rangle|^2)$$

$$\times \text{Im}\int_0^\infty\frac{d\omega}{2\pi}\frac{\varepsilon(\omega)-1}{\varepsilon(\omega)+1}\frac{2\omega_{\text{eg}}}{\omega_{\text{eg}}^2 - i\Gamma_{\text{eg}}\omega - \omega^2}, \qquad (5.15)$$

where $\varepsilon(\omega)$ is the dielectric function of the surface material. The summation runs over all excited states that are connected with electric dipole transitions to the ground state, with the D1/D2 transitions giving the most significant contribution [1, 9]. In Chapter 4, Eq. (4.49) provides the generalization to surfaces with a permeability $\mu(\omega) \neq 1$, but is restricted to an isotropic polarizability.

An electromagnetic surface mode occurs at a frequency given by $\varepsilon(\omega) = -1$, called surface plasmon resonance for a metal. Equation (5.15) suggests that this frequency (of the order of the metal's plasma frequency) gives a large contribution to the van der Waals shift. This turns out to be a too simple picture because for ground-state atoms, the shift is always due to a broad continuum of non-resonant virtual photons. For the same reason, it also turns out that even in optical traps, electric dipoles in ground-state atoms have quantum fluctuations that give a much larger contribution to $\langle d^2 \rangle$ than the dipole moment induced by the trapping lasers – the latter can be neglected for the calculation of the atom–surface interaction. Resonant interactions with surface plasmon modes have been observed for electronically excited states where real photons emitted by the atom give rise to forces as well [10–12]. Similarly, there is a resonant force from the absorption of thermal photons, see Chapter 4 and [13].

The van der Waals energy attracts a ground-state atom to the surface and provides a significant distortion of atom chip traps at distances $z \leq \lambda_{eg}$, about 1 μm and below. A typical order of magnitude is

$$V_{vdW}(z) \sim -\frac{\hbar \Gamma_{eg}}{(z/\lambda_{eg})^3(1+z/\lambda_{eg})}, \tag{5.16}$$

where Γ_{eg} is the linewidth of the dominant electric dipole transition. The power law in Eq. (5.16) changes to $1/z^4$ when the distance becomes large compared to the transition wavelength $\lambda_{eg} = 2\pi c/\omega_{eg}$. This effect is due to the non-zero time delay required by light for traveling from the atom to the surface and back.

5.2.3
Casimir–Polder Potential

The potential in the retarded range, $\lambda_{eg} \ll z$, is called the Casimir Polder potential. Its asymptotic expression involves the atomic polarizability and the dielectric function of the chip surface at zero frequency, with an angular average of surface reflection coefficients

$$V_{CP}(z) = -\frac{3c}{4\pi^2 \varepsilon_0 (2z)^4} \sum_e \frac{\langle g|d_i|e\rangle \langle e|d_j|g\rangle}{\omega_{eg}}$$

$$\times \int_0^{\pi/2} d\psi \sin\psi \left\{ -\cos^2\psi \, \Delta_{ij} \, r_{TE}(\varepsilon_{stat}) \right.$$

$$\left. + (2\sin^2\psi \, \delta_{iz}\delta_{jz} + \Delta_{ij}) \, r_{TM}(\varepsilon_{stat}) \right\}, \tag{5.17}$$

$$r_{TE}(\varepsilon_{stat}) = \frac{1 - \sqrt{\varepsilon_{stat}\cos^2\psi + \sin^2\psi}}{1 + \sqrt{\varepsilon_{stat}\cos^2\psi + \sin^2\psi}}, \tag{5.18}$$

$$r_{TM}(\varepsilon_{stat}) = \frac{\varepsilon_{stat} - \sqrt{\varepsilon_{stat}\cos^2\psi + \sin^2\psi}}{\varepsilon_{stat} + \sqrt{\varepsilon_{stat}\cos^2\psi + \sin^2\psi}}. \tag{5.19}$$

This formula is valid at $T = 0$ and can be derived from the results of [7]. We have denoted $\Delta_{ij} = \delta_{ij} - \delta_{iz}\delta_{jz}$. For a perfect conductor and more generally for metals with a non-zero DC conductivity, $\varepsilon_{stat} = i\infty$, and therefore $r_{TE} \equiv -1$ and $r_{TM} \equiv +1$. In that case, the ψ-integral in Eq. (5.17) gives $4/3\delta_{ij}$. For an explicit formula for all ε_{stat}, see [14]. Equation (5.17) is complementary to Eq. (4.48) of Chapter 4 where a surface permeability $\mu_{\text{stat}} \neq 1$ is allowed for, but the polarizability is assumed isotropic.

If the distance exceeds the wavelength $\lambda_T = 2\pi\hbar c/(k_B T)$ of thermal photons (a few micrometers at $T = 300$ K), the Casimir–Polder potential is reduced by a factor z/λ_T. Equation (5.16) with its $1/z^3$ power law thus applies again if it is multiplied with a small prefactor of the order of $k_B T/\hbar\omega_{eg} \ll 1$ (see [13, 15]). This regime typically holds very well for electric dipole transitions in the visible range and room temperature. The Casimir–Polder potential has been measured in the distance range $z = 5$–10 µm, using the change in the oscillation frequency of a Bose–Einstein condensate (BEC) in a harmonic trap near a surface [16], see Eq. (4.53) in Chapter 4.

5.2.4
Recent Developments

Temperature Dependence Atom–surface interactions have also been studied in recent years for non-equilibrium situations. The case of an atom in a well-defined state (non-thermal populations) has been analyzed in [7, 10, 12, 17], see also Section 4.4 and Eq. (4.32) in Chapter 4. Another example is a surface that is warmer than its surroundings where the field itself is not in equilibrium [22, 23]. In that case, one may use the fluctuation electrodynamics developed by Rytov and coworkers [18, 19] that has been turned into a consistent scheme for field quantization in dispersive and absorbing media (see Chapter 4). This approach ascribes to the surface material thermally excited sources (a current density or a polarization field) with an oscillator strength fixed by the electric and magnetic susceptibilities. These sources radiate fields into the vacuum that add incoherently to those generated by the surroundings (photons incident on the surface and reflected by it) [20]. The resulting atom–surface interaction need not be describable in terms of a potential [21]; the basic quantity is then the radiation force F given by

$$F(r) = \langle d_i \nabla E_i(r) \rangle, \tag{5.20}$$

where the average must take into account both the fluctuating dipole and the fluctuating field [22–24]. Repulsive forces may occur in this context in the form of radiation pressure from the warm surface. The dependence on the two temperatures has been experimentally verified using the trap oscillations of a condensate [16, 25].

Potential Probing with Atomic de Broglie Waves Experiments at shorter distances exploit the distortion of the trapping potential. The main effect is then that atoms are lost from the trap by "spilling over" the potential barrier to the surface. This has been demonstrated in [26] using cold atoms reflected from an evanescent wave mirror, and in [27] where an atom chip trap has been moved towards the surface.

A very different behavior occurs for ultra-cold atoms whose de Broglie wavelength is comparable to the scale on which the atom–surface interaction potential varies. In that case, an incident matter wave can be "quantum reflected", with a non-zero probability, even if the potential is attractive. (The same effect occurs routinely for light at an interface.) First experiments with ultra-cold hydrogen atoms have been performed in [28]. Quantum reflection is related to the breakdown of the WKB or short-wavelength approximation, since the latter predicts that the matter wave follows the corresponding classical particle path [29]. For the typical power-law potentials occurring in atom–surface interactions, the WKB approximation breaks down in a limited range of distances: in this domain the potential varies rapidly on a scale set by the incident kinetic energy [30–32]. When the surface potential is weaker, the "bad lands", where the WKB approximation fails, approach the surface and, what is more important, they become more pronounced. This has been exploited in experiments with cold atomic beams to enhance the efficiency of quantum reflection. The surface density of a silicon wafer has been reduced by writing a relief grating, and this improved the reflectivity for metastable neon atoms [33, 34]. It has been studied in detail to which extent the atom–surface interaction actually plays a role in this context [35, 36]. The authors of [37, 38] have demonstrated the quantum reflection of a rubidium BEC from a silicon wafer. A reasonable agreement has been found with a theory for non-interacting atoms, using a potential that interpolates between the van der Waals and Casimir–Polder limits. Measurements of this kind, using ultra-cold atoms, hence provide the possibility to explore the weak long-distance tail of atom–surface interactions [31, 32, 39, 40].

Geometry Dependence Let us finally comment on the behavior of the van der Waals potential in other geometries. The following estimates apply, strictly speaking, only in the short-distance regime where retardation can be ignored. A Table that also lists the retarded range can be found in Chapter 4 (Table 4.1). The $1/z^3$ law can be obtained, with an error of about 15%, by integrating a $1/r^6$ van der Waals pair potential between the atomic constituents of the surface and the "test atom" [15, 41]. This method can be easily applied to bodies with other geometries: for an atom near a thin sheet of polarizable matter, the potential scales like $1/z^4$, near a thin rod, like $1/z^5$. The discrepancy with respect to exact calculations is due to screening effects that make the interaction nonadditive in the amount of polarizable matter. An experiment with ultra-cold atoms has been recently suggested to measure a lateral Casimir–Polder force near a corrugated surface, with the aim to demonstrate its nonadditivity [42]. Carbon nanotubes provide in this context a very clean model system for thin rods [43–47]. The challenge here is to approach the nanotube so closely (below 1 μm) that the atom–nanotube interaction becomes a relevant energy. Naturally, it is quite small compared to a bulk surface, given the strongly reduced amount of material.

5.3
Surface-Induced Atomic Transitions

We now turn to transitions between atomic states that are induced by the presence of the atom chip surface. These transitions tend to drive the atom towards thermal equilibrium – where it is eventually lost since the surface temperature is much larger than typical atom chip trapping potentials. The useful time scale for stable trapping and coherent manipulation is thus set by the rate constants for transitions that drive the atom into non-trapped states. The rates are, by Fermi's Golden Rule, proportional to the spectral densities of electric and magnetic fields, evaluated at the atom position and at the relevant transition frequency. In this kind of loss processes, the atom thus acts as a quasi point-like *detector* of the local electromagnetic field spectrum.

The sources of electromagnetic field fluctuations in atom chips can be roughly categorized as follows.

(a) Vacuum (also called zero point) fluctuations of field modes and, at lower frequencies, their thermal counterpart:

(a-1) One type of field modes are scattering states labeled by plane wave photons incident from infinity (or from the walls of the experimental setup). They give rise to a spatially modulated field pattern upon reflection from the chip structures.

(a-2) A second type of modes is provided by the radiation due to polarization charges or currents inside the chip material. These modes have to be taken into account separately when the chip material shows absorption. At subwavelength distances, their non-propagating near-fields actually dominate the field fluctuations and increase them orders of magnitude beyond the Planck blackbody spectrum.

(b) The fluctuations of the external currents and fields that provide the trapping potential. In a first step, they can be calculated along the same lines as the static fields that form the trap, using a linearization procedure. This works provided retardation is negligible at the relevant frequencies. Current fluctuations at the shot noise level require a more careful description of electron transport through the chip wires (see [48] for a simple ballistic model).

5.3.1
Visible Frequencies: Spontaneous Emission

Let us begin with electric dipole transitions in the visible and near infrared. They occur at a rate given by Fermi's Golden Rule that can be written in the form

$$\gamma_{i \to f} = \frac{1}{\hbar^2} \sum_{i,j} \langle i|d_i|f\rangle \langle f|d_j|i\rangle S^E_{ij}(\mathbf{r},\mathbf{r};\omega_{fi}) \,. \tag{5.21}$$

Here, $|i, f\rangle$ are the initial and final states of the atom, ω_{fi} is the corresponding transition frequency, and $S^E_{kl}(\mathbf{r},\mathbf{r};\omega)$ is the local spectral density of the electric field, defined in Eq. (5.6). We recall that it gives the strength of field fluctuations at the frequency ω. It is usually expressed as the local density of photon states LDOS [49–51]. Because of an asymmetry between electric and magnetic fields in the near field, this has to be done with some care, as explained in [52]: we need here, of course, the local spectrum of electric fields. Using the fluctuation–dissipation theorem (5.10), one finds that $\gamma_{i \to f}(\mathbf{r})$ is proportional to the imaginary part of the electric Green tensor, $\operatorname{Im} G_{ij}(\mathbf{r}, \mathbf{r}, \omega)$, projected onto the direction of the transition dipole, $\langle i | \mathbf{d} | f \rangle$ (a "projected local density of states" [53]). The temperature dependence simplifies here as $1/(1 - e^{-\beta \omega}) \mapsto 1$ because of the large electric-dipole transition frequencies. This holds no longer for polar molecules whose vibrational and rotational resonance frequencies are lower and where absorption and stimulated emission of thermal photons plays a role, see Section 4.3.5.

The LDOS can also be split into radiative and non-radiative parts – this is similar to the distinction made above between photon scattering states and material radiation. Above a structured substrate, the decay rate $\gamma_{i \to f}(\mathbf{r})$ is position-dependent and allows one to infer local optical properties with a sub-wavelength spatial resolution [54, 55]. The enhanced resolution is essentially due to the non-radiative LDOS where evanescent fields carry information about the surface on small scales. This can be used in scanning near-field microscopy, the advantage being that $\gamma_{i \to f}(\mathbf{r})$ is independent of illumination and detection conditions. Some evidence for a distance-dependent decay rate $\gamma_{i \to f}(z)$ close to a planar dielectric has been found in [56]: cold atoms dropped onto a glass surface have been detected via the absorption of an evanescent wave coating the surface.

5.3.2
Thermal Frequencies: Spin-Flips

Consider now lower Bohr frequencies, as they occur among the hyperfine and Zeeman levels of the electronic ground state. Thermal occupation plays a prominent role, and the replacement $1/(1 - e^{-\beta \omega}) \mapsto k_B T/(\hbar \omega) \gg 1$ in Eq. (5.21) is a good approximation. In addition, traps at micrometer distances from an atom chip probe the extreme near-field of the surface, as the resonance wavelengths are of the order of centimeters (hyperfine splitting) or larger. For both reasons, the transition rates are enhanced by several orders of magnitude compared to spontaneous emission into a "cold, empty space". We recall that this was precisely the motivation of Purcell's seminal paper on cavity-enhanced decay rates [57]: the simple-minded application of the free-space formula for spontaneous emission due to magnetic dipole coupling gives a lifetime far too long compared to experimental observations (exceeding the age of the Universe, see Chapter 4 after Eq. (4.25)). Given the large wavelengths, one must take into account that the electromagnetic modes are strongly modified by nearby objects (i.e., closer than λ_{fi}).

The transition rate $\gamma_{i \to f}$ due to magnetic dipole coupling can be found from Eq. (5.21) with the replacements $\mathbf{d} \mapsto \boldsymbol{\mu}$ and $S^E_{ij} \mapsto S^B_{ij}$. Above a planar surface, the result takes the form

$$\gamma_{i \to f} = \frac{\mu_0 k_B T}{2 \pi \hbar^2 \omega_{fi}} |\langle i | \mu_z | f \rangle|^2 \operatorname{Im} \int_0^\infty \frac{K \, dK}{(-i k_z)} e^{2 i k_z z} K^2 r_{TE}(K, \omega_{fi}) \,. \tag{5.22}$$

We have calculated the magnetic noise spectrum with the rules spelled out after Eq. (5.11). The transition dipole is assumed perpendicular to the surface only; a similar formula holds with the replacement $K^2 r_{TE} \mapsto \frac{1}{2}[(\omega_{fi}/c)^2 r_{TM} + (K^2 - (\omega_{fi}/c)^2) r_{TE}]$ for a parallel transition dipole. The approximation $k_z \approx iK$ holds very well in the relevant range of K-vectors ($K \gg \omega/c$, corresponding to $z \ll \lambda_{fi}$. For a normally conducting metal surface, the reflection coefficient r_{TE} depends in this regime only on the product $K\delta$ of the K-vector and the skin depth $\delta = \delta(\omega_{fi})$. One can derive the following interpolation formula [5] (see also Chapter 4, Eq. (4.25))

$$\gamma_{i \to f}(z) \approx \frac{\mu_0 k_B T \left(|\langle i|\mu_\parallel|f\rangle|^2 + 2|\langle i|\mu_z|f\rangle|^2\right)}{16\pi \hbar^2 \omega_{fi} \delta^2 z \left(1 + \frac{2}{3}(z/\delta)^3\right)} . \tag{5.23}$$

This interpolation correctly reproduces the limits of distance z very small or very large compared to the skin depth δ. It is roughly a factor three too small in the transition region $z \sim \delta$, as can be seen by performing numerically the integration in Eq. (5.22). Observe in Eq. (5.23) how the thermal occupation number increases the very small characteristic energy Eq. (5.5) associated with the magnetic dipole. For a good conductor like copper and transition frequencies in the MHz range, the skin depth δ is in the 100 µm range. Spin-flip transitions between different Zeeman sublevels then occur at rates $100-10\,\mathrm{s}^{-1}$ for distances between 1–10 µm. Since the trapping potential in a magnetic microtrap is proportional to the magnetic (Zeeman) quantum number, one or two magnetic transitions can put the atom into a state that is no longer trapped. This becomes observable as a limitation of the trap lifetime at small atom–surface distances [27, 58–61].

The physics behind these magnetic fluctuations can be understood from the thermal excitation of low-frequency currents that are coupled to the magnetic field (eddy currents). In the metal, these currents show spatial diffusion, and the skin depth emerges as a characteristic length scale from forced diffusive waves, $\delta = (2D/\omega)^{1/2}$, with diffusion constant $D = 1/(\mu_0 \sigma_0)$. The fluctuating magnetic field penetrates through the surface of the metal into the vacuum.

The dependence on the conductivity of the chip material is encoded in Eq. (5.23) via the skin depth δ. The worst case at fixed z occurs for $\delta \sim z$. Shorter skin depths can be realized with superconductors, see [62, 63] and Chapters 4 [Eq. (4.26)] and 10. With semiconductors, δ gets larger. Noise reduction can also be achieved by cooling, but the chip must then use suitable alloys where $T\sigma(T)$ is not constant [64]. This approach is limited to weak wire currents where Joule heating of the wires can be neglected; for more details, see [61, 65].

An alternative route is to reduce the amount of metallic material using thin films or narrow wires on a substrate with a smaller conductivity [27, 61]. Approximate calculations for these geometries show indeed a change in the power-law dependence on distance from $1/z$ (half-space) through $1/z^2$ (layer thinner than z) to $1/z^3$ (thin cylinder). The idea behind these estimates is to sum up incoherently over the magnetic noise generated by all current elements in the metallic object [66]. One makes an error of about a factor two or three (nonadditivity) and one does not capture the anisotropy of the field polarization, as gauged by numerical calculations of the magnetic Green function [67]. The power law for the half-space can be easily understood from this picture: a metallic volume element with a given current density creates a magnetic field that decays with distance as a $1/r^2$ law. Average the square to get the magnetic noise spectrum ($\sim 1/r^4$) and integrate over the metallic half-space: the distance to the surface defines a characteristic volume $\sim z^3$ so that the weak dependence $1/z$ remains. Thin films or wires have smaller characteristic volumes ($\sim z^2$ or $\sim z$) which explains the faster decay with distance. More accurate calculations based on the Green tensor for the cylinder geometry have been performed [68] and compare well to the experiment of [59]. A numerical analysis of rectangular wires, but restricted to two dimensions, can be found in [67].

5.3.3
Trap Heating

The magnetic states considered so far differ in their magnetic (or spin) quantum numbers, and the matrix element $\langle i|\boldsymbol{\mu}|f\rangle$ is a transition magnetic dipole. In a typical magnetic trap on a chip, however, atoms also have a permanent magnetic moment $\langle i|\boldsymbol{\mu}|i\rangle$ (parallel to the static trapping field) that makes them sensitive to field fluctuations even when no transitions between Zeeman levels are involved. We consider here the frequency range where the resonances of the center-of-mass motion are located. A typical resonance is the trap oscillation, with frequency in the kHz to MHz range for the tightly confined geometries of atom chips. To couple different center-of-mass quantum states, one needs a spatially inhomogeneous magnetic field or, equivalently, a magnetic force. In a sense, we are dealing with a random Stern–Gerlach experiment where the de Broglie wave of the atom is scattered. The main consequence is the heating of an atom sample that is initially prepared in the ground state of the trap.

It turns out that the dominant source of trap heating are "technical" fluctuations in the currents flowing through the chip or of the bias field: they generate magnetic fluctuations that turn into a force when they combine with the (static) bias field. Estimates for the transition rate between the two lowest eigenstates in a one-dimensional harmonic trap have been worked out in [48, 69, 70]. Consider the simplest case of a thin wire trap in a homogeneous bias field (see Chapter 2). The noise spectrum $S_I(\omega)$ of the current translates into magnetic field noise at the height z above the wire as

$$S_B(\omega) = \left(\frac{\mu_0}{2\pi z}\right)^2 S_I(\omega), \qquad (5.24)$$

where the frequency ω is assumed sufficiently small that the magnetic noise fields can still be calculated within the quasi-static approximation (roughly, $\omega \ll c/z$ to avoid retardation). Above a wire on a chip, this field is oriented parallel to the surface and shifts the position where it cancels with the bias field. The trap center is moving around randomly which is equivalent to a fluctuating force acting on the atom, $\delta F_z = M\Omega^2 \delta z$, where $M\Omega^2$ is the trap spring constant. The excitation rate for the vibrational motion perpendicular to the surface (angular frequency Ω), from the ground-state $|0\rangle$ to the first excited state $|1\rangle$, can be estimated from the force fluctuation spectrum $S_{zz}^F(\omega)$ as [48]

$$\gamma_{0\to 1} = \frac{1}{2\hbar M \Omega} S_{zz}^F(-\Omega) = \frac{e}{2\hbar} \frac{M\Omega^3 z^2}{I} \frac{S_I(-\Omega)}{SN_I}, \quad (5.25)$$

where M is the atom mass, I the wire current, and the current noise spectrum $S_I(\omega)$ is normalized to the shot noise level $SN_I \equiv eI$. The product $\hbar \Omega \gamma_{0\to 1}$ gives the heating rate in terms of absorbed power; this quantity can be compared to experiments even if the atoms are not prepared in the trap ground state. Typical parameters (100 amu atom mass, 100 kHz, 10 μm, 1 mA, current spectrum at the shot noise level) give a quite significant heating rate $\gamma_{0\to 1} \sim 10\,\text{s}^{-1}$, well within the reach of current experiments. We note that an alternative mechanism for trap heating involves flips between spin states that move in magnetic potentials of different curvature, see Chapter 4.

Indications for a heating process from technical noise have been observed in the experiments of [71] and [59]: current supplies with parasitic noise reduce the lifetime of a BEC in a surface microtrap. Low-noise electronics and shielding results in significant improvements down to the level where surface-induced spin-flips became relevant for the trap lifetime.

We note that the picture of trap heating becomes subtle when we analyze the atoms in a frame co-moving with the randomly shifted trap. It can be shown that for a strictly harmonic trap, no heating occurs in this frame [72]. What happens to a BEC is that the phase of its macroscopic wave function ("order parameter") acquires a finite coherence length, once the average over the Galilean gauge transformations back into the laboratory frame is done. This reduced phase coherence can be made visible by letting two halves of a split condensate interfere. The challenge of this experiment is to reduce the intrinsic phase noise due to the low-lying excitation modes of the BEC [73, 74].

The heating due to thermal magnetic fields radiated by the surface is much smaller, the main effect being trap loss, as discussed in Section 5.3.2. It has been shown in [48, 75] that thermal magnetic fields radiated by the surface are spatially "rough" on a scale given by the trap height. This roughness scale determines the magnetic forces. The excitation rate of higher trap eigenstates then scales with $(\Delta x/z)^2$ times the loss rate due to a spin-flip, where Δx ($\ll z$) is the typical size of the trap ground state.

5.3.4
Atom Chips and Decoherence

One of the key peculiarities of quantum mechanics are linear superpositions of quantum states. They are also at the heart of the exceptionally fast "parallelism" in quantum computers compared to classical ones. In a quantum system coupled to an environment, superpositions are destroyed because the different quantum states involved get entangled with environment states that rapidly become orthogonal [76, 77]. "Open quantum systems", as they have been dubbed, can no longer be described by Hilbert space vectors. One has to use (reduced) density matrices that are akin to correlation functions of state vectors and that permit one to retrieve all observables pertaining to the system alone, once we admit that we do not have access to the environment observables. Superpositions are characterized by density matrices with non-vanishing off-diagonal elements, also called "coherences", if written in the basis spanned by the states involved in the superposition. Decoherence is the process in which a particular basis exists (also known as "pointer basis") in which off-diagonal elements decay to zero [77]. This leads to a density matrix that can be interpreted as a mixture of these pointer states, weighted with classical probabilities: from being "here and there", the system has come to be "here or there".

Spin Decoherence A typical result of the theory of open quantum systems is that environment-induced transitions between basis states also suppress coherence. For a closed two-state spin system $|g\rangle, |e\rangle$, for example, the off-diagonal element of the density matrix ρ evolves like

$$\frac{d\rho_{eg}}{dt} = -\frac{i}{\hbar} \langle e| [H, \rho] |g\rangle - \frac{1}{2} \left(\gamma_{e \to g} + \gamma_{g \to e} \right) \rho_{eg} , \qquad (5.26)$$

where we can identify a "decoherence rate" $\frac{1}{2} \left(\gamma_{e \to g} + \gamma_{g \to e} \right)$. If transitions to other levels occur, that rate is further increased.

There are interactions that speed up decoherence without inducing transitions, a process called "dephasing". An example are random fluctuations $\delta\omega_{eg}(t)$ of the transition frequency of the two-level system. They randomize the relative phase between the components of a superposition state, and the superposition gets lost after averaging. This model for an open qubit system can actually be solved analytically [78]. It can be used to estimate the impact of magnetic field fluctuations polarized along the static trapping field (and hence along the permanent magnetic moment of the trapped atoms). If the fluctuations have a correlation time much shorter than other experimental time scales (i.e., for a white noise spectrum), an additional decay is found for off-diagonal elements of the density matrix with a rate [79]

$$\gamma_{deph} = \frac{\Delta\mu^2}{2\hbar^2} S_{nn}^B (\mathbf{r}, \mathbf{r}; \omega \to 0) , \qquad (5.27)$$

where $\Delta\mu$ is the difference in the magnetic moment between the two basis states and $S_{nn}^B = n_i n_j S_{ij}^B$ is the noise spectrum for the field component along the static trapping field (unit vector $\mathbf{n}$). Thermal surface noise gives from Eq. (5.27) a dephasing rate comparable to the rate for spin-flip processes: note that $\delta^2\omega = (\mu_0\sigma_0)^{-1}$ and $\delta \to \infty$ in the static limit $\omega \to 0$. The same power laws apply as in Eq. (5.23) above. Wire current fluctuations show strongly anisotropic noise for thin wires (mainly azimuthally polarized), and give a decoherence rate scaling like $1/z^4$, as shown in [80].

Recent experiments have exploited the suppression of dephasing between states with the same magnetic moment ($|\Delta\mu| \ll \mu_B$), in order to demonstrate an atomic clock in a chip microtrap [81]. The two states are in different hyperfine manifolds and only the small nuclear magnetic moment distinguishes their coupling to magnetic fields. One can even switch off the dependence on magnetic fields to first order by working at a particular value of the trapping field ("magic field"). Coherent superpositions between the two hyperfine states are created in a Ramsey–Bordé interferometer scheme (see Chapters 7 and 8). The dephasing between the states did not show any distance-dependent enhancement.

Technical noise in the wire currents and bias fields from the external electronics is likely to give the dominant contribution to dephasing because of its increase at low frequencies (see Eq. (5.27)), in particular if $1/f$ noise is present. In that limit, the noise spectrum in Eq. (5.27) has to be evaluated at a small cutoff frequency set by the duration of the experiment, as shown in [82].

De Broglie Wave Decoherence Atom chips can test the (de)coherence of matter waves by letting different parts of a spatially delocalized state interfere. Indeed, a delocalized wave in a potential, for example, is in a continuous superposition of position eigenstates. The degree of "quantum-ness" of this wave can be characterized from the correlation or coherence function $\langle \psi^*(x, t)\psi(x', t)\rangle$. This quantity is related to the contrast of the interference generated by a fictitious double-slit experiment (the slits being placed at x and x') [83]. The range in $x - x'$ over which the correlation function decays to zero is called the correlation (or coherence) length l_{coh}. Spatial decoherence is the decrease of l_{coh} due to coupling to an environment. This involves in particular random scatterings that broaden the momentum distribution, since it can be shown that $l_{coh} \sim \hbar/\Delta p$ [77].

Model calculations for cold atoms interacting with a fluctuating, rough potential indicate that the atomic coherence length is reduced to the correlation length of the potential, after a time set by the power spectrum of the fluctuations [66, 84]. These calculations are based on a semi-classical Boltzmann equation, neglecting matter wave localization. In [85, 86], the decoherence of a trapped condensate that scatters non-condensed atoms has been considered, including spinor condensates. In the latter case, robust spin coherence can be maintained if the atom interactions do not break a $SU(2)$ symmetry. The condensate superfluidity which is due to atom–atom interactions, reduces the decoherence rate, as shown by model calculations in [87]. This is because condensate excitations at long wavelengths are mainly phase waves (phonons) and are inefficiently excited by a fluctuating potential.

In order to test matter wave coherence on an atom chip, an interferometer setup is the method of choice (see Chapter 7). Time-dependent schemes where a trapped cloud is split and recombined, have been suggested [88–90] and realized in [73, 74] to probe the intrinsic phase dynamics of a BEC. These experiments probe the phase coherence between two parts of a condensate via the interference pattern formed when the two start to overlap in free fall. Recent experiments have been reported for elongated clouds that have been split along their axial direction by Bragg diffraction [91] and by a magnetic grating made from a wire array [92]. Splittings in the radial direction have been reported with an optically created double well [93] and with a two-wire scheme on a chip [94, 95]. The relative phase can be locked when tunneling through the barrier is still possible. Still, quantum and thermal fluctuations of the atom number difference between the two parts let the relative phase diffuse [73, 96–98]. If the barrier height exceeds the chemical potential of the condensate, a rapid phase randomization has been observed [95]. At the time of writing, we are not aware of experiments that have detected surface-induced decoherence in on-chip atom interferometry.

5.4 Perspectives

Recent work on miniaturized atom traps has improved our understanding of atom–surface interactions in a regime where, compared to the chip geometry, optical transition wavelengths are "small" or "relevant", but hyperfine or vibrational wavelengths are "large". The fact that cold atoms can be maintained at distances of a few micrometers from a macroscopic, hot surface, highlights the weakness of the atom–surface coupling via the electromagnetic field. At the same time, these interactions do determine limits for miniaturization and coherent manipulation. Strategies to circumvent these limits exist, and first experimental demonstrations have been reported, for example an integrated atom clock with "decoherence-free" hyperfine states [81] (Chapter 8).

Experiments have probed atom–surface interactions by measuring the motion of ultra-cold samples or transitions between atomic sublevels in the trapping potentials. Both, the atom–surface potential (electric dipole coupling) and trap loss due to thermally excited magnetic near-fields are now quantitatively understood. Ultracold matter wave dynamics thus has emerged as a sensitive tool to probe the surface interaction. More work is required to achieve a quantitative understanding in more complex geometries (wires of finite cross-section, chips with different materials, anharmonic traps). But we can say that the precise control over the center of mass motion afforded by atom chips has opened new routes for atomic spectroscopy.

The next steps in atom chip development are likely to involve optimized materials and geometries for magnetic noise reduction. This should enable in the near future the observation of surface interactions using the interference of matter waves. The impact of atom–atom interactions has already been studied, and we expect to see precision measurements of surface potentials, of surface-induced decoherence and dephasing.

On the theoretical side, much work has been invested into detailed models for atom chip materials and geometries, including absorption, dispersion, and finite temperature. The unusual regimes of cavity quantum electrodynamics provided by atom chips have been identified and put to use in efficient calculations. The relevant scaling laws have been spelled out and are being used for future atom chip designs. Versatile computational schemes for magnetic near-field noise calculations in three dimensions will be developed in the near future. The complexity of predicting the van der Waals–Casimir–Polder potential in a generic geometry is likely to become manageable using clever approximation schemes and tools from computational electrodynamics. (For an example dealing with the macroscopic Casimir interaction, see [99].) We also anticipate an improved understanding of mesoscopic physics by studying condensate dynamics in small-scale structures near surfaces.

Other challenges that are currently being addressed are electromagnetic noise spectra near superconducting atom chips (see Chapters 4 and 10), and optimized materials based on an anisotropic conductivity [100] or based on artificial nanostructures (metamaterials) [101]. Finally, one may contemplate self-contained traps that do not need any electromagnetic trapping fields because the confinement of ultra-cold matter waves is achieved by quantum reflection [102].

References

1 Johnson, W.R., Dzuba, V.A., Safronova, U.I., and Safronova, M.S. (2004) Finite-field evaluation of the Lennard-Jones atom-wall interaction constant $C3$ for alkali-metal atoms. *Phys. Rev. A* **69**, 022508.

2 Drexhage, K.H. (1974) Interaction of light with monomolecular dye layers, in (ed. E. Wolf), *Progress in Optics XII*, North-Holland, Amsterdam, pp. 163–232.

3 Chance, R.R., Prock, A., and Silbey, R. (1978) Molecular fluorescence and energy transfer near interfaces, in (eds I. Prigogine and S.A. Rice), *Advances in Chemical Physics XXXVII*, Wiley & Sons, New York, pp. 1–65.

4 Agarwal, G.S. (1975) Quantum electrodynamics in the presence of dielectrics and conductors. I. Electromagnetic-field response functions and black-body fluctuations in finite geometries. *Phys. Rev. A* **11**, 230.

5 Henkel, C., Pötting, S., and Wilkens, M. (1999) Loss and heating of particles in small and noisy traps. *Appl. Phys. B* **69**, 379.

6 Callen, H.B. and Welton, T.A. (1951) Irreversibility and generalized noise. *Phys. Rev.* **83**, 34.

7 Wylie, J.M. and Sipe, J.E. (1984) Quantum electrodynamics near an interface. *Phys. Rev. A* **30**, 1185.

8 Scheel, S. and Buhmann, S.Y. (2008) Macroscopic Quantum Electrodynamics – Concepts and Applications. *Acta Phys. Slov.* **58**, 675.

9 Caride, A.O., Klimchitskaya, G.L., Mostepanenko, V.M., and Zanette, S.I. (2005) Dependences of the van der Waals atom-wall interaction on atomic and material properties. *Phys. Rev. A* **71**, 042901.

10 Wylie, J.M. and Sipe, J.E. (1985) Quantum electrodynamics near an interface. II. *Phys. Rev. A* **32**, 2030.

11 Hinds, E.A. and Sandoghdar, V. (1991) Cavity QED level shifts of simple atoms. *Phys. Rev. A* **43**, 398.

12 Gorza, M.-P., Saltiel, S., Failache, H., and Ducloy, M. (2001) Quantum theory of van der Waals interactions between excited atoms and birefringent dielectric surfaces. *Eur. Phys. J. D* **15**, 113.

13 Mendes, T.N.C. and Farina, C. (2007) Atom–wall dispersive forces from the master equation formalism. *J. Phys. A: Math. Gen.* **40**, 7343.

14 Antezza, M., Pitaevskii, L.P., and Stringari, S. (2004) Effect of the Casimir–Polder force on the collective oscillations of a trapped Bose–Einstein condensate. *Phys. Rev. A* **70**, 053619.

15 Parsegian, V.A. (2006) *Van der Waals Forces – A Handbook for Biologists, Chemists, Engineers, and Physicists*, Cambridge University Press, New York.

16 Harber, D.M., Obrecht, J.M., McGuirk, J.M., and Cornell, E.A. (2005) Measurement of the Casimir–Polder force through center-of-mass oscillations of a Bose–Einstein condensate. *Phys. Rev. A* **72**, 033610.

17 Buhmann, S.Y. and Scheel, S. (2008) Thermal Casimir vs Casimir–

Polder forces: Equilibrium and nonequilibrium forces. *Phys. Rev. Lett.* **100**, 253201.

18 Rytov, S.M., Kravtsov, Y.A., and Tatarskii, V.I. (1989) Elements of Random Fields. *Principles of Statistical Radiophysics*, vol. 3, Springer, Berlin.

19 Lifshitz, E.M. and Pitaevskii, L.P. (1980) Statistical Physics (Part 2). *Landau and Lifshitz, Course of Theoretical Physics*, vol. 9, Pergamon, Oxford, 2nd edn..

20 Henry, C.H. and Kazarinov, R.F. (1996) Quantum noise in photonics. *Rev. Mod. Phys.* **68**, 801.

21 Buhmann, S.Y., Dung, H.T., Knöll, L., and Welsch, D.-G. (2004) Casimir–Polder forces: A non-perturbative approach. *Phys. Rev. A* **70**, 052117.

22 Henkel, C., Joulain, K., Mulet, J.-P., and Greffet, J.-J. (2002) Radiation forces on small particles in thermal near fields. *J. Opt. A: Pure Appl. Opt.* **4**, S109.

23 Antezza, M., Pitaevskii, L.P., and Stringari, S. (2005) New asymptotic behaviour of the surface-atom force out of thermal equilibrium. *Phys. Rev. Lett.* **95**, 113202.

24 Buhmann, S.Y. and Welsch, D.-G. (2007) Dispersion forces in macroscopic quantum electrodynamics. *Progr. Quantum Electron.* **31**, 51.

25 Obrecht, J.M., Wild, R.J., Antezza, M., Pitaevskii, L.P., Stringari, S., and Cornell, E.A. (2007) Measurement of the temperature dependence of the Casimir–Polder force. *Phys. Rev. Lett.* **98**, 063201.

26 Landragin, A., Courtois, J.-Y., Labeyrie, G., Vansteenkiste, N., Westbrook, C.I., and Aspect, A. (1996) Measurement of the van der Waals force in an atomic mirror. *Phys. Rev. Lett.* **77**, 1464.

27 Lin, Y.-J., Teper, I., Chin, C., and Vuletić, V. (2004) Impact of Casimir–Polder potential and Johnson noise on Bose–Einstein condensate stability near surfaces. *Phys. Rev. Lett.* **92**, 050404.

28 Berkhout, J.J., Luiten, O.J., Setija, I.D., Hijmans, T.W., Mizusaki, T., and Walraven, J.T.M. (1989) Quantum reflection: Focusing of hydrogen atoms with a concave mirror. *Phys. Rev. Lett.* **63**, 1689.

29 Berry, M.V. and Mount, K.E. (1972) Semiclassical approximations in wave mechanics. *Rep. Prog. Phys.* **35**, 315.

30 Maitra, N.T. and Heller, E.J. (1996) Semiclassical perturbation approach to quantum reflection. *Phys. Rev. A* **54**, 4763.

31 Segev, B., Côté, R., and Raizen, M.G. (1997) Quantum reflection from an atomic mirror. *Phys. Rev. A* **56**, R3350.

32 Friedrich, H., Jacoby, G., and Meister, C.G. (2002) Quantum reflection by Casimir–van der Waals potential tails. *Phys. Rev. A* **65**, 032902.

33 Shimizu, F. (2001) Specular reflection of very slow metastable Neon atoms from a solid surface. *Phys. Rev. Lett.* **86**, 987.

34 Shimizu, F. and Fujita, J.-I. (2002) Giant quantum reflection of Neon atoms from a ridged silicon surface. *J. Phys. Soc. Japan* **71**, 5.

35 Oberst, H., Kouznetsov, D., Shimizu, K., Fujita, J.-I., and Shimizu, F. (2005) Fresnel Diffraction mirror for an atomic wave. *Phys. Rev. Lett.* **94**, 013203.

36 Kouznetsov, D. and Oberst, H. (2005) Scattering of atomic matter waves from ridged surfaces. *Phys. Rev. A* **72**, 013617.

37 Pasquini, T.A., Shin, Y.-I., Sanner, C., Saba, M., Schirotzek, A., Pritchard, D.E., and Ketterle, W. (2004) Quantum reflection of atoms from a solid surface at normal incidence. *Phys. Rev. Lett.* **93**, 223201.

38 Pasquini, T.A., Saba, M., Jo, G.-B., Shin, Y., Ketterle, W., Pritchard, D.E., Savas, T.A., and Mulders, N. (2006) Low velocity quantum reflection of Bose–Einstein condensates. *Phys. Rev. Lett.* **97**, 093201.

39 Druzhinina, V. and DeKieviet, M. (2003) Experimental observation of quantum reflection far from threshold. *Phys. Rev. Lett.* **91**, 193202.

40 Kallush, S., Segev, B., and Côté, R. (2005) Manipulating atoms and molecules with evanescent-wave mirrors. *Eur. Phys. J. D* **35**, 3.

41 Langbein, D. (1974) Theory of Van der Waals attraction. *Springer Tracts in Modern Physics*, vol. 72, Springer, Berlin.

42 Moreno, G.A., Dalvit, D.A.R., and Calzetta, E. (2010) Bragg spectroscopy for measuring Casimir–Polder interac-

tions with Bose–Einstein condensates above corrugated surfaces *New J. Phys.* **12**, 033009.

43 Bondarev, I.V. and Lambin, P. (2004) van der Waals energy under strong atom-field coupling in doped carbon nanotubes. *Solid State Commun.* **132**, 203.

44 Peano, V., Thorwart, M., Kasper, A., and Egger, R. (2005) Nanoscale atomic waveguides with suspended carbon nanotubes. *Appl. Phys. B* **81**, 1075.

45 Blagov, E.V., Klimchitskaya, G.L., and Mostepanenko, V.M. (2005) Van der Waals interaction between microparticle and uniaxial crystal with application to hydrogen atoms and multiwall carbon nanotubes. *Phys. Rev. B* **71**, 235401.

46 Fermani, R., Scheel, S., and Knight, P.L. (2007) Trapping cold atoms near carbon nanotubes: Thermal spin flips and Casimir–Polder potential. *Phys. Rev. A* **75**, 062905.

47 Petrov, P.G., Machluf, S., Younis, S., Macaluso, R., David, T., Hadad, B., Japha, Y., Keil, M., Joselevich, E., and Folman, R. (2009) Trapping cold atoms using surface-grown carbon nanotubes. *Phys. Rev. A* **79**, 043403.

48 Henkel, C., Krüger, P., Folman, R., and Schmiedmayer, J. (2003) Fundamental limits for coherent manipulation on atom chips. *Appl. Phys. B* **76**, 173.

49 Barnes, W.L. (1998) Fluorescence near interfaces: the role of photonic mode density. *J. mod. Opt.* **45**, 661.

50 Chicanne, C., David, T., Quidant, R., Weeber, J.C., Lacroute, Y., Bourillot, E., Dereux, A., C. des Francs, G., and Girard, C. (2002) Imaging the local density of states of optical corrals. *Phys. Rev. Lett.* **88**, 097402.

51 Colas des Francs, G., Girard, C., and Dereux, A. (2002) Theory of near-field optical imaging with a single molecule as a light source. *J. Chem. Phys.* **117**, 4659.

52 Joulain, K., Carminati, R., Mulet, J.-P. and Greffet, J.-J. (2003) Definition and measurement of the local density of electromagnetic states close to an interface. *Phys. Rev. B* **68**, 245405.

53 Savasta, S., Stefano, O.D., Pieruccini, M., and Girlanda, R. (2004) Comment on 'Imaging the local density of states of optical corrals'. *Phys. Rev. Lett.* **93**, 069701.

54 Henkel, C. and Sandoghdar, V. (1998) Single-molecule spectroscopy near structured dielectrics. *Opt. Commun.* **158**, 250.

55 Parent, G., Labeke, D.V., and Barchiesi, D. (1999) Fluorescence lifetime of a molecule near a corrugated interface: Application to near-field microscopy. *J. Opt. Soc. Am. A* **16**, 896.

56 Ivanov, V.V., Cornelussen, R.A., van Linden van den Heuvell, H.B., and Spreeuw, R.J.C. (2004) Observation of modified radiative properties of cold atoms in vacuum near a dielectric surface. *J. Opt. B: Quantum Semiclass. Opt.* **6**, 454.

57 Purcell, E.M. (1946) Spontaneous emission probabilities at radio frequencies. *Phys. Rev.* **69**, 681.

58 Fortágh, J., Ott, H., Kraft, S., and Zimmermann, C. (2002) Surface effects on a Bose–Einstein condensate in a magnetic microtrap. *Phys. Rev. A* **66**, 041604(R).

59 Jones, M.P.A., Vale, C.J., Sahagun, D., Hall, B.V., and Hinds, E.A. (2003) Spin coupling between cold atoms and the thermal fluctuations of a metal surface. *Phys. Rev. Lett.* **91**, 080401.

60 Harber, D.M., McGuirk, J.M., Obrecht, J.M., and Cornell, E.A. (2003) Thermally induced losses in ultra-cold atoms magnetically trapped near room-temperature surfaces. *J. Low Temp. Phys.* **133**, 229.

61 Zhang, B., Henkel, C., Haller, E., Wildermuth, S., Hofferberth, S., Krüger, P., and Schmiedmayer, J. (2005) Relevance of sub-surface chip layers for the lifetime of magnetically trapped atoms. *Eur. Phys. J. D* **35**, 97.

62 Sidles, J.A., Garbini, J.L., Dougherty, W.M., and Chao, S.-H. (2003) The classical and quantum theory of thermal magnetic noise, with applications in spintronics and quantum microscopy. *Proc. IEEE* **91**, 799.

63 Scheel, S., Rekdal, P.-K. Knight, P.L., and Hinds, E.A. (2005) Atomic spin decoherence near conducting and su-

perconducting films. *Phys. Rev. A* **72**, 042901.

64 Dikovsky, V., Japha, Y., Henkel, C., and Folman, R. (2005) Reduction of magnetic noise in atom chips by material optimization. *Eur. Phys. J. D* **35**, 87.

65 Groth, S., Krüger, P., Wildermuth, S., Folman, R., Fernholz, T., Schmiedmayer, J., Mahalu, D., and Bar-Joseph, I. (2004) Atom chips: Fabrication and thermal properties. *Appl. Phys. Lett.* **85**, 2980.

66 Henkel, C. and Pötting, S. (2001) Coherent transport of matter waves. *Appl. Phys. B* **72**, 73.

67 Zhang, B. and Henkel, C. (2007) Magnetic noise around metallic microstructures. *J. Appl. Phys.* **102**, 084907.

68 Rekdal, P.-K., Scheel, S., Knight, P.L., and Hinds, E.A. (2004) Thermal spin flips in atom chips. *Phys. Rev. A* **70**, 013811.

69 Savard, T.A., O'Hara, K.M., and Thomas, J.E. (1997) Laser-noise-induced heating in far-off resonance optical traps. *Phys. Rev. A* **56**, R1095.

70 Gehm, M.E., O'Hara, K.M., Savard, T.A., and Thomas, J.E. (1998) Dynamics of noise-induced heating in atom traps. *Phys. Rev. A* **58**, 3914.

71 Hänsel, W., Hommelhoff, P., Hänsch, T.W., and Reichel, J. (2001) Bose–Einstein condensation on a microelectronic chip. *Nature* **413**, 498.

72 Japha, Y. and Band, Y.B. (2002) Motion of a condensate in a shaken and vibrating harmonic trap. *J. Phys. B: Atom. Mol. Opt. Phys.* **35**, 2383.

73 Jo, G.-B., Choi, J.-H. Christensen, C.A., Lee, Y.-R., Pasquini, T.A., Ketterle, W., and Pritchard, D.E. (2007) Matter–Wave interferometry with phase fluctuating Bose–Einstein condensates. *Phys. Rev. Lett.* **99**, 240406.

74 Hofferberth, S., Lesanovsky, I., Fischer, B., Schumm, T., and Schmiedmayer, J. (2007) Non-equilibrium coherence dynamics in one-dimensional Bose gases. *Nature* **449**, 324.

75 Henkel, C., Joulain, K., Carminati, R., and Greffet, J.-J. (2000) Spatial coherence of thermal near fields. *Opt. Commun.* **186**, 57.

76 Stern, A., Aharonov, Y., and Imry, Y. (1990) Phase uncertainty and loss of interference: A general picture. *Phys. Rev. A* **41**, 3436.

77 Zurek, W.H. (1991, August) Decoherence and the transition from quantum to classical. *Phys. Today* **44**, 36.

78 Unruh, W.G. (1995) Maintaining coherence in quantum computers. *Phys. Rev. A* **51**, 992.

79 Folman, R., Krüger, P., Schmiedmayer, J., Denschlag, J.H., and Henkel, C. (2002) Microscopic atom optics: from wires to an atom chip. *Adv. At. Mol. Opt. Phys.* **48**, 263.

80 Schroll, C., Belzig, W., and Bruder, C. (2003) Decoherence of cold atomic gases in magnetic microtraps. *Phys. Rev. A* **68**, 043618.

81 Treutlein, P., Hommelhoff, P., Steinmetz, T., Hänsch, T.W., and Reichel, J. (2004) Coherence in microchip traps. *Phys. Rev. Lett.* **92**, 203005.

82 Shnirman, A., Makhlin, Y., and Schön, G. (2002) Noise and decoherence in quantum two-level systems. *Phys. Scr.* **T102**, 147.

83 Goodman, J.W. (2000) *Statistical Optics*, Wiley-Interscience, New York.

84 Jayannavar, A.M. and Kumar, N. (1982) Nondiffusive quantum transport in a dynamically disordered medium. *Phys. Rev. Lett.* **48**, 553.

85 Kuklov, A.B. and Birman, J.L. (2000) Implications of SU(2) symmetry on the dynamics of population difference in the two-component atomic vapor. *Phys. Rev. Lett.* **85**, 5488.

86 Kuklov, A.B., Chensincki, N., and Birman, J.L. (2002) Symmetries and decoherence of two-component confined atomic clouds: study of the atomic echo in the two-component Bose–Einstein condensate. *Laser Phys.* **12**, 127.

87 Henkel, C. and Gardiner, S.A. (2004) Decoherence of Bose–Einstein condensates in microtraps. *Phys. Rev. A* **69**, 043602.

88 Cahn, S.B., Kumarakrishnan, A., Shim, U., Sleator, T., Berman, P.R., and Dubetsky, B. (1997) Time-domain de Broglie wave interferometry. *Phys. Rev. Lett.* **79**, 784.

89 Hinds, E.A., Vale, C.J., and Boshier, M.G. (2001) Two-wire waveguide and interferometer for cold atoms. *Phys. Rev. Lett.* **86**, 1462.

90 Hänsel, W., Reichel, J., Hommelhoff, P., and Hänsch, T.W. (2001) Trapped-atom interferometer in a magnetic microtrap. *Phys. Rev. A* **64**, 063607.

91 Wang, Y.-J., Anderson, D.Z., Bright, V.M., Cornell, E.A., Diot, Q., Kishimoto, T., Prentiss, M., Saravanan, R.A., Segal, S.R., and Wu, S. (2005) Atom Michelson interferometer on a chip using a Bose–Einstein condensate. *Phys. Rev. Lett.* **94**, 090405.

92 Günther, A., Kraft, S., Kemmler, M., Koelle, D., Kleiner, R., Zimmermann, C., and Fortágh, J. (2005) Diffraction of a Bose–Einstein condensate from a magnetic lattice on a micro chip. *Phys. Rev. Lett.* **95**, 170405.

93 Shin, Y., Saba, M., A. Pasquini, T., Ketterle, W., Pritchard, D.E., and Leanhardt, A.E. (2004) Atom interferometry with Bose–Einstein condensates in a double-well potential. *Phys. Rev. Lett.* **92**, 050405.

94 Schumm, T., Hofferberth, S., Andersson, L.M., Wildermuth, S., Groth, S., Bar-Joseph, I., Schmiedmayer, J., and Krüger, P. (2005) Matter–wave interferometry in a double well on an atom chip. *Nature Physics* **1**, 57.

95 Shin, Y., Sanner, C., Jo, G.-B., Pasquini, T.A., Saba, M., Ketterle, W., Pritchard, D.E., Vengalattore, M., and Prentiss, M. (2005) Interference of Bose–Einstein condensates on an atom chip. *Phys. Rev. A* **72**, 021604(R).

96 Javanainen, J. and Wilkens, M. (1997) Phase and phase diffusion of a split Bose–Einstein condensate. *Phys. Rev. Lett.* **78**, 4675.

97 Pezzé, L., Smerzi, A., Berman, G.P., Bishop, A.R., and Collins, L.A. (2005) Dephasing and breakdown of adiabaticity in the splitting of Bose–Einstein condensates. *New J. Phys.* **7**, 85.

98 Gati, R., Hemmerling, B., Fölling, J., Albiez, M., and Oberthaler, M.K. (2006) Noise thermometry with two weakly coupled Bose–Einstein condensates. *Phys. Rev. Lett.* **96**, 130404.

99 Rodriguez, A., Ibanescu, M., Iannuzzi, D., Capasso, F., Joannopoulos, J.D., and Johnson, S.G. (2007) Computation and Visualization of Casimir Forces in Arbitrary Geometries: Nonmonotonic lateral-wall forces and the failure of proximity-force approximations. *Phys. Rev. Lett.* **99**, 080401.

100 David, T., Japha, Y., Dikovsky, V., Salem, R., Henkel, C., and Folman, R. (2008) Magnetic interactions of cold atoms with anisotropic conductors. *Eur. Phys. J. D* **48**, 321.

101 Sarychev, A.K. and Shalaev, V.M. (2007) *Electrodynamics of Metamaterials*, World Scientific, Singapore.

102 Jurisch, A. and Friedrich, H. (2006) Realistic model for a quantum reflection trap. *Phys. Lett. A* **349**, 230.

Part Three   Coherence on Atom Chips

6
Diffraction and Interference of a Bose–Einstein Condensate Scattered from an Atom Chip-Based Magnetic Lattice

A. Günther, T.E. Judd, J. Fortágh, and C. Zimmermann

6.1
Introduction

Atomic Bose–Einstein condensates (BECs) are intensely discussed because of their fascinating prospects for fundamental many-body physics. At the same time, the condensates we generate in the laboratory are different from typical systems studied in condensed matter physics. Atomic condensates are tiny clouds of only several hundred thousand atoms. They are subject to gravity and inertial forces, can be mechanically manipulated with magnetic and optical fields, and they can be photographed and even filmed. Just as other macroscopic quantum systems, atomic condensates exhibit a well-defined quantum mechanical phase. This gives rise to interference phenomena and a number of interesting applications. The quantum mechanical properties of a condensate become particularly interesting if it is brought in contact with other fundamental physical systems such as light in cavities, surfaces, current conductors or even macroscopic quantum systems such as superconducting circuits or liquid helium. During the last few years there have been a number of pioneering experiments with condensates in the vicinity of surfaces. Van der Waals forces have been studied this way as well as static and dynamic magnetic fields generated by inhomogeneous currents inside the conductor (for a review see [1]). More recently, condensates near superconductors [2–4] and silicon cantilevers [5] have also been studied. It is too early to speculate about the future of such hybrid systems but the experimental synthesis of atomic and solid-state systems is fascinating enough to be explored in detail.

Many fundamental properties of condensates in microtraps depend on interatomic interactions. In structures smaller than the condensate's healing length, the atoms tend to behave as independent particles obeying Schrödinger's equation. If, however, the potential structures reach the micrometer scale the condensate behaves as a macroscopic wave function and the atom–atom interaction gives rise to an extra mean-field term proportional to the atomic density. This non-linear Schrödinger equation includes a wealth of new physics, but it is difficult to solve even numerically, especially for fully 3D problems without any symmetry to reduce

Atom Chips. Edited by Jakob Reichel and Vladan Vuletić
Copyright © 2011 WILEY-VCH Verlag GmbH & Co. KGaA, Weinheim
ISBN: 978-3-527-40755-2

the complexity of the problem. Nonetheless, many interesting scenarios are of that kind. This is especially true for hybrid systems.

In this chapter we describe the particular example of a condensate reflected from a surface that carries a micro-fabricated magnetic lattice [6, 7]. The lattice diffracts the condensate and splits it into a number of momentum classes. The experiment is elementary because it realizes the matter wave analog of an optical reflecting diffraction grating. These are extensively used, for example, in optical spectrometers. Such an atom interferometer is even sensitive to the exact position of the condensate relative to the micro-fabricated lattice. Thus, atomic matter wave optics of this kind require precise control over the position and the velocity of the atoms. This can be achieved with micron-size current conductors on a chip. They generate a suitable magnetic field structure and with the resulting mechanical potential the atoms can be moved over several millimeters and positioned with a reproducibility on a length scale of tens of nanometers [8]. This remarkable stability has been crucial in almost all atom chip experiments and may eventually turn out to be one of the most important advantages of chip-based atom traps. Here, we use a chip trap to precisely adjust the velocity of the condensate at which it hits the magnetic lattice. Controlled diffraction takes place only in a small velocity range between 5.7 and 7.2 mm s^{-1}. With a lattice constant of $a = 4$ μm the chemical potential of the atom–atom interaction energy exceeds the recoil energy of the lattice ($E_r = 2\pi^2\hbar^2/ma^2$) by far, thus mean-field effects are important. This may appear as an extra unwanted complication, but it turns out that in some regimes diffraction can be described by the very simple model of phase imprinting.

We begin this chapter with a technical description of the experimental apparatus and the magnetic grating for diffraction experiments. Special emphasis is given to the magnetic field configuration near meandering conductors and to the wave mechanics of diffraction. We describe experimental data of BEC diffraction and interference and analyze them with an analytic phase-imprinting model. A full three-dimensional simulation of the Gross–Pitaevskii equation [9] is used to verify the regimes in which this model is valid.

6.2
Experimental Setup

6.2.1
The BEC Apparatus

Ultra-cold atoms are prepared using a compact experimental setup, with all field-generating elements like coils, wires and microconductors placed within a vacuum chamber at a pressure of 10^{-11} mbar (Figure 6.1). Two magnetic coils with a vertical distance of 42 mm are driven in anti-Helmholtz configuration and generate the magnetic field for the operation of the magneto-optical trap (MOT) at the center between the two coils. The MOT is operated with three pairs of counter-propagating laser beams, one in each spatial direction, with the laser wavelength

Figure 6.1 Cold atom setup placed inside the vacuum chamber at a pressure of 10^{-11} mbar. Rubidium atoms are emitted from dispensers and loaded into a magneto-optical trap (MOT) operated by the MOT-coils and six laser beams. The transfer coils and the Ioffe wire allow for a subsequent transport of the ultracold atoms to the atom chip surface.

red detuned to the rubidium cooling transition $5S_{1/2}, F = 2 \rightarrow 5P_{3/2}, F = 3$. Rubidium dispenser sources are mounted directly next to the MOT-coils and generate thermal atoms which are collected by the MOT. The dispensers are heated at a current of 6 A for several seconds [10]. Two shielding wires in front of the dispensers prevent direct impacts between the dispenser-emitted atoms and the MOT atoms. The MOT is then loaded by the rubidium background pressure. This shielding efficiently reduces the MOT decay channel due to non-Rubidium atoms emitted by the dispensers. A total number of about 3×10^8 atoms can be collected in the MOT at a temperature of about 100 µK. After additional molasses cooling to 30 µK and optical pumping to the spin-polarized $F = 2, m_F = 2$ state, the atoms are transferred to a spherical quadrupole trap generated by the MOT-coils. A second pair of magnetic coils – the transfer coils (see left side of Figure 6.1) – is used to shift the atoms in the horizontal plane towards the atomic microchip. Once the atoms are trapped between the transfer coils, the activation of the Ioffe-wire (see Figure 6.1) transforms the spherical magnetic quadrupole trap to a parabolic Ioffe-type trap[5]. This parabolic trap is characterized by axial and radial trap frequencies of $\omega_z = 2\pi \times 14$ Hz and $\omega_r = 2\pi \times 120$ Hz. In this trap the atoms are evaporatively cooled within 20 s to a temperature of about 5 µK with 2×10^7 remaining atoms. By lowering the current in the upper transfer coil the atoms are shifted towards the carrier chip surface.

Figure 6.2 shows a chip used in our experiments for nano-positioning atoms near fabricated surfaces. We refer to it as the "carrier chip" because other chips can be fixed on its surface. Micron-scaled wires are electroplated on both sides of a 250-µm-thick aluminum oxide substrate. The wires allow not only for the generation of a purely chip-based magnetic trap but also for the precise spatial and momentum

---

5) While a spherical quadrupole trap shows a vanishing magnetic field at the trap center and a radially increasing field amplitude from the center, the Ioffe trap is characterized by a harmonic potential with a non-vanishing magnetic field at the trap center. The finite magnetic field suppresses atom losses at cold temperatures due to Majorana spin-flips [11].

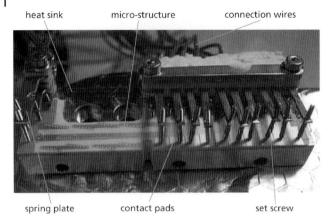

Figure 6.2 Atom chip for nano-positioning atoms near surfaces ("carrier chip"). The chip contains all microwires for the generation and μm-precise positioning of a condensate. Electronic contact to the Kapton-isolated connection wires is made via spring plates, pressing the set screws onto the contact pads. The chip is mounted on a copper heatsink.

control of atom clouds [8]. Electric contact to the microwires is realized via the contact pads, which are linked to the connection wires by spring plates and set screws (see Figure 6.2). The microstructure itself is glued on top of a heatsink, which is mounted upside down at the upper transfer coil (see Figure 6.1). By controlling the currents in the micron wires, the thermal cloud located between the transfer coils can now be transferred to the chip-based magnetic microtrap. At the chip, the cloud can be shifted towards the desired position where it is evaporatively cooled to a Bose–Einstein condensate with up to 5×10^5 atoms [12].

While the carrier chip features all the magnetic fields for trapping and positioning of the ultra-cold atomic ensemble, the coherent manipulation of the condensate is done with additional "atom-optics chips" which are glued on top of the carrier

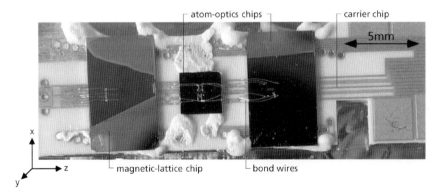

Figure 6.3 The atom-optics chips are fixed on top of the carrier chip and are electronically connected by bonding wires. The nano-scaled wires on the atom-optics chips allow for coherent manipulation of the condensate wave function.

chip (see Figure 6.3). The atom-optics chips carry nano-scaled conductors produced by electron-beam lithography on top of a 250-μm-thick silicon substrate. By using this kind of combined chip, the condensate can be kept far away from the wires of the carrier chip which otherwise would cause problems due to condensate fragmentation [13–16] or thermally induced spin-flips [17, 18]. With the atom-optics chips on top of the carrier chip these problems can be avoided while, at same time, the condensate can be positioned very close (μm-regime) to the nano-scaled wires on the atom-optics chips. As these wires are produced with electron-beam lithography and contain much less material than the carrier chip wires, the expected fragmentation problems and spin-flip losses are minimized [18]. This opens the door for coherent matter wave experiments near the surface of magnetic microchips. Because of the carrier chip geometry the condensates are cigar shaped with the long axis pointing along the z-direction as shown in Figure 6.3. This has to be taken into account for the design of the nano-scaled devices such as the lattice pattern described next.

6.2.2
The Magnetic Lattice Chip

For diffraction and interference experiments, we implement a wire pattern on an atom-optics chip which generates a magnetic lattice potential with a lattice constant of $a = 4$ μm. Figure 6.4 shows a schematic drawing of the micro-conductor geometry. Two nested wires with 1 μm width, carrying a current of I_{M1} and I_{M2} respectively, form a meandering pattern. This creates a set of 372 parallel wires with a length of 100 μm each and a clearance of 1 μm. Choosing the current direction as shown in Figure 6.4 the parallel wire array features a 4 μm periodicity with respect to the current direction. Figure 6.5 shows two microscope images of the fabricated lattice-

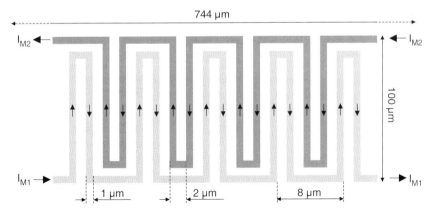

Figure 6.4 Schematic drawing of the nested meander pattern (not to scale). The pattern creates a set of parallel conductors with 100 μm length. Neighboring wires are spaced by 2 μm and have alternating current directions for $I_{M1} = I_{M2}$. This way a 4 μm periodic wire structure is created.

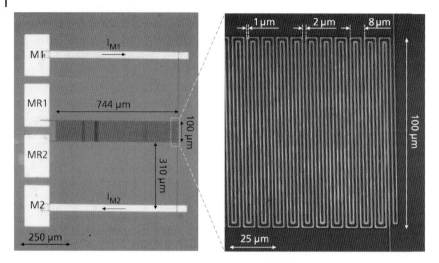

Figure 6.5 Microscope images of the lattice chip featuring the wire pattern for the magnetic lattice potential. Two independent wires are meander shaped and nested into each other. Electrical contact is made via the contact pads M_i and MR_i ($i = 1, 2$).

chip. Each of the meander wires is 1 μm in width, 150 nm in height, and 2 mm in length. The maximum current in the wires of 15 mA corresponds to a current density of 1×10^7 A/cm². The lattice chip is glued with an ultra-high vacuum compatible glue [19] to the surface of the carrier chip (left chip in Figure 6.3). The lattice wires are oriented perpendicular to the condensate's long axis (cf. Figures 6.5 and 6.3).

6.3
The Magnetic Lattice Potential

6.3.1
Infinite Lattice

Keeping the double meander structure in mind, we calculate the magnetic potential generated from a lattice consisting of an infinite number of infinitely long parallel wires with negligible radius (see Figure 6.6). The wires are spaced by a distance of $a/2 = 2$ μm and carry opposite currents in neighboring wires. Each wire is referenced by an index number n. The coordinate system is chosen such that the wire pattern is located in the x–z plane while the wire with $n = 0$ is aligned along the x-axis (see Figure 6.6). The current in the nth wire is given by $I_n = (-1)^n I$ and generates a circular magnetic field [20]:

$$\boldsymbol{B}_n(y, z) = \frac{K}{2\pi} \frac{(-1)^n}{(2y/a)^2 + (2z/a - n)^2} \begin{pmatrix} 0 \\ -(2z/a - n) \\ 2y/a \end{pmatrix} \quad (6.1)$$

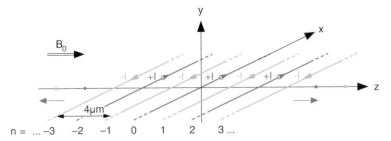

Figure 6.6 Infinite lattice of parallel conductors, separated by $a/2 = 2$ μm. Each wire carries a current I and the current direction alternates between neighboring wires. In addition to the magnetic field from this pattern of wires, there is a homogeneous field B_0 along the z-direction.

with $K := 2\mu_0 I/a$, given by the wire geometry. In our case $K/I = 6.8$ G/mA. The total magnetic field is given by the superposition of all individual lattice-site fields and can be written in terms of trigonometric and hyperbolic functions [21]:

$$B_y(y,z) = \sum_{-\infty}^{+\infty} B_{n,y}(y,z) = -K \frac{\cosh(k_0 y) \cdot \sin(k_0 z)}{\cosh(2k_0 y) - \cos(2k_0 z)} \quad (6.2)$$

$$\xrightarrow{y > a/4} -K e^{-k_0 y} \sin(k_0 z) , \quad (6.3)$$

$$B_z(y,z) = \sum_{-\infty}^{+\infty} B_{n,z}(y,z) = +K \frac{\sinh(k_0 y) \cdot \cos(k_0 z)}{\cosh(2k_0 y) - \cos(2k_0 z)} \quad (6.4)$$

$$\xrightarrow{y > a/4} +K e^{-k_0 y} \cos(k_0 z) . \quad (6.5)$$

Here $k_0 := 2\pi/a$ is the lattice vector corresponding to the lattice period a. The magnetic field is translationally invariant along the x-direction and the field amplitude is given by:

$$B(x,y,z) := |\mathbf{B}(x,y,z)| = K e^{-k_0 y} \left(1 - 2e^{-2k_0 y} \cos(2k_0 z) + e^{-4k_0 y}\right)^{-1/2} . \quad (6.6)$$

Figure 6.7a,b shows the resulting potential of the lattice pattern. The potential is only slightly modulated along the z-axis and drops off exponentially with increasing distance from the patterned surface. This can be seen by expanding Eq. (6.6) in powers of $\exp(-k_0 y)$ up to third order

$$B(x,y,z) = K e^{-k_0 y} \left(1 + e^{-2k_0 y} \cos(2k_0 z) + O(a^4)\right) . \quad (6.7)$$

The magnetic field modulus is thus given by the superposition of an unmodulated part $K \exp(-k_0 y)$ and a periodic modulated part $K \exp(-3k_0 y) \cos(2k_0 z)$ with periodicity[6] of $a/2$ (see Figure 6.7a and b). With increasing distance from the wire

---
6) Although the wire pattern shows a periodicity of 4 μm with respect to the current direction, the field modulus does not depend on the current direction and is thus 2 μm periodic.

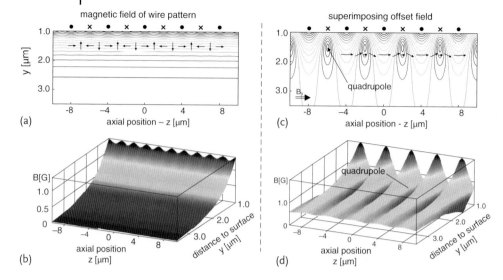

Figure 6.7 (a) Equipotential lines ($\Delta B = 85$ mG) and field modulus (b) of the periodic wire pattern for $I = 1$ mA. For distances $y > a/4$ the field shows almost no modulation along the axial direction of the lattice. The potential is dominated by an exponentially decaying field amplitude depending only on the distance from the wire pattern. Although the magnetic field changes its direction with respect to the axial position z, its amplitude stays constant. Atoms entering the potential from large distances see a magnetic mirror potential rather than a modulated field amplitude. The symbols × and ● show the position of the lattice bars with the corresponding current direction. Superimposing an offset field $B_0 = 0.6$ G in the z-direction changes the equipotential lines (c) and the magnetic field potential (d) drastically. Because of the offset field the total magnetic field is enhanced underneath the ● marked wires and attenuated underneath the × wires. The total field thus shows a 4 µm periodicity along z. Quadrupole waveguides parallel to the lattice bars appear at positions where the magnetic field of the wire pattern is compensated by the offset field. The total potential is now dominated by a periodic part, which looks suitable for diffraction and interference experiments.

pattern the modulated field term decays three-times faster than the unmodulated one. Thus, for distances $y > a/4 = 1$ µm the field is completely dominated by the unmodulated term decaying exponentially on a length scale of $1/k_0 = 0.64$ µm. To an incoming atomic matter wave such a potential acts more like a magnetic mirror rather than a diffraction grating.

The potential shape changes dramatically if the magnetic field from the lattice structure is superimposed with a homogeneous field B_0 along the z-direction (see Figure 6.7c and d). The field modulus now becomes

$$\tilde{B}(x, y, z) = \sqrt{B_y^2 + (B_z + B_0)^2} = B_0\sqrt{1 + \frac{2B_z}{B_0} + \left(\frac{B}{B_0}\right)^2} \qquad (6.8)$$

$$\approx B_0 + B_z + \frac{B_y^2}{2B_0}, \qquad (6.9)$$

which can be written for $y > a/4$ with Eqs. (6.3) and (6.5) as:

$$\tilde{B} \approx B_0 + Ke^{-k_0 y} \cos(k_0 z) + \frac{K^2}{4B_0} e^{-2k_0 y} (1 - \cos(2k_0 z)). \tag{6.10}$$

The magnetic field is now dominated by the homogeneous offset field B_0 which does not act on the atoms. The modulated part $Ke^{-k_0 y} \cos(k_0 z)$ is now dominant, showing a periodicity[7] of 4 µm. The potential modulation amplitude decays exponentially but now three-times slower than in the case of the infinite lattice without offset field. The non-modulated field part, however, now decays much faster than the modulated one and can thus be neglected. As seen from Figure 6.7 the superposition of the offset field changes the original mirror potential to a magnetic lattice potential suitable for diffraction and interference experiments.

6.3.2
Finite Size Effects

Unlike an infinite lattice, a real lattice is finite and characterized by the number of lattice sites N and their length L. The total width of the pattern is thus given by $2w = (N-1)a/4$ with $a/2$ being the distance between neighboring lattice sites (see Figure 6.8). The magnetic potential of the infinite lattice thus needs to be corrected for finite size effects.

Finite Length of the Wires Although the magnetic field direction of a single wire is independent of its length, the field amplitude B above the center is changed by a

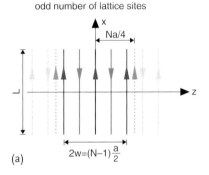
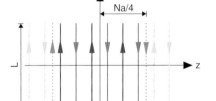

Figure 6.8 Finite lattice of N parallel conductors with length L (solid lines). Near the lattice center, the magnetic field of the finite lattice can be described as the superposition of the infinite lattice field (solid and dashed lines) and two correction wires (dotted lines) at the position $z = \pm Na/4$. The current direction in these additional wires is parallel for odd N (a) and antiparallel for even N (b).

---

7) The offset field B_0 breaks the field symmetry above two neighboring wires and is thus responsible for the 4 µm periodicity of the resulting potential.

factor γ with respect to the field B_∞ of an infinitely long wire:

$$B = \gamma B_\infty \quad \text{with} \quad \gamma := \frac{1}{\sqrt{1 + \left(\frac{2r}{L}\right)^2}}. \quad (6.11)$$

For distances $r \ll L$ the field is independent of the wire length ($\gamma \approx 1$). However, it decays with $1/r^2$ for distances $r \gg L$ compared with $1/r$ for infinitely long wires ($\gamma \sim 1/r$). As the total field of the lattice structure decays exponentially with at least $e^{-k_0 y}$ all experiments need to be carried out at distances $y \approx a$, that is very close to the lattice surface. In this region the field of the nearby wires is not affected by their finite length because $L \gg a$. Furthermore the field contributions from the wires far away are barely changed, because these wires always appear in pairs with opposite current. As the field of such a wire pair decays with $1/r^2$, their field is almost negligible for the atoms and the additional field reduction by a factor $\gamma \sim 1/r$ can be neglected.

Finite Number of Lattice Sites The magnetic field near the center of a finite lattice can be described by a superposition of an infinite lattice field and the field of two additional wires placed a quarter lattice period outside the lattice (see Figure 6.8) [22]. These additional wires at $z = \pm Na/4$ are driven with half the lattice current in a direction given by the outermost wires. Depending on the total number of lattice sites the currents in these additional wires are parallel (N odd) or antiparallel (N even). For an odd number of lattice sites these two wires cause an additional magnetic field gradient at the center of the lattice:

$$\partial_y B = \partial_z B = \frac{\mu_0 I}{2\pi w^2} = \frac{8\mu_0 I}{\pi a^2} N^{-2}. \quad (6.12)$$

For an even number of lattice sites they produce a homogeneous field along y and a field curvature near the lattice center:

$$B_y = \frac{2\mu_0 I}{\pi a} N^{-1}, \quad \partial_y^2 B = \partial_z^2 B = \frac{\mu_0 I}{\pi w^3} = \frac{64 \mu_0 I}{\pi a^3} N^{-3}. \quad (6.13)$$

6.3.3
The Double Meander Potential

Using the results from the last two sections we concentrate on the potential created by the lattice pattern of our double meander pattern. It consists of $N = 372$ parallel wires with lengths of $L = 100$ μm (see Figure 6.5). Neighboring wires are separated by a distance of 2 μm yielding a 4 μm periodic current pattern for $I_{M1} = I_{M2}$. Close to the center of the wire pattern the magnetic field is well approximated by the infinite lattice field. Together with an offset field B_0 the wire pattern generates a suitable diffraction potential. According to Eq. (6.13) the finite size effects of the

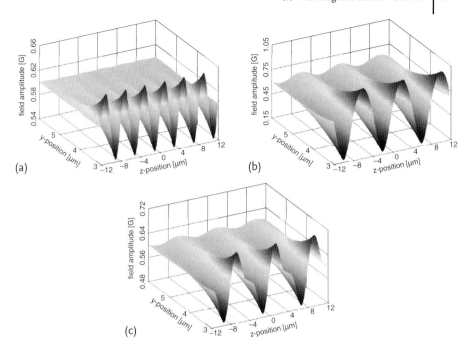

Figure 6.9 Magnetic potential of the double meander structure for $B_0 = 0.6$ G and (a) $I_{M1} = I_{M2} = 1$ mA, (b) $I_{M1} = -I_{M2} = 1$ mA and (c) $I_{M1} = 1$ mA, $I_{M2} = 0.5$ mA.

real structure causes an additional offset field and a field curvature. The offset field $B_y/I = 5.4$ mG/mA can be neglected with respect to the external offset field of typically > 600 mG. As for all diffraction experiments the condensate is placed in a parabolic trap close to the lattice pattern, the additional field curvature $\partial_z^2 B/I = \partial_y^2 B/I = 7.77$ G mA^{-1}/cm^2 causes a small change in the trap frequencies which can also be neglected.

Figure 6.9 shows the numerically simulated field of the double meander structure for a lattice current of 1 mA and an offset field of 0.6 G. Choosing $I_{M1} = I_{M2}$ a 4 µm lattice potential appears close to the wire pattern surface (see Figure 6.9a). For this case the numerical field simulations are in perfect agreement with the analytic formulas from the previous section. For $I_{M1} = -I_{M2}$ the wire pattern and thus the magnetic potential becomes 8 µm periodic with respect to the current direction (see Figure 6.9b). For $I_{M1} \neq I_{M2}$ asymmetric potentials can be created which could be useful for the generation of quantum ratchets (see Figure 6.9c) [23].

The potential of the magnetic grating can be further shaped by applying oscillating magnetic fields on top of the static magnetic field configurations. Such radio-frequency (rf) controlled adiabatic potentials are used for generating high-reflectivity diffraction gratings (see Section 6.6).

6.4
Diffraction and Interference

6.4.1
Diffraction Scheme

Since the magnetic lattice potential decays exponentially with increasing distance from the lattice chip surface, the atoms have to be brought very close to the surface before diffraction and interference occur. The experimental steps used for the lattice experiments are shown in Figure 6.10a. At first a Bose–Einstein condensate is prepared 30 μm underneath the lattice chip in a magnetic trap generated by the carrier chip. At this position the lattice potential is negligible and the harmonic trapping potential is characterized by an axial and radial trap frequency of $\omega_z = 2\pi \times 16$ Hz and $\omega_r = 2\pi \times 76$ Hz respectively. The offset field at the trap center amounts to $B_0 = 0.87$ G and is oriented along the axial direction of the trap. The long axis of the condensate is oriented perpendicular to the lattice wires, that is parallel to the z-direction. The lattice chip produces a periodic potential with an exponentially decaying modulation amplitude along the y-direction. The condensate is now excited to a controlled center-of-mass oscillation inside the parabolic harmonic trap. During this oscillation the condensate approaches the lattice and interacts with the lattice potential for a short time at the upper turning point. After the condensate is reflected from the lattice, all magnetic fields are turned off and the condensate ballistically expands under gravity. During this expansion the internal momentum distribution of the condensate is converted to a spatial distribution revealing diffraction and interference effects.

Figure 6.10b shows a characteristic potential in which the condensate approaches the lattice chip. The potential is created by the superposition of the lattice potential and the magnetic trapping potential with the trap center placed at a distance of 15 μm from the lattice chip surface. For distances $y > 8$ μm the potential structure is dominated by the magnetic trap. At distances $y \sim 8$ μm the periodic potential becomes significant and its modulation amplitude increases rapidly for even shorter distances. An oscillating Bose–Einstein condensate can deeply penetrate the modulated potential region. A controlled oscillation is initialized by instantaneously shifting the magnetic trap center, originally placed at $y_0 = 30$ μm, towards the chip surface. As shown in Figure 6.10c the condensate acquires potential energy during the trap displacement and starts an oscillation towards the chip surface. With increasing displacement d the upper turning point approaches the lattice chip and eventually enters the region where it interacts with the periodic lattice potential. The interaction strength is thus well controlled by changing the initial trap displacement d.

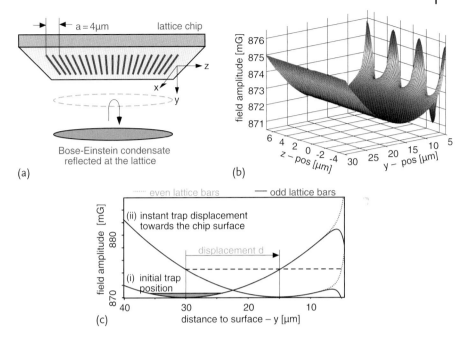

Figure 6.10 (a) Experimental steps to diffract a condensate from the lattice potential. In a controlled oscillation the condensate approaches the lattice and enters the periodic modulated potential region of the lattice chip. After the interaction the BEC moves back to its starting position and is released from the magnetic trap. During the ballistic expansion the condensate's momentum structure is revealed. (b) Superposition of the lattice potential with the harmonic trapping potential centered at a distance of 15 μm from the lattice surface. The trap parameters are chosen as in the experiment. Atoms oscillating in this potential deeply penetrate the modulated potential region of the lattice pattern. (c) Trap potential with superimposed lattice potential in the direction perpendicular to the chip surface. Near the chip surface the potential increases exponentially above every second wire (dotted lines). Above those wires in between, the potential is lowered (solid lines). The difference between the solid and dotted line is the potential modulation. Initially the condensate is placed in a harmonic trap at a distance of 30 μm from the surface (i). Afterwards the trap center is shifted instantaneously by a controlled displacement d towards the chip surface. (ii) shows this for d = 15 μm. The dashed line shows the potential energy of the condensate after displacement and specifies how deeply the condensate penetrates the lattice potential.

6.4.2
Theoretical Model for the Interaction

The interaction of the condensate with the magnetic lattice potential can be described with a simple one-dimensional model. In its center-of-mass frame the condensate interacts with an axially modulated potential $V(z, t)$ which we assume to be constant across the radial extent of the condensate. In the following we neglect the inter-atomic interactions. The influence of the mean-field is described in Section 6.5. Without interactions the problem reduces to solving the one-dimensional time-dependent Schrödinger equation for a 4 µm periodic potential with time-dependent amplitude:

$$i\hbar \partial_t \Psi(z,t) = -\frac{\hbar^2}{2m}\nabla^2 \Psi(z,t) + V(z,t)\Psi(z,t) \tag{6.14}$$

with $\quad V(z,t) = U_0(t)\cos(k_0 z)$. (6.15)

The time-dependent modulation amplitude could be set to $U_0(t) = K\exp(-k_0 Y(t))$ with $k_0 = 2\pi/a$, $a = 4$ µm, and $Y(t)$ being the center-of-mass distance between the condensate and the surface. However, the determination of the exact form of the function $Y(t)$ requires a detailed simulation of the condensate's center-of-mass motion during the interaction with the lattice. Instead we use a simplified ansatz for the time-dependent potential strength $U_0(t)$, acting on the condensate during diffraction

$$U_0(t) = \begin{cases} 0: & t < t_a, t > t_b \\ U: & t_a < t < t_b . \end{cases} \tag{6.16}$$

Here, we assume that the interaction is activated instantaneously at $t = t_a$, kept constant for the interaction time $t_i = t_b - t_a$ and instantaneously deactivated at $t = t_b$ (see Figure 6.11). The interaction is divided into three time intervals. The first interval accounts for the condensate's oscillation towards the chip surface without lattice interaction. During the second interval the condensate interacts with the lattice for a time t_i at a modulation amplitude U. During the third interval the condensate oscillates back in the absence of the lattice potential.

Equation (6.14) can now be solved by independently solving the time-dependent Schrödinger equation for the three intervals I–III and using a sudden approximation to connect the wave functions in the different time intervals. As the potential is time-independent in all three intervals, it is sufficient to solve the time-independent Schrödinger equation for the eigenfunctions Ψ_i and eigenenergies E_i. By expanding the condensate's wave function $\Psi = \sum_i \Psi_i$ the temporal propagation in each time region can easily be determined: $\Psi(t) = \sum_i \Psi_i \exp(-i/\hbar E_i t)$.

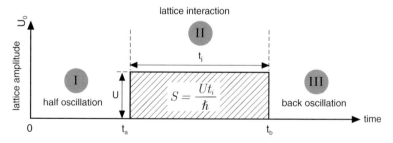

Figure 6.11 The interaction of the condensate with the periodic potential is divided into three regions. In region I the atoms are moving towards the chip surface and do not yet feel an interaction potential. At $t = t_a$ the interaction is turned on instantaneously and kept constant till $t = t_b$ when it is deactivated again.

Region I We calculate the diffraction for atoms initially at rest in the axial direction. The initial wave function is thus given by a plane wave function with momentum zero:[8]

$$|\Psi(0)\rangle = |0\rangle. \tag{6.17}$$

As this function is an eigenfunction of the Schrödinger equation in region I with the eigenenergy $E_0 = 0$, the wave function at $t = t_a$ is given by

$$|\Psi(t_a)\rangle = \exp\left(-\frac{i}{\hbar} E_0 t_a\right) = |0\rangle. \tag{6.18}$$

Region II The eigenfunctions in region II are no longer pure momentum states, but rather superpositions of plane waves with wave vectors differing by an integer multiple of the lattice vector k_0. For every $k \in [-k_0/2, k_0/2]$ the eigenfunctions of the time-independent Schrödinger equation can be written as Bloch functions [24]:

$$\left|\Psi_k^b\right\rangle = \sum_{n=-\infty}^{\infty} C_b(k - nk_0) |k - nk_0\rangle, \quad b \in \mathbb{Z} \tag{6.19}$$

with the band index b. The coefficients $C_b(k - nk_0)$ and the eigenenergies $E_b(k)$ of the eigenstates $|\Psi_k^b\rangle$ depend on the modulation amplitude U and can easily be calculated [24].

At $t = t_a$ the periodic potential is activated instantaneously. The wave function $|\Psi(t_a)\rangle$ is thus projected onto the new eigenstates $|\Psi_k^b\rangle$:

$$|\Psi(t_a)\rangle = \underbrace{\sum_{k,b} \left|\Psi_k^b\right\rangle\left\langle\Psi_k^b\big|0\right\rangle}_{1} = \sum_b \left|\Psi_0^b\right\rangle\underbrace{\left\langle\Psi_0^b\big|0\right\rangle}_{C_b^*(0)} = \sum_b C_b^*(0) \left|\Psi_0^b\right\rangle.$$

$$\tag{6.20}$$

The time evolution of these eigenstates $|\Psi_0^b\rangle$ is given by the energy E_0^b, and the atomic wave function at the end of region II becomes:

$$|\Psi(t_b)\rangle = \sum_b C_b^*(0) e^{-\frac{i}{\hbar} E_0^b t_i} \left|\Psi_0^b\right\rangle. \tag{6.21}$$

Region III At $t = t_b$ the periodic potential is deactivated suddenly and the wave function $|\Psi(t_b)\rangle$ is projected back to the momentum states $|k\rangle$ (eigenstates in region III). Therefore, we expand $|\Psi(t_b)\rangle$ into momentum eigenfunctions by replacing the state $\left|\Psi_0^b\right\rangle$ in Eq. (6.21) by its Fourier expansion (Eq. (6.19))

$$|\Psi(t_b)\rangle = \sum_b C_b^*(0) e^{-\frac{i}{\hbar} E_0^b t_i} \sum_n C_b(0 - nk_0) |-nk_0\rangle \tag{6.22}$$

$$= \sum_n \underbrace{\sum_b C_b^*(0) C_b(0 - nk_0) e^{-\frac{i}{\hbar} E_0^b t_i}}_{A_{-n}(U, t_i)} |-nk_0\rangle \tag{6.23}$$

---

8) The state $|k\rangle$ denotes plane waves with momentum $\hbar k$: $\langle x | k \rangle \sim \exp(ikx)$.

$$= \sum_n A_n(U, t_i) |nk_0\rangle. \tag{6.24}$$

Thus, the condensate's wave function after the interaction with the lattice is given by a superposition of plane waves with discrete wavevectors $k_n = nk_0, n \in \mathbb{Z}$. During the back oscillation of the condensate not only does this momentum distribution stay unchanged, but also the occupation probability for each of these wave vectors k_n:

$$p_n = |A_n(U, t_i)|^2 = \left| \sum_b C_b^*(0) C_b(0 + nk_0) e^{-\frac{i}{\hbar} E_0^b t_i} \right|^2. \tag{6.25}$$

The interaction with the lattice thus causes a formation of discrete momentum classes (diffraction orders) inside the condensate. During the subsequent ballistic expansion this momentum distribution is converted to a spatial distribution usually known as a diffraction pattern.

To characterize the occupation of the different diffraction orders for various interaction times t_i and modulation amplitudes U, we define a parameter $S := U \times t_i / \hbar$, which, besides a factor $\hbar$, is the time integral of $U_0(t)$ in Figure 6.11. Thus S characterizes the interaction strength of the condensate with the lattice potential. Figure 6.12 shows the occupation of the different diffraction orders (black lines) versus the interaction strength S for three different interaction times ranging from 10 to 0.1 ms. As the interaction time becomes smaller and smaller, the occupation probability becomes independent of the interaction time and solely depends on the interaction strength S. In fact the occupation probabilities for small interaction times are perfectly described by the Bessel functions J_n of the first kind: $p_n = |J_n(S)|^2$ (see gray lines in Figure 6.12). This model holds for short interaction times and directly shows the connection between the occupation probabilities and the Bessel functions (cf. Section 6.4.3).

The critical value S up to which the diffraction process can be described by the Bessel functions is given by the energy-time uncertainty. This uncertainty allows for the occupation of higher diffraction orders, which would otherwise violate the

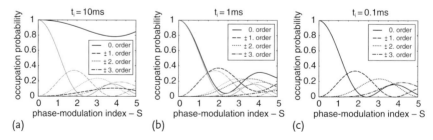

Figure 6.12 Occupation probabilities of the four lowest diffraction orders (black lines) for different interaction times: (a) 10 ms, (b) 1 ms, (c) 0.1 ms. The probabilities are plotted versus the interaction strength S. For smaller and smaller interaction times the occupation probabilities become independent of the interaction time and can be described by Bessel functions of the first kind $p_n \sim |J_n(S)|^2$ (gray lines).

energy conservation law. For a given interaction strength S, the occupation probability becomes independent of t_i if the interaction time is so small that the corresponding energy uncertainty allows the occupation of the maximum expected diffraction order (typically $\sim S$). With our experimental parameters we achieve good agreement with the Bessel functions up to $S = 5$ for interaction times $t_i < 0.5$ ms. We will further investigate the interaction time in Section 6.7 where we also take inter-atomic interactions into account.

6.4.3
Diffraction in the Raman–Nath Regime

In this section we will have a closer look at the condensate's wave function after diffraction in the limit of small interaction times. The condensate wave function before the lattice interaction is given by its ground-state wave function in a harmonic trap:

$$\Psi_0(r) = \sqrt{n_0(r)} e^{i\Theta_0} \tag{6.26}$$

Θ_0 is the constant phase across the condensate and can be set to zero. As before, we describe the lattice interaction in the condensate's center-of-mass frame with an interaction potential $V(z,t) = U_0(t) \cos(k_0 z + \varphi)$ which is constant over the radial extension of the condensate. The additional phase φ is used to characterize the relative position of the lattice with respect to the condensate's. A center-of-mass shift of the condensate by Δz corresponds to a phase shift $\Delta \varphi = 2\pi \Delta z / a$. With $Y(t)$ being the condensate's center-of-mass distance to the surface, the modulation amplitude (see Eq. (6.10)) can be written as $U_0(t) = K \exp(-k_0 Y(t))$. In the framework of a semi-classical description, the atomic wave function during the interaction can be calculated by solving the action integral along the classical trajectory [25]:

$$\Theta(t) = \frac{1}{\hbar} \int \mathcal{L}(t) dt = \frac{1}{\hbar} \int \left(E_{\text{kin}} - E_{\text{pot}} \right) dt . \tag{6.27}$$

In the approximation of short interaction times – known as Raman–Nath approximation – the motion of the atoms during the interaction is neglected. The density of the condensate directly after the interaction is thus unchanged from the initial density distribution. The phase, however, is changed by the lattice potential according to Eq. (6.27). With $E_{\text{kin}} = 0$ and $E_{\text{pot}} = V(z,t)$ the wave function of the condensate directly after the lattice interaction thus reads:

$$\Psi(r,t) = \sqrt{n_0(r)} \exp\left(-\frac{i}{\hbar} \int_0^{t_i} V(r,t) dt \right) \tag{6.28}$$

$$= \sqrt{n_0(r)} \exp\left(-i \cos(k_0 z + \varphi) \underbrace{\frac{1}{\hbar} \int_0^{t_i} U_0(t) dt}_{S} \right) . \tag{6.29}$$

In the Raman–Nath regime the interaction of the condensate with the lattice thus imprints a periodic phase pattern onto the condensate's wave function. The amplitude of the phase modulation is given by the phase-modulation index S (see Eq. (6.29)) which is identical to the diffraction strength parameter introduced in the last section. The condensate wave function after phase imprinting then reads:

$$\Psi(r) = \sqrt{n_0(r)} \exp\left(-iS \cos(k_0 z + \varphi)\right). \tag{6.30}$$

While the period of the phase imprinting is given by the 4 μm period of the magnetic lattice potential, the phase-modulation index S depends on the strength of the interaction which can be tuned by the initial trap displacement d. The momentum distribution of the phase-modulated wave function, which is essential for the cloud propagation during ballistic expansion, can easily be obtained by using the Jakobi–Anger representation [26] which allows one to expand a modulated phase term in Bessel functions of the first kind

$$e^{-iS\cos(k_0 z + \varphi)} = \sum_{n=-\infty}^{+\infty} (-i)^n J_n(S) e^{in(k_0 z + \varphi)} \tag{6.31}$$

$$\Rightarrow \Psi(r) = \sqrt{n_0(r)} \sum_{n=-\infty}^{+\infty} e^{in(\varphi - \pi/2)} J_n(S) e^{ink_0 z} \tag{6.32}$$

$$= \sum_{n=-\infty}^{+\infty} \Psi_n \tag{6.33}$$

with $\quad \Psi_n = \sqrt{n_0(r)} J_n(S) e^{in(\varphi - \pi/2)} e^{ink_0 z} . \tag{6.34}$

As in the case of arbitrary interaction times, the condensate's wave function after the interaction with the lattice is given by a discrete superposition of momentum eigenstates, that is plane waves with wavevectors $k_n = nk_0$. The occupation probability of each diffraction order, however, depends only on the interaction strength and is given by $p_n = |J_n(S)|^2$. The gray lines in Figure 6.12a,b and the black lines in Figure 6.12c show this occupation probability in the Raman–Nath regime for phase-modulation indices up to $S = 5$. Without lattice interaction ($S = 0$) only the zeroth order is populated referring to the undiffracted condensate. However, for increasing interaction strength S, higher and higher diffraction orders are occupied. For a given strength S the occupation probability is thus purely described by the Bessel functions $J_n(S)$.

6.4.4
Evolution of the Wave Function after the Lattice Interaction

The detection of the diffracted condensate is done by absorption imaging. A resonant laser beam illuminates the condensate and the transmitted light is captured with a CCD-camera. This absorption image reveals the atomic density distribution integrated along the imaging axis (x).

In the Raman–Nath regime the density of the condensate remains unchanged during the interaction with the lattice. Thus, the diffracted condensate cannot be imaged in situ, that is directly after interaction. The condensate remains in the trap for an additional $\tau_1 = 8\,\text{ms}$ (back oscillation to the initial BEC position) before it is released for $\tau_2 = 20\,\text{ms}$ of ballistic expansion. During this time the momentum distribution of the condensate leads to a spatial separation of the diffraction orders and with it a change of the density distribution. To compare the observed density distributions with the theory, the wave functions of the different diffraction orders have to be modeled up to the imaging time $T = \tau_1 + \tau_2$: $\Psi(t) = \sum_n \Psi_n(t)$. This includes the density distribution $|\Psi_n(t)|^2$ as well as the corresponding phase evolution $\Theta_n(t)$. To simplify matters we restrict our analysis to the line density profiles which result from the integration of the three-dimensional density distribution along the radial x- and y-directions. The remaining line profile describes the density of the BEC in the direction of the modulated lattice potential (z).

Density Evolution The condensate's line density before the lattice interaction is given by an inverted parabola:

$$n_0(z) = \frac{\mu \pi r_r^2}{2g} \cdot \text{Max}\left\{0, 1 - \frac{z^2}{r_z^2}\right\}^2 \tag{6.35}$$

with g being the coupling constant, μ being the chemical potential, and r_r, r_z being the Thomas–Fermi radii of the condensate.[9]

While the condensate remains in the parabolic potential, the density distribution of the diffraction orders does not change. However, when the condensate is released from the trap, the diffraction orders ballistically expand while keeping their functional form of an inverted parabola [28]. At the imaging time $T = \tau_1 + \tau_2$ the density distribution of the nth diffraction order in the center-of-mass (COM) frame is thus given by:

$$\rho_{n,\text{COM}}(T) = |J_n(S)|^2 \times \frac{\mu \pi r_r^2 r_z}{2g\,R_z(\tau_2)} \cdot \text{Max}\left\{0, 1 - \frac{z^2}{R_z^2(\tau_2)}\right\}^2. \tag{6.37}$$

In the following we assume that each diffraction order expands independently of the other orders. This ad hoc assumption is further discussed in Section 6.5. For an

---

9) The density distribution of a ground-state condensate in a harmonic trap is given by the solution of the time-independent Gross–Pitaevskii equation [27] which adds a non-linear term to the well-known Schrödinger equation, accounting for the atomic interactions:

$$\mu\phi(r) = \left[-\frac{\hbar^2}{2m}\nabla^2 + U(r) + g|\phi(r)|^2\right]\phi(r). \tag{6.36}$$

In the Thomas–Fermi limit of strong interactions, the kinetic energy term $-\hbar^2\nabla^2/(2m)$ can be neglected leaving an analytic expression for the condensate's density $\rho(r) = |\phi(r)|^2 = 1/g \times \text{Max}\{0, \mu - U(r)\}$. This density directly maps the underlying potential structure up to a given chemical potential which is fixed by the total atom number. With a cigar-shaped parabolic potential $U(r) = 1/2 m\omega_r^2 (x^2 + y^2) + 1/2 m\omega_z^2 z^2$ the outer shape of the condensate becomes an ellipsoid with a semi-major and semi-minor axis given by the Thomas–Fermi radii $r_z = \sqrt{2\mu/m}\,1/\omega_z$ and $r_r = \sqrt{2\mu/m}\,1/\omega_r$, respectively. The line density profile in Eq. (6.35) is obtained by integrating this density along the radial directions.

individual diffraction order the time evolution of the Thomas–Fermi radius $R_z(t)$ is given by the scaling theory [28] and can be written in the limit $\tau_2 > 1/\omega_r = 2.1$ ms as:

$$R_z(\tau_2) = r_z \left(1 + \frac{\omega_z^2}{\omega_r} \frac{\pi}{2} \tau_2\right). \tag{6.38}$$

After the interaction with the lattice the diffraction orders move with a velocity $v_n = n \times \hbar k_0/m$ and continuously separate from each other. During the first 8 ms this happens in a parabolic potential of axial frequency ω_z causing a separation of neighboring diffraction orders by 8.2 μm and a deceleration of the nth diffraction order from $n \times 1.14$ μm/ms to $n \times 0.79$ μm/ms. During the subsequent ballistic expansion the diffraction orders separate at constant velocity resulting in a final spacing Δz between the center-of-mass position of neighboring diffraction orders:

$$\Delta z = \frac{\hbar k_0}{m\omega_z}(\sin(\omega_z\tau_1) + \omega_z \cos(\omega_z\tau_1)\tau_2) = 24 \text{ μm}. \tag{6.39}$$

The density distribution of the nth diffraction order at the imaging time is thus given by:

$$\rho_n(T) = |J_n(S)|^2 \cdot n(z - n\Delta z) \tag{6.40}$$

$$\text{with} \quad n(z) = \frac{\mu \pi r_r^2 r_z}{2g R_z(\tau_2)} \cdot \text{Max}\left\{0, 1 - \frac{z^2}{R_z^2(\tau_2)}\right\}. \tag{6.41}$$

Phase Evolution The starting point for the phase evolution of the different diffraction orders is given by Eq. (6.34):

$$\Theta_n(0) = n\left(\varphi - \frac{\pi}{2}\right) + nk_0 z. \tag{6.42}$$

After the interaction with the lattice, each diffraction order accumulates a phase due to its center-of-mass motion along a classical trajectory. This homogeneous phase contribution is constant across the extension of the diffraction order and given, in analogy to Eq. (6.27), by:

$$\Delta\Theta_{\text{hom},n}(T) = \frac{1}{\hbar} \int_0^T \mathcal{L}(t) dt = \frac{1}{\hbar} \int_0^T (E_{\text{kin}} - E_{\text{pot}}) dt \sim n^2. \tag{6.43}$$

As the diffraction orders expand during the ballistic drop, an additional inhomogeneous phase contribution, that is a phase contribution changing across the diffraction order, has to be taken into account. In fact the spatially varying phase gradient is connected to a velocity field causing the expansion of the cloud. The phase structure of an expanding condensate is parabolic and given by [28]:

$$\Delta\Theta_{\text{inhom},n}(T) = \frac{m}{2\hbar} \frac{\dot{R}_z(T)}{R_z(T)}(z - n\Delta z)^2. \tag{6.44}$$

The total phase accumulated by the nth diffraction order thus reads:

$$\Theta_n(T) = \frac{m}{2\hbar} \frac{\dot{R}_z(T)}{R_z(T)}(z - n\Delta z)^2 + n(\varphi - \pi/2)$$
$$+ nk(\tau_1)(z - n\Delta z) + \Delta\Theta_{\text{hom},n}(T). \tag{6.45}$$

Total Wave Function Using Eq. (6.40) and Eq. (6.45) the wave function of the nth diffraction order at the imaging time is given by:

$$\Psi_n(T) = \sqrt{n(z - n\Delta z)} J_n(S) \times \exp[i\Theta_n], \tag{6.46}$$

$$\Theta_n = \frac{m}{2\hbar} \frac{\dot{R}_z}{R_z} (z - n\Delta z)^2 + n(\varphi - \pi/2)$$
$$+ nk(\tau_1)(z - n\Delta z) + \Delta\Theta_{\text{inhom},n}(T), \tag{6.47}$$

$$n(z) = \frac{\mu \pi r_r^2 r_z}{2g R_z(\tau_2)} \cdot \text{Max}\left\{0, 1 - \frac{z^2}{R_z^2(\tau_2)}\right\}. \tag{6.48}$$

The model function for the axial line density profile of the diffracted condensate after the ballistic expansion thus reads:

$$\rho_{\text{coh}}(z) = \left|\sum_n \Psi_n(T)\right|^2 \otimes f_{\text{opt}}(z) = \left[\sum_n \sum_m \Psi_m^*(T) \Psi_n(T)\right] \otimes f_{\text{opt}}(z). \tag{6.49}$$

The convolution with the function $f_{\text{opt}}(z)$ is necessary to account for the limited optical resolution of the imaging system. It is chosen as gaussian with a $1/e$ width of 13 µm corresponding to our optical resolution. The density distribution in Eq. (6.49) accounts for the coherent superposition of different diffraction orders including interferences between them. This is significantly different from the incoherent superposition:

$$\rho_{\text{incoh}}(z) = \left[\sum_n |\Psi_n(T)|^2\right] \otimes f_{\text{opt}}(z) \tag{6.50}$$

which accounts only for the density distribution of the different diffraction orders and completely neglects coherent effects like interference. Although the incoherent superposition depends only on the interaction strength S it should be well suited to describing the overall shape of the expected line density profile. The coherent superposition depends additionally on the relative phase φ and describes interference effects in the final line density profile. Figure 6.13a shows the expected coherent (solid black line) and incoherent (dotted black line) line density profiles for four different relative phases between the condensate and the lattice structure. Figure 6.13b shows the corresponding interference pattern as the difference between the coherent and incoherent line profiles. For the illustration the interaction strength was set to $S = 1.3$. Figure 6.13a shows in addition the line density profiles of the individual diffraction orders (zeroth order – solid gray line, first order – dashed gray line, second order – dotted gray line). The clear overlap of the diffraction orders is the reason for the emergence of the interference pattern in the coherent line profile. In the example shown, interference mainly occurs between the zeroth and ±first diffraction order. As expected, the incoherent line profiles do not

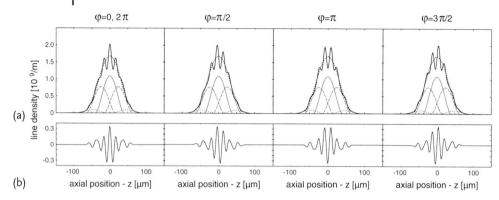

Figure 6.13 (a) Coherent (solid black line) and incoherent (dotted black line) diffraction patterns for four different relative phases φ at an imprint strength of $S = 1.3$. The phase φ characterizes the relative position of the condensate with respect to the lattice. With increasing φ the interference pattern shifts to the left before it is reproduced for $\varphi = 2\pi$. (b) Corresponding interference pattern deduced from the difference of the coherent and incoherent model function.

change with increasing phase φ. However, the interference pattern in the coherent line profile is shifted to the left for increasing phase φ. For $\varphi = 2\pi$ the original line density is restored.[10] The position of the interference fringes can thus be used to derive the position of the condensate relative to the magnetic lattice. Any force that changes this position shows up as a shift of the interference pattern. The scheme may be interpreted as a sensitive force sensor.

6.5
Ballistic Expansion and Phase Imprinting

We have assumed that during ballistic expansion the diffraction orders expand as if they were individual condensates. Although confirmed by the experimental observations, this assumption is somewhat surprising since on one hand the mean-field is strong enough to cause a significant overlap between the expanding diffraction orders, on the other hand there is no observable mixing between the orders due to mean-field nonlinearities. Here, some further insight is gained if we extend the phase-imprinting interpretation to the ballistic expansion right after the trap has been turned off. For a single condensate at rest in a trap the density dependent mean-field cancels the trapping potential. Only then is the phase of the condensate homogeneous and constant in time. If the trapping potential is suddenly switched off, the mean-field is uncompensated and starts to imprint an inhomogeneous phase structure $\phi(r)$ onto the condensate wave function. As a consequence the condensate starts to expand with a velocity given by the local phase gradient. Dur-

---

10) A phase shift $\varphi = 2\pi$ corresponds to a relative displacement of the initial condensate by a full lattice period. Therefore, the original density profile is restored for $\varphi = 2\pi$.

ing this expansion the density rapidly drops and with it the mean-field. It is thus a reasonable approximation to assume that the strongest phase structure (parabolic phase) is imprinted during the relatively short time interval after the trap has been turned off and before the atomic density starts changing. In other words, most of the velocity is acquired before the atoms actually start moving. For a diffracted condensate the mean-field phase structure $\phi(r)$ adds to the already imprinted periodic phase due to the lattice. At the time when the trap has been turned off, that is the time when the imprinting by the lattice and the mean-field is completed but before the expansion starts, the total wave function is thus proportional to:

$$\Psi(r, T) \sim \exp\left(-iS\cos(k_0 z + \varphi) + i\phi(r)\right) \quad (6.51)$$

$$\sim \exp(i\phi(r)) \cdot \exp(-iS\cos(k_0 z + \varphi)) \quad (6.52)$$

$$\sim \exp(i\phi(r)) \sum_n (-i)^n J_n(S) \exp(in(k_0 z + \varphi)) \quad (6.53)$$

$$\sim \sum_n (-i)^n J_n(S) \exp(in(k_0 z + \varphi) + i\phi(r)). \quad (6.54)$$

One obtains a sum of diffraction orders, each one modulated with the same phase structure $\phi(r)$ that governs the subsequent expansion dynamics. The above assumption is well justified in a 3D scenario where the density drops rapidly right at the beginning of the ballistic expansion. According to the three-dimensional simulation of Section 6.7, the mean-field does not affect the phase profile after ~ 1 ms of ballistic expansion. If the diffracted cloud were kept in a waveguide where it can only axially expand, the mean-field potential would be active over a longer time interval [10] and may influence the number of atoms in the various diffraction orders. Since the mean-field energetically favors homogeneous densities one may expect that diffraction will be suppressed by the atomic interactions. However, the velocity pattern imprinted by the lattice transfers a certain amount of kinetic energy to the atoms which will be released during expansion in arbitrary directions. It thus might be interesting to study such momentum splitting of condensates also in respect to number squeezing and coherence properties.

6.6
Experimental Results

For the realization of the experiment we prepare a condensate of 1.2×10^5 rubidium atoms at a distance of 30 µm from the lattice chip surface (see Figure 6.10). At this position the atoms are trapped in a cigar-shaped magnetic trap generated by the carrier chip and characterized by axial and radial trap frequencies of $\omega_z = 2\pi \times 16$ Hz and $\omega_r = 2\pi \times 76$ Hz, respectively. The offset field at the trap center, $B_0 = 0.87$ G, is oriented along z. The long axis of the condensate is also oriented along the z-direction, that is parallel to the chip surface and perpendicular to the lattice sites. The Thomas–Fermi radii of the condensate are $r_z = 26.2$ µm and $r_r = 5.5$ µm.

As the atoms are relatively far away from the chip surface, the lattice potential can be turned on without influencing the condensate. This is done by driving the current in the double meander pattern with typically 0.2 mA. The interaction with the lattice potential is now initiated by instantaneously displacing the trap center towards the chip surface. As described in Section 6.4.1, this causes the condensate to oscillate against the chip surface. By changing the initial displacement d, we can influence the interaction strength and thus the amplitude of the phase modulation experienced by the condensate. After the reflection of the condensate from the lattice (i.e., 12 ms after the initial trap displacement) all magnetic fields are turned off and the condensate is dropped for ballistic expansion. The diffraction and interference pattern of the condensate is imaged after 20 ms of free expansion. The imaging beam is oriented parallel to the x-axis which yields a two-dimensional density profile of the condensate.

Figure 6.14 shows the diffracted condensate for three different displacements d of 13, 14, and 14.6 μm. The left images show the two-dimensional density distributions as taken from the CCD-camera.[11] The corresponding vertical integrated line density profiles are shown on the right-hand side. Obviously the axial width of the density distribution increases with initial displacement d. This is to be expected from the diffraction model because with larger displacements the imprint strength on the condensate increases, leading to the occupation of higher and higher diffraction orders. This causes an increased width of the density profile, due to the 24 μm separation of the diffraction orders at the imaging time. In fact the overall shape of the measured density profiles (black dots) can be very well described by the incoherent superposition (solid black lines) of different diffraction orders. Equation (6.50) is fitted to the experimental data by adjusting the imprint strength S. As this is the only free parameter, S fixes the relative strength of the diffraction orders (gray lines) and thus the overall shape of the line profiles (solid black lines). An additional Gaussian distribution accounts for the thermal atoms (dashed black line).[12] Even for large imprint strengths this model agrees very well with the experiment, as can be seen from Figure 6.14 for occupations up to the fifth diffraction order.

For each initial trap displacement d one can now extract the corresponding phase-modulation index S. As this parameter characterizes the interaction strength of the condensate with the lattice, it is strongly correlated with the displacement d, as shown in Figure 6.15. The interaction with the lattice starts for displacements of about 12 μm. For larger displacements the interaction strength increases steadily and reaches $S = 5$ for a displacement of $d = 15$ μm. It should be noted that the imprint strength can be tuned between $S = 0$ and $S = 5$ by changing the displacement by only 3 μm. Without a sub-μm precision positioning system as

---

11) The color coding is chosen so that the density distribution increases from black to white. Areas of very high atomic density are shown in black again. This reveals the clouds' internal structure with high contrast.

12) The width of the Gaussian function was set to 135 μm and extracted from the $d = 0$ measurement, where no interaction with the lattice took place. This corresponds to a temperature of 125 nK for the thermal part of the atoms. The fraction of thermal atoms was fixed to 30 % of the total atom count.

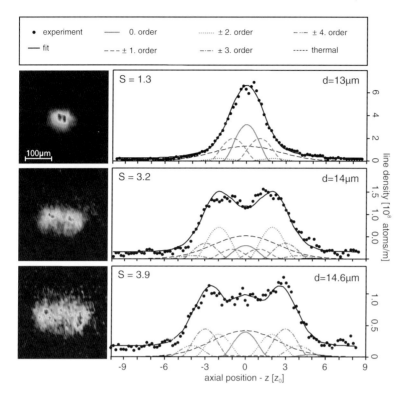

Figure 6.14 Absorption images (for color coding see Footnote 12) and vertical integrated line density profiles after 20 ms of ballistic expansion for three different displacements d of 13, 14, and 14.6 μm. The distance $z_0 = 24$ μm corresponds to the separation of neighboring diffraction orders at the imaging time. The measured density profile (black dots) can be described by the sum (solid black line) of overlapping diffraction orders (gray lines). An additional Gaussian distribution (dashed black line) accounts for the thermal atoms. The relative strength of each diffraction order is given by the interaction strength S which can be derived by fitting the incoherent model function from Eq. (6.50) to the experimental data.

supplied by our carrier chip, a specific choice of the interaction strength would be impossible. This is mainly because the modulation amplitude of the periodic lattice potential increases exponentially on a length scale of $1/k_0 = 0.64$ μm with decreasing distance from the surface.

Diffraction on a static magnetic lattice is, however, associated with a significant loss of atoms (see Figure 6.16). The loss can be understood by considering the steep magnetic field gradients inside the potential wells of the lattice (see Figure 6.10). Atoms entering the wells are lost due to Majorana spin-flips which result from the sudden break down of the adiabatic condition along the atoms' trajectories [11]. This limitation can be eliminated by introducing RF-controlled adiabatic potentials [29, 30] to "plug" the Majorana hole. Figure 6.16 demonstrates the enhanced reflectivity when the RF plug is turned on: An oscillating magnetic field of

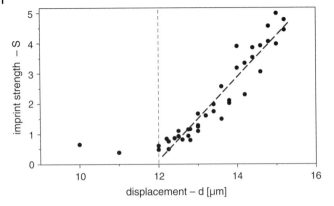

Figure 6.15 For different initial trap displacements d, the incoherent model function (Eq. (6.50)) was fitted to the measured line profiles. From each of those fits, the imprint strength S was extracted. The interaction with the lattice starts at about $d = 12\,\mu m$ and rises continuously for increasing d.

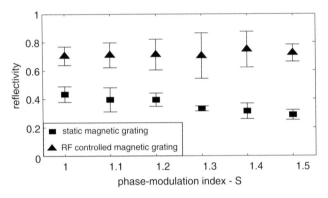

Figure 6.16 Reflectivity of the static (squares) and the RF-controlled (triangles) magnetic lattice. The reflectivity is plotted against the phase-modulation index S characterizing the interaction strength of the condensate with the lattice.

10 mG amplitude is superimposed on the static field of the lattice. The polarization of the RF field is parallel to the lattice conductors and the frequency (585 kHz) is below the Larmor frequency (610 kHz) corresponding to the offset field B_0 in the trap center. The resulting adiabatic lattice potential features wells of limited depth and the initially steep, attractive potential slopes are converted into repulsive potential walls (for details see [7]). The enhanced reflectivity indicates that the motion of atoms becomes adiabatic in the RF controlled lattice potential.[13] The high reflectivity diffraction grating allows phase coherent manipulation of Bose–Einstein condensates and the realization of a matter wave interferometer on a chip.

13) For these measurements the RF field was activated simultaneously to the displacement and deactivated prior to turning off the magnetic trap for ballistic expansion.

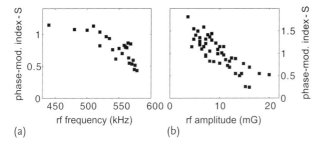

Figure 6.17 Controlling the diffraction strength by means of the RF magnetic field. The diffraction is initiated by a displacement d of the trap center towards the lattice. (a) The diffraction is controlled by the frequency (constant RF amplitude of 10 mG, $d = 14$ μm). (b) The diffraction is controlled by the amplitude of the RF field (constant frequency of 585 kHz, $d = 14.5$ μm).

Using this technique the phase-modulation index and thus the population of the diffraction orders can be controlled not only by the displacement d but also by the RF field. The frequency and amplitude of the oscillating field determines the coupling between internal atomic states [31] and with it the depth and shape of the potential wells. Radio-frequency control of diffraction, as demonstrated in Figure 6.17, may be particularly convenient for integrated atom-optical devices when sequences of diffraction pulses are required. The phase-modulation index can be electronically changed while the condensate performs a constant center-of-mass oscillation in the trap and interacts with the grating in subsequent oscillation periods.

Although the envelope of the measured line density profiles can be very well described by the incoherent superposition, a closer look at the line profiles shows deviations. Figure 6.14a, for instance, shows a mismatch between the theoretical and experimental profile in the overlapping region between the zeroth and ±first diffraction order. This is because the interference between these diffraction orders was neglected in the incoherent model (Eq. (6.50)). The coherent model function Eq. (6.49) should solve this problem and yield better agreement with the experimental data. As the complexity of the interference pattern depends on the number of overlapping diffraction orders, interference effects are best studied in the case of only a few overlapping orders. We thus restrict our analysis to phase modulation indices between 1 and 1.5 where mainly the zeroth and ±first diffraction orders are occupied and thus interfere. Figure 6.18 shows the absorption images and line density profiles for an undiffracted condensate (left) and two typical imprint strengths in this regime. The phase-modulation index S can be determined by fitting the incoherent superposition (dotted black line) of the diffraction orders to the measured line profile (black dots). As mentioned before this fit describes the overall shape of the line profile quite well but shows significant discrepancies in the overlapping region of the diffraction orders where interference appears. Keeping S fixed we improve the theoretical description by fitting the coherent model function to the experimental data. This results in an almost perfect agreement with the experimental line profiles and returns a value for the relative phase φ between the condensate and the lattice structure with φ being the only free-fitting parameter.

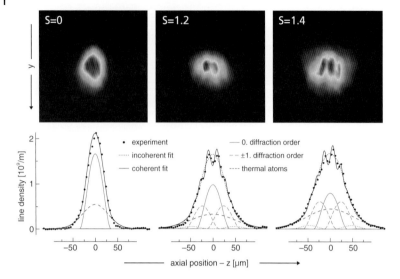

Figure 6.18 Absorption images (for color coding see Footnote 6.6) and line density profiles after 20 ms of ballistic expansion. In the overlapping region between the zeroth and ±first diffraction order, the incoherent superposition (dotted black line) of the diffraction orders (gray lines) can not describe the measured data (black dots). The interference between diffraction orders is taken into account by the coherent superposition (solid black line), which fits the experimental data much better.

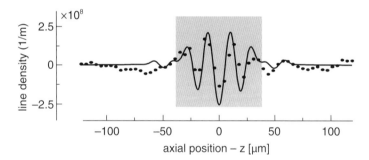

Figure 6.19 Interference pattern calculated from the deviation of the experimental data from the incoherent fit function for $S = 1.2$. The coherent model function (solid curve) describes the interference pattern very well and yields the relative phase of the condensate to the lattice structure $\varphi = 3.2$ rad.

To visualize the interference structure the incoherent fit can be subtracted from the experimental data. This is shown in Figure 6.19 for the central image of Figure 6.18 with an imprint strength of $S = 1.2$. The remaining density profile shows a clear interference pattern (dots) localized in the overlapping region of the zeroth and ±first diffraction order (shaded rectangle). The coherent model function (solid line) fitted to the experimental data, shows very good agreement with the observed interference structure for $\varphi = 3.2$ rad.

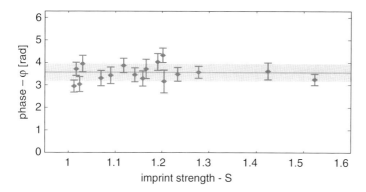

Figure 6.20 Phase φ extracted from fitting the density profiles with the coherent model function for different S ranging from 1 up to 1.5. Each data point is a single measurement. The error bar indicates the confidence interval of the fit. Averaging all data the phase amounts to $\overline{\varphi} = 3.57$ rad with a phase uncertainty of 21°.

To analyze the experimental data we concentrate on the reproducibility of the interference pattern for different realizations of the experiment. This directly depends on the reproducibility of the condensate's position relative to the lattice structure. All other phase contributions in Eq. (6.49) are fixed by the experimental parameters and should not change during successive experimental realizations. Figure 6.20 shows the extracted phase information for different interaction strengths S. As expected, φ does not show a dependance on the phase-modulation index S and amounts to an averaged value of $\overline{\varphi} = (3.57 \pm 0.37)$ rad. The phase spread of 21° could be based on a position uncertainty of about 250 nm. However, it is more likely that the phase spread is caused by the finite length of the condensate. With an axial extension of 50 μm the condensate illuminates only a limited number of $j \approx 12$ lattice sites. This constrains the momentum resolution of the lattice to $\Delta k/k_0 = 1/j$ which corresponds to a phase resolution of $\Delta \varphi = 2\pi/j \approx 0.5$ rad. The observed phase fluctuation is thus probably purely due to the finite length of the condensate.

The reproducibility of the interference pattern proves the phase coherent splitting of the condensate at the lattice. The control over the experimental parameters is good enough to observe a deterministic, predictable phase relation between neighboring diffraction orders. The interference at the lattice can thus be used as an integrated matter wave interferometer on an atom chip. In contrast to other interferometers it depends on a single diffraction pulse which phase coherently splits the condensate in a magnetic waveguide. During the final ballistic expansion the interferometric path is closed due to the released inter-atomic interaction energy. The analysis of the interference pattern reveals the relative phase between the different diffraction orders. Besides an initial shift of the condensate's position relative to the lattice structure, a magnetic field gradient should be detectable as this changes the relative phase accumulated during the motion in the waveguide and the ballistic expansion.

6.7
Effect of Atomic Interactions

Up until now we have analyzed the experiments using a quasi non-interacting theory. This explains the data very well in most of the circumstances we consider. However, in certain regimes, interactions are expected to be significant and will distort the behavior of an interferometer based on these techniques. It is also unclear if interactions will cause any systematic shift in the estimation of the phase imprint strength.

By undertaking full three-dimensional simulations of the Gross–Pitaevskii equation, we can investigate these issues and find "safe" regimes where interferometry can be performed without fear of atomic interactions distorting the data. In this section, we describe how in some regimes, atomic interactions can broaden diffraction orders and depopulate higher ones. In the context of the experiments here, atoms in the first diffraction order have a kinetic energy comparable with the interaction energy. Hence, if the majority of the atoms are in the first diffraction order the interactions significantly affect the phase evolution. Otherwise, the non-interacting theory generally provides a good description.

Our calculations show good agreement with the diffraction patterns observed in the experiments in all regimes, as long as we allow for interactions when the imprinting strength S is low. Obviously it is convenient to work in a regime where interaction effects are small and we can predict the interferometer's behavior on the basis of simple analytical formulas. We find this can be done provided that the kinetic energy corresponding to the BECs separating diffraction orders is larger than the peak inter-atomic interaction energy. If this is not the case, BECs respond in a complicated and somewhat counterintuitive way to spatially varying phase patterns. It is still possible to perform atom interferometry in this interaction-dominated regime but a thorough calibration of the device will be needed.

6.7.1
Modeling BEC Surface Diffraction

As we described in Section 6.4.1, the diffraction process effectively has two stages. First there is the imprinting stage where the BEC reflects from the magnetic diffraction grating and is phase imprinted. The trap is then switched off and we move to the expansion stage during which the diffracted BEC undergoes time-of-flight expansion. We explore the role of atomic interactions during both stages. We first consider the phase imprinting stage. To simplify our analysis and reduce computational effort to an acceptable level we then perform new simulations to describe the expansion stage using an idealized starting configuration.

We use the Gross–Pitaevskii equation

$$-\frac{\hbar^2}{2m}\nabla^2 \Psi + V_T \Psi + \frac{4\pi\hbar^2 a_s}{m}|\Psi|^2 \Psi = i\hbar\frac{\partial \Psi}{\partial t}, \qquad (6.55)$$

where V_T is the combined potential of the trap and magnetic lattice, and a_s the atomic scattering length. We first prepare a ground-state BEC with the same parameters as in the experiment. We then use a Biot–Savart solver to model the magnetic potential landscape from the grating and combine this with the harmonic trap potential.

The theoretical BEC initially starts 30 µm from the surface as in the experiment and we begin the imprinting simulations at time $t = 0$ by displacing the trap by 11–15 µm in the y-direction to swing the BEC towards the magnetic lattice. Atom losses are simulated with a linear imaginary potential beginning at $y = 26$ µm. The strength of the imaginary term is adjusted to give rough agreement with experimentally measured reflection probabilities.

Equation (6.55) is solved using the Crank–Nicolson method [32]. The potential landscape contains sharp, small-scale features, hence a high resolution grid is required. It is, therefore, necessary to employ parallel programming on high-performance computers.

6.7.2 Density Profile Dynamics

We begin by considering what effect the reflection process has on the density profile of the BEC. We can also use these simulations to extract information about the phase imprint. In this case we use a trap displacement of 13 µm. Figure 6.21 shows the density of the BEC through its middle ($x = 0$ plane) as it interacts with the grating in this simulation. Figure 6.21a shows the initial position and shape of the BEC. After $t \approx 5.6$ ms, the BEC has just made contact with the lattice potential and after 7.6 ms the BEC has begun to move away (Figure 6.21b). Because we consider a relatively low speed in this example, the atoms interact with the lattice for about $t_i \simeq 1$–2 ms. This does not quite satisfy the short interaction time criterium discussed earlier in the chapter but the approximation improves at higher velocity. After interaction with the lattice, both density and phase imprints are observed in the BEC (see Figure 6.21b including the inset). At $t = 12.0$ ms (the end of the reflection stage, just before expansion begins) there is still a clear density imprint

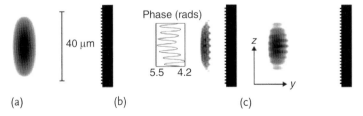

Figure 6.21 Slices through the middle of the diffracting BEC ($x = 0$ plane, dark indicates high atom density). Axes shown in (c). Timings are $t =$ (a) 0.0 ms, (b) 7.6 ms, and (c) 12.0 ms. Black shapes are sketches of the atom chip and grating, showing approximate positions. The inset in (b) shows the phase through the middle of the BEC along z at $t = 7.6$ ms.

(Figure 6.21c). The phase and density imprints seen in the simulations are approximately sinusoidal. This confirms that the interactions do not prevent the formation of a clean phase imprint and justifies the assumption made in Eq. (6.30). This is the case for all trap displacements considered here.

We do not observe the formation of topological excitations which have been observed in some previous studies of reflecting BECs [32]. This is because the time required for sound waves to reach the edge of the condensate and thus trigger the formation of excitations [32] is longer than the time it takes for the BEC to interact with the lattice. We have also repeated our simulations including quantum fluctuations which are known to cause decoherence and scattering halos in certain circumstances. These fluctuations have very little effect in the cases we consider here. This is primarily due to the geometry of the experiment [33] which is very suitable for reflection. We, therefore, do not need to use advanced variants of the Gross–Pitaevskii equation to describe the results.

6.7.3
Phase Modification by Inter-Atomic Interactions

The effect of atomic interactions on the phase profile of a diffracting BEC has been studied in detail elsewhere [9] by examining the early stages of the BEC's dynamics shortly after a phase imprint. Here we summarize the key results.

We note firstly that interaction effects occur on a timescale given by the correlation time (the BEC's healing length divided by the maximal speed of sound in the condensate). For the experiments, this is approximately 0.14 ms. If the BEC's interaction with the lattice lasts for less than this correlation time, we may speak of a clean phase imprint where the BEC's density profile is not modulated during the interaction. If, however, the interaction time lasts for longer, the density will become modulated and we can no longer speak of a pure phase imprint. It is not possible to distinguish density modulations and phase modulations from the expansion images in the experiment. However, it is possible to heuristically quantify an "overall" imprint strength in this regime. This means the experimental techniques are still perfectly satisfactory for use as interferometers, although individual calibration of the device may be required if the lattice interaction time is long.

Secondly, atomic interactions tend to broaden diffraction orders and *depopulate* the higher diffraction orders. Some 2 ms after the application of a pure phase imprint to a BEC, sharp density peaks appear in the BEC's density profile due to interference between separating diffraction orders (see [9] for further details). Because of this process, the kinetic energy that the atoms possess due to the diffraction (which, of course, has an associated direction along the z-axis), is transferred into mean-field internal energy in the BEC. As the expansion continues, the density peaks relax and the internal energy is transferred back to kinetic energy. However, this time there is no preferred direction for this energy, which scatters evenly in all directions. In this way, atomic interactions remove energy from the z-direction.

Rather than slowing the speed of separating diffraction orders (which interestingly remain constant) the energy is removed by scattering atoms out of higher diffraction orders. The BEC naturally resists attempts to form modulations with wavelengths smaller than its healing length, hence short wavelength interference fringes due to the separation of higher diffraction orders are strongly suppressed.

These interaction effects can be reduced if the kinetic energy associated with the diffraction can be increased, either by using higher impact velocities or a grating with a smaller periodicity. At slow BEC speeds, it becomes difficult to associate a phase amplitude S with the process, both because of the long lattice interaction time, and also because interaction effects are able to significantly affect the momentum distribution. However, as we noted, it is still possible to associate S with an overall imprint strength in any regime and at higher S values, the analytical non-interacting theory can be used to extract accurate phase amplitudes.

6.7.4
Comparison of the Interacting Theory with Experiment

Up until now we have only considered the dynamics of the cloud shortly after imprinting when the far-field diffraction pattern has not yet formed. To compare the interacting theory with experiment we must now simulate the expansion of an imprinted BEC. To simplify the analysis and the computation we start with an idealized situation with perfect artificial phase and density imprints, similar to those shown in Figure 6.21b. We apply such imprints to the basic ground-state solution of the harmonic trap, and the BEC is left in the trap for 6 ms. The trap is then turned off, and the BEC is allowed to expand for 20 ms. We show the results of this process for a non-interacting matter wave and a BEC in Figure 6.22a,b. In the case of the non-interacting atom cloud (Figure 6.22a), we see a typical far-field diffraction pattern with separated peaks. The zeroth, first, and second diffraction orders (arrowed) all have significant populations. However, in the case of the interacting BEC (Figure 6.22b), the diffraction orders have not separated after 20 ms and hence rapid modulations due to interference between separating orders are observed. In a non-interacting system, the nodes in this pattern would drop to zero but we see here, as discussed in the previous section, that the interactions reduce the fringe amplitude.

We now compare an experimental expansion image with simulation for $S = 1.4$. Figure 6.23a shows the experimental data. Figure 6.23b shows the corresponding theoretical density profile. The 3D wave function has been integrated along the x-axis, after which a low-pass filter was applied (this filter has the same resolution

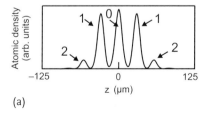
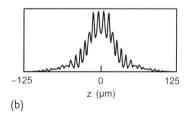

Figure 6.22 Density profiles along the z-axis after a sinusoidal phase imprint and 6 + 20 ms expansion time for the non-interacting (a) and interacting (b) clouds.

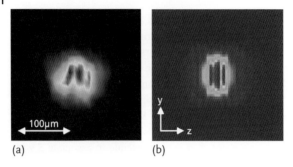

Figure 6.23 Comparison of experimental (a) and theoretical (b) two-dimensional density profiles after 6 + 20 ms time-of-flight for $S = 1.4$. The field of view is 250 × 250 μm² (for color coding see Footnote 12).

as the imaging camera used in the experiment). Like the experimental density profile, the simulated profile contains three small peaks and the envelope function is roughly the same. The peaks are interference fringes between separating diffraction orders, aliased by the filter. As described earlier in this section, these peaks are far weaker than the non-interacting case where nodes are observed. The simulated cloud profile is slightly thinner than the experimental data due to the depopulation of higher diffraction orders and the lack of a thermal cloud. However, since our zero-temperature mean-field theory captures the main features of the experiment, we draw the important conclusion that heating of the atom cloud by the experimental procedure is negligible.

6.7.5
Locating the Low-Interaction Regime

Thus far, we have considered regimes where interactions noticeably affect the diffraction. How can we recover the "analytic" behavior and identify the parameter regimes in which the expansion simulations give approximately non-interacting behavior? We employ two different strategies, repeating our expansion simulations firstly with a larger recoil energy and secondly, with a lower initial density to decrease the interaction strength. All other parameters remain the same, including $S = 1.4$. Figure 6.24 shows the final expansion images from these two strategies after the same expansion procedure as before.

The simulation results approximate non-interacting behavior better and better as the kinetic energy associated with the first diffraction order $E_r = 2\pi^2\hbar^2/ma^2$ (i.e., the recoil energy) increases beyond the maximum interaction energy $E_I = 4\pi\hbar^2 a_s n_0/m$. As before, we have defined a as the grating periodicity, n_0 as the peak atom density, and a_s as the atomic scattering length. We can satisfy this condition for E_I by reducing the grating period to $a = 2$ μm. Now, unlike the original case with $a = 4$ μm, the diffraction pattern reveals separated peaks (see Figure 6.24a) although we have not quite reached the far-field. Satisfying this energy condition also implicitly helps us satisfy the short interaction time criterium.

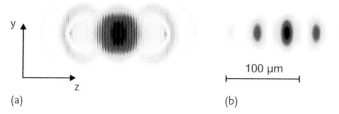

Figure 6.24 Density profiles for expanded BECs with improved diffraction peaks (dark indicates high atom density). (a) is for four times the original recoil energy and (b) is with 1% of the original atom number.

Alternatively, we can satisfy the low-interaction condition $E_r > E_l$ by retaining $a = 4\,\mu$m but reducing the BEC's density. By reducing the number of atoms in the BEC from 10^5 to 10^3 (so the peak density drops to $n_0 \approx 1.4 \times 10^{19}$ atoms m^{-3}) but using the original grating, we see a good approximation to the single atom picture. In this case, 20 ms expansion is long enough for us to see a far-field distribution with fully separated peaks (see Figure 6.24b). This occurs because the BEC's initial size is small (due to the low interactions) and hence, the BEC reaches the far-field sooner. This may be an important consideration when designing experiments.

6.8 Conclusion

We have described the interaction of a condensate with a magnetic lattice at the surface of a microchip. In particular, we describe diffraction of BECs in a regime where the chemical potential is larger than the recoil energy associated with the lattice. An important consequence is that the diffraction orders overlap and form a characteristic interference pattern. The position of the interference fringes depends sensitively on the location of the condensate relative to the lattice, while the resolution of the interference fringes is only limited by the size of the condensate. For sufficiently short interaction times with the lattice a simplified phase-imprinting model turned out to be applicable. On the other hand the three-dimensional Gross–Pitaevskii equation identifies regimes in which atomic interactions become dominant, giving rise to non-linear phenomena. Thus, the construction of atom interferometers based on diffraction techniques requires careful consideration of interatomic interactions.

Since the reflected condensate encodes the information about the lattice in its phase distribution, the experiment can be interpreted as a holographic sensor for mechanical potentials near surfaces. This opens new ways to investigate dispersive surface forces caused by van der Waals and Casimir–Polder potentials. By combining materials with different dielectric functions versatile potential landscapes for the atoms may be engineered. In principle, such landscapes can be made very steep such that the quantum character of the trapped atoms becomes apparent already at temperatures of several Kelvin. With such atomic "nanotraps" one could start to do quantum optics near surfaces with unexplored possibilities for fundamental research and for quantum technology.

Acknowledgments

We would like to thank C. Vale for early contributions to the experiment and M. Kemmler, D. Koelle, and R. Kleiner for the production of the magnetic lattice chip. We also wish to thank R. Scott and T.M. Fromhold for contributions to the theoretical part of this work. This work was supported by the Deutsche Forschungsgemeinschaft, the Landesstiftung Baden-Württemberg, the European Union, EPSRC and Nottingham HPC facilities.

References

1 Fortágh, J. and Zimmermann, C. (2007) Magnetic microtraps for ultracold atoms. *Rev. Mod. Phys.* **79**, 235.

2 Cano, D., Kasch, B., Hattermann, H., Kleiner, R., Zimmermann, C., Koelle, D., and Fortágh, J. (2008) Meissner effect in superconducting microtraps. *Phys. Rev. Lett.* **101**, 183006.

3 Roux, C., Emmert, A., Lupascu, A., Nirrengarten, T., Nogues, G., Brune, M., Raimond, J.-M., and Haroche, S. (2008) Bose–Einstein condensation on a superconducting atom chip. *Europhys. Lett.* **81**, 56004.

4 Mukai, T., Hufnagel, C., Kasper, A., Meno, T., Tsukada, A., Semba, K., and Shimizu, F. (2007) Persistent supercurrent atom chip. *Phys. Rev. Lett.* **98**, 260407.

5 Treutlein, P., Hunger, D., Camerer, S., Hänsch, T.W., and Reichel, J. (2007) Bose–Einstein condensate coupled to a nanomechanical resonator on an atom chip. *Phy. Rev. Lett.* **99**, 140403.

6 Günther, A., Kraft, S., Kemmler, M., Koelle, D., Kleiner, R., Zimmermann, C., and Fortágh, J. (2005) Diffraction of a Bose–Einstein condensate from a magnetic lattice on a micro chip. *Phys. Rev. Lett.* **95**, 170405.

7 Günther, A., Kraft, S., Zimmermann, C., and Fortágh, J. (2007) Atom interferometer based on phase coherent splitting of Bose–Einstein condensates with an integrated magnetic grating. *Phys. Rev. Lett* **98**, 140403.

8 Günther, A., Kemmler, M., Kraft, S., Vale, C.J., Zimmermann, C., and Fortágh, J. (2005) Combined chips for atom optics. *Phys. Rev. A* **71**, 063619.

9 Judd, T.E., Scott, R.G., and Fromhold, T.M. (2008) Atom-chip diffraction of Bose–Einstein condensates: The role of interatomic interactions. *Phys. Rev. A* **78**, 053623.

10 Fortágh, J., Ott, H., Kraft, S., Günther, A., and Zimmermann, C. (2003) Bose–Einstein condensates in magnetic waveguides. *Appl. Phys. B* **76**, 157.

11 Sukumar, C.V. and Brink, D.M. (1997) Spin-flip transitions in a magnetic trap. *Phys. Rev. A* **56**, 2451.

12 Ott, H., Fortágh, J., Schlotterbeck, G., Grossmann, A., and Zimmermann, C. (2001) Bose–Einstein condensation in a surface microtrap. *Phys. Rev. Lett.* **87**, 230401.

13 Fortágh, J., Ott, H., Kraft, S., Günther, A., and Zimmermann, C. (2002) Surface effects in magnetic microtraps. *Phys. Rev. A* **66**, 041604.

14 Leanhardt, A.E., Chikkatur, A.P., Kielpinski, D., Shin, Y., Gustavson, T.L., Ketterle, W., and Pritchard, D.E. (2002) Propagation of Bose–Einstein

condensates in a magnetic waveguide. *Phys. Rev. Lett.* **89**, 040401.

15 Estève, J., Aussibal, C., Schumm, T., Figl, C., Mailly, D., Bouchoule, I., Westbrook, C.I., and Aspect, A. (2004) Role of wire imperfections in micromagnetic traps for atoms. *Phys. Rev. A* **70**, 043629.

16 Wildermuth, S., Lesanovsky, S.H.I., Haller, E., Andersson, L., Groth, S., Bar-Joseph, I., Krüger, P., and Schmiedmayer, J. (2005) Microscopic magnetic-field imaging. *Nature* **435**, 440.

17 Jones, M., Vale, C., Sahagun, D., Hall, B., and Hinds, E. (2003) Spin coupling between cold atoms and the thermal fluctuations of a metal surface. *Phys. Rev. Lett.* **91**, 080401.

18 Henkel, C., Krüger, P., Folman, R., and Schmiedmayer, J. (2003) Fundamental limits for coherent manipulation on atom chips. *Appl. Phys. B* **76**, 173.

19 Resbond 920, Contronics Corp. 3379 Shorw Pkwy. Brooklyn, New York.

20 Jackson, J.D. (1974) *Classical Electrodynamics*, John Wiley & Sons, New York, 2nd edn.

21 Gradsteyn, I.S. and Ryzhik, I.M. (1965) *Table of Integrals, Series and Products*, Academic Press, New York and London.

22 Lau, D., Sidorov, A., Opat, G., McLean, R., Rowlands, W., and Hannaford, P. (1999) Reflection of cold atoms from an array of current-carrying wires. *Eur. Phys. J. D* **5**, 193.

23 Reimann, P., Grifoni, M., and Hänggi, P. (1997) Quantum Ratchets. *Phys. Rev. Lett.* **79**, 10.

24 Ashcroft, N.W. and Mermin, N.D. (1976) *Solid state physics*, Brooks Cole, New York.

25 Henkel, C., Courtois, J.-Y., and Aspect, A. (1994) Atomic diffraction by a thin phase grating. *J. Phys. II France* **4**, 1955.

26 Weisstein, E.W. *Jacobi–Anger Expansion*, From MathWorld – A Wolfram Web Resource, http://mathworld.wolfram.com.

27 Dalfovo, F., Giorgini, S., Pitaevskii, L., and Stringari, S. (1999) Theory of Bose–Einstein condensation in trapped gases. *Rev. Mod. Phys.* **71**, 463.

28 Castin, Y. and Dum, R. (1996) Bose–Einstein condensates in time dependent traps. *Phys. Rev. Lett.* **77**, 5315.

29 Schumm, T., Hofferberth, S., Andersson, L.M., Wildermuth, S., Groth, S., Bar-Joseph, I., Schmiedmayer, J., and Krüger, P. (2005) Matter-wave interferometry in a double well on an atom chip. *Nature Physics* **1**, 57.

30 Courteille, P.W., Deh, B., Fortágh, J., Günther, A., Kraft, S., Marzok, C., Slama, S., and Zimmermann, C. (2006) Highly versatile atomic micro traps generated by multifrequency magnetic field modulation. *J. Phys. B: At. Mol. Opt. Phys.* **39**, 1055.

31 Lesanovsky, I., Schumm, T., Hofferberth, S., Andersson, L.M., Krüger, P., and Schmiedmayer, J. (2006) Adiabatic radio frequency potentials for the coherent manipulation of matter waves. *Phys. Rev. A* **73**, 033619.

32 Scott, R.G., Martin, A.M., Fromhold, T.M., and Sheard, F.W. (2005) Anomalous quantum reflection of Bose–Einstein condensates from a silicon surface: The role of dynamical excitations. *Phys. Rev. Lett.* **95**, 073201.

33 Scott, R.G., Hutchinson, D.A.W., and Gardiner, C.W. (2006) Disruption of reflecting Bose–Einstein condensates due to interatomic interactions and quantum noise. *Phys. Rev. A* **74**, 053605.

7
Interferometry with Bose–Einstein Condensates on Atom Chips

Thorsten Schumm, Stephanie Manz, Robert Bücker, David A. Smith, and Jörg Schmiedmayer

7.1
Introduction

Wire traps have been shown to be a versatile tool for robust trapping and manipulation of ultra-cold atomic gases [1, 2] as described in Chapter 2 of this book. The integration and miniaturization of current-carrying wires onto atom chips using methods from microelectronics enabled the realization of strongly confining magnetic traps. This yields a significant simplification of the creation of Bose–Einstein condensation (BEC) [3–5] in only a few seconds. Additionally, and maybe more importantly, the proximity of the atomic quantum gas to complex field-creating structures raises the possibility of manipulating the condensate wave function on its intrinsic length scale [1, 2, 6].

The ability to shape trapping potentials on a microscopic scale makes atom chips ideal candidates for implementing atom-optical elements. One of the long-standing goals is the realization of a complete integrated matter-wave interferometer using Bose–Einstein condensates on a single chip. On-chip BEC creation was first achieved in [3, 4] and successively realized in many labs worldwide. The concept of the wire trap itself (see Chapter 2) automatically provides a guiding mechanism for matter-waves in analogy to optical fibers and controlled transport of wave packets was realized in conveyor-belt structures [4, 7]. Other elements known from laser optics like mirrors [8, 9], gratings [10], and lenses [11] have equally been implemented individually. Even the detection of atomic samples [12] down to the single particle level has recently been realized on an atom chip [13–15].

However, the key element of an interferometer, a phase-preserving splitter and combiner, turned out to be surprisingly complicated to realize with static magnetic field configurations, both for technical and conceptual reasons [16, 17]. Despite the enormous freedom in designing potential "landscapes" on an atom chip, Maxwell's equations of electrodynamics prevent the splitting of a single magnetic trap of lowest (quadrupole) order in the transverse, strongly confining direction of a magnetic trap [18]. Only the hexapole configuration (which can be construct-

Atom Chips. Edited by Jakob Reichel and Vladan Vuletić
Copyright © 2011 WILEY-VCH Verlag GmbH & Co. KGaA, Weinheim
ISBN: 978-3-527-40755-2

ed as two spatially coalescing quadrupoles) allows splitting into two traps. In the hexapole configuration atomic confinement is weak and the configuration is highly sensitive to feeble magnetic field fluctuations [19]. Implementing this "transverse" splitting scheme requires strong gradients and absolute field stability to reduce the influence of fluctuations. It has been shown that this can only be realized with small atom chip structures on the micrometer level, which makes the system difficult to fabricate and vulnerable to surface effects [19]. A splitting of the wave function along the trapping wire and hence the weakly confining direction of a static magnetic wire trap circumvents topological problems [20] but suffers from even weaker atomic confinement and hence comparable stability issues. As before, only very small trapping structures will be able to provide sufficient stability to enable coherent manipulation and splitting of the wave function [21]. As a consequence, despite numerous attempts and approaches, no phase-preserving splitter for Bose–Einstein condensates could so far be realized using static magnetic fields only.

This problem could be overcome by the introduction of dressed-state adiabatic potentials [22–25]. Here, internal Zeeman states of atoms trapped in static magnetic atom chip traps are strongly coupled by means of oscillating magnetic fields in the radio-frequency (RF) range. This leads to new "dressed" eigenstates, which can be described as acting in an effective dressed potential. These dressed potentials are not bound by Maxwell's equations and hence enable new topologies. Most importantly in the present context, dressed potentials allow the splitting of a single atom trap of quadrupole geometry to a double-well potential while maintaining the strong atomic confinement provided by the atom chip. This renders the system stable against fluctuations and interferometry can be performed with comparatively large trapping structures at considerable distances (both of the order of 100 μm), mitigating malicious surface effects. As a consequence, phase-coherent splitting and interferometry on an atom chip could be demonstrated in [25] and has now become a standard technique in many atom chip experiments [26–30].

In this chapter we will briefly review schemes and efforts towards atom chip interferometry based on static magnetic and electric fields in Section 7.2. In Section 7.3 we will then give an introduction to dressed-state adiabatic potentials and describe the formation of double-well potentials and further possible trapping topologies not accessible with static fields alone. We will then review recent experiments performing interferometry with Bose–Einstein condensates in Section 7.4. Here, we focus on coherence properties and different ways to measure the relative phase of coherently split or independently created condensates. We will investigate fundamental decoherence processes, namely phase diffusion, in three-dimensional Bose condensates and discuss the recombination of BECs after a coherent splitting. Performing matter-wave interferometry with one-dimensional samples allows one to probe static and dynamic properties of thermal and quantum phase fluctuations and will be discussed in Section 7.5.

In the following we will focus on the most simple implementation of an atom-chip BEC splitter, based on transforming a single trap containing a ground-state matter-wave function into a double-well potential with wave packets localized at the respective trap minima. This scheme can either be implemented in the time domain by dynamically deforming a three-dimensional single trapping potential into a double trap or in the spatial domain, by realizing a Y-shaped trapping geometry and have the atoms propagate through the junction. The recombination of the wave packets is realized by either inverting the splitting process or by overlapping both wave packets in free expansion after release from the atom chip trap.

We will focus on interferometry in real space in contrast to interferometry in momentum space as realized with optical Bragg pulses [10, 31–33] or by phase imprinting using magnetic lattices (see Chapter 6 of this book). This essentially means that the interferometer sequence can in most cases be stopped during the split configuration, allowing for variable hold times, differential phase evolution and the study of decoherence processes as described in the following sections.

Double-well potentials will be discussed for their application as BEC splitters. Other fascinating aspects like the observation of Josephson tunneling [34, 35] or the implementation of quantum phase gates with state selective potentials [21, 36, 37] will only be mentioned briefly. We will also assume the wave packets to be in a single internal quantum state, interferometry with internal states on an atom chip has been demonstrated in [38] and is discussed in Chapter 8 of this book.

7.2
Atom Chip BEC Splitters Based on Static Fields

In this section we will concentrate on beam-splitting schemes and proposals based on *static* magnetic and electric fields. Static in this context means that potential variations take place on a timescale $\tau \gg \omega_L^{-1}$ where $\omega_L = |m_F g_F \mu_B B_{\text{trap}}|/\hbar$ is the Larmor frequency of the magnetically trapped atoms at any point in the trap. This is to be seen in contrast to schemes based on adiabatic potentials, where different atomic Zeeman or hyperfine states are coupled by means of (near-resonant) rapidly oscillating magnetic fields (see Section 7.3). Slow potential variations, that do not lead to spin-flips or coupling may however be necessary for the following schemes for example for dynamic BEC splitting in the time domain.

7.2.1
Transverse Splitting

By *transverse* splitting we denote splitting along the strongly confined direction of the initial single magnetic atom chip trap. In most realizations of atom chip traps, the strong confinement is implemented by a side wire guide (see Chapter 2), *transverse* hence also refers to the direction of the wire. Along the line connecting both potential minima of the double well after splitting, the magnetic field changes in both amplitude and direction.

7.2.1.1 The Two-Wire Splitter

The simplest way to realize a double-well potential is based on the magnetic fields of two parallel side wire guides together with a common external bias field B_{bias} (see Chapter 2). It can be shown that in a planar wire configuration, the number of necessary wires is equal to or larger than the number of potential minima [39] one wants to create. The so-called *two-wire splitter* was first proposed in [40, 41] for parallel current-carrying wires and [36, 42] proposed similar geometries based on periodically magnetized magnetic films.

We assume two parallel wires separated by $2d$, carrying an equal current I as indicated in Figure 7.1. For an external bias field B_{bias} smaller than a critical value B_c two quadrupole traps are formed on a vertical line between the two wires. For $B_{bias} = B_c$, the two quadrupoles coalesce to form a field of hexapole shape with a single potential minimum at a distance d from the wire surface. For $B_{bias} > B_c$ the hexapole again splits into two quadrupole traps, located on a circle of radius d. As in most atom chip experiments the chip surface is mounted horizontally, the horizontal splitting obtained for $B_{bias} > B_c$ is favorable, as both wells are equally affected by gravity.

For the theoretical description of the system, it is convenient to consider the situation $B_{bias} = B_c$ and add a small additional homogeneous field b to describe the splitting of the hexapole. Around the hexapole minimum, the magnetic field is

$$B_x = A\left(y^2 - x^2\right) \quad \text{and} \quad B_y = 2Axy, \tag{7.1}$$

where we have positioned the reference system at the coalescence point, and

$$A = \frac{\mu_0 I}{4\pi d^3} \tag{7.2}$$

describes the strength of the hexapole field in analogy to the gradient describing the strength of a quadrupole field (see Chapter 2). The critical bias field B_c to superimpose both quadrupole traps at the coalescence point is

$$B_c = \frac{\mu_0 I}{2\pi d} = Ad^2. \tag{7.3}$$

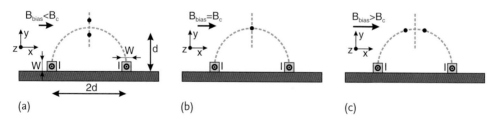

Figure 7.1 Schematic representation of the two-wire splitter/double-well scheme proposed by [40]: for a small external bias field B_{bias}, two trap minima form along a vertical line (a), as indicated by black dots. They merge at the coalescence point located at a distance d from the surface for $B_{bias} = B_c$ (b) and split horizontally for larger bias field (c). For $B_{bias} > B_c$, both wells are located on a half circle of radius d.

7.2 Atom Chip BEC Splitters Based on Static Fields

To prevent Majorana spin-flip losses at the potential minima, one superimposes a homogeneous longitudinal ("Ioffe") field $B_z = B_I$. Along the main axes of the hexapole (e.g., $y = 0$) the magnetic field is then

$$|B(r)| = \sqrt{B_I^2 + A^2 r^4} \approx B_I \left(1 + \frac{1}{2}\left(\frac{A}{B_I}\right)^2 r^4\right) = B_I \left(1 + \frac{1}{2}\left(\frac{B_c}{B_I}\right)^2 \left(\frac{r}{d}\right)^4\right). \tag{7.4}$$

At the coalescence point, the atomic confinement hence relies on a quartic magnetic field dependence, where the strength is determined by the ratio B_c/B_I.

Adding a small homogeneous field b to B_c will split the hexapole into two quadrupole traps separated (to first order) by $2r_0$, where

$$r_0 = \sqrt{\frac{b}{A}} = d\sqrt{\frac{b}{B_c}}. \tag{7.5}$$

If the field b is adjusted with an angle θ with respect to the horizontal x-axis, the two trap minima will be located on a line which makes an angle $-\theta/2$ with the x-axis. By rotating the external field b, the minima of the double well can be rotated on a circle of radius r_0 (see Figure 7.2). Applying for example a field orthogonal to the chip surface will create a double well tilted by 45°. In the following we assume a reference system rotated by the angle $-\theta/2$, so that the new x' axis contains both trap minima.

In the limit $B_I \gg b$ and using the newly defined axes $\{x', y'\}$, we can develop the 2D double-well potential to

$$V(x', y') = \frac{m\omega_0^2}{4} y'^2 + \frac{m\omega_0^2}{4x_0'^2} x'^2 y'^2 + \frac{m\omega_0^2}{8x_0'^2} \left(x'^2 - x_0'^2\right)^2, \tag{7.6}$$

where $x_0' = r_0 = \sqrt{b/A}$ is the position of the trap minima and

$$\omega_0 = \sqrt{\frac{4 m_F g_F \mu_B A b}{m B_I}}. \tag{7.7}$$

The two-wire configuration seems in principle well suited for the realization of a matter-wave splitter. In practice, this setup is very sensitive to noise in the involved

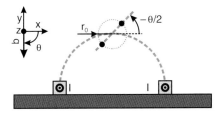

Figure 7.2 Tilting the double well in the two-wire scheme: a small magnetic field b splits the double well to a radius $r_0 = \sqrt{A/b}$, rotating this field by θ rotates the double well by $-\theta/2$. A field b orthogonal to the chip surface rotates the double well to 45°.

currents and magnetic fields: as an example we will analyze fluctuations in the trap distance and hence tunnel coupling due to instabilities in the external field b effectuating the splitting. From Eq. (7.3) and Eq. (7.5) we derive

$$\frac{\delta r_0}{r_0} = \frac{1}{2}\frac{\delta b}{b} = \frac{1}{2}\frac{d^2}{r_0^2}\frac{\delta b}{B_c}. \tag{7.8}$$

This illustrates that stability can be increased using small wire separations d (comparable to the desired double-well separation r_0) and large external fields B_c (and hence high wire currents, see Eq. (7.2)).

As an example we consider two wires of $W = 500$ nm square cross-section, separated by $2d = 2.5$ µm [19]. To estimate the required stability, we set the parameters (coalescence field $B_c = 16$ G, splitting field $b = 183$ mG) to obtain a tunnel frequency of $2\pi \times 100$ Hz (for a single particle) between the wells. A magnetic field fluctuation of 2 mG would be sufficient to change the tunnel coupling by 10%, not accounting for possible fluctuations in the wire currents. As a consequence, the total magnetic field of 16 G has to be controlled at least on the 2 mG level (relative stability of $\approx 10^{-4}$) which is at the limit of conventional current sources and will probably require additional active or passive magnetic shielding. An experimental realization of a two-wire splitter configuration with $r_0/d \approx 0.1$ ($2d = 300$ µm, $W = 50$ µm) succeeded to observe interference between split condensates, however technical fluctuations lead to a non-reproducible relative phase in successive experiments [43].

7.2.1.2 The Five-Wire Splitter

To circumvent these stability problems, the Orsay/Palaiseau group proposed the use of a system where the magnetic fields forming the hexapole trap are entirely generated by current-carrying chip wires [19]. In analogy to the self-sustaining three-wire traps [44] consisting of a main side guide wire and two parallel wires creating the "external" bias field, a *five-wire geometry* can create a stable hexapole field (see Figure 7.3). The final splitting into the double-well potential is still performed by adding a small homogeneous external field b, which now only has to be stable on the percent level to realize a stable double-well potential. As the same current will be used in all five wires, a high level of noise rejection can be obtained.

The five-wire structure can be understood as two close-by three-wire traps, where the two central wires were merged into one in order to approach the two wells. It is interesting to note, that a hexapole magnetic field cannot be created by less than five wires in a planar geometry [39]. As the wire system will be symmetric, only the two distances d and D completely determine the pattern, as indicated in Figure 7.3.

We are now interested in calculating the distances for which the same current I in all wires creates a single trap of hexapole shape. For symmetry reasons, this trap will be located on the y-axis above the central wire. Along this axis the magnetic field is

$$B_x = \frac{\mu_0 I}{2\pi}\left(\frac{2y}{D^2 + y^2} - \frac{2y}{d^2 + y^2} + \frac{1}{y}\right). \tag{7.9}$$

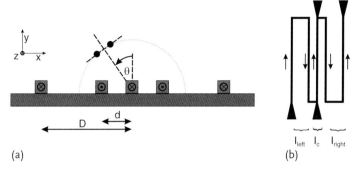

Figure 7.3 Five-wire configuration to realize a stable matter-wave splitter (a). Equal currents in all wires create a single trap directly above the central wire. Additional contacts on the wires in series enable control of the individual currents in the outer two branches and the central wire (b). By unbalancing these currents the hexapole can be moved on a circle of radius $\sqrt{3}d$ (a). The splitting can be either performed by an additional external field b or by increasing the central wire current.

This field has a double zero which indicates the hexapole configuration for $D = 3d$. The height of the hexapole trap above the central wire will be $y_0 = \sqrt{3}d$. The position of this coalescence point is exclusively fixed by the wire pattern. It does not depend on the current in the wires, which enhances stability.

An expansion of the magnetic field around the hexapole position gives

$$B_x = A'\left(x^2 - y^2\right) \tag{7.10}$$

$$B_y = -2A'xy\,, \tag{7.11}$$

where $A' = (4/\sqrt{3})\,A$ and A is defined as in (7.2). Apart from the factor $4/\sqrt{3}$, the five-wire configuration with wire separations d and $3d$ creates an identical magnetic field as two wires separated by $2d$ together with an external magnetic field $B_{\text{bias}} = B_c$. The potential around the hexapole configurations is well described by (7.6). However, the problem of magnetic field fluctuations is transferred to stability constraints on the wire current and the precision of the lithographic process. Using actively stabilized current drivers, it is relatively easy to obtain a current stability of $\delta I/I < 10^{-4}$. Also, standard micro-fabrication techniques easily reach 10^{-4} precision. The quality of the micro-fabricated wires themselves still represents a critical issue as outlined in Section 3 of this book.

The splitting of the magnetic hexapole into a double well is realized by adding a small homogeneous field b, analogous to the two-wire system. A field stability of 1 % is sufficient to implement a stable double-well potential.

Splitting of a thermal atomic cloud has been realized using this scheme, as illustrated in Figure 7.4 [19]. Atoms were loaded into the "upper" well of a vertically split double well potential, generated by adding an external magnetic field parallel to the chip surface with equal current in all wires. By reducing and then inverting this field, the two wells were approached, merged at the coalescence point, and then split horizontally. By adding a small offset field orthogonal to the atom chip

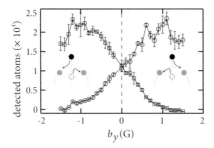

Figure 7.4 Splitting of a thermal cloud in the five-wire scheme. By adding a small additional field orthogonal to the chip surface, the two vertical traps (the one coming from the chip surface being empty) slightly "avoided" each other in the merging sequence, which allowed one to adjust the populations in the horizontally split double well. Atom loss for strongly imbalanced splitting was explained by a strong compression at the end of the sequence, "boiling" atoms out of the trap (figure from [19]).

surface, the spatial overlap of the two traps in the merging-splitting sequence could be adjusted. This allowed one to control the relative population of the horizontally split traps from 50 %/50 % to 100 %/0 %. Although the structures were miniaturized to 700 nm wires with 2.5 μm gaps to maintain strong atomic confinement at the coalescence point (see Section 7.2.1.1), a significant heating of the sample was observed when going through the splitting (not observed when the two traps entirely avoided each other). Because of this heating, the sequence could not be performed with a BEC.

7.2.1.3 The Y Splitter

The splitting schemes described above are designed to dynamically deform a single atom trap into a double-well potential and thus realize a BEC splitting in the time domain. To realize a splitting of atoms propagating along an axis, wire structures have to be designed accordingly to implement the so-called Y splitter scheme. Here, a region providing a single (transverse) trap and a region of double-well shape (again in the transverse direction) coexist at the same time at different (longitudinal) positions on the chip.

Figure 7.5 illustrates several realizations of a Y splitter scheme. The most simple realization, indicated in Figure 7.5a, consists of a single side wire guide, forking into two outgoing wires [18]. Similar geometries have been used in photon and electron interferometers [1, 45]. The resulting trapping potential features one main input port and two output waveguides. An additional "loss channel" port emerges from the atom chip surface and connects to the coalescence point. The position of the trapping potential minima can be described by

$$y_\pm = \frac{1}{2}\left(d_\text{split} \pm \sqrt{d_\text{split}^2 - d^2}\right) \quad (7.12)$$

where d is the distance between the outgoing wires and

$$d_\text{split} = \frac{\mu_0}{\pi} \frac{I}{B_\text{bias}}. \quad (7.13)$$

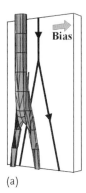 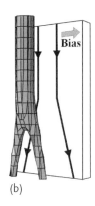 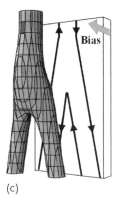 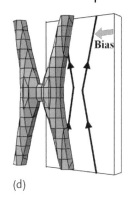

(a) (b) (c) (d)

Figure 7.5 Different realization of the Y splitter scheme: Indicated are the wire arrangements on the surface of an atom chip, the direction of current flow, and the direction of the external bias field. Also shown are typical equipotential surfaces to illustrate the shape of the resulting potential. (a) A simple Y splitter consisting of a single wire that is split into two: The output side guides are more confining and closer to the surface than the input guide. Note that a second minimum closer to the chip surface occurs in the region between the wire splitting and the actual split point of the potential. (b) A two-wire guide split into two single-wire guides does not exhibit this "loss channel" if the external field is tuned to the coalescence field for the input channel. (c) Here, the output guides have the same characteristics as the input guide minimizing the back-scattered amplitude. The vertical orientation of the bias field ensures exact symmetry of the two output guides. (d) In an X-shaped wire pattern the splitting occurs because of tunneling between two side guides in the region of close approach of the two wires (figure from [1]).

Note that the two waveguides merge at $d = d_{split}$ and not at the forking point of the wire structure. Here, the coalescence condition (7.3) is fulfilled and one locally recovers the physics of the two-wire splitter described in Section 7.2.1.1.

The simple Y splitter scheme base on a Y-shaped wire has several drawbacks. First, as current is halved in the outgoing leads, the output ports are more strongly confined and located closer to the atom chip surface, which can lead to back reflections [18]. Second the "loss channel" towards the surface inhibits ground-state splitting and leads to excitations and loss. So far, only splitting of a ballistically expanding thermal atom cloud could be demonstrated, as indicated in Figure 7.6. If employed in a Ioffe–Pritchard configuration with an additional longitudinal "Ioffe" field B_I, the two output ports are slightly imbalanced due to the different angle of the respective wires. This can be partially compensated by adjusting the current in the output leads (see Figure 7.6c).

To circumvent the "loss channel", a two-wire configuration as shown in Figure 7.5b can be employed. Here, the input port is realized by a two-wire guide with a bias field B_c, realizing the coalescence situation as described in 7.2.1.1. Note, that as equally described above, this configuration is very vulnerable to magnetic or current fluctuations. Still, the two output ports show stronger confinement than the input port (which now relies on a hexapole configuration). A way to solve this is to use a two-wire configuration with opposite currents and a bias field orthogonal to the atom chip plane [46, 47] (Figure 7.5c) realizing a truly symmetric splitter.

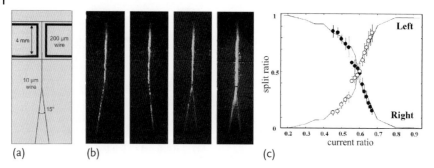

Figure 7.6 A Y splitter on an atom chip. (a) Wire layout realizing a simple Y splitter by forking a side guide wire, initial trapping is realized by a double U configuration. (b) Fluorescence images of thermal atoms propagating along the Y splitter for different settings of the relative currents in the output ports. (c) Splitting ratios obtained for different current ratios. The solid lines indicate a classical Monte Carlo simulation [18].

We find it interesting to note that while most beam-splitting schemes rely on single-mode splitting of Bose–Einstein condensates, a back-to-back Y splitter interferometer can in principle operate with many transverse modes. This is based on the fact that in a symmetric Y beam splitting, two adjacent symmetric and antisymmetric wave functions become degenerate in the limit of strong splitting,[14] realizing disjunct interferometers for each pair of vibrational states. The output of each of these interferometers can be in-phase at specific times, allowing high-contrast interference with a thermal sample occupying many transverse levels [48]. However, this scheme still remains to be realized experimentally.

The simple Y splitter scheme depicted in Figure 7.5a has been implemented experimentally with free-standing wires [41] and on atom chips [18]. In the atom chip case ^{7}Li atoms were loaded from a single straight side wire guide combined with two large U-shaped wires (Figure 7.6a). The atoms were released into the guide and pushed towards the Y splitter with an additional longitudinal gradient (Figure 7.6b). By adjusting the relative currents in the two output ports, the relative population in the output ports could be controlled between 50%/50% and 85%/15% (Figure 7.6c).

The X junction illustrated in Figure 7.5d has been experimentally demonstrated in [49] for guided thermal atoms using two square 100 μm-wide conductors with equal and parallel currents in a bias field. One of the arms was loaded with atoms. Depending on the current distribution in the wires, the ratio of atom numbers in the two output ports was varied between 100%/0% and 15%/85%, respectively. The splitting ratio also depends on the bias field and the bending of the conductors, as these parameters determine the overlap of the waveguides.

---

14) For optimal operation, the transverse trapping frequency in the split double-well situation should be twice the trap frequency of the single input channel.

7.2.2
Longitudinal Splitting

An alternative scheme to realize a deformation of a single trap into a double-well potential and hence a splitting of a BEC into two has been proposed in [20]. This scheme relies on crossed wires, where the longitudinal "Ioffe" field of a side wire guide trap is spatially deformed to a double-well shape (see Figure 7.8a). Here, the magnetic fields along the line connecting the trap minima only change in amplitude, changes in direction can be neglected. Note that this splitting scheme only allows splitting and interferometry in the time domain.

A current I_0 in a main trapping wire together with an orthogonal bias field B_{bias} creates the main wire trap as indicated in Figure 7.7. Two crossing wires carrying a current I_{ext} provide longitudinal confinement along the side wire guide, acting as a magnetic "dimple". An additional central crossing wire (I_c) with opposite current is used to erect a potential barrier. A detailed theoretical analysis of the effectively (longitudinal) 1D dynamics has shown that excitations of higher vibrational states can be suppressed to below 1% when using optimized splitting ramps [20]. Because of the fact that the double-well potential is constructed in the weakly confining direction, the spatial separation on the split wave packets is comparable to the distances between the wires. To realize rapid tunneling or to achieve sufficient overlap of the wave functions in order to observe interference effects in expansion, the wire structure shown in Figure 7.7 has to be miniaturized. Recent proposals following this scheme employ wire structures below the micrometer level [21, 37, 50, 51].

A conceptually similar scheme has been experimentally realized in [52]. Here, two longitudinally shifted U-shaped wires were used to replace a wire crossing, which is often difficult to fabricate in a planar atom chip geometry (see Figure 7.8). Current in the two outer leads of the U wires provided the longitudinal confinement, similar to the basic Z wire trap described in Chapter 2, replacing the two outer confinement wires in Figure 7.7. The inner leads realized a magnetic field comparable to the central crossing wire in Figure 7.7, that implements the double-well barrier. Transverse confinement was provided by a standard side wire guide carrying a current I_0. Lowering this current allowed the magnetic trap to approach the chip, deforming the single well (with the effect of the barrier averaged out) to

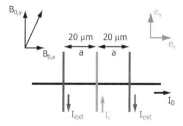

Figure 7.7 Schematic representation of the crossing wire layout to realize a double well in the longitudinal direction of the main trapping side guide wire (I_0) (figure from [20]).

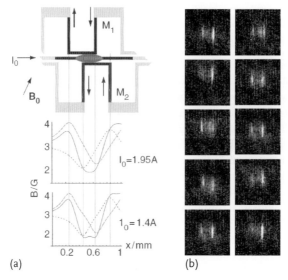

Figure 7.8 (a) Top: Schematic wire scheme to implement a longitudinal double-well potential without the need for wire crossings. Bottom: Deformation of a single trapping potential to a double well by decreasing the current in the main trapping wire and hence approaching the atomic sample from 126 µm to 87 µm to the chip surface. (b) Successive absorption images of split Bose–Einstein condensates released from the double well. Each condensate contains between 500 and 1500 atoms (figure from [52]).

a double-well potential (see Figure 7.8a) with a separation of 135 µm between the wells.

Bose–Einstein condensates could successfully be split in this geometry, Figure 7.8b shows successive time-of-flight (TOF) absorption images after performing the experimental sequence. BECs could clearly be identified in both wells, however the large splitting distance made the system vulnerable to fluctuations, which lead to instabilities in the relative population. The condensate wave functions did not overlap in expansion, hence no interference could be observed.

7.2.3
Electrostatic Splitter

An interesting approach to realizing a double-well potential is to make use of local DC Stark shifts generated by electric fields on the atom chip, which additionally pattern and deform a static magnetic trap [53]. The interaction of a neutral but polarizable (polarizability α) atom with an electrostatic field E is described by the interaction potential

$$U_{el} = -\frac{\alpha}{2} E^2 \, . \tag{7.14}$$

Equation (7.14) shows that atoms will always be attracted to regions of high electric field. As a consequence, the minimum of a purely electrostatic trapping potential will always be located at the (local) maximum of E. According to Earnshaw's the-

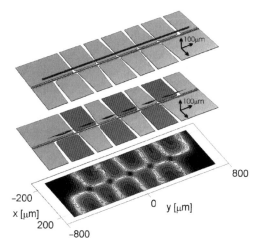

Figure 7.9 Combining electric and magnetic interaction on an atom chip. The magnetic trap of a straight side wire guide is longitudinally patterned by electric fields generated by on-chip electrodes. Typical voltages across the gaps are 475 V (figure from [53]).

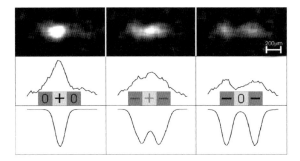

Figure 7.10 Dynamic splitting of a thermal atom cloud using electric potentials created by three electrodes on an atom chip (figure from [53]).

orem [54], such a maximum can not exist in free space, hence purely electrostatic trapping is not possible. However, the (attractive) electric potential can be combined with a (repulsive) magnetic potential to realize a three-dimensional confinement above the atom chip. Figure 7.9 shows equipotential lines of such a combined magneto-electrical trap. A standard side wire guiding potential is longitudinally patterned by electrical fields, generated by on-chip electrodes.

This combination of electrostatic fields and static magnetic traps can be employed to realize an atom chip BEC splitter as depicted in Figure 7.10. By dynamically adjusting control voltages of three electrodes, a single well could be dynamically deformed to a double-well potential. This has been qualitatively demonstrated with thermal atomic clouds in [53]. It shall be pointed out, that the electrostatic and magnetic potential add up (to first order) independently and hence the splitting can be performed in any spatial direction (longitudinal and transverse) with an appropriately designed atom chip layout.

7.3
Atom Chip BEC Splitters Based on Dressed Adiabatic Potentials

Combining static magnetic trapping techniques with rapidly oscillating magnetic fields significantly enhances the flexibility in designing potential landscapes on atom chips. Coupling internal Zeeman or hyperfine states of the trapped atoms leads to new "dressed" eigenstates of the system which can be expressed as effective adiabatic potentials. The employed oscillating fields are in the radio-frequency (RF) regime for coupling Zeeman transitions or in the microwave (MW) regime for coupling hyperfine transitions[15] and can be created with sizable amplitude using atom chip "antenna" wires[16]. In designing the wire patterns one obtains a high level of control over the amplitude and polarization of the oscillating fields which directly reflects in an enhanced flexibility in the design of dressed adiabatic potentials.

It shall be pointed out that the effective adiabatic potential originates from an *effective* magnetic "vector" field and hence allows the design of trapping geometries not realizable with static fields alone. Most notably in the present context, dressed adiabatic potentials enable the deformation of a single static magnetic trap into a double-well potential and hence the dynamic splitting of a Bose–Einstein condensate (both in space and in the time domain) for interferometry. Maintaining strong atomic confinement throughout the splitting (see Section 7.3.5), dressed adiabatic potentials overcome most of the difficulties in atom chip BEC splitter schemes described in the previous section and consequently enabled the first realization of coherent BEC interferometry on an atom chip [25].

Altering energy eigenstates by using dressed states was first employed in neutron physics [55]. First experiments with neutral atoms were performed in [56] where a macroscopic trap was generated by a standing MW field in a resonator. Combining static magnetic trapping with dressed adiabatic potentials was first proposed in 2001 by Zobay and Garraway [22]. Their approach relies on a generic static trap of Ioffe–Pritchard configuration. Combination with an oscillating RF magnetic field realizes closed-shell 3D trapping potentials of "bubble" geometry [23] that may be exploited to realize two-dimensional Bose–Einstein condensates [22]. Pioneering experiments realizing 2D thermal atomic clouds based on a macroscopic QUIC trap were carried out by the Villetaneuse group [24].

As pointed out in [17, 25], the relative orientation of the static magnetic trapping fields and the polarization of the oscillating fields have to be taken into account to compute the arising dressed adiabatic potentials (see Section 7.3.2). Designing the vector coupling by patterning wires on an atom chip allows for a large variety in

---

15) In the following we will consider mainly RF transitions between adjacent Zeeman levels which are of σ^+ or σ^- character, depending on the Landé g-factor of the trapped atomic state. Note that for MW transitions between different hyperfine states, σ^+, σ^-, and π transitions are possible, leading to different couplings and topologies than discussed here.

16) Note, that the considered oscillating RF fields are non-radiative near-fields, which for all purposes can be seen as DC (wire) fields with a time modulation $\sin \omega t$.

possible trapping topologies, such as 2D traps [23], ring traps [57], Mach–Zehnder geometries [17, 30], torus traps [28] and many more. Modulating dressed adiabatic potentials in time (time-averaged potentials) additionally enhances the flexibility [58]. As the coupling depends on the polarization of the RF field, the arising potentials depend on the Landé g-factor of the trapped atomic state. Hence, dressed double-well potentials can be made state-selective, which is an important prerequisite for the implementation of quantum information schemes [21, 36, 37, 57].

In the following we will briefly describe the concept of dressed-state adiabatic potentials and then focus on the implementation of a coherent splitter for Bose–Einstein condensates on an atom chip.

7.3.1
Dressed Adiabatic State Potentials

The power of RF-dressed state potentials arises from the combination of an RF oscillating magnetic field with an inhomogeneous static magnetic trap. We begin by introducing a time-independent static magnetic field $B_S(r)$ that constitutes the static trap. We will also introduce a time-dependent RF magnetic field $B_{RF}(r, t)$ which can be composed of several RF components:

$$B_{RF}(r, t) = \sum_n B_n^{RF}(r) \cos(\omega t - \theta_n) , \quad (7.15)$$

where ω and θ_n are the frequency and relative phase of the RF field component $B_n^{RF}(r)$.[17] As such, the total magnetic field is given by the vector sum of $B_S(r)$ and $B_{RF}(r, t)$:

$$B_{total}(r, t) = B_S(r) + B_{RF}(r, t) . \quad (7.16)$$

Before writing down the Hamiltonian of the system, we first make some assumptions and clarifications. First, we assume the oscillating magnetic field to be of high amplitude and classical, so that we do not account for quanta of photons being removed from or added to the RF field(s). Next, we assume that we are in the regime where the magnetic sub-level splitting due to the Zeeman effect is linear in the magnetic field (i.e., F is still a meaningful quantum number) such that we can write

$$H = \frac{p^2}{2m} + g_F \mu_B F \cdot B_{total}(r, t), \quad (7.17)$$

where μ_B is the Bohr magneton and g_F is the Landé g-factor.

We now make a transformation from the laboratory x, y, z coordinate system to a coordinate system where the new $\tilde{z}$-axis is parallel to $B_S(r)$.[18] Here, we take $B_S(r)$ as our *local* quantization axis. The stress here is on the word *local* as $B_S(r)$ spatially changes in direction and magnitude. We can use a unitary transformation $U_S(r)$ such that [17]

$$U_S^\dagger(r) F U_S(r) \cdot B_S(r) = F_{\tilde{z}} |B_S(r)| , \quad (7.18)$$

where $F_{\tilde{z}}$ is a diagonal matrix with values of $\tilde{m}_F$ from $-F$ to F.

---

17) Here we consider a single RF frequency ω, the case of multiple frequencies is discussed in [59].
18) Parameters denoted $\tilde{x}$, $\tilde{y}$, or $\tilde{z}$ are used to differentiate from parameters in the laboratory frame.

We can now write down the Hamiltonian in this new basis

$$\tilde{H} = \frac{1}{2m}\left[p + A(r,t)\right]^2 + g_F\mu_B |B_S(r)| F_{\tilde{z}} + g_F\mu_B B_{\tilde{z}}^{RF}(r,t) F_{\tilde{z}}$$
$$+ g_F\mu_B B_{\tilde{x}}^{RF}(r,t) F_{\tilde{x}} + g_F\mu_B B_{\tilde{y}}^{RF}(r,t) F_{\tilde{y}}, \tag{7.19}$$

where B_i^{RF} corresponds to the projections of the RF field onto the denoted axis. The gauge field $A(r,t)$ emerges due to the change in coordinate system. Note that the RF field is now split into components parallel and perpendicular to the quantization axis $\tilde{z}$ (i.e., parallel and perpendicular to $B_S(r)$), such that

$$B_{\parallel}^{RF}(r,t) = B_{\tilde{z}}^{RF}(r,t) \tag{7.20}$$

$$B_{\perp}^{RF}(r,t) = \sqrt{B_{\tilde{x}}^{RF}(r,t)^2 + B_{\tilde{y}}^{RF}(r,t)^2}. \tag{7.21}$$

It would now be useful to remove the time dependencies contained within the Hamiltonian in Eq. (7.19). The method for this removal is described in detail in [17] and will not be rewritten here. Briefly, we apply a unitary transformation to enter into a frame rotating at the RF frequency ω,

$$U_R = \exp\left[-i\frac{g_F}{|g_F|} F_{\tilde{z}}\omega t\right] \tag{7.22}$$

and apply a series of unitary transformations to remove unwanted terms, followed by three approximations (see below). We arrive at the now stationary Hamiltonian

$$H_{\text{approx}} = \frac{p^2}{2m} + g_F\mu_B |B_{\text{eff}}(r)| F_{\tilde{z}} \tag{7.23}$$

where the effective field is defined as[19]

$$B_{\text{eff},\tilde{x}}(r) = \frac{1}{2}\sum_n (\cos(\gamma_n), -\sin(\gamma_n), 0) \cdot B_n^{RF}(r)$$

$$B_{\text{eff},\tilde{y}}(r) = \frac{1}{2}\sum_n (\sin(\gamma_n), \cos(\gamma_n), 0) \cdot B_n^{RF}(r)$$

$$B_{\text{eff},\tilde{z}}(r) = |B_S(r)| - \frac{\hbar\omega}{|g_F|\mu_B} \tag{7.24}$$

with

$$\gamma_n = -\frac{g_F}{|g_F|}\theta_n. \tag{7.25}$$

The three approximations used to obtain Eq. (7.23) are as follows.

- First, we assume that the Larmor frequency associated with the $\tilde{z}$-component of the RF field(s) is much lower than the RF frequency ω itself. This assumption

[19] Again parameters denoted $\tilde{x}$, $\tilde{y}$, or $\tilde{z}$ are used to differentiate from parameters in the laboratory frame.

allows us to neglect the $\tilde{z}$-component of the RF field, that is that dynamics due to the RF field occur in the $\tilde{x}$–$\tilde{y}$ plane and the $\tilde{z}$-component of the total B-field is dominated by $\boldsymbol{B}_S(\boldsymbol{r})$.

- The second approximation is the rotating-wave approximation (RWA), which consists of neglecting the non-resonant processes (i.e., those that are not near the RF frequency ω) that occur in the atom-field system [60]. The Hamiltonian in Eq. (7.23) is, relative to the laboratory frame, in a frame rotating with the RF frequency ω around the $\tilde{z}$-axis. In this rotating frame, we apply the RWA by excluding all terms that oscillate with frequency 2ω. The rotation Eq. (7.22), therefore, predisposes which terms are neglected and which ones are selected by Eq. (7.22) depending on the sign of the Landé g-factor g_F. Equation (7.22) ensures that the rotation is in the direction of the precession of the atomic spin around the quantization axis.
- The third approximation is the adiabatic approximation [61, 62]. This consists of neglecting the gauge term arising from the change in reference frame, which is essentially assuming that the orientation between the total spin vector and the magnetic field remains constant during the atoms motion in the trap.

We have now arrived at the Hamiltonian Eq. (7.23) and will discuss its components in more detail. Equations (7.24) define an effective magnetic field $\boldsymbol{B}_{\text{eff}}(\boldsymbol{r})$. It shall be pointed out that $\boldsymbol{B}_{\text{eff}}(\boldsymbol{r})$ is not a "true" magnetic field, in the sense that it is constructed from static and time-dependent components. This, however, is not a disadvantage since it means that RF-dressed state potentials are not subject to Maxwell's equations or Earnshaw's theorem [54], meaning that a large variety of potentials are possible. $\boldsymbol{B}_{\text{eff}}(\boldsymbol{r})$ represents the magnetic field to which the magnetic quantum numbers in the dressed state potential can be related. For $B_{\text{eff},\tilde{z}}(\boldsymbol{r}) \gg B_{\text{eff},\tilde{x}}(\boldsymbol{r})$, $B_{\text{eff},\tilde{y}}(\boldsymbol{r})$, this quantization axis is in the direction of the static field of the trap $\boldsymbol{B}_S(\boldsymbol{r})$. In addition, Eq. (7.25) defines an angle of rotation γ that is dependent upon the sign (but not the magnitude) of the Landé g-factor g_F and also the phase of the particular RF field component θ. The rotation due to γ is in the $\tilde{x}$–$\tilde{y}$ plane, that is around the direction of the static field direction. The effective magnetic field $\boldsymbol{B}_{\text{eff}}(\boldsymbol{r})$ lies in the plane spanned by the $\tilde{z}$-axis and γ. It is tilted away from the static quantization axis $\boldsymbol{B}_S(\boldsymbol{r})$ by an angle α with

$$\tan \alpha = \frac{\Omega(\boldsymbol{r})}{\Delta(\boldsymbol{r})} \tag{7.26}$$

defining the detuning term Δ as

$$\Delta(\boldsymbol{r}) = \frac{|g_F|\mu_B}{\hbar} B_{\text{eff},\tilde{z}}(\boldsymbol{r}) = \frac{|g_F|\mu_B}{\hbar}|\boldsymbol{B}_S(\boldsymbol{r})| - \omega \tag{7.27}$$

and the Rabi frequency Ω as

$$\Omega(\boldsymbol{r}) = \frac{|g_F|\mu_B}{\hbar}\sqrt{B_{\text{eff},\tilde{x}}(\boldsymbol{r})^2 + B_{\text{eff},\tilde{y}}(\boldsymbol{r})^2} \, . \tag{7.28}$$

The detuning Δ relates the RF frequency ω to the Lamor frequency of an atom at a specific position in the trap and Ω represents the coupling between the Zeeman sublevels at this position.[20] Both Δ and Ω are *local* values since they are spatially dependent, which is fundamental to the functionality of the RF-dressed state potentials. This spatial dependence in Δ and Ω is what leads to the deformation of the dressed adiabatic potential compared to the static one.

In taking the adiabatic approximation [61, 62], we can take Eq. (7.23) a step further and define the adiabatic potential seen by the atoms:[21]

$$H_{\text{approx}} = \frac{p^2}{2m} + g_F \mu_B |\mathbf{B}_{\text{eff}}(\mathbf{r})| F_z = \frac{p^2}{2m} + V_{\text{ad}}(\mathbf{r}) \tag{7.29}$$

where

$$V_{\text{ad}}(\mathbf{r}) = \tilde{m}_F g_F \mu_B \sqrt{B_{\text{eff},\tilde{z}}(\mathbf{r})^2 + B_{\text{eff},\tilde{y}}(\mathbf{r})^2 + B_{\text{eff},\tilde{x}}(\mathbf{r})^2}$$

$$= \tilde{m}_F \hbar \sqrt{\Delta(\mathbf{r})^2 + \Omega(\mathbf{r})^2} \,. \tag{7.30}$$

7.3.2
A BEC Splitter Based on Dressed Adiabatic State Potentials

Although a large variety of trapping geometries can be realized using dressed states, we will concentrate on RF-induced double-well potentials, which have been used for the interferometry experiments detailed in the later parts of this chapter. For the creation of this double-well potential, we require only a single linear RF field component, which we will define as having frequency ω,

$$\mathbf{B}_{\text{RF}}(t) = B_{\text{RF}} \cos(\omega t) \mathbf{e}_x \,. \tag{7.31}$$

For simplicity, we consider the RF field to have a constant amplitude over the region of trapped atoms. We assume a simple generic form for the static trapping field,

$$\mathbf{B}_s(r, \phi, z) = G r (\cos \phi \, \mathbf{e}_x - \sin \phi \, \mathbf{e}_y) + B_I(z) \mathbf{e}_z \,, \tag{7.32}$$

(i.e., cylindrically symmetric) where $r = \sqrt{x^2 + y^2}$, G is the gradient of the trap, and $\tan \phi = y/x$. Here we are concentrating on the 2D x–y plane in the region of the center of the trap ($z = 0$). We can then write the coupling term Ω as [17]

$$\Omega(\mathbf{r}) = \frac{|g_F| \mu_B}{\hbar} \sqrt{[B_{\text{RF}}/(2|\mathbf{B}_s(\mathbf{r})|)]^2 (B_I^2 + G^2 r^2 \sin^2 \phi)} \tag{7.33}$$

and, as before,

$$\Delta(\mathbf{r}) = \frac{|g_F| \mu_B}{\hbar} |\mathbf{B}_s(\mathbf{r})| - \omega \,.$$

For given magnetic fields and an RF frequency ω, the minima of this potential are located at $\phi = 0, \pi$, that is at two distinct points in space, giving rise to a

---

20) More precisely it relates ω to the Lamor frequency of an atom with $m_F = 1$.
21) Note that interesting effects like the appearance of geometric phases may be observed, when adiabaticity is not ensured [63].

double-well structure. Atoms close to the center of the trap ($Gr \ll B_I$) experience a potential (from Eq. (7.30))

$$V_{ad}(r, \phi = 0, \pi) = \tilde{m}_F g_F \mu_B \sqrt{\frac{G^4}{4B_I^2}\left(r^2 - r_0^2\right)^2 + B_0^2} \quad (7.34)$$

where

$$r_0 = \frac{\sqrt{B_{RF}^2 - B_C^2}}{\sqrt{2}G} \quad (7.35)$$

is the position of the potential minima and we define the critical field

$$B_C = 2\sqrt{\frac{\hbar \Delta}{|g_F|\mu_B} B_I}. \quad (7.36)$$

The condition $B_{RF} > B_C$ must be satisfied for the creation of a double-well structure. The magnetic field at position r_0, we define as B_0, where

$$B_0 \approx \frac{B_{RF}}{2}\sqrt{\frac{\hbar \Delta}{|g_F|\mu_B B_I} + 1}. \quad (7.37)$$

Also, the assumption $Gr \ll B_I$ leaves us with a detuning term

$$\Delta \approx \frac{|g_F|\mu_B}{\hbar} B_I - \omega. \quad (7.38)$$

The potential in Eq. (7.34) describes a double-well potential split along the x-axis, since this is the direction in which the RF was chosen to oscillate, and which coincides with one axis of the transverse quadrupole (see Figure 7.11). The minima occur where the components of $\boldsymbol{B}_{RF}$ and $\boldsymbol{B}_S$ are parallel in the plane perpendicular to the axis of the static trap. A rotation of the RF field in the x–y plane affectuates a rotation of the double well in the opposite direction (see Figure 7.12c) [57].

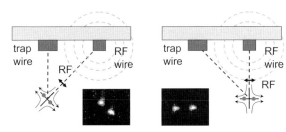

Figure 7.11 Balancing an RF-induced double-well potential. The spatial orientation of the double well is determined by the relative orientation of the static magnetic trapping field and the oscillating magnetic RF field; potential minima form where both fields are parallel, which results in a reduced local coupling of the atomic Zeeman states. A balanced double-well potential, allowing equal splitting of the condensate, can be realized for several static trap positions between the trapping wire and the RF antenna.

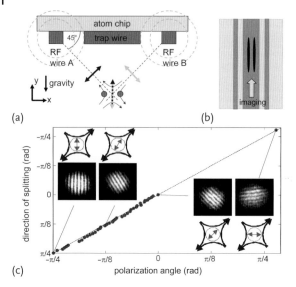

Figure 7.12 Atom chip setup for RF potentials. (a) Side view of the three-wire setup. A broad (100 μm) central wire creates the static trapping potential. Smaller (10 μm) wires on each side provide the oscillating magnetic fields that create the adiabatic RF potentials. The static trap is positioned such that the two RF fields are perpendicular at its center. (b) Top view of the atom chip showing the relevant wires. The longitudinal confinement of the static trap is provided by the outer leads of the Z-wire. Additional U-wires on the side can be used to increase this confinement. (c) Orientation of the double-well potential as a function of the plane of polarization in the case of a linearly polarized RF field. The double-well orientation is inferred from interference patterns. The observed fringes are always perpendicular to the double-well axis. Coherent splitting is possible in all directions.

7.3.3
Beyond the Rotating-Wave Approximation

The effective adiabatic dressed state potential (see Eqs. (7.30) and (7.34)) is derived under several assumptions detailed in Section 7.3.1. Specifically, one assumes the Rabi frequency of the RF driving field Ω to be much smaller than the driving frequency ω and one assumes "near resonant coupling" meaning $\hbar\Delta \ll |g_F|\mu_B|\mathbf{B}_S(0)|$ where $\mathbf{B}_S(0)$ refers to the magnetic field at the center of the static magnetic trap.

In experimental realizations of dressed state potentials on atom chips, owing to the strong amplitude of the oscillating RF field, these assumptions may not hold. In this case, the time-dependent Hamiltonian Eq. (7.19) has to be solved numerically, for example using the Floquet formalism [64]. Here, the matrix ($m_F \times m_F$) representation of the time-dependent Hamiltonian is transferred to a time-independent ($\infty \times \infty$)-matrix which can be truncated and diagonalized numerically to obtain new eigenenergies of the dressed states.

Taking beyond-RWA effects into account has two important consequences: first, the obtained eigenenergies can differ from the ones calculated solving the simplified Hamiltonian Eq. (7.23). In the example of RF-induced dressed double wells, this significantly changes the system parameters such as tunneling rates, which depend exponentially on the details of the effective potential. Furthermore, additional "multi-photon" transitions between different manifolds of dressed states (characterized by RF photon number) become possible. Both effects have been observed spectroscopically in [65].

7.3.4
Implementation on an Atom Chip

Radio-frequency-induced double-well potentials were first implemented in [25] using a "one-plus-one-wire scheme" as depicted in Figure 7.11, similar schemes were then adapted in [26, 29, 30]. It consists of a trapping wire (e.g., Z-shaped wire trap as described in Chapter 2) carrying a DC current, which, together with a homogeneous external magnetic field, provides a static magnetic trap of Ioffe–Pritchard type. The static trap can be rotated around the trapping wire by changing the angle of the external field, changing it's magnitude will change the distance from the wire. An additional "RF wire" carrying an AC current creates the required oscillating magnetic field. It shall be pointed out, that the created oscillating fields are non-radiative near-fields, which for all purposes can be calculated as DC fields with a time modulation $\sin \omega t$.

As indicated in Figure 7.11, balanced double wells, yielding 50%/50% splitting of cold atomic clouds, can be realized at several positions on the atom chip. However, as ultra-cold atoms or Bose–Einstein condensates are extremely sensitive to small energy gradients (order $h \times 100$ Hz), balancing the double well is an issue. The following three processes contribute to double-well imbalance (ordered according to their influence in typical experimental realizations):

- Inhomogeneity of the RF field. The oscillating magnetic field varies in direction and amplitude over the size of the double well due to the $1/r$ dependence of the field amplitude and the finite size of the field-generating wire. This leads to a stronger coupling on the double well site closer to the RF wire and atoms favor the averted well. RF field inhomogeneity is the dominant imbalancing effect in most experimental realizations. It may be reduced or eliminated by using RF antenna wires with increased distance from the atoms, for example by fabricating them on the backside of the atom chip, using macroscopic wire structures or even coils [66].
- Gravity obviously creates a potential gradient and drags atoms into the lower well. To split a cloud of 100 nK temperature or $h \times 2$ kHz chemical potential, the double well has to be balanced to a height difference better than 500 nm. The influence of gravity is usually weaker than the effect of the $1/r$ dependence of the RF field amplitude.
- Inhomogeneity of the static trapping potential. As the static trapping potential is not perfectly rotational symmetric (the magnetic field gradient being slightly higher towards the trapping wire when moving away from the minimum) the

oscillation frequencies of the two double-well sites are not fully identical for widely split double-well potentials. This increases the energy of the ground-state (BEC) wave function, even when the potential minima are perfectly balanced. This effect is usually the weakest contribution to imbalance and can often be neglected.

As the above effects do not necessarily go in the same direction, they can be used to cancel each other out. Most importantly, the $1/r$ dependence of the RF field amplitude can be used to cancel the influence of gravity, allowing the realization of tilted (see Figure 7.11) or even entirely vertical double-well orientations. Configurations with a high degree of symmetry are easier to balance and less vulnerable to fluctuations. Initial balancing can be performed by simply looking at the relative atom population in the two wells. Fine tuning is achieved by adjusting the relative phase evolution in an interferometer sequence (see Section 7.4.2.3) [25].

Although the one-plus-one-wire scheme presented above allows the realization of a stable double-well potential and the implementation of a coherent splitter for Bose–Einstein condensates, the flexibility in choosing different RF polarizations is strongly limited. A more versatile "one-plus-two-wire-layout" is depicted in Figure 7.12. It makes use of one DC trapping wire and two RF wires, symmetrically fabricated on either side. By adjusting the relative phase and amplitude of the AC currents in the RF wires, any polarization of the RF field at the position of the atoms can be realized. Linear polarization of adjustable angle allows one to turn the double well at will, balancing is realized by adjusting the height of the static trap above the DC wire (see Figure 7.12). Most importantly, a stable balanced vertical double well can be realized which allows absorption imaging orthogonal to the elongated atomic clouds, parallel to the atom chip surface. This enables interferometric detection of longitudinal phase dynamics in 1D condensates as described in Section 7.5.

Applying a circularly polarized RF field[22] will create a ring-shaped trap. However, where balancing a double-well potential (e.g., two distinct points in space) is already experimentally demanding, balancing an RF-induced ring potential (e.g., a connected line in space) is extremely difficult and necessitates a more complex and dedicated atom chip design [66].

7.3.5
Advantages of RF-Induced Splitters over Static Splitters

RF-induced double-well potentials overcome many of the topological problems of static schemes described in Section 7.2 as they permit the true splitting of a single atom trap into a double-well potential without the presence of loss channels or asymmetries (see Figure 7.13). Furthermore, dressed adiabatic state double wells maintain tight atomic confinement throughout the splitting, making them less vulnerable to fluctuations in magnetic fields or wire currents. In consequence, the RF

---

22) The helicity depends on the sign of the g-factor of the trapped atomic state.

static 2-wire beam splitter 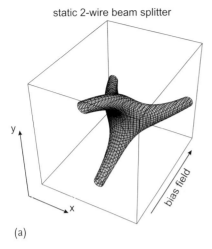   RF-induced beam splitter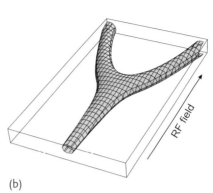

(a)   (b)

Figure 7.13 Equipotential surfaces for a static two-wire interferometer (a) and an interferometer based on dressed adiabatic potentials (b) in the vicinity of the splitting region. The advantage of the RF interferometer is the splitting of a true single input into two output ports. In contrast, the static implementation generates two input and two output ports.

double-well scheme can be implemented using comparably large wire structures (see Section 7.2.1.1). This on one side simplifies the fabrication of the atom chip, on the other side it permits experiments at large distances from the atom chip surface, mitigating malicious surface effects such as heating and fragmentation.

To illustrate the scaling of BEC splitter schemes based on static fields and dressed state potentials, we consider wire structures of a characteristic size d. We assume the static trap to be provided by a basic side wire guide, with also the distance from the atom chip surface being d. For the static trap we have in mind the two-wire scheme depicted in Figure 7.1 with a separation $2d$ between the wires, both running a DC current I. For the RF double-well scheme we compare to the one-plus-two-wire scheme depicted in Figure 7.12. Assuming equal currents I, equal double-well separations r_0, equal B_1 and $\Delta = 0$ we find

$$\frac{\omega_{RF}}{\omega_{static}} \propto \frac{d}{r_0} \sqrt{\frac{B_{RF}}{B_1}} \tag{7.39}$$

where ω_{RF} and ω_{static} are the oscillation frequencies of the double-well sites of the RF or static double well, respectively. As RF amplitudes can be significant in atom chip setups, we can set $B_{RF} \approx B_1$. Hence the ratio of the double-well separation r_0 to the structure size d describes the gain in trap frequency for the RF-induced scheme. As double-well separations are typically micrometers and structure sizes often are around 100 μm, this gain can be easily two orders of magnitude compared to the static two-wire scheme.

To estimate the required stability of experimental parameters, we perform an analysis similar to Section 7.2.1.1 for wires of different structure size d (d also being the distance of the static trap minimum from the wire). We again fix the tunnel coupling (for a single particle) in the double well to $2\pi \times 100$ Hz and furthermore enforce a single-well trap frequency of $2\pi \times 1$ kHz to fulfil the two-modes approximation. For wire sizes between 1 and 500 µm, we can always find a set of experimental parameters (regarding the RF amplitude and frequency) that meets these conditions, leading to double-well separations r_0 between 0.5 and 5 µm. To maintain a stable tunnelling within 10 %, only a relative stability of 1 % in *all* experimental parameters is required (the most sensitive being the RF amplitude) [16].

In conclusion, stable double-well potentials for splitting or tunnelling experiments can be implemented for a large range of structure sizes up to 500 µm with typical experimental stabilities using RF-dressed potentials. To realize similar experiments using static magnetic fields, the trapping structure has to be miniaturized down to a few micrometers (preferably making use of a noise-rejecting configuration, see Section 7.2.1.2), the atom–surface distance being of the same order. Furthermore, the experimental stability has to be excellent, necessitating active or passive magnetic field stabilization or shielding. It hence seems that splitters based on RF-induced adiabatic potentials are easier to implement and may be more robust against external perturbations. So far, RF-induced splitting has mainly been performed in the time domain, to which extent a full spatial interferometer (e.g., of Mach–Zehnder type) can be implemented is a question of ongoing research.

7.4
Matter–Wave Interferometry with Bose–Einstein Condensates

Interferometry with Bose–Einstein condensates is a powerful tool to probe properties like phase dynamics and (de-) coherence of interacting many-particle quantum systems. In this section, we briefly resume the theory describing interference of two spatially separated Bose–Einstein condensates. Each atomic sample will be treated as a coherent matter-wave of spatially uniform phase, which is common for the case of three-dimensional systems. Lower dimensional systems show thermal and quantum fluctuations of the phase, which will be described in Section 7.5. We will in particular review how interference patterns of Bose condensed atoms released from a double-well potential give information about the splitting process itself, the phase dynamics in a double well, and the merging of condensates after reversing the splitting process.

7.4.1
Theoretical Aspects

Before we start analyzing the outcome of interferometry with Bose–Einstein condensates, we have to describe what is input into the interferometer. Using ultra-cold gases, the objects which are brought into contact can consist of up to millions of atoms. Still, in the quantum degenerate regime, the interference does not have to be described on the single particle level, but as interference of macroscopic wave functions with a global probability distribution and phase. The theoretical descrip-

7.4 Matter–Wave Interferometry with Bose–Einstein Condensates

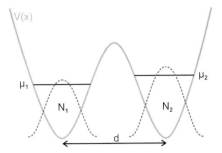

Figure 7.14 Schematic of two wave packets confined to a double-well potential after splitting of a Bose–Einstein condensate. The double-well separation d determines the fringe spacing obtained after release and expansion, a difference in well population $N_{1,2}$ or chemical potential $\mu_{1,2}$ influences the evolution of the relative phase.

tion for a weakly interacting, degenerate Bose condensate can be carried out in close analogy to classical electro-magnetic waves, treating interactions in a mean-field approach. The existence (or emergence) of a global phase in a single Bose–Einstein condensate has been discussed in [67], where the occurrence of a phase is explained by a broken symmetry mechanism [68, 69], and shall not be repeated here.

The wave function of a single condensate can be written as

$$\Psi(r) = \sqrt{n_0(r)} e^{i\phi} , \qquad (7.40)$$

with a global phase ϕ and density $n_0(r)$. The wave function is normalized to the total number of atoms by $\int |\Psi|^2 \, dr = N$.

In many interferometer sequences, a condensate is created in a single-well trap, and then split dynamically by deforming the trapping potential into a double well. The final situation is schematically depicted in Figure 7.14. Because of the splitting, the two wave packets in the double-well trap are prepared in a *coherent state*, each of which can be described by a single-particle wave function $\Psi_{1,2}(r, t)$.[23] The total wave function describing the system is then

$$\Psi(r, t) = \sqrt{N_1}\, \Psi_1(r, t) + \sqrt{N_2}\, \Psi_2(r, t) \qquad (7.41)$$

for given expectation values of atom number N_1 and N_2 in each well.

The experiments described in this section are mainly of the following type (with the exception of Section 7.4.2.4): A sample of Bose-condensed atoms is split spatially using one of the schemes described above. Then the atoms are released from the trapping potential and fall freely due to gravity. During time-of-flight, the atoms expand according to their initial momentum distribution and interaction energy. The which-way information is lost completely in the expansion and overlapping of the two clouds, as the final width of the clouds is much larger than the spatial separation in the trap. Therefore, we can think of the interferometer as a Young

23) We are here neglecting thermal or quantum fluctuations in $\Psi_{1,2}$ which will be discussed in Section 7.4.2.3.

double slit-type for coherent (light-) fields if the interference pattern is detected after a sufficiently long time-of-flight.

The atomic density after expansion is simply given by the square of the wave function:

$$n(r, t) = \left|\sqrt{N_1}\,\Psi_1(r, t) + \sqrt{N_2}\,\Psi_2(r, t)\right|^2$$
$$= N_1 |\Psi_1(r, t)|^2 + N_2 |\Psi_2(r, t)|^2 + 2\sqrt{N_1 N_2}\, \mathrm{Re}\left\{\Psi_1(r, t)\Psi_2^*(r, t)\right\}. \quad (7.42)$$

The last term in Eq. (7.42) describes the first-order correlations and gives rise to spatial interference.

Let us assume the initial clouds to be of a Gaussian shape of width σ_0. The condensates are confined to a double well with wells centered at $r = \pm d/2$ and have initial phases ϕ_1 and ϕ_2. The trap is switched off and the condensates expand freely (neglecting interactions). During time-of-flight t_{TOF} the size of the clouds increases:

$$\sigma_t = \sqrt{\sigma_0^2 + \left(\frac{\hbar t_{\mathrm{TOF}}}{m\sigma_0}\right)^2}, \quad (7.43)$$

where m denotes the mass of the atoms.

In the absence of interactions, the interference term in Eq. (7.42) takes the shape

$$2\sqrt{N_1 N_2}\,\mathrm{Re}\left\{\Psi_1(r, t)\Psi_2^*(r, t)\right\} \propto E \cos\left(\frac{\hbar}{m}\frac{rd}{\sigma_0^2\sigma_t^2}t + \phi_1 - \phi_2\right), \quad (7.44)$$

which describes the density modulation of the interference pattern under an envelope E. The density modulation manifests as fringes orthogonal to the vector connecting the trap minima of the double well, which shows the close analogy to a double-slit experiment.

From such an interference pattern the following information can be extracted (e.g., by fitting a modulated cosine function or a more advanced profile obtained by numerical simulation):

- *Fringe spacing:*
  Assuming ballistic expansion of the atomic clouds, the fringe spacing of the interference pattern is given by [70]:

$$\lambda = \frac{h t_{\mathrm{TOF}}}{md}, \quad (7.45)$$

where t_{TOF} is the expansion time, which is typically of the order of tens of milliseconds. For an initial spacing on the order of a few micrometers one expects a fringe spacing of several tens of micrometers which can be easily detected using standard absorption imaging. This simple expression will change in the presence of interactions during the early stage of the expansion, which leads to a density-dependent increase of fringe spacing for the case of repulsive interactions, as suggested in [25].

7.4 Matter–Wave Interferometry with Bose–Einstein Condensates

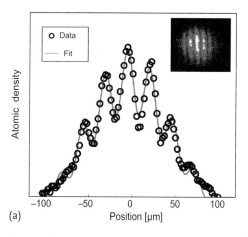
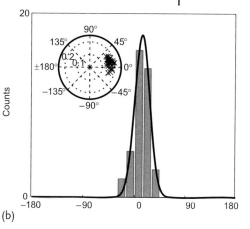

Figure 7.15 (a) A cosine function with Gaussian envelope is fitted to an integrated atomic density profile, obtained from absorption images of matter-wave interference patterns (inset). Information on fringe spacing, contrast and relative phase is extracted from each fit. (b) Contrast and relative phase for 40 realizations of the same experiment are plotted in a polar diagram (inset). A histogram of the same data shows a very narrow distribution of the differential phase ($\sigma = 13°$) directly after separating the clouds.

- *Relative phase*:
  By comparing the fringe position of the interference pattern within the envelope, one can detect a relative phase $\phi_r = \phi_1 - \phi_2$ between the two condensates. As shown in Figure 7.15, one can often use a Gaussian or parabolic envelope modulated with a cosine function to fit a typical interference pattern.
- *Contrast*:
  The degree of coherence of an interfering system (see Section 7.5) can often be quantified by the visibility of the interference pattern:

$$V = \frac{n(r, t_{\text{TOF}})_{\max} - n(r, t_{\text{TOF}})_{\min}}{n(r, t_{\text{TOF}})_{\max} + n(r, t_{\text{TOF}})_{\min}}. \tag{7.46}$$

  In the case of equal atom numbers and spatially uniform phase of the interfering wave packets one can expect a visibility of 1. Yet the contrast can be reduced by atom number imbalance ($V \propto 2\sqrt{N_1 N_2}/N$) and finite resolution of the imaging system. Taking these effects into account, the visibility may be used as a measure for the coherence of the system, as described in Section 7.5.
- *Atom number*:
  Since the number of atoms is conserved, one can detect the total number of atoms by integrating over the spatial profile.

In the first experimental realizations of interference with Bose–Einstein condensates [71], interference was observed even for independently prepared wave functions, not obtained by a splitting of a single source sample (see Figure 7.16). In this situation, since it is in principle possible to know the exact atom number in each ensemble (neglecting finite temperature and interaction effects), one has to treat

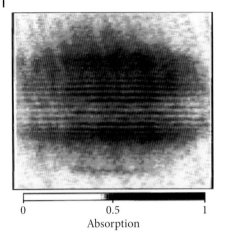

Figure 7.16 First demonstration of interference of Bose–Einstein condensates (figure from [71]).

the atoms as being in a *number state* or *Fock state*. In contrast to a coherent state, a number state does not possess a well-defined phase, which is a result of the phase-number uncertainty. Thus, also no relative phase in the sense of Eq. (7.44) can be defined, as long as the condensates are completely separate. However, it was shown in [72–74], that for an initial number state a (random) relative phase will build up during the measurement, the system dynamically evolves into a coherent state. Yet, quantum mechanics will make a prediction only for the expectation value that is observed in a sufficiently long series of measurements. Averaging over the interference patterns of completely independent condensates (with random relative phases) will lead to vanishing visibility. Hence, to ascertain whether a system has been prepared in a number state or a coherent state, an experimental interferometer sequence has to be repeated many times. Accordingly, a BEC beam-splitting sequence is termed *coherent* only when it yields a reproducible, non-random phase in an arbitrary number of repetitions (see Figure 7.15b).

7.4.2
Experimental Realizations

The goal of experiments with interfering Bose–Einstein condensates is often to explore phase dynamics. As outlined above, it is, therefore, important that the splitting process provides a reproducible relative phase between the two atomic ensembles (given that they are in a coherent state) after the splitting process and that the phase evolution is not disturbed by the splitting process itself. In this section we will briefly review the first experiments demonstrating the interference of ultracold bosonic clouds. Then we will focus on the coherence of the splitting and the phase evolution of split condensates.

The first interference of Bose–Einstein condensates was demonstrated in 1997 by the MIT group [75]. A condensate of Sodium atoms was prepared in a magnetic cloverleaf trap [71]. The condensate was split by focusing a blue-detuned light sheet into the center of the trap, generating a localized repulsive potential barrier. The resulting double-well potential was switched off and interference could be observed after 40 ms of free expansion. Figure 7.16 shows a raw data image: in the center, where the two clouds overlap, interference fringes are visible with a contrast between 20 and 40 %. Because of technical instabilities, the splitting was not coherent in a sense that the relative phase showed significant shot-to-shot fluctuations.

Coherent splitting and reduced fluctuations in relative phase have been demonstrated in a more stable setup based on purely optical trapping [76]: a Bose–Einstein condensate was confined in an focused dipole trap. The trap was spatially deformed into a double well until the condensate was split into two wave packets with a potential barrier high enough to prevent tunneling or locking of phases. The double well was created by splitting the optical trap radially using an acousto-optical-modulator (AOM) driven by two frequencies. The radial separation d was controlled by the frequency difference, and the trap depth of each single well could be adjusted by changing the averaged laser power in each beam. After releasing the atoms from the double-well trap, interference patterns were recorded by absorption imaging. By evaluating the interference patterns as described above, the relative phase was extracted from each single picture.

A low spread of the measured relative phases was shown for short hold times in the double well, proving a reproducible relative phase in repetitions of the experiment. The standard deviation of the relative phase for short times after the splitting was found to be less than $40°$, which is significantly smaller than the expected standard deviation of more than $100°$ for a random phase distribution. It was also observed, that the relative phase evolved linearly with longer hold times in the double-well trap. The phase sensitivity of the condensate interferometer was demonstrated by variations of the trap depth in each single well: by changing the imbalance between the wells, the temporal evolution of the relative phase could be controlled. After 5 ms hold time, the phase coherence was lost, and the interference patterns showed a substantial curvature due to residual excitations created in the splitting process.

7.4.2.1 Coherent Splitting on Atom Chips

We will now concentrate on coherent splitting making use of atom chip traps as described in [25]. In this experiment a standard magnetic microtrap, generated by the combined fields of a current-carrying trapping wire and an external bias field was used; a static magnetic field minimum formed where atoms in low-field-seeking states could be trapped. The BEC of up to 10^5 rubidium-87 atoms was prepared in the $F = m_F = 2$ state. A sinusoidally alternating current through an additional chip wire provided the magnetic RF field that generated the dressed adiabatic double-well potential (see Section 7.3.1) used to implement the BEC splitter. The RF field was held at a constant frequency slightly below the Larmor frequency of the atoms at the minimum of the static trap. The amplitude of the RF field was ramped up from zero to its final value to perform the dynamic splitting. The splitting was performed transversely to the long axis of the trap, as shown in Figure 7.11.

The spatial separation of the atomic wave packets after the splitting process was expected to be high enough such that tunneling was inhibited and the condensates could be expected to be completely independent. The splitting process itself took place on a timescale of ≈ 0.5 ms, which is also faster than the expected tunnel dynamics through the barrier. We will have a closer look at the timescales of the dynamic splitting of a condensate in Section 7.4.2.3.

To study the coherence of the splitting process the clouds were recombined in time-of-flight expansion after a non-adiabatic fast (< 50 μs) extinction of the double-well potential. The transverse density profile was detected along the weak trapping direction, that is, integrating over the long axis of the elongated clouds. The fringe spacing λ and the relative phase ϕ_r were extracted by fitting a cosine function with a Gaussian envelope to the measured profiles, as shown in Figure 7.15a. The relative phase has been repeatedly measured. A narrow distribution of relative phase was found, indicating a reproducible and hence coherent splitting of the condensate. The distribution is plotted in Figure 7.15b, showing a phase spread of 13°, which is clearly nonrandom.

7.4.2.2 Interference of Independent Condensates

Complementary experiments have been performed in [57]. Here, a thermal cloud above the critical condensation temperature ($T_C \approx 400$ nK) was dynamically split using the above-described experimental setup and techniques. A high potential barrier of $U \approx 4$ μK was erected to clearly separate the two ensembles. Forced evaporative cooling in the RF-dressed double-well potential was applied to create two completely separated Bose–Einstein condensates in the respective wells. After release from the trapping potential, the overlapping clouds showed high-contrast interference but with a relative phase randomly distributed between 0 and 360° as expected from a number state (see Section 7.4.1).

7.4.2.3 Phase Dynamics of Split Condensates

In the experiment described in the previous section, the measured relative phase was recorded during a dynamic splitting sequence, meaning that the phase evolution for different times after the complete splitting was observed. The width of the relative phase distribution has also been monitored throughout an entire splitting sequence as shown in Figure 7.17. No phase evolution was found for small double well separations, where tunneling locked the relative phase to zero. At a separation of 3.5 μm the splitting was assumed to be complete. The experiment was carried out for different realizations of the double-well potential, where a slight imbalance of the double well was controlled by the current in the DC-trapping wire. Depending on the strength of the imbalance, a differential evolution of the relative phase could be observed. This phase evolution can be used to fine-tune the balancing of the double-well potential (see Section 7.3.4). Along with the directed evolution of the relative phase, the measurement also showed a continuous increase of phase spread until the relative phase was almost random for the largest splitting distances. At the same time a significant loss of contrast in individual images was observed. The timescale of phase randomization and complete loss of contrast was

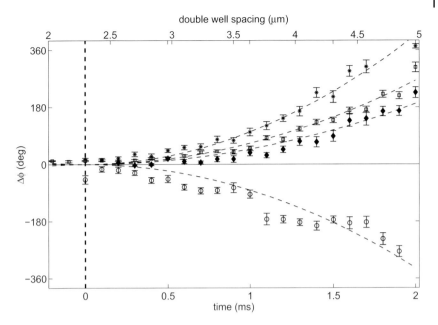

Figure 7.17 Top: evolution of the relative phase throughout the splitting process (15 ms constant ramp time, increasing final splitting distance). The dashed vertical line indicates the critical trap separation of 3.34 μm from which both condensates are fully separated. To affect the phase evolution, the position of the static magnetic trap has been slightly changed, deliberately introducing a double-well imbalance (different trap positions are distinguished by marker shape). The error bars indicate the statistical error of the mean measured relative phase. The dashed lines show the result of a numerical simulation of the splitting process, where the initial double-well imbalance has been adjusted to fit the data.

found to be 2.5 ms, two orders of magnitudes shorter than the lifetime of the condensates in the double-well potential themselves (>500 ms).

The vanishing phase signal could later be explained by longitudinal phase fluctuations, which are a typical feature of elongated condensates in three and one dimension. A quantitative analysis of this effect will be discussed in Section 7.5.

We will now take a closer look at the timescales of splitting a Bose–Einstein condensate, as they are crucial for the fundamental process of *phase diffusion*. To obtain adiabatic splitting (with respect to connecting the single-well ground state before the splitting to the symmetric ground state of the double-well potential) the splitting has to take place on a timescale longer than the inverse of the involved trap frequencies. Optimal splitting ramps that avoid excitations have been derived in [77]. During the finite splitting time, two effects determine the further dynamics of the relative phase: *tunneling* of atoms between the two wells is possible, as long as the atomic wave functions in the two wells spatially overlap, and *interactions* (which will be considered to be repulsive for our purpose) introduce an additional energy scale to the system. In a theoretical description, the splitting can be modeled as a two-step process [78]: as long as tunneling between the wells is allowed, we will make use of the Josephson formalism. In the second step, the two wave packets will be treated as independent atomic samples with a fixed atom number and individual phase evolution.

The Josephson Hamiltonian includes tunneling and interactions and for the near-equilibrium situation can be written as

$$H \approx \frac{E_C}{2} N_r^2 + \frac{E_J}{2} \phi_r^2 , \qquad (7.47)$$

where $E_C = d\mu/dN$ is the charging energy, E_J is the Josephson tunneling energy [79], $N_r = N_1 - N_2$ is the relative atom number, and $\phi_r = \phi_1 - \phi_2$ the relative phase in the two condensates. The Hamiltonian is of a harmonic oscillator-type with the characteristic (plasma) frequency $\omega_J = \sqrt{E_C E_J}/\hbar$. During a dynamic splitting process the potential barrier height in the double well will rise and the distance between the condensates will increase, which makes ω_J time-dependent. The characteristic splitting time T_s hence has to be compared to this frequency ω_J. For $\hbar\omega_J(t) > T_s^{-1}$ the system will stay in the ground state of the Josephson Hamiltonian, and the splitting can be expected to be nearly adiabatic (now with respect to the tunneling dynamics). As soon as $\hbar\omega_J(t) < T_s^{-1}$ tunneling is inhibited and the splitting is complete.

As we will see below, fluctuations in relative atom number will play an important role for phase diffusion. Since the atom number difference N_r and the relative phase ϕ_r are canonically conjugate variables, they have to fulfill the relation:

$$\Delta N_r \Delta \phi_r \geq 1/2 , \qquad (7.48)$$

which highlights the interplay between fluctuations in phase and atom number during the splitting. In particular, a sudden splitting would prepare the system in a coherent state with exact zero relative phase and $\sqrt{N}$ fluctuations in N_r. The variance of relative atom number depends on the splitting time:

$$(\Delta N_r)^2 = \hbar N/\mu T_s . \qquad (7.49)$$

Once the two condensates are split entirely, the relative phase between them is no longer locked, and the individual phases can evolve independently. The phase evolution of each single condensate is determined by atomic scattering interactions (among other factors) and hence scales with the number of atoms $N_{1,2}$. Thus, the evolution of the relative phase between the condensates will depend on the relative atom number [74, 80]. The energy of the split system can be expanded as

$$E(N_r) = E(0) + \frac{1}{2} E_C N_r^2 . \qquad (7.50)$$

Because of the interactions, the charging energy E_C is nonzero, and the N_r^2 term will lead to a dispersion of the phase (phase diffusion). The time until this dispersion leads to a phase spread on the order of 2π is given by the *coherence time*

$$\tau = \frac{2\hbar}{E_C \Delta N_r} . \qquad (7.51)$$

After this time the phase coherence between the condensates in the double well is lost. The time-dependent variance of the phase is written as:

$$\Delta\phi(t)^2 = \Delta\phi(0)^2 + \left(\frac{\Delta N_r E_C}{2\hbar}\right)^2 t^2. \tag{7.52}$$

At much larger timescales the original spread $\Delta\phi(0)$ revives and broadens again within τ. The revival happens on a typical timescale of tens or hundreds of seconds and might, therefore, be hard to observe experimentally [69].

The phase diffusion of two coherently split Bose–Einstein condensates has been experimentally investigated in [29] by the MIT group. Jo et al. demonstrated reproducible splitting and high coherence times up to 200 ms. The long coherence time was shown to be an indicator of a high *squeezing*[24] of atom number uncertainty throughout the splitting process.

The system is modeled as described above. In addition, a *squeezing factor* ξ is introduced, defined by:

$$\frac{1}{\xi} = \frac{\Delta N_r}{\sqrt{N}} \tag{7.53}$$

which quantifies the reduction of relative atom number fluctuations compared to $\sqrt{N}$. The squeezing factor is related to the splitting time and chemical potential by the relation $\xi = \sqrt{\mu T_s}$ [78]. This means that very slow splitting will lead to a small variation in relative atom number, resulting in a high ξ, whereas faster splitting will lead to number fluctuations on the order of $\Delta N_r \approx \sqrt{N}$, as expected for a coherent state with perfectly defined relative phase. Complex non-linear splitting ramps to achieve maximal squeezing have been calculated in [81] using optimal control theory.

Once the splitting is complete, the two separate Bose–Einstein condensates are described as a superposition of many relative number states $|N_r = N_1 - N_2\rangle$, where N_1 and N_2 are the number of atoms occupying each well. The interaction energy in each well depends on the number of atoms, so the phase evolution in each well will be different for $N_r \neq 0$. The superposition in relative number states hence transforms into a superposition of relative phases measured in the experiment, which leads to a diffusion of the relative phase with time.

For the quantification of this process, it is convenient to resort to the *coherence factor* known from circular statistics [82] to give a measure for the correlation of the repeated measurements.

$$\Psi(t) = \frac{1}{M}\left|\sum_M e^{i\phi(t)_r}\right|, \tag{7.54}$$

where M denotes the number of measurements and $\phi_r(t)$ the obtained relative phase between both condensates after a hold time t. For M measurements on a random phase distribution, the coherence factor $\Psi(t)$ would have an expectation

24) By squeezing we denote any realization of relative atom number fluctuations $\Delta N_r < \sqrt{N}$.

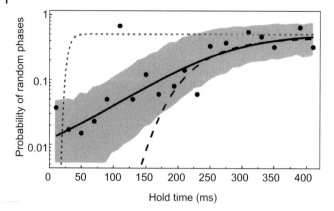

Figure 7.18 Phase diffusion and number squeezing. The randomness probability of ten measurements of the relative phase is displayed up to 400 ms after splitting. The dotted curve (dashed curve) shows a simulation for a phase-coherent state (number squeezed state with $\xi = 10$), which have negligible initial phase uncertainty. The solid line includes an initial phase uncertainty of 0.28π. The shaded region represents the window where ten data points from the sample with the given phase uncertainty would fall with 50 % probability (figure from [29]).

value of $\sqrt{M}$. This value has to be compared to the coherence factor acquired from the measurements. The difference gives the probability for the randomness of the phase distribution (see Figure 7.18).

Phase diffusion has been studied experimentally by the MIT group in a setup comparable to the one depicted in Figure 7.11. Bose–Einstein condensates of $\approx 4 \times 10^5$ sodium-23 atoms in the $|F = 1, m_F = -1\rangle$ state were loaded into a static magnetic wire trap on an atom chip ($\omega_\perp = 2\pi \times 2.1\,\text{kHz}$). A radio-frequency-induced splitter was used to split the atom cloud into a double-well potential (separation $2r_0 = 8.7\,\mu\text{m}$, barrier height $U = h \times 30\,\text{kHz}$, chemical potential per sample $\mu = h \times 6\,\text{kHz}$). The condensates were held in the trap for different hold times t up to 400 ms, which was small compared to the condensate lifetime of 1.8 s. Then the atoms were released from the trap by switching of the trapping potential within 30 μs. The spatial interference pattern was recorded, and the relative phases extracted from the images.

The measured phase diffusion is depicted in Figure 7.18 which shows the randomness of relative phases plotted for different hold times t, defined as the deviation of the coherence factor from $\sqrt{M}$. For the first 200 ms the phase diffusion is clearly nonrandom. The following function was used to fit the data points, each of which represents the mean value of ten measurements:

$$\Delta \phi_r(t)^2 = \left(\frac{\xi}{\Delta N_r} \right)^2 + (Rt)^2 . \tag{7.55}$$

A squeezing of $\xi > 10$ was inferred from the data and fit, including technical shot-to-shot variations and thermal fluctuations. $\xi = 10$ corresponds to very low relative atom number fluctuations of $\pm 0.03\,\%$ corresponding to ± 50 atoms on a total of 4×10^5 atoms. Even with this squeezing factor only being an estimation, the long phase coherence time itself implies that the initial state must have been strongly number squeezed. A non-squeezed number state would have shown a much faster randomization of relative phases on the order of a few ms. Data and simulations for both a phase-coherent state and a number-squeezed state are compared in Figure 7.18. The observed number squeezing was associated to repulsive interactions, which make number imbalance energetically unfavorable [83].

7.4.2.4 Merging of Split Condensates

All interferometry experiments described above used time-of-flight expansion to recombine atomic clouds after the splitting. Once released from the confining potential, the wave packets propagate according to their momentum distribution (and under the influence of interactions) and are lost for further manipulation. Merging the atoms after the splitting, while keeping them trapped in a confining potential on the chip, closes the interferometer sequence and avoids this disadvantage.

An exact theoretical description for such a process exists only for two special cases of the relative phase in the interferometer arms, $\phi_r = 0$ and $\phi_r = \pi$. For the merging of two ensembles of *non-interacting* atoms with relative phase $\phi_r = 0$, one can expect an evolution into the ground state of the single-well potential, if the recombination happens in an adiabatic fashion [20, 40, 48]. For $\phi_r = \pi$, the wave function evolves into the first anti-symmetric excited state of the single well. The system acquires then an excitation energy of $N\hbar\omega$. For intermediate relative phases, the final state will be a superposition of ground and first excited single particle state.

The merging of *interacting* atomic ensembles with $\phi_r = 0$ results again in the Thomas–Fermi ground state of the single well. For a relative phase of $\phi_r = \pi$ a static dark soliton will form at the position of the former barrier [84, 85]. Intermediate relative phases will lead to complex dynamics on the microscopic scale [84] and may be related to an overall heating after some relaxation time [86, 87]. The presence of phase fluctuations in elongated or one-dimensional condensates further complicates a quantitative analysis [88].

The merging of Bose condensates was experimentally investigated by the MIT group in [29]. Sodium-23 condensates were coherently split with an RF-induced BEC splitter and then recombined in a single-well trap by inverting the splitting procedure. The atoms were released from the trap and the density profiles were detected with standard absorption imaging. To get a measure for the number of atoms in the condensate fraction of each sample, the number of atoms in a fixed region in the center of the cloud was recorded. A loss of atoms in that central part indicated heating of the sample. This loss was compared to measurements without merging of the clouds. The fractional loss of atoms was measured for different hold times t before recombination.

The dependence of the relative phase between the two split condensates on t was additionally observed by interference experiments as described in Section 7.4.2.3. A strong correlation between the two measurements was found, indicating that heating and atom loss may both be used as indicators of relative phase. A larger

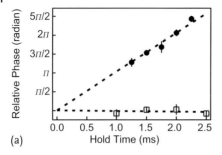 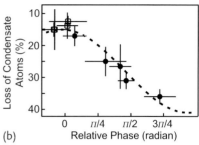

Figure 7.19 Phase-sensitive recombination of two separate condensates. (a) The relative phase of two split condensates was monitored for various hold times after splitting by suddenly releasing the two condensates and observing interference fringes. For the independent condensates (solid circle), the evolution rate of the relative phase was determined from the linear fit to be 500 Hz. For the weakly coupled condensates (open square), the relative phase did not evolve. At 0 ms hold time, the relative phase was set to zero for both cases. (b) For the same range of delay times as in (a), the condensate atom loss after in-trap recombination was determined. The relative phase (x axis) was obtained from interference patterns as in (a). The merging time was 5 ms (figure from[29]).

relative phase leads to an increased atom loss in the center of the condensate fraction (see Figure 7.19). This loss can be interpreted either as heating, in case the soliton decays, or as additional momentum contribution due to the presence of the soliton itself.

7.5
Interferometry with 1D quasi condensates

Atom chip wire traps enable the creation of highly elongated "cigar shaped" Bose–Einstein condensates with aspect ratios reaching from 10–10^4 [89]. They are hence ideally suited to generate and study one-dimensional Bose gases [90]. In the following, we will focus on *1D quasi condensates*, as introduced in Chapter 11 of this book. As 1D quasi condensates we consider any realization of a quantum degenerate system with $\mu, k_B T \leq \omega_\perp$ and $T_\phi < T < T_{\mathrm{deg}}$. In this regime, density fluctuations of the wave function are suppressed, however, the phase field shows enhanced thermal and quantum fluctuations, both in the time and the spatial domain [90]. The *coherence* or *correlation length* l_c of phase fluctuations is given by $l_c \approx L(T_\phi/T)$. Here,

$$k_B T_\phi \approx \frac{15 \left(\hbar \omega_\parallel\right)^2 N}{32 \mu} \tag{7.56}$$

describes the temperature for which the thermal coherence length is comparable to the overall length of the system L and thermal phase fluctuations can be considered as damped out. Equation (7.56) is valid also for elongated 3D condensates. In the 1D limit ($\mu \approx \hbar \omega_\perp$) we obtain $k_B T_\phi \approx (\hbar \omega_\parallel)^2 N/(\hbar \omega_\perp)$. It shall be pointed out, that no phase transition occurs at T_ϕ. Note also, that even at $T = 0$ no true long-range order can be found in the one-dimensional system. *Quantum fluctuations* are always present and lead to a power-law decay of the correlation function [91]. Here, we will exclusively consider the temperature regime $T_\phi < T < T_{\text{deg}}$, a detailed overview on regimes of quantum degeneracy in 1D can be found in Chapter 11 of this book. Whether the above-listed criteria for one-dimensionality are sufficient is the subject of ongoing discussions. In the following, we focus on long-wavelength excitations of the system, beyond the healing length, which are clearly of 1D character. Other, more short-range properties like collisions may still show 3D behavior within the given parameters [92, 93].

Although one-dimensional with respect to excitations, any real system still has a finite radial extension, given by $\sigma_\perp = a_\perp(1 + 4a_0 n_{1D})^{1/4}$ with $a_\perp = \sqrt{\hbar/m\omega_\perp}$ the single particle transverse ground-state size [94]. If this extension is larger than the characteristic length scale of the inter-atomic potential a_0 (s-wave scattering length), collisions remain three-dimensional and one can derive an effective 1D coupling constant g_{1D} by integrating over the radial wave function [95, 96]

$$g_{1D} = \frac{2\hbar^2 a_0}{m a_\perp^2} = 2\hbar\omega_\perp a_0. \tag{7.57}$$

The condition $\mu \approx \hbar\omega_\perp$ for the crossover into the 1D regime imposes a limit on the maximum particle number in the system. Rewriting this limitation in terms of linear 1D density yields $n_{1D} \approx 1/a_0$, implying that the line density of a 1D quasi condensate is limited to approximately one atom per scattering length, independent of the strength of radial confinement. For rubidium-87 atoms, this amounts to $n_{1D} \approx 200$ atoms/μm.

In the following we will present experiments, which combine the methods of RF-induced matter-wave interferometry and 1D Bose gases on atom chips to investigate equilibrium and non-equilibrium phase properties. These experiments are complementary to efforts in optical lattices, where thousands of 1D samples are created in parallel, providing access to averaged quantities only. Often, we will employ Luttinger liquid (LL) theory for uniform 1D systems as the theoretical framework [91, 97].

7.5.1
Coherently Split 1D BECs: Coherence Dynamics

In this section we study the time-evolution of a 1D system prepared by dynamic splitting of a single sample (see Section 7.4.2.1). Directly after the splitting, the relative phase between the two condensates is zero over the entire length of the system, that is, the splitting creates two identical copies of the phase fluctuation pattern at the moment of the splitting. This situation is a highly non-equilibrium state of the split system, which will accordingly relax towards equilibrium over time. The equilibrium is given by completely uncorrelated phase fluctuations in each condensate. Consequently, the relaxation results in a randomization of the local relative phase $\phi_r(z, t)$.

This decoherence process was studied experimentally in [27]. A single one-dimensional quasi condensate was created on an atom chip and coherently split

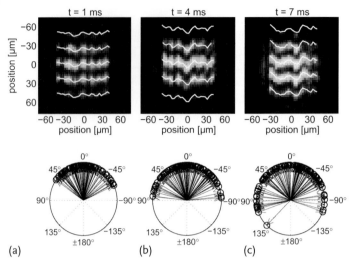

Figure 7.20 Decoherence of the relative phase between two uncoupled 1D Bose condensates after $t = 1, 4, 7$ ms ((a–c), respectively). The upper row shows example interference images. The decoherence dynamics could be identified as an increasing undulation in the fringe pattern, reflecting the dephasing in the transverse coordinate $\phi_r(z, t)$. The lower row shows relative phase vectors $e^{i\phi_r}$ normalized to zero global phase. Here, decoherence showed up as a spread in the distribution of vectors, which lead to a decay of the coherence factor $\Psi(t)$ with time.

(to zero tunneling) in the vertical direction, employing the setup depicted in Figure 7.12. To study the dynamics of the phase relaxation process, the two BECs were held in the double well for a varying time t before they were released and the interference pattern was recorded. Figure 7.20 shows example images for holding times $t = 1, 4, 7$ ms for a pair of uncoupled condensates. From a single image, a vector of the local relative phases $\phi_r(z, t)$ along the system was extracted by individually analyzing the fringe pattern in each vertical pixel slice.

The relaxation towards equilibrium resulted in an increasing spread of the phases in each such vector, which was quantified (for a single image) by the coherence factor (see Section 7.4.2.3)

$$\Psi(t) = \frac{1}{N} \cdot \left| \sum_j e^{i\phi(z_j, t)} \right| \tag{7.58}$$

where N is the number of considered pixel slices. This analysis included only the central slices in the range $L < 1/2 L_{\text{tot}}$, where $L_{\text{tot}} \approx 80\text{--}100$ μm was the total length of the 1D condensates. In this central region, the line density could be considered as constant, that is as independent of the longitudinal confinement: $g_{1D} n_{1D} = \mu$ (see Chapter 11).

The time evolution of the coherence factor has been theoretically studied by Burkov et al. [78] based on a path integral approach to the Luttinger liquid description of the phase fluctuations. They obtained the following result for zero temperature:

$$\Psi_d(t) \propto \exp\left[-\frac{\mu^2 t^2}{2\hbar^2 N \xi^2}\right] \times \exp\left[\frac{-\mu t}{2\pi\hbar K \xi^2}\right], \quad (7.59)$$

where ξ is the factor that describes the squeezing arising in the dynamic splitting (see Section 7.4.2.3), and $K \approx \pi/\sqrt{\gamma}$ is the Luttinger liquid parameter [68, 78] with $\gamma = mg_{1D}/\hbar^2 n_{1D}$.

The first of the two exponents describes the "global" phase diffusion due to atom number uncertainty, while the second one is due to the quantum phase fluctuations present in a uniform 1D system even at $T = 0$.

The thermal part of the phase fluctuations can only be neglected for times $t < \hbar/k_B T$. To observe a significant effect on the coherence factor due to the quantum fluctuations in this time, $\mu/2\pi K\xi^2 k_B T$ must be large. As expected, quantum effects become more important for small K and T.

For $t > \hbar/k_B T$, the thermal fluctuations have to be included in the calculations. The coherence factor is then dominated by a decay of the form

$$\Psi(t) \propto \exp\left[-\left(\frac{t}{t_0}\right)^{2/3}\right], \quad (7.60)$$

with the time constant $t_0 = ?\,61\pi\hbar\mu K/(k_B T)^2$.[25] The important feature of this result is the non-analytic time dependence, which is a general property of 1D thermal decoherence dynamics. This is due to the fact that in 1D liquids, damping at finite T is always nonhydrodynamic [98]. Furthermore, at finite T there is an anharmonic coupling of the two degrees of freedom of the system, namely the relative phase $\phi_r = \phi_1 - \phi_2$ and the "center-of-mass" total phase $\Phi = \phi_1 + \phi_2$.

7.5.1.1 Decoherence of Uncoupled 1D Systems

To produce a pair of uncoupled 1D systems after a coherent splitting, a sufficiently high and wide potential barrier such that any tunnel coupling could be neglected was introduced in experiments described in [27]. This was verified by numerical two-mode model calculations of the tunnel coupling. As discussed above, for the experimentally achieved temperatures $T \approx 100$ nK, one expects the phase dynamics to be dominated by thermal phase fluctuations. The corresponding theoretical prediction for the time evolution of the coherence factor $\Psi_d(t)$ is given by Eq. (7.60). Taking the logarithm twice, this expression becomes

$$\ln(-\ln(\Psi_d(t))) \propto \frac{2}{3}\ln(t) + \text{const.} \quad (7.61)$$

Figure 7.21 shows six different measurement series of $\Psi(t)$ in a double logarithmic plot. Each data set corresponds to a different combination of initial temperature T, line density n_{1D}, and transverse confinement $\omega_\perp$ in the experiment [27].

---

[25] The scaling of the decay time constant t_0 with temperature is an issue of ongoing discussion, see [92].

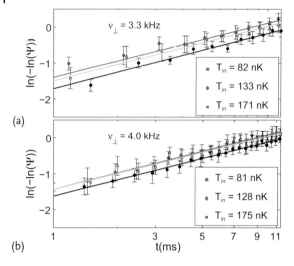

Figure 7.21 Double logarithmic plot of measured coherence factors over time for six different combinations of trapping frequencies (top $\omega_\perp = 2\pi \times 3.3$ kHz, bottom $\omega_\perp = 2\pi \times 4.0$ kHz), line densities, and temperatures. The solid lines are linear regression fits yielding slopes of 0.67±0.06, 0.65±0.07, 0.64±0.04 for datasets in (a) (ascending in T) and 0.65±0.03, 0.66±0.06, 0.64±0.06 for (b). All experimental parameters and results are summed in Table 7.1.

The solid lines in the plots are linear fits to the data, the slope directly yields the exponent of Eq. (7.60). The exact values of the fitted slopes are given in the figure caption and in Table 7.1.

For all data sets the results of the fits were in very good agreement with the theoretical prediction of 2/3 for the slope. This confirmed the non-trivial time dependence of the thermal phase decoherence predicted by Burkov et al. [78]. It was considered the first experimental test (and verification) of a prediction based on the Luttinger liquid description for non-equilibrium dynamics in weakly interacting 1D Bose gas.

Table 7.1 Measured decoherence exponents and decay time constants t_0 after the complete splitting of a 1D quasi condensate for the data shown in Figure 7.21.

T_{in} [nK]	n_{1d} [1/μm]	μ/h [kHz]	$\nu_\perp$ [kHz]	decay exponent	t_0 [ms]
82(28)	20(4)	0.7(1)	3.3	0.64(8)	9.0(4)
133(25)	34(5)	1.2(2)	3.3	0.65(7)	5.5(3)
171(19)	52(4)	1.8(1)	3.3	0.64(4)	6.4(3)
81(31)	22(4)	0.9(2)	4.0	0.65(3)	8.1(2)
128(23)	37(4)	1.5(2)	4.0	0.66(3)	5.9(2)
175(20)	51(5)	2.1(2)	4.0	0.64(6)	6.1(4)

7.5.1.2 Coherence Dynamics for Coupled 1D Condensates

When splitting 1D condensates to a distance/barrier height, where a residual tunnel coupling J between the samples remains, the decoherence process described above is partially counterbalanced by phase locking. The equilibrium situation of two coupled 1D quasi condensates has been studied by Whitlock and Bouchoule [99]. Expressing their results in terms of the (equilibrium) coherence factor, one obtains

$$\Psi(T, J) = \exp\left[-\frac{\sqrt{mg_{1D}}}{4\pi\hbar\sqrt{n_{1D}}} F(b)\right] \times \exp\left[-\frac{\sqrt{mg_{1D}}}{2\pi\hbar\sqrt{n_{1D}}} S(b)\right] \quad (7.62)$$

with the abbreviation $b = J/g_{1D}n_{1D}$. The auxiliary function $S(b)$ is given by

$$S(b) = \int da \frac{\sqrt{\frac{1}{(a^2+b)} + 1}}{e^{\frac{\sqrt{(a^2+b+1)(a^2+b)}}{k_B T}} - 1} \quad (7.63)$$

and $a^2 = \frac{\hbar^2 k^2}{2m}$. The function $F(b)$ is given by

$$F(b) = \sqrt{\frac{1}{(a^2+b)} + 1} - 1. \quad (7.64)$$

In this form, $S(b)$ and $F(b)$ can be integrated numerically in a straight-forward manner. The two exponentials on the right-hand side of Eq. (7.62) correspond to the quantum and thermal contributions to the coherence factor. Equation (7.62) allows for a direct extraction of the coupling energy J from measurements of the equilibrium coherence factor $\Psi(T, J)$.

Experimentally, coupled 1D systems were prepared following the procedure described in Section 7.5.1.2 but ending the splitting process with a non-negligible overlap between the wave packets, implementing an (unknown) tunnel coupling J. Figure 7.22 indicates the experimental observations: the coherence factor showed an initial decay on a timescale between 5–10 ms and then leveled off at a constant value. Although there was no formal proof, that this indicated indeed the equilibrium situation of the system, the levels with the coherence factor $\Psi(T, J)$ as in Eq. (7.62) were used to extract the tunnel coupling J (starting from known temperatures T). Note that in the case of strong tunnel coupling (upper curve in Figure 7.22) the coherence factor did essentially not decay, in other words, decoherence due to thermal phase fluctuation was completely inhibited by locking of the transverse phase. Tunnel couplings derived from Figure 7.22 in units of interaction energies were $J/g_{1D}n_{1D} = 0.2, 0.01, 0.003$, respectively. This indicated that the experiments were performed deeply in the Josephson regime [99, 100], the corresponding plasma frequencies were $\omega_J = 2\pi \times 900, 200$, and 80 Hz. These values agreed to a numerical calculation based on a beyond-RWA calculation of the double-well potentials (see Section 7.3.3) and an improved two-modes model [101] within a factor of 2.

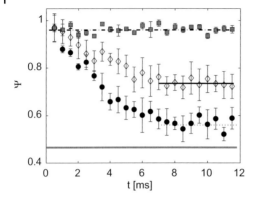

Figure 7.22 Initial decay and leveling off of the coherence factor for the case of coupled 1D quasi condensates. The constant coherence factor after an initial decay indicates an equilibrium situation, where decoherence due to (longitudinal) thermal phase fluctuations is counterbalanced by (transverse) phase locking due to tunneling. For the three shown datasets, tunneling energies of $J/g_{1D} n_{1D} = 0.2, 0.01, 0.003$ can be inferred, respectively.

7.5.2
Independent 1D BECs: Noise Statistics of Interference Amplitude

We will now consider interference of two independent 1D quasi condensates at equilibrium. The phase fluctuations in both systems are assumed to be entirely uncorrelated. We assume both condensates to be identical in their physical parameters like temperature T, line density n_{1D}, and the characteristics of the trapping potentials. They hence show identical statistical behavior concerning their phase properties.

Interference patterns obtained from independent 1D samples show a periodic density modulation at the interference wave vector $Q = md/\hbar t$ (see Section 7.4.1), where d is the separation of the two 1D systems in the trap, and t is the time of free propagation (see Figure 7.23). Assuming ballistic expansion, the (complex) amplitude of the density modulation after integration over a length L is given by [102, 103]

$$A_Q(L) = \int_{-L/2}^{L/2} dz\, a_1^\dagger(z) a_2(z) , \tag{7.65}$$

where $a_{1,2}$ are the annihilation operators within the two 1D BECs before the expansion. $A_Q(L)$ is called *interference amplitude*, it is related to the *interference contrast* $C(L)$ of (integrated) interference images as given by $C(L) = A_Q(L)/n_{1D} L$. The phase of $A_Q(L)$ describes the position of interference fringes and is determined by the relative phase between the two condensates averaged over the interval $-L/2$ and $+L/2$.

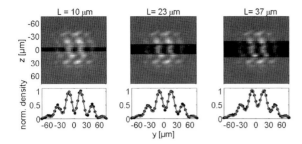

Figure 7.23 Example image of interference of two independent 1D quasi condensates. Integrating the image over different lengths L yields the interference amplitude $A_Q(L)$, its scaling with L contains information on phase fluctuations and correlations in the original 1D samples.

7.5.2.1 Average Interference Amplitude Square

In interference experiments with independent 1D quasi condensates, the phase of A_Q is random, and the expectation value $\langle A_Q \rangle$ is zero. This does not imply the absence of fringes in individual measurements but shows the unpredictable random phase in individual interference patterns [104–106] (see Section 7.4.1). As a consequence, to obtain meaningful information on the contrast statistics of the interference patterns, one has to consider the quantity $\langle |A_Q(L)|^2 \rangle$. This quantity is independent of the global phase difference but is strongly affected by local phase twisting within each condensate. We will refer to it as *average interference amplitude square*.

For ideal condensates and in the absence of phase fluctuations one expects perfect interference contrast/amplitude independent of the system length L considered. This implies $\langle |A_Q(L)|^2 \rangle \propto L^2$. In the opposite limit of short-range phase correlations with finite correlation length ξ_ϕ, the interference pattern arises from adding up fringes in L/ξ_ϕ uncorrelated domains. In this case the obtained interference contrast is washed out and appears only as a linear dependence in the average contrast amplitude square, $\langle |A_Q(L)|^2 \rangle \propto L\xi_\phi$.

More precisely, $\langle |A_Q(L)|^2 \rangle$ is determined by the integral of the two-point correlation function:

$$\langle |A_Q(L)|^2 \rangle = \int_{-L/2}^{L/2} dz_1 \int_{-L/2}^{L/2} dz_2 \, \langle a_1^\dagger(z_1) a_1(z_2) \rangle \langle a_2^\dagger(z_2) a_2(z_1) \rangle .$$

Using a standard expression for the two-point correlation function in Luttinger liquid theory [91, 107, 108] we can express the average contrast as

$$\langle |A_Q(L)|^2 \rangle = n_{1D}^2 L^2 \left(\frac{\xi_h}{L} \right)^{1/K} f\left(\frac{\xi_\phi(T)}{KL}, K \right). \tag{7.66}$$

Here, $K = \pi\hbar\sqrt{n_{1D}/g_{1D}m}$ is the Luttinger parameter for the weakly interacting 1D Bose gas, $g_{1D} = 2\hbar\omega_\perp a_0$ the effective 1D coupling constant, and a_0 the s-wave scattering length. $\xi_h = \hbar/\sqrt{mgn_{1D}}$ is the healing length and

$\xi_\phi(T) = \hbar^2 n_{1D}\pi/mk_B T$ is the thermal correlation length of the 1D condensates in the weakly interacting regime. The function $f(x, K)$ is given by

$$f(x, K) = \int_0^1 \int_0^1 du\, dv \left(\frac{\pi}{x \sinh\left(\frac{\pi|u-v|}{x}\right)} \right)^{1/K}. \tag{7.67}$$

Equation (7.66) provides us with the scaling of the average interference amplitude square for the considered system length L. We will first consider the regime of low temperatures and/or short system sizes ($L/\xi_\phi(T) \ll 1$). Here we only need to consider quantum fluctuations, which originate from interactions between atoms. Non-interacting bosons at zero temperature have no phase fluctuations, their interference pattern provides perfect contrast, leading to $\langle|A_Q(L)|^2\rangle \propto L^2$. In the other extreme, impenetrable bosons (Tonks–Girardeau gas [95, 109]) show very strong fluctuations (even at $T = 0$) and their interference pattern corresponds to short-range correlations, $\langle|A_Q(L)|^2\rangle \propto L$. For finite interaction strengths we find an intermediate scaling with length L, resulting in $\langle|A_Q(L)|^2\rangle \propto L^{2-1/K}$.

Finite temperature introduces thermal fluctuations, which creates phase fluctuations with a temperature-dependent correlation length $\xi_\phi(T)$. When $L > \xi_\phi(T)$ thermal fluctuations dominate and the interference amplitude scales as $\langle|A_Q(L)|^2\rangle \propto L\xi_\phi(T)$.

Interference of independent 1D quasi condensates was studied experimentally in [110]. In contrast to previous experiments, a cold thermal gas was (vertically) split in an RF-induced double-well potential and both halves were successively and independently cooled to quantum degeneracy. Both samples where recombined in free expansion after switching off the double-well potential, where the interference was recorded by standard absorption imaging.

Figure 7.23 shows an example interference image. To extract $|A_Q|$ from the observed interference patterns, transverse density profiles were integrated over the longitudinal direction for different lengths L. By fitting a cosine function with a Gaussian envelope to the resulting fringe profiles, the relative phase and the interference amplitude $|A_Q|$ were extracted as a function of L. Again, the analysis was restricted to the central 50 % of the system to ensure approximately uniform 1D density.

To compare to the scaling of Eq. (7.66), the average interference amplitude square $\langle|A_Q|^2\rangle$ was plotted over system length for three different temperatures, with the density $n_{1D} = 50(4)\,\mu m^{-1}$ and the transverse trapping frequency $\omega_\perp = 2\pi \times 3020(10)$ Hz (chemical potential $\mu \approx h \times 1.5$ kHz, Luttinger parameter $K = 42$, and healing length $\xi_h = 0.3\,\mu m$) identical for all three data sets, as depicted in Figure 7.24. Expression (7.66) was fitted to the obtained data with the temperature T as a free parameter. The functional behavior of the measured contrasts appeared to be in very good agreement with the theoretical predictions. This is of particular interest as both, quantum and thermal fluctuations contribute to the average interference amplitude as discussed above. For integration lengths longer than 20–30 µm, a linear dependence of $\langle|A_Q(L)|^2\rangle$ on L was observed. This corresponds to the $L \gg \xi_\phi(T)$ regime where thermal fluctuations dominate.

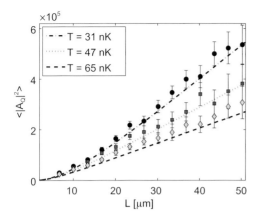

Figure 7.24 Scaling of the average interference amplitude square $\langle|A_Q(L)|^2\rangle$ with system length L for three different temperatures. The markers indicate the experimental measurements, lines correspond to fits according to expression (Eq. (7.66)) leaving the temperature as a free parameter.

Quantum fluctuations become important for shorter sample lengths L. However, the analysis presented in Figure 7.24 was not able to clearly identify quantum fluctuations. The experimentally implemented Luttinger parameter was $K = 42$, and it was hence impossible to observe the $L^{-1/K}$ correction to the ideal case power law L^2 in the limited range of lengths available due to finite imaging resolution.

From the fits, temperatures $T = 33(7), 47(6), 68(6)$ nK were obtained for 0, 50, 100 ms waiting time in the double-well potential, respectively. The increase of the temperature with longer waiting times was consistent with the heating rates observed in the experimental setup [110]. No alternative verification of the obtained temperatures was possible, a reliable detection of the thermal background was possible only down to $T \approx 80$ nK.

7.5.2.2 Full Counting Statistics of Interference Amplitude

The interference patterns obtained in interferometry with independent 1D quasi condensates contain more information than just the average interference amplitude square $\langle|A_Q(L)|^2\rangle$. We now consider the shot-to-shot variations of individual measurements of $|A_Q(L)|^2$ around the expectation value, which are characterized by the higher moments $\langle|A_Q|^{2n}\rangle$ and ultimately by the *full distribution function* $W(|A_Q(L)|^2)$. For quantifying the shot-to-shot fluctuations of the average interference amplitude square, it is convenient to introduce the normalized variable $\alpha(L) = |A_Q(L)|^2/\langle|A_Q(L)|^2\rangle$ and its distribution function $W(\alpha(L))$. The importance of the higher moments $\langle|A_Q|^{2n}\rangle$ is that they are directly related to the corresponding higher-order correlation functions of the 1D interacting Bose gas [102].

We will briefly summarize the qualitative scaling of the distribution function, a more theoretical approach can be found in [110]. For non-interacting ideal 1D condensates at $T = 0$, that is perfect interference patterns in each experimental realization, the distribution function $W(\alpha(L))$ is described by a delta function centered on one. Taking interactions into account (but still at $T = 0$), we expect a narrow distribution of width $1/K$ and $W(\alpha(L))$ to approach a universal Gumbel-like distribution [103, 111].

For long integration lengths, $L \gg \xi_\phi(T)$ and finite temperature, thermal fluctuations are dominant. In this case the interference pattern arises from adding local interference patterns from many uncorrelated domains resulting in a Poissonian distribution function centered on zero.

Figure 7.25 indicates histograms of experimentally measured distribution functions $W(\alpha(L))$ for four different length scales L and two different temperatures, $T = 31(6)\,\text{nK}$ (upper row, Figure 7.25a) and $T = 60(5)\,\text{nK}$ (lower row, Figure 7.25b), obtained using the average interference amplitude method. For these data sets, density and transverse confinement were $n_{1d} = 59(5)\,\mu\text{m}^{-1}$ and $\omega_\perp = 2\pi \times 3020(10)\,\text{Hz}$, resulting in $\mu \approx h \times 1.9\,\text{kHz}$ and $K = 46$.

Figure 7.25 equally shows numerically calculated distribution functions for the corresponding experimental parameters for each histogram [112]. Note that once the temperature was derived following the above-described measurement of the av-

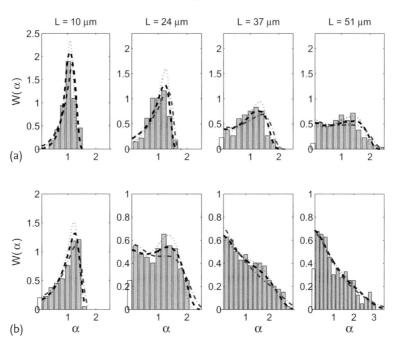

Figure 7.25 Experimentally measured distribution functions $W(\alpha(L))$ for $\alpha(L) = |A_Q(L)|^2/\langle |A_Q(L)|^2 \rangle$ for different system lengths. The upper row (a) corresponds to $T = 31(6)\,\text{nK}$, the lower row (b) to $T = 60(5)\,\text{nK}$. The dashed lines correspond to a numerical calculation of $W(\alpha/L)$ for the given experimental parameters, the dotted and dash-dotted lines indicate the same calculation taking into account the uncertainty in obtaining the system temperature.

erage interference amplitude square, no free parameters remained in the analysis as all remaining quantities were determined independently.

One clearly observes the transition from a narrow, asymmetric distribution centered on $\alpha = 1$ at low temperatures and small L to a broad Poissonian distribution around $\alpha = 0$ for higher temperatures and larger L, passing by an intermediate regime characterized by a double-peak structure. Whether this transition is sufficient to identify and separate the quantum and thermal contribution to phase fluctuations in 1D samples is an issue of ongoing discussions.

Quantum noise analysis, looking at the full distribution function of a single-shot (nonaveraged) observable, is a powerful new tool to characterize complex many-body states, most probably not only in the atomic system. A comparable analysis has recently been performed to detect the Berezinskii–Kosterlitz–Thouless transition in a two-dimensional quantum gas [113].

7.6
Summary and Outlook

We have shown that atom chips are a promising approach to interferometry with Bose–Einstein condensates. They have been successfully employed to characterize fundamental properties of the matter-wave source such as long-range coherence and have been established as a powerful tool to characterize complex many-body quantum systems, novel quantum phases and non-equilibrium decoherence dynamics in 1D, 2D, and 3D.

The atom chip has been identified as a highly suitable technological platform with the potential to integrate all components of a matter-wave interferometer in a single robust and transportable device. In recent years, significant advances have been made on many fronts; some of which are described in this manuscript. Although technological at first sight, these advances touch the most fundamental questions of modern many-particle quantum physics such as the role of nonlinear interactions, properties of low-dimensional systems, fundamental decoherence processes, squeezing and entanglement.

Whether an integrated matter-wave interferometer can be a useful device for precision measurements such as inertial sensing or gravity wave detection is today an open question, both experimentaly and theoretically. Along with other directions, future research will be encompassed by the search for further atom-optical elements on atom chips:

- A DC matter-wave source. Currently, experiments are performed in a pulsed manner with single Bose–Einstein condensates consisting of a finite, limited number of atoms. This leads to an inhomogeneous density profile of the matter-wave, dispersion, and simply limited flux. Experimental sequences take several seconds to minutes, at least one order of magnitude longer than any interferometer sequence so far. Drifts of trapping potentials are hence difficult to trace from one experimental cycle to another. A continuous matter-wave source would allow active stabilization which is required for true metrological applications. At the same time, it will enable a multitude of fundamental studies on transport properties, quantum reflections, or localization effects.

- Monomode guiding: transport of atoms along "atom fibers" on atom chips has been demonstrated [114], but in contrast to monomode optical fibers, atom chip traps are multimode in the first place. Strong transverse confinement increases the level separation, reduces thermal excitations and renders the system more robust. However, to increase confinement, the atom chip trap needs to be brought closer to the surface, rendering the system vulnerable to potential corrugation and heating. Furthermore, restricting the system to the vibrational ground state leads to one-dimensional quantum gases. This represents a fascinating research area in itself, whether low-dimensional systems are well suited for interferometry is the subject of ongoing discussion and study. Confining the matter-wave in the longitudinal direction maintains the system in the 3D regime and enables controlled transport as demonstrated in conveyor belt experiments [52, 115]. This however again leads to density variations and a pulsed operation of the interferometer.
- The BEC splitter: although coherent splitting of BEC on an atom chip could be demonstrated and is now routinely employed in many labs worldwide, many questions remain to be studied. Most interesting may be the role of interactions in guided matter-wave interferometry. As outlined in this manuscript, interactions lead to the fundamental process of phase diffusion which complicates interferometry. On the other hand, interactions also lead to squeezing and a resulting reduction of phase diffusion. Strategies to make optimal use of squeezing to reach Heisenberg-limited interferometry will have to be devised [116]. Alternatively, methods to alter the strength and sign of interactions (e.g., through Feshbach resonances) compatible with atom chip technology should be investigated.
- A matter-wave combiner: a phase-sensitive recombination (by reversing a splitting process) has been demonstrated. However, complex dynamics take place on the microscopic scale in the merging and the analysis is nontrivial. An alternative approach based on controlled tunneling instead of splitting could be investigated. This brings about the interesting physics of bosonic Josephson junctions [34, 35, 116] which has not been discussed here.
- Detection: Most information about coherence or phase dynamics is so far derived from absorption imaging performed after releasing the atomic system from the trapping potential and allowing for ballistic expansion. As this detection technique is destructive, it again enforces pulsed operation of the interferometer. In situ detection of atoms on the single particle level on atom chips has already been demonstrated [14, 15] and could be combined with interferometer setups.

Also on the technological side impressive progress has been made in recent years, probably highlighted by the realization of transportable cold atom setups which have been successfully tested in parabola flights [117], drop towers [118] and soon in space missions. Industrial companies have started to cooperate with research groups, funding agencies have identified quantum-based metrology as one of the key emerging technologies of the forthcoming decade. In conclusion, research on integrated matter-wave interferometry and in particular with Bose–Einstein condensates on atom chips will remain an exciting and stimulating field at the interface of fundamental research and applied quantum technology.

References

1 Folman, R., Krüger, P., Schmiedmayer, J., Denschlag, J., and Henkel, C. (2002) Microscopic atom optics: from wires to an atom chip. *Adv. At. Mol. Opt. Phys.* **48**, 263.

2 Fortágh, J. and Zimmermann, C. (2007) Magnetic microtraps for ultracold atoms. *Rev. Mod. Phys.* **79**, 235.

3 Hänsel, W., Hommelhoff, P., Hänsch, T.W., and Reichel, J. (2001) Bose–Einstein condensation on a microelectronic chip. *Nature* **413**, 498.

4 Ott, H., Fortágh, J., Schlotterbeck, G., Grossmann, A., and Zimmermann, C. (2001) Bose–Einstein condensation in a surface microtrap. *Phys. Rev. Lett.* **87**, 230401.

5 Schneider, S., Kasper, A., Hagen, C.V., Bartenstein, M., Engeser, B., Schumm, T., Bar-Joseph, I., Folman, R., Feenstra, L., and Schmiedmayer, J. (2003) Bose–Einstein condensation in a simple microtrap. *Phys. Rev. A* **67**, 023612.

6 Reichel, J. (2002) Microchip traps and Bose–Einstein condensation. *Appl. Phys. B* **74**, 469.

7 Hänsel, W., Reichel, J., Hommelhoff, P., and Hänsch, T.W. (2001) Magnetic conveyor belt for transporting and merging trapped atom clouds. *Phys. Rev. Lett.* **86**, 608.

8 Savalli, V., Stevens, D., Estève, J., Featonby, P.D., Josse, V., Westbrook, N., Westbrook, C.I., and Aspect, A. (2002) Specular reflection of matter-waves from a rough mirror. *Phys. Rev. Lett.* **88**, 250404.

9 Rosenbusch, P., Hall, B., Hughes, I., Saba, C., and Hinds, E. (2000) Manipulation of cold atoms by an adaptable magnetic reflector. *Appl. Phys. B: Lasers Opt.* **70**, 709.

10 Günther, A., Kraft, S., Zimmermann, C., and Fortágh, J. (2007) Atom interferometer based on phase coherent splitting of Bose–Einstein condensates with an integrated magnetic grating. *Phys. Rev. Lett.* **98**, 140403.

11 Smith, D.A., Arnold, A.S., Pritchard, M.J., and Hughes, I.G. (2008) Experimental single-impulse magnetic focusing of launched cold atoms. *J. Phys. B: At. Mol. Opt. Phys.* **41**, 125302.

12 Quinto-Su, P.A., Tscherneck, M., Holmes, M., and Bigelow, N.P. (2004) On-chip optical detection of laser cooled atoms. *Opt. Express* **12**, 5098.

13 Wilzbach, M., Haase, A., Schwarz, M., Heine, D., Wicker, K., Liu, X., Brenner, K.-H., Groth, S., Fernholz, T., Hessmo, B., and Schmiedmayer, J. (2006) Detecting neutral atoms on an atom chip. *Fortschr. Phys.* **54**, 746.

14 Colombe, Y., Steinmetz, T., Dubois, G., Linke, F., Hunger, D., and Reichel, J. (2007) Strong atom-field coupling for Bose–Einstein condensates in an optical cavity on a chip. *Nature* **450**, 272.

15 Wilzbach, M., Heine, D., Groth, S., Liu, X., Raub, T., Hessmo, B., and Schmiedmayer, J. (2009) Simple integrated single-atom detector. *Opt. Lett.* **34**, 259.

16 Schumm, T., Krueger, P., Hofferberth, S., Lesanovsky, I., Wildermuth, S., Groth, S., Bar-Joseph, I., Andersson, L.M., and Schmiedmayer, J. (2006) A double well interferometer on an atom chip. *Quantum Inf. Process.* **5**, 537.

17 Lesanovsky, I., Schumm, T., Hofferberth, S., Andersson, L.M., Krüger, P., and Schmiedmayer, J. (2006) Adiabatic radio frequency potentials for the coherent manipulation of matter-waves. *Phys. Rev. A* **73**, 033619.

18 Cassettari, D., Hessmo, B., Folman, R., Maier, T., and Schmiedmayer, J. (2000) Beam splitter for guided atoms. *Phys. Rev. Lett.* **85**, 5483.

19 Esteve, J., Schumm, T., Trebbia, J.-B., Bouchoule, I., Aspect, A., and Westbrook, C.I. (2005) Realizing a stable magnetic double-well potential on an atom chip. *Eur. Phys. J. D* **35**, 141.

20 Hänsel, W., Reichel, J., Hommelhoff, P., and Hänsch, T.W. (2001) Trapped-atom interferometer in a magnetic microtrap. *Phys. Rev. A* **64**, 063607.

21 Treutlein, P., Hänsch, T.W., Reichel, J., Negretti, A., Cirone, M.A., and Calarco, T. (2006) Microwave potentials and optimal control for robust quantum gates on an atom chip. *Phys. Rev. A* **74**, 022312.

22 Zobay, O. and Garraway, B.M. (2001) Two-dimensional atom trapping in field-induced adiabatic potentials. *Phys. Rev. Lett.* **86**, 1195.

23 Zobay, O. and Garraway, B.M. (2004) Atom trapping and two-dimensional Bose–Einstein condensates in field-induced adiabatic potentials. *Phys. Rev. A*. **69**, 023605.

24 Colombe, Y., Knyazchyan, E., Morizot, O., Mercier, B., Lorent, V., and Perrin, H. (2004) Ultracold atoms confined in RF-induced two-dimensional trapping potentials. *Europhys. Lett.* **67**, 593.

25 Schumm, T., Hofferberth, S., Andersson, L.M., Wildermuth, S., Groth, S., Bar-Joseph, I., Schmiedmayer, J., and Krüger, P. (2005) Matter wave interferometry in a double well on an atom chip. *Nat. Phys.* **1**, 57.

26 Extavour, M.H.T., LeBlanc, L.J., Schumm, T., Cieslak, B., Myrskog, S., Stummer, A., Aubin, S., and Thywissen, J.H. (2006) *Dual-Species Quantum Degeneracy of ^{40}K and ^{87}Rb on an Atom Chip*, in (eds C. Roos, H. Häffner, and R. Blatt), ATOMIC PHYSICS 20: XX International Conference on Atomic Physics – ICAP 2006, Vol. **869**, 241–249, AIP.

27 Hofferberth, S., Lesanovsky, I., Fischer, B., Schumm, T., and Schmiedmayer, J. (2007) Non-equilibrium coherence dynamics in one-dimensional Bose gases. *Nature* **449**, 324.

28 Fernholz, T., Gerritsma, R., Kruger, P., and Spreeuw, R.J.C. (2007) Dynamically controlled toroidal and ring-shaped magnetic traps. *Phys. Rev. A* **75**, 063406.

29 Jo, G.-B., Shin, Y., Will, S., Pasquini, T.A., Saba, M., Ketterle, W., Pritchard, D.E., Vengalattore, M., and Prentiss, M.(2007) Long phase coherence time and number squeezing of two Bose–Einstein condensates on an atom chip. *Phys. Rev. Lett.* **98**, 030407.

30 J. P. van Es, J., Whitlock, S., Fernholz, T., H. van Amerongen, A., and J. van Druten, N. (2008) Longitudinal character of atom-chip-based RF-dressed potentials. *Phys. Rev. A* **77**, 063623.

31 Stenger, J., Inouye, S., Chikkatur, A.P., Stamper-Kurn, D.M., Pritchard, D.E., and Ketterle, W. (1999) Bragg spectroscopy of a Bose–Einstein condensate. *Phys. Rev. Lett.* **82**, 4569.

32 Hagley, E.W., Deng, L., Kozuma, M., Trippenbach, M., Band, Y.B., Edwards, M., Doery, M., Julienne, P.S., Helmerson, K., Rolston, S.L., and Phillips, W.D. (1999) Measurement of the coherence of a Bose–Einstein condensate. *Phys. Rev. Lett.* **83**, 3112.

33 Wang, Y.-J., Anderson, D.Z., Bright, V.M., Cornell, E.A., D., Q., Kishimoto, T., Prentiss, M., Saravanan, R.A., Segal, S.R., and Wu, S. (2005) Atom Michelson interferometer on a chip using a Bose–Einstein condensate. *Phys. Rev. Lett.* **94**, 090405.

34 Albiez, M., Gati, R., Fölling, J., Hunsmann, S., Cristiani, M., and Oberthaler, M.K. (2005) Direct observation of tunneling and nonlinear self-trapping in a single bosonic Josephson junction. *Phys. Rev. Lett.* **95**, 010402.

35 Levy, S., Lahoud, E., Shomroni, I., and Steinhauer, J. (2007) The a.c. and d.c. Josephson effects in a Bose–Einstein condensate. *Nature* **449**, 579.

36 Calarco, T., Hinds, E.A., Jaksch, D., Schmiedmayer, J., Cirac, J.I., and Zoller, P. (2000) Quantum gates with neutral atoms: controlling collisional interactions in time-dependent traps. *Phys. Rev. A* **61**, 022304.

37 Böhi, P., Riedel, M.F., Hoffrogge, J., Reichel, J., Hansch, T.W., and Treutlein, P. (2009) Coherent manipulation of Bose–Einstein condensates with state-dependent microwave potentials on an atom chip. *Nat. Phys.* **5**, 592.

38 Treutlein, P., Hommelhoff, P., Steinmetz, T., Hänsch, T.W., and Reichel, J. (2004) Coherence in microchip traps. *Phys. Rev. Lett.* **92**, 203005.

39 Davis, T.J. (2002) 2D magnetic traps for ultra-cold atoms: A simple theory using complex numbers. *Eur. Phys. J. D* **18**, 27.

40 Hinds, E.A., Vale, C.J., and Boshier, M.G. (2001) Two-wire waveguide and interferometer for cold atoms. *Phys. Rev. Lett.* **86**, 1462.

41 Cassettari, D., Chenet, A., Folman, R., Haase, A., Hessmo, B., Krüger, P., Maier, T., Schneider, S., Calarco, T., and Schmiedmayer, J. (2000) Micromanipulation of neutral atoms with nanofabricated structures. *Appl. Phys. B* **70**, 721.

42 Zobay, O. and Garraway, B.M. (2000) Controllable double waveguide for atoms. *Opt. Commun.* **178**, 93.

43 Shin, Y., Sanner, C., Jo, G.-B., Pasquini, T.A., Saba, M., Ketterle, W., Pritchard, D.E., Vengalattore, M., and Prentiss, M. (2005) Interference of Bose–Einstein condensates split with an atom chip. *Phys. Rev. A* **72**, 021604.

44 Dekker, N.H., Lee, C.S., Lorent, V., Thywissen, J.H., Smith, S.P., Drndić, M., Westervelt, R.M., and Prentiss, M. (2000) Guiding neutral atoms on a chip. *Phys. Rev. Lett.* **84**, 1124.

45 Buks, E., Schuster, R., Heilblum, M., Mahalu, D., and Umansky, V. (1998) Dephasing in electron interference by 'which-path' detector. *Nature* **391**, 871.

46 Lou, X., Krüger, P., Klein, M.W., Brugger, K., Wildermuth, S., Groth, S., Bar-Joseph, I., Folman, R., and Schmiedmayer, J. (2004) An atom fiber for omnidirectional guidung of cold neutral atoms. *Opt. Lett.* **29**, 2145.

47 Lou, X., Krüger, P., Klein, M.W., Brugger, K., Wildermuth, S., Groth, S., Bar-Joseph, I., Folman, R., and Schmiedmayer, J. (2005) Two wire guides and traps with vertical bias field on atom chips. *Phys. Rev. A* **72**, 023607.

48 Andersson, E., Calarco, T., Folman, R., Andersson, M., Hessmo, B., and Schmiedmayer, J. (2002) Multi mode interferometer for guided matter-waves. *Phys. Rev. Lett.* **88**, 100401.

49 Müller, D., Cornell, E.A., Prevedelli, M., Schwindt, P.D.D., Zozulya, A., and Anderson, D.Z. (2000) Waveguide atom beam splitter for laser-cooled neutral atoms. *Opt. Lett.* **25**, 1382.

50 Cirone, M., Negretti, A., Calarco, T., Krüger, P., and Schmiedmayer, J. (2005) A simple quantum gate with atom chips. *Eur. Phys. J. D* **35**, 165.

51 Charron, E., Cirone, M.A., Negretti, A., Schmiedmayer, J., and Calarco, T. (2006) Theoretical analysis of a realistic atom-chip quantum gate. *Phys. Rev. A* **74**, 012308.

52 Hommelhoff, P., Hänsel, W., Steinmetz, T., Hänsch, T.W., and Reichel, J. (2005) Transporting, splitting and merging of atomic ensembles in a chip trap. *New J. Phys.* **7**, 3.

53 Krüger, P., Luo, X., Klein, M.W., Brugger, K., Haase, A., Wildermuth, S., Groth, S., Bar-Joseph, I., Folman, R., and Schmiedmayer, J. (2003) Trapping and manipulating neutral atoms with electrostatic fields. *Phys. Rev. Lett.* **91**, 233201.

54 Wing, W.H. (1984) On neutral particle trapping in quasistatic electromagnetic fields. *Prog. Quantum Electron.* **8**, 181.

55 Muskat, E., Dubbers, D., and Schärpf, O. (1987) Dressed neutrons. *Phys. Rev. Lett.* **58**, 2047.

56 Spreeuw, R.J.C., Gerz, C., Goldner, L.S., Phillips, W.D., Rolston, S.L., Westbrook, C.I., Reynolds, M.W., and Silvera, I.F. (1994) Demonstration of neutral atom trapping with microwaves. *Phys. Rev. Lett.* **72**, 3162.

57 Hofferberth, S., Lesanovsky, I., Fischer, B., Verdu, J., and Schmiedmayer, J.

57 (2006) Radio-frequency dressed state potentials for neutral atoms. *Nat. Phys.* **2**, 710.
58 Lesanovsky, I. and von Klitzing, W. (2007) Time-averaged adiabatic potentials: Versatile matter–wave guides and atom traps. *Phys. Rev. Lett.* **99**, 083001.
59 Courteille, P.W., Deh, B., Fortágh, J., Günther, A., Kraft, S., Marzok, C., Slama, S., and Zimmermann, C. (2006) Highly versatile atomic micro traps generated by multifrequency magnetic field modulation. *J. Phys. B: At. Mol. Opt. Phys.* **39**, 1055.
60 Cohen-Tannoudji, C., Dupont-Roc, J., and Grynberg, G. (1992) *Atom-Photon Interactions*, Wiley, New York.
61 Majorana, E. (1932) Majorana flops. *Nuovo Cimento* **9**, 43.
62 Gov, S., Shtrikman, S., and Thomas, H. (2000) Magnetic trapping of neutral particles: Classical and quantum-mechanical study of a Ioffe–Pritchard type trap. *J. Appl. Phys.* **87**, 3989.
63 Zhang, P. and You, L. (2007) Geometric phase of an atom inside an adiabatic radio-frequency potential. *Phys. Rev. A* **76**, 033615.
64 Shirley, J.H. (1965) Solution of the Schrödinger equation with a Hamiltonian periodic in time. *Phys. Rev.* **138**, B979.
65 Hofferberth, S., Fischer, B., Schumm, T., Schmiedmayer, J., and Lesanovsky, I. (2007) Ultracold atoms in radio-frequency dressed potentials beyond the rotating-wave approximation. *Phys. Rev. A* **76**, 013401.
66 Göbel, M. (2008) *Low-Dimensional Traps for Bose-Fermi Mixtures*, Ph.D. thesis, University of Heidelberg.
67 Leggett, A. and Sols, F. (1991) On the concept of spontaneously broken gauge symmetry in condensed matter physics. *Found. Phys.* **21**, 353.
68 Leggett, A.J. and Sols, F. (1998) Comment on "*Phase and phase diffusion of a split Bose–Einstein condensate*". *Phys. Rev. Lett.* **81**, 1344.
69 Sols, F. (1998) Bose–Einstein Condensation in Atomic Gases. *Enrico Fermi School Lectures*.
70 Pethick, C.J. and Smith, H. (2002) *Bose–Einstein Condensation in Dilute Gases*, Cambridge University Press.
71 Mewes, M.O., Andrews, M.R., J. van Druten, N., Kurn, D.M., Durfee, D.S., and Ketterle, W. (1996) Bose–Einstein condensation in a tightly confining DC magnetic trap. *Phys. Rev. Lett.* **77**, 416.
72 Javanainen, J. and Yoo, S.M. (1996) Quantum phase of a Bose–Einstein condensate with an arbitrary number of atoms. *Phys. Rev. Lett.* **76**, 161.
73 Naraschewski, M., Wallis, H., Schenzle, A., Cirac, J.I., and Zoller, P. (1996) Interference of Bose condensates. *Phys. Rev. A* **54**, 2185.
74 Castin, Y. and Dalibard, J. (1997) Relative phase of two Bose–Einstein condensates. *Phys. Rev. A* **55**, 4330.
75 Andrews, M.R., Townsend, C.G., Miesner, H.-J., Durfee, D.S., Kurn, D.M., and Ketterle, W. (1997) Observation of interference between two Bose condensates. *Science* **275**, 637.
76 Shin, Y., Saba, M., A. Pasquini, T., Ketterle, W., Pritchard, D.E., and Leanhardt, A.E. (2004) Atom interferometry with Bose–Einstein condensates in a double-well potential. *Phys. Rev. Lett.* **92**, 050405.
77 Hohenester, U., Rekdal, P.K., Borzì, A., and Schmiedmayer, J. (2007) Optimal quantum control of Bose–Einstein condensates in magnetic microtraps. *Phys. Rev. A* **75**, 023602.
78 Burkov, A.A., Lukin, M.D., and Demler, E. (2007) Decoherence dynamics in low-dimensional cold atom interferometers. *Phys. Rev. Lett.* **98**, 200404.
79 Zapata, I., Sols, F., and Leggett, A.J. (1998) Josephson effect between trapped Bose–Einstein condensates. *Phys. Rev. A* **57**, R28.
80 Lewenstein, M. and You, L. (1996) Quantum phase diffusion of a Bose–Einstein condensate. *Phys. Rev. Lett.* **77**, 3489.
81 Grond, J., Schmiedmayer, J., and Hohenester, U. (2009) Optimizing number squeezing when splitting a mesoscopic condensate. *Phys. Rev. A* **79**, 021603.
82 Fisher, N.I. (1993) *Statistical analysis of circular data*, Cambridge University Press.

83 Menotti, C., Anglin, J.R., Cirac, J.I., and Zoller, P. (2001) Dynamic splitting of a Bose–Einstein condensate. *Phys. Rev. A* **63**, 023601.

84 Weller, A., Ronzheimer, F.P., Gross, C., Esteve, J., Oberthaler, M.K., Frantzeskakis, D.J., Theocharis, G., and Kevrekidis, P.G. (2008) Experimental observation of oscillating and interacting matter-wave dark solitons. *Phys. Rev. Lett.* **101**, 130401.

85 Negretti, A. and Henkel, C. (2004) Enhanced phase sensitivity and soliton formation in an integrated BEC interferometer. *J. Phys. B: At. Mol. Opt. Phys.* **37**, L385.

86 Scott, R.G., Judd, T.E., and Fromhold, T.M. (2008) Exploiting soliton decay and phase fluctuations in atom chip interferometry of Bose–Einstein condensates. *Phys. Rev. Lett.* **100**, 100402.

87 Scott, R.G., Hutchinson, D.A.W., Judd, T.E., and Fromhold, T.M. (2009) Quantifying finite temperature effects in atom chip interferometry of Bose–Einstein condensates, arXiv:0901.4602.

88 Grandi, C.D., Barankov, R.A., and Polkovnikov, A. (2008) Adiabatic non-linear probes of one-dimensional Bose gases. *Phys. Rev. Lett.* **101**, 230402.

89 Reichel, J. and Thywissen, J.H. (2004) Using magnetic chip traps to study Tonks–Girardeau quantum gases. *J. Phys. France* **116**, 265.

90 Petrov, D.S., Shlyapnikov, G.V., and Walraven, J.T.M. (2000) Regimes of quantum degeneracy in trapped 1D gases. *Phys. Rev. Lett.* **85**, 3745.

91 Giamarchi, T. (2003) *Quantum Physics in One Dimension*, Oxford University Press.

92 Mazets, I.E., Schumm, T., and Schmiedmayer, J. (2008) Breakdown of integrability in a quasi-1D ultracold bosonic gas. *Phys. Rev. Lett.* **100**, 210403.

93 Yurovsky, V.A., Olshanii, M., and Weiss, D.S. (2008) Collisions, correlations, and integrability in atom wave guides. *Adv. At. Mol. Opt. Phys.*, **55**, 61.

94 Gerbier, F. (2004) Quasi-1D Bose–Einstein condnesates in the dimensional crossover regime. *Europhys. Lett.* **66**, 771.

95 Olshanii, M. (1998) Atomic scattering in the presence of an external confinement and a gas of impenetrable bosons. *Phys. Rev. Lett.* **81**, 938.

96 Kim, J.I., Schmiedmayer, J., and Schmelcher, P. (2005) Quantum scattering in quasi-one-dimensional cylindrical confinement. *Phys. Rev. A* **72**, 042711.

97 Bloch, I., Dalibard, J., and Zwerger, W. (2008) Many-body physics with ultracold gases. *Rev. Mod. Phys.* **80**, 885.

98 Andreev, A.F. (1980) The hydrodynamics of two-dimensional and one-dimensional fluids. *Sov. Phys. JETP* **51**, 1038.

99 Whitlock, N.K. and Bouchoule, I. (2003) Relative phase fluctuations of two coupled one-dimensional condensates. *Phys. Rev. A* **68**, 053609.

100 Smerzi, A., Fantoni, S., Giovanazzi, S., and Shenoy, S.R. (1997) Quantum coherent atomic tunneling between two trapped Bose–Einstein condensates. *Phys. Rev. Lett.* **79**, 4950.

101 Ananikian, D. and Bergeman, T. (2006) The Gross–Pitaevskii equation for Bose particles in a double well potential: Two mode models and beyond. *Phys. Rev. A* **73**, 013604.

102 Polkovnikov, A., Altman, E., and Demler, E. (2006) Interference between independent fluctuating condensates. *Proc. Natl. Acad. Sci. USA* **103**, 6125.

103 Imambekov, A., Gritsev, V., and Demler, E. (2007) *Fundamental noise in matter interferometers*, in (eds M. Inguscio, W. Ketterle, and C. Salomon), *Proceedings of the 2006 Enrico Fermi Summer School on Ultracold Fermi gases*.

104 Castin, Y. and Dalibard, J. (1997) Relative phase of two Bose–Einstein condensates. *Phys. Rev. A* **55**, 4330.

105 Javanainen, J. and Wilkens, M. (1997) Phase and phase diffusion of a split Bose–Einstein condensate. *Phys. Rev. Lett.* **78**, 4675.

106 Leggett, A.J. (2001) Bose–Einstein condensation in the alkali gases. *Rev. Mod. Phys.* **73**, 307.

107 Haldane, F. (1981) Effective harmonic-fluid approach to low-energy properties of one-dimensional quantum fluids. *Phys. Rev. Lett.* **47**, 1840.

108 Cazalilla, M. (2004) Bosonizing one-dimensional cold atomic gases. *J. Phys. B: At. Mol. Opt. Phys.* **37**, S1.

109 Girardeau, M. (1960) Relationship between systems of impenetrable bosons and fermions in one dimension. *J. Math. Phys.* **1**, 516.

110 Hofferberth, S., Lesanovsky, I., Schumm, T., Imambekov, A., Gritsev, V., Demler, E., and Schmiedmayer, J. (2008) Probing quantum and thermal noise in an interacting many-body system. *Nat. Phys.* **4**, 489.

111 Gumbel, E.J. (1958) *Statistics of Extremes*, Columbia University Press.

112 Imambekov, A., Gritsev, V., and Demler, E. (2008) Mapping of Coulomb gases and sine-Gordon models to statistics of random surfaces. *Phys. Rev. A* **77**, 063606.

113 Hadzibabic, Z., Krüger, P., Cheneau, M., Battelier, B., and Dalibard, J. (2006) Berezinskii–Kosterlitz–Thouless crossover in a trapped atomic gas. *Nature* **441**, 1118.

114 Brugger, K., Krüger, P., Luo, X., Wildermuth, S., Gimpel, H., Klein, M.W., Groth, S., Folman, R., Bar-Joseph, I., and Schmiedmayer, J. (2005) Two-wire guides and traps with vertical bias fields on atom chips. *Phys. Rev. A* **72**, 023607.

115 Günther, A., Kemmler, M., Kraft, S., Vale, C.J., Zimmermann, C., and Fortágh, J. (2005) Combined chips for atom-optics. *Phys. Rev. A* **71**, 063619.

116 Pezzé, L., Smerzi, A., Berman, G.P., Bishop, A.R., and Collins, L.A. (2006) Nonlinear beam splitter in Bose–Einstein-condensate interferometers. *Phys. Rev. A* **74**, 033610.

117 Stern, G., Battelier, B., Geiger, R., Varoquaux, G., Villing, A., Moron, F., Carraz, O., Zahzam, N., Bidel, Y., Chaibi, W., Pereira Dos Santos, F., Bresson, A., Landragin, A., and Bouyer, P. (2009) Light-pulse atom interferometry in microgravity. *Eur. Phys. J. D* **53**, 353.

118 Könemann, T., Brinkmann, W., Göklü, E., Lämmerzahl, C., Dittus, H., van Zoest, T., Rasel, E., Ertmer, W., Lewoczko-Adamczyk, W., Schiemangk, M., Peters, A., Vogel, A., Johannsen, G., Wildfang, S., Bongs, K., Sengstock, K., Kajari, E., Nandi, G., Walser, R., and Schleich, W. (2007) A freely falling magneto-optical trap drop tower experiment. *Appl. Phys. B: Lasers Opt.* **89**, 431.

8
Microchip-Based Trapped-Atom Clocks

Vladan Vuletić, Ian D. Leroux, and Monika H. Schleier-Smith

8.1
Basic Principles

A two-level quantum system in a vacuum – in the absence of any perturbing fields – constitutes an ideal clock whose oscillation frequency $\omega = E/\hbar$ is given by the energy difference E between the two levels $|1\rangle, |2\rangle$. In a single measurement of duration T performed on a single particle this frequency can be determined with an uncertainty $\Delta\omega = 1/T$. If the measurement is performed simultaneously on N independent identical particles, and this measurement is repeated with a cycle time $T_c > T$ for a total averaging time $\tau > T_c$, the fundamental quantum uncertainty of the frequency determination is given by the standard quantum limit [1]

$$\frac{\Delta\omega}{\omega} = \frac{1}{\omega T}\sqrt{\frac{T_c}{N\tau}}. \tag{8.1}$$

The $N^{-1/2}$ scaling arises because the quantities to be measured are the non-zero probabilities to find a particle in either of the clock states $|1\rangle, |2\rangle$: when the independent particles in the ensemble are read out, the observed populations of the two clock states are binomially distributed, leading to so-called projection noise on the estimation of those probabilities [2, 3]. The $\sqrt{T_c/\tau}$ scaling arises because the sequential measurement repeated τ/T_c times using N particles each is equivalent to a single measurement using $N\tau/T_c$ atoms. For a given total measurement time τ the stability improves as the duration T of the single measurement is increased. The latter is limited by the coherence time of the transition. While the absolute stability does not depend on the transition frequency ω, the fractional stability improves with higher transition frequency.

8.2
Atomic-Fountain versus Trapped-Atom Clocks

In the absence of other fields, particles fall under gravity, which sets a practical limit on the single-measurement time T. Atomic fountains [4, 5], where an ensemble

Atom Chips. Edited by Jakob Reichel and Vladan Vuletić
Copyright © 2011 WILEY-VCH Verlag GmbH & Co. KGaA, Weinheim
ISBN: 978-3-527-40755-2

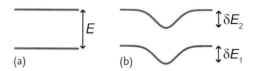

Figure 8.1 Illustration of free-space clock (a) and trapped-atom clock (b). A trapped-atom clock can be operated with small linewidth if the difference $\delta E = \delta E_2 - \delta E_1$ between the energy shifts $\delta E_1, \delta E_2$ of the trapped states is small.

of atoms is launched upwards into a ballistic flight region, allow one to increase the measurement time. A vacuum apparatus with a height of 1 m yields a typical interaction time of $T \sim 0.7$ s. While larger systems are under construction for the measurement of weak gravitational effects [6], substantial further increase of measurement time for freely falling atoms is impractical – unless working in a micro-gravity environment [7, 8] – in view of the quadratic dependence of apparatus height on measurement time.

Achieving long measurement times T with reasonable-sized apparatus therefore requires the use of trapped atoms, held against gravity by some externally applied force. Such a force necessarily perturbs the atomic energy levels E_1 and E_2, but this perturbation can be tolerated provided that the *differential* energy shift between the two clock levels $\delta E = \delta E_1 - \delta E_2$ is sufficiently small (see Figure 8.1). Under such circumstances the external field can be used to provide a trap for the particles without compromising clock stability or accuracy. An example is the electrostatic Coulomb force used to trap ions, which produces only a small differential shift of the clock transition [9, 10]. In such systems, a coherence time of 10 min has been reported between hyperfine ground states using Be^+ ions [9]. To date, a single Al^+ ion trapped in a Paul trap constitutes the best clock, with current fractional accuracy now exceeding 10^{-17} using an optical transition [10, 11]. Similar non-perturbing traps can be crafted for hyperfine transitions in neutral atoms using magnetostatic forces [12–14] provided the linear Zeeman shift is the same in both clock states [15, 16]. For electronic transitions at optical frequencies, traps based on the optical dipole force (AC Stark shift) [17, 18] can be used at certain "magic wavelengths" where the polarizability of the two clock states is the same [19]. This approach, developed independently for studies of cavity quantum electrodynamics by Kimble and coworkers [20, 21] and for optical-transition clocks by Katori and coworkers [22–24], has enabled many of the recent successes using magic-wavelength optical traps for neutral atoms [25–34].

Aside from allowing long interrogation times, the confinement of atoms to a small volume, typically of millimeter to micrometer size, allows better control of perturbing external fields than can be achieved over the meter-scale flight region of an atomic fountain. On the other hand, this confinement also leads to a higher atomic density for a given atom number, resulting in larger collision shift of the clock transition frequency than in the dilute clouds used in fountain clocks. These collision shifts can be suppressed for trapped-atom clocks at the expense of an increase in overall trapping volume, by choosing periodic confining potentials with less than one atom per site so that atoms never collide [26–31, 35]. However, even without such a suppression the achievable accuracy of trapped-atom clocks is interesting for many commercial applications. Microchip-based atom traps, which allow compact experimental setups with modest power requirements, might thus allow the construction of robust, portable trapped-atom secondary frequency standards that would be technologically valuable even if they do not exceed the absolute performance of fountain clocks in the laboratory [16, 36]. The hope of making the stability of cold-atom clocks available in the field has fueled much of the commercial and experimental interest in chip clocks.

8.3
Optical-Transition Clocks versus Microwave Clocks

Since the fractional accuracy of an atomic clock improves with increasing transition frequency, it is natural to consider the use of optical-frequency electronic transitions rather than microwave-frequency hyperfine transitions as the basis for a clock. Clocks operating on optical transitions have already surpassed the much more mature hyperfine-transition atomic-fountain clocks in accuracy [25–32, 35], and further improvements are expected.

An optical clock could be constructed on an atom chip, using integrated fiber optics for the optical trapping and probing fields [37]. However, the measurement of optical transition frequencies currently requires bulky and vibration-sensitive laser systems with ultra-stable reference cavities and optical frequency combs, negating the advantages of compactness and robustness that make chip clocks interesting in the first place. It is, therefore, likely that work on chip clocks will concentrate, for the medium term, on hyperfine transitions in the microwave region of the electromagnetic spectrum.

8.4
Clocks with Magnetically Trapped Atoms: Fundamental Limits to Performance

Microwave clocks cannot use the same optical dipole traps as optical-transition clocks due to the lack of suitable "magic wavelengths" for hyperfine transitions (but see [38] for a recently proposed workaround which exploits the tensor polarizability). Instead, a "magic"-confinement approach is available for hyperfine clocks using magnetic trapping. While fountain clocks use magnetically untrapped $|F, m_F = 0\rangle, |F + 1, m_F = 0\rangle$ states for which the linear Zeeman shift vanishes at zero magnetic field, it is also possible to find magnetically trappable states where the *difference* between the Zeeman shift for the two states $|1\rangle, |2\rangle$ for some magnetic-field value B_0 varies only quadratically around B_0. Thus, it is possible to realize a hyperfine clock with magnetically trapped atoms that has similar sensitivity to external magnetic fields as a fountain clock [15, 16, 36].

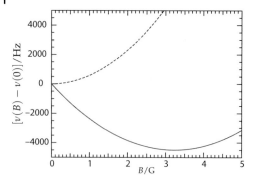

Figure 8.2 Transition frequency shift vs. magnetic field for the standard clock transition $|F = 1, m_F = 0\rangle \rightarrow |F = 2, m_F = 0\rangle$ (dashed line), and for the "magic" magnetic-trap transition $|F = 1, m_F = -1\rangle \rightarrow |F = 2, m_F = 1\rangle$ in ^{87}Rb (solid line). For the former, there is no linear Zeeman shift at zero magnetic field, while for the latter, the linear Zeeman shift vanishes at a field of 3.23 G.

Figure 8.2 shows the dependence of the hyperfine transition frequency on magnetic fields for a standard ^{87}Rb fountain clock that uses the transition $|F = 1, m_F = 0\rangle \rightarrow |F = 2, m_F = 0\rangle$, and for a trapped-atom clock that uses the transition $|1\rangle \equiv |1, -1\rangle \rightarrow |2\rangle \equiv |2, 1\rangle$. For the latter, the coefficient for the quadratic variation

$$\nu(B) = \nu_0 + \beta(B - B_0)^2 \qquad (8.2)$$

around the magic field $B_0 = 3.228\,917(3)$ G is given by $\beta = 431.359\,57(9)$ Hz/G^2 for ^{87}Rb [15], which is a little smaller than for the fountain-clock transition $|1, 0\rangle \leftrightarrow |2, 0\rangle$, where it amounts to $\beta' = 575.14$ Hz/G^2 [39]. The transition frequency at the minimum of the $|1, -1\rangle \leftrightarrow |2, 1\rangle$ transition is given by $\nu_0 = 6\,834\,678\,113.59(2)$ Hz for ^{87}Rb [40], slightly smaller than the hyperfine splitting and transition frequency of the $|1, 0\rangle \leftrightarrow |2, 0\rangle$ transition at zero field, $\nu'_0 = 6\,834\,682\,610.904\,32(2)$ Hz [5].

However, in a fountain clock one can apply a uniform magnetic field, whereas the magic-magnetic-trap approach requires the atoms to experience spatially varying magnetic field magnitude for trapping. This means that the thermal motion of the atoms in the trap will necessarily cause them to sample regions of the potential with different magnetic fields, and thus different residual quadratic Zeeman shift. The range of magnetic fields sampled by the atoms, and the resulting broadening and shift of the clock transition, increase as the cloud's temperature increases, so it seems advantageous to operate the clock with an ensemble of atoms that is as cold as possible. However, cooling the sample increases its density (for fixed atom number) and thus leads to a higher collision shift. Since the two perturbations have opposite temperature dependences, a compromise must be found. Here we analyze these two dominant line shift mechanisms, following the arguments of references [15, 36], in order to find the achievable performance for a magnetic-trap hyperfine clock as first demonstrated by Harber and coworkers in a macroscopic trap [15], pioneered for a microchip trap by Treutlein et al. [16], and recently upgraded to a precision device by Ramírez–Martínez et al. [41].

8.4 Clocks with Magnetically Trapped Atoms: Fundamental Limits to Performance

We restrict the analysis to a magnetically trapped ^{87}Rb hyperfine clock and do not consider a clock using ^{133}Cs. First, unlike ^{87}Rb where the density shift is independent of temperature for the temperature range of interest (i.e., temperatures lower than ~ 100 µK), the density shift for ^{133}Cs remains temperature-dependent down to nK temperatures [42] due to a multitude of low-field Feshbach resonances [43–48]. Related, but more important, is the fact that the density shift in ^{133}Cs, at typical temperatures of interest, is about two orders of magnitude larger than in ^{87}Rb, so that microchip magnetic traps, with their relatively high atomic densities, would yield very poorly performing ^{133}Cs clocks with large line shift and broadening.

As an atom moves in the trap, it experiences a time-varying magnetic field $B(x(t))$ that constitutes the source of its potential energy $U(B(x))$. This potential energy in a trap with a field minimum B_{min} is given by

$$U(B) = g_F m_F \mu_B (B - B_{min}) = \Upsilon (B - B_{min}), \quad (8.3)$$

where g_F is the atomic Landé factor, μ_B the Bohr magneton, and $\Upsilon/h \approx 0.70$ MHz/G for ^{87}Rb for the trapped states $|F = 1, m_F = -1\rangle, |F = 2, m_F = 1\rangle$ [39].

It then follows that if we choose the minimum trap field equal to the magic field, $B_{min} = B_0$, the average quadratic Zeeman shift of the clock transition is given by

$$\langle \delta \nu \rangle_Z = \frac{\beta}{\Upsilon^2} \langle U^2 \rangle. \quad (8.4)$$

For a thermal distribution of temperature T in a three-dimensional harmonic trap the clock frequency shift can be written as

$$\langle \delta \nu \rangle_Z |_T = \frac{15\beta}{4\Upsilon^2} (k_B T)^2 = \zeta T^2. \quad (8.5)$$

Thus, we find that the clock Zeeman shift is quadratic in the atomic temperature with a coefficient $\zeta = 1.43$ Hz/µK^2 for ^{87}Rb. We have assumed that the temperature is large compared to the trap oscillation frequency, so that the trap zero-point energy can be ignored. This is a good approximation for the optimum range of temperatures once the density shift is taken into account.

The Zeeman frequency shift and broadening due to the thermal energy of the gas in the trap must be traded off with the collision shift, which for fixed atom number and trap frequency increases as the atoms get colder. For a given atomic density n and equal populations of the two clock states the frequency shift is given by the expression [15, 36, 49]

$$\delta \nu_C = \frac{2\hbar}{m} n (a_{22} - a_{11}). \quad (8.6)$$

Here a_{22} and a_{11} are the scattering lengths for $|2\rangle + |2\rangle$ and $|1\rangle + |1\rangle$ collisions, respectively. Thus, the frequency shift is proportional to the atomic density and the difference in scattering lengths for the different clock states. For ^{87}Rb a serendipitous degeneracy leads to almost identical scattering lengths $a_{11} = 100.44 a_0$ and $a_{22} = 95.47 a_0$ [40, 50], where a_0 is the Bohr radius.

The density shift for a thermal cloud of N atoms in a harmonic trap can be written as

$$\langle \delta v \rangle_C = \frac{2\hbar}{m}(a_{22} - a_{11})\langle n \rangle = -\chi \frac{N\bar{\omega}^3}{T^{3/2}} \tag{8.7}$$

with $\bar{\omega} = \sqrt[3]{\omega_x \omega_y \omega_z}$ the geometric mean of the trap frequencies along the three axes, and $\chi = 9.2 \times 10^{-15}$ Hz s^3 μK$^{3/2}$.

To maximize coherence time in the magnetic trap, it is convenient to choose parameters such that the Zeeman shift and collision shift are of the same magnitude. Since they have approximately the same position dependence in the trap, the inhomogeneous shifts which they respectively impose on the cloud will then cancel each other, considerably extending the coherence time of the Ramsey fringes [36]. We therefore consider clock operation at a temperature such that $\langle \delta v \rangle_Z$ and $\langle \delta v \rangle_C$ are of the same magnitude and sum to zero. Figure 8.3 shows the combined quadratic Zeeman shift and density shift of the clock transition as a function of temperature for different total trapped atom numbers. For $N = 10^6$ atoms in a trap with vibration frequencies $\omega_x = \omega_y = \omega_z = 2\pi \times 100$ Hz the shift goes through zero at a temperature of $T_0 = (\chi \bar{\omega}^3 N/\zeta)^{2/7} = 1.1$ μK. Under these conditions the rms cloud size in each dimension is $\sqrt{k_B T_0/m\bar{\omega}^{-1}} = 17$ μm and the shift from each effect individually is 1.9 Hz. Note that $T_0/T_c = (0.3\,\text{mK}/T_c)^{1/7}$, where $k_B T_c = 0.94\hbar\bar{\omega} N^{1/3}$ is the critical temperature for Bose–Einstein condensation [51]. For the (low) densities of interest for clock operation, $T_c \ll 0.3$ mK so that the zero-shift temperature T is always above the critical temperature, and the clock always stays in the classical (non-degenerate) regime. Assuming that the atom number can be measured and controlled to 5%, a fractional accuracy of 10^{-11} on the hyperfine transition at 6.8 GHz appears possible in such a trap. A more detailed analysis, including the effects of motional and collisional averaging [15, 52], and assuming better control of the atom number by measuring it at

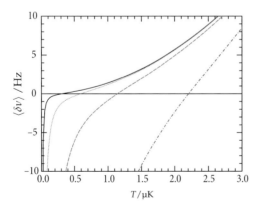

Figure 8.3 The total (Zeeman and collision) frequency shift of the magic-field clock transition $|1,-1\rangle \to |2,1\rangle$ for an ensemble of magnetically trapped ^{87}Rb atoms vs. ensemble temperature. A trap with a vibration frequency $\bar{\omega} = 2\pi \times 100$ Hz is assumed. The curves from left to right correspond to 10^4, 10^5, 10^6, and 10^7 trapped atoms.

the end of each experimental cycle, suggests that fractional stability of 10^{-12} and fractional accuracy at the level of 10^{-13} might be attainable [36].

Such accuracy is much worse than the state of the art in hyperfine atomic fountains that have reached 10^{-16} fractional accuracy [5]. Nevertheless, the small size of the microchip clock in combination with its accuracy renders it interesting as a secondary standard for commercial or other applications. Also it may be possible to improve upon the parameters discussed above. For instance, a trap with a vibration frequency as low as 10 Hz has been demonstrated in the study of quantum reflection from a surface [53]. In such a trap and with $N = 10^5$ atoms, the optimum temperature is around 80 nK, reducing the quadratic Zeeman and collision shifts to below 10 mHz.

The density shift could be avoided altogether if the atoms were confined in an array of traps, such that at most one atom is trapped in each individual trap, as with optical lattices. It is possible to make an array of magnetic traps on the microchip [54], but the individual trap size is larger than for optical lattices, resulting in a larger total trap volume. Nevertheless, with a 30 μm lattice period a planar array of 10^5 traps could be created within a 1 cm^2 area.

An additional challenge that is specific to microchip clocks is that they are operated in close vicinity to the chip surface, where electric and magnetic fields are enhanced. Electric field effects are negligible for the hyperfine spin states of interest for clock operation, while magnetic field effects are understood [55, 56] and have been quantified in experiments (see Chapters 4 and 5). In particular, there are increased magnetic field fluctuations near conducting surfaces that are due to Johnson-noise-induced currents in the conductor [57]. However, such fields can be reduced by using a chip design where the trap is located close to a non-conducting surface, but at relatively large distance from any conductor [57].

8.5
Clocks with Magnetically Trapped Atoms: Experimental Demonstrations

The tradeoff between Zeeman shift and collision shift was investigated experimentally by Eric Cornell's group at Boulder in a standard setup for the creation of Bose–Einstein condensates, a macroscopic magnetic trap [15]. An ensemble of typically 10^6 ^{87}Rb atoms was prepared by a combination of laser and evaporative cooling at a typical temperature of 500 nK for thermal clouds, or as a Bose–Einstein condensate with large condensate fraction. Starting from the state $|1\rangle = |F = 1, m_F = -1\rangle$, the atoms were then prepared in a superposition of states $|1\rangle$ and $|2\rangle \equiv |F = 2, m_F = 1\rangle$ using a two-photon microwave transition. The transition frequency and clock coherence were then investigated using the Ramsey technique of separated oscillatory fields as a function of both atomic density and offset magnetic field B_0 at the bottom of the magnetic trap.

For the transition frequency ν_{12} as a function of magnetic field, the expected quadratic variation of Eq. (8.2) was observed with a quadratic coefficient $\beta \approx$

431 Hz/G^2. In such a setup, control of the magnetic field to a level of 10^{-3} (3 mG) is straightforward, which in itself would permit a clock accuracy of 4 mHz, corresponding to a fractional stability of 10^{-12}. With magnetic shielding in combination with interspersed measurement of the magnetic field on a transition with linear Zeeman shift, and corresponding field compensation, one would expect to be able to achieve a field stability on the order of 10 µG, which would limit the clock accuracy only at the level of 10^{-17}. However, as is discussed in Section 8.4, collision shifts limit the clock accuracy long before this level is reached.

Section 8.4 shows that the quadratic coefficient β also quantifies the frequency shift and decoherence that arise from the sample's small but non-zero temperature. For the trap parameters of the experiment by Harber et al. [15] the quadratic Zeeman shift of the clock transition amounted to $\sim$1 Hz at a sample temperature of 500 nK. For low-density clouds coherence times in excess of 2 s were observed, which demonstrated that in terms of coherence times trapped atoms can compete with atomic fountains. More recently, coherence times approaching one minute have been observed with atoms trapped on a microchip by a collaboration led by Jakob Reichel (ÉNS) and Peter Rosenbusch (SYRTE) [52].

The relatively high density in a trapped-atom clock then makes collision shifts easily observable. The density shift becomes particularly large when a Bose–Einstein condensate is formed, at much larger density than the associated thermal cloud. For the experiment by Harber et al. [15] with trap vibration frequencies $(\omega_x, \omega_y, \omega_z)$ of $2\pi \times (230, 230, 7)$ Hz, the transition frequency shift in a non-degenerate gas at a temperature of $T = 500$ nK amounted to $-3.9(3) \times 10^{-13}$ Hz cm^3, or typically $\langle \delta \nu \rangle_C = -5$ Hz. For the same trap parameters the density shift in the pure condensate with $N = 10^6$ atoms was as large as $\langle \delta \nu \rangle_C = -25$ Hz.

An intriguing feature of the collision shift in a degenerate quantum gas is the factor of two appearing in Eq. (8.6), associated with exchange symmetry for bosons [15], and a similar effect for fermions that could be dubbed the "factor of (not) zero" [33]. (See also the work by Kurt Gibble [49] for a unified approach for bosons [15, 58] and fermions [59, 60].) For identical bosons, the s-wave collision cross-section is given in terms of the s-wave scattering length a by $\sigma_{id} = 8\pi a^2$, whereas the corresponding expression for distinguishable particles (e.g., bosons in different quantum states) with the same scattering length is two times smaller, $\sigma_{dist} = 4\pi a^2$. The difference arises from the two different possible paths in the collision that give rise to the same final state, such that the amplitudes for the processes add, giving the larger value for identical bosons. For a coherently prepared sample, all particles are in the same internal state, and the larger cross-section for identical bosons applies, as verified experimentally by Harber et al. [15].

Another curious observation made in this experiment was that collisions actually serve to lengthen the coherence time of the sample. Indeed, the observed decay time of the Ramsey oscillations could be up to eight times longer than what would have been predicted based on the quadratic Zeeman broadening of the transition in a simple collision-free model. The authors concluded that the collisions, by randomly exchanging the velocities of pairs of particles in the cloud, cause them all to sample the same magnetic field environment. Thus, all the atoms in the cloud experience substantially the same average Zeeman shift, even if the shifts are different for each atom at any given instant. The long coherence times observed by Deutsch et al. in their chip trap are also believed to be due in part to collisions, but the hypothesized mechanism is different: under certain conditions colliding identical atoms can coherently exchange their phase, so that an atom that has precessed farther than its collision partner finds its phase set back, rather as though a spin-echo procedure were continually being applied to the sample [36, 52].

The experiment by Harber et al. [15] demonstrated a Ramsey clock with cold atoms and verified the parameters determining clock performance, but used a macroscopic magnetic trap rather than a microchip. Two years later, Treutlein et al. applied this approach to microchips, demonstrating a clock with a coherence time of 1 s, and verified that clock operation was possible within 5 μm of the chip surface [16].

Treutlein et al. used a silver-coated microchip where the magnetic trap was loaded from a mirror magneto-optical trap (MOT), as in the first demonstration of Bose–Einstein condensation on a microchip [61, 62]. The loading was followed by RF-induced evaporation. In this way an ultra-cold atom cloud containing typically 10^4 ^{87}Rb atoms was prepared at a typical temperature of 0.6 μK in a magic-field magnetic trap with frequencies of $\omega = 2\pi \times (50, 350, 410)$ Hz, comparable to the example studied in Section 8.4. The trap's position relative to the chip surface could be calculated from the known wire geometry and currents, and verified experimentally by comparing the absorption image of the atoms in the trap to the reflected image seen in the coated chip surface [63] or, for very short distances, by studying the rate of surface-induced trap loss [57]. The two-photon microwave transition from $|F = 1, m_F = -1\rangle$ to $|F = 2, m_F = 1\rangle$ was driven by a microwave photon from an external antenna, detuned by 1.2 MHz from the intermediate state, and an RF photon supplied directly by an on-chip wire, yielding a Rabi frequency for the clock transition of ~ 500 Hz. The population of the clock states could then be read out by absorption imaging after a 4 ms time-of-flight, unfortunately with a signal-to-noise ratio of only 6, substantially worse than the projection noise limit of ~ 100 for the system (see Section 8.6). To study the effect of the chip surface on atomic coherence, the authors performed Ramsey spectroscopy (i.e., operated a clock) using atomic samples trapped at varying distances from the surface. At distances below 5 μm, the trap loss induced by the chip surface limited the lifetime of the atomic population to a second or less, which is undesirable for clock operation. Beyond this range, however, no effect of the chip on clock performance was observed: the coherence time – as determined from the decay of the Ramsey contrast – and the phase noise – as measured by the signal-to-noise ratio – were invariant within experimental error for atom-chip distances from 5 to 132 μm.

Working at a distance of 54 μm from the chip surface, with a 23-s experimental cycle time and a 1-s Ramsey interrogation time, Treutlein et al. successfully operated a clock with a stability of $\sigma(\tau) = 1.7 \times 10^{-11} \sqrt{s}\tau^{-1/2}$, limited by fluctuations of

24 mHz in the quadratic Zeeman shift due to magnetic field noise in the laboratory. The long-term fractional stability was limited to $\sim 10^{-12}$ (after 10 min of integration time) by slow drifts of their microwave oscillator reference. Since the readout noise, duty cycle, and magnetic shielding can all be improved well beyond the levels used in this experiment, substantial improvements in chip clock performance beyond that demonstrated in this work are realizable. Already, Ramírez–Martínez et al. have improved the short-term stability to $1.5 \times 10^{-12} \sqrt{s}\tau^{-1/2}$ [41].

8.6
Readout in Trapped-Atom Clocks

The signal-to-noise ratio of clocks operating with ensembles of independent two-level atoms is limited by the projection noise associated with the independent measurement outcomes for the individual two-level atoms. Atomic-fountain clocks operate at this limit [1], while magic-wavelength optical clocks are approaching it [32]. To achieve projection-noise-limited readout, the number of photons detected per atom must exceed one. For absorption imaging, as used by Harber et al. and Treutlein et al., it is difficult to achieve projection-noise-limited detection in view of beam intensity fluctuations and interference fringes on the camera, and both experiments were substantially above the atom projection noise limit. For instance, in the work of Treutlein et al., the observed signal-to-noise ratio was around 6, while projection-noise-limited measurement with their $N > 10^4$ atoms would have allowed a ratio of $\sqrt{N} > 100$ [16]. However, recent experiments have demonstrated absorption imaging at the projection noise limit [64, 65].

Better state readout can be obtained if the cloud is placed inside an optical resonator that serves to enhance the signal by inducing repeated atom–light interaction [37, 66, 67], or using on-chip large-aperture fiber optics [68]. It is possible to use either fluorescence measurements or dispersive measurements, with dispersive measurements providing better signal-to-noise ratio at large atom number [66].

Besides projection-noise-limited resolution, another desirable feature of the clock's readout is that it be nondestructive. Any trapped-atom clock will suffer from dead time while fresh atoms are cooled and loaded into the trap, an operation that normally takes several seconds. During this dead time the noise of the local microwave oscillator is left uncorrected (Dick effect). If a given sample of cold trapped atoms can be reused for multiple successive frequency measurements, then the clock can spend less of its time loading fresh atoms and more of it measuring frequency, suppressing local oscillator noise and improving stability. Furthermore, by reducing the number of new atoms loaded for each measurement, such reuse can also reduce the effects of trap loading noise, which is typically well above the atom-number shot noise.

For this reason, Lodewyck et al. have built a non-destructive dispersive readout system into their optical clock apparatus [34]. They trap around 10^4 ^{87}Sr atoms in the ground state of an optical lattice, obtaining a cloud with a radius of 10 μm. The trap is placed in one arm of a Mach–Zehnder interferometer (Figure 8.4) with a

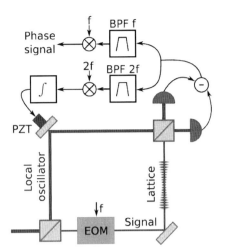

Figure 8.4 Setup for non-destructive detection of atomic-state populations in an ^{87}Sr optical lattice clock (from [34]). The phase shift of the RF component at the modulation frequency f is proportional to the number of atoms in the optical lattice. The harmonic at frequency $2f$ is used to lock the phase of the interferometer. This setup allows one to measure atomic-state populations with high signal-to-noise ratio without losing the atoms from the lattice. This enables the repeated use of the same atoms for several clock cycles, thereby improving the clock duty cycle and stability.

probe beam waist of 37 μm. The weak (∼ nW) probe beam undergoes a dispersive phase shift due to the presence of the atoms, which is detected by beating it against the strong (∼ mW) optical local oscillator at the final beam-splitter of the interferometer. The two output ports of this beam-splitter are detected on fast photodiodes with an overall quantum efficiency of 43 % and the signals subtracted, to obtain a signal that is insensitive to laser power drift to first order. Rather than using the DC signal, which measures the phase between probe and local oscillator and would require careful stabilization of the relative path length of the two interferometer arms, the detector makes use of a sideband modulation scheme reminiscent of the Pound–Drever–Hall frequency locking technique for optical resonators. An electro-optic crystal in the probe arm of the interferometer modulates the probe beam phase at an RF frequency $f = 90$ MHz, thus generating a pair of sidebands to either side of the atomic resonance, detuned by approximately three atomic linewidths. Since the sidebands have opposite detuning, they are subjected to opposite dispersive phase shifts from the atoms, and this *differential* phase shift is detectable as a phase shift of the RF component at frequency f of the interferometer output. Fluctuations in the path length difference of the interferometer impart a *common* phase shift to the two sidebands and do not contribute to this signal to first order. Path length fluctuations only affect the amplitude of the atomic signal, and a servo loop based on the $2f$ component of the output is used to detect such fluctuations and lock the interferometer at the position of maximum signal. The net result is a photon-shot-noise-limited determination of the dispersive atomic phase shift, yielding a measurement of the atom number

Figure 8.5 Optical resonator mounted on microchip [66, 69, 70] for quantum non-demolition measurements of the clock states and spin squeezing. The resonator mode is aligned at a height of 200 μm above the chip.

in one of the two clock states at the projection noise level for 10^4 atoms. Up to 95 % of the atoms in the sample remain in the trap after such a readout and can be reused for the next cycle after a few ten ms of recooling. The authors foresee realistic clock operation with only 100 ms of dead time between Ramsey interrogations, improving the duty factor of the clock from the typical value of 10 % to over 80 %.

Another approach that has been successful in chip-based experiments employs dispersive state measurement [71–77] via an optical resonator, which has been used to realize both projection-noise-limited readout and spin squeezing (see Section 8.7) for an atomic clock operated on a microchip [69, 70, 78]. In the demonstration of reference [69], an ensemble of up to 10^5 ^{87}Rb atoms was prepared in an elongated trap overlapped with an optical resonator mode 200 μm from the surface of a microchip, and transferred into a far-detuned optical trap formed inside the resonator (Figure 8.5). The deep standing-wave dipole trap ensured that the sample with radial extension of 8.1(8) μm was located well within the resonator mode with a waist of 56.9(4) μm. The cavity length was adjusted such that one optical resonance lay between the optical resonance frequencies for the two hyperfine clock states $|F = 1, m_F = 0\rangle$ and $|F = 2, m_F = 0\rangle$. The state-dependent index of refraction of the atomic sample then changed the cavity resonance frequency by an amount proportional to the population difference between the clock states. Measuring this shift provided a resolution of ~ 30 atoms, substantially better than the projection-induced fluctuations of over one hundred atoms. Much less than one photon per atom was scattered into free space during the measurement so that the atomic sample was little heated and could be reused.

8.7
Spin Squeezing

Even in the absence of technical readout noise, the standard quantum limit of Eq. (8.1) places a bound on the achievable signal-to-noise ratio of a clock operated with an ensemble of N independent atoms. This fundamental instability – due to quantum projection noise on the final state readout – can in principle be made arbitrarily small by increasing the total atom number N. In practice, however, the allowable atom number is severely limited by the onset of collision shifts which, as we have seen, are the bane of compact trapped-atom clocks. Further improvements to chip clock stability must, therefore, be made at fixed N by overcoming the projection noise limit. This can be achieved by treating the N atoms not as independent particles but as a single (entangled) ensemble, with quantum correlations between atoms. Such correlations can be used to generate a Ramsey fringe that oscillates faster with frequency than a single-atom Ramsey fringe, increasing the size of the signal for constant projection noise. This approach has been demonstrated in trapped-ion clocks [79], but it is not clear how to engineer the required entangled states in large ensembles such as those envisaged for chip-based neutral-atom clocks. Alternatively, quantum correlations can be used to reduce the projection noise of the ensemble below the limit for independent particles, an approach known as spin squeezing. Spin squeezing of superpositions of clock states is an active research area, with several recently demonstrated implementations [64, 65, 69, 70, 80, 81], and more expected in the near future.

So far, three spin squeezing methods have been demonstrated in experiments on neutral atomic ensembles. The first employs a non-destructive measurement of the ensemble to prepare it in a squeezed initial state [69, 71, 81–83]. The essential idea is to measure the quantum projection noise separately, before the clock is operated, so that it may be removed from the final Ramsey signal. As long as less than one photon per atom is scattered into free space, measurement of the ensemble state with a single spatial mode of probe light does not reveal the states of the individual atoms, but merely how many are in each of the clock levels. Therefore, a measurement that resolves the atom number in each clock state to better than the projection noise limit can entangle the atoms, where the entangled state of the ensemble is conditioned on the outcome of the measurement. The outcome of a later readout can then be predicted more precisely than would be possible for independent particles. The cavity-enhanced state readout described in Section 8.6 has been used to implement such measurement-based squeezing [69]. By comparing this reduction of readout noise to the reduction of coherence (and hence clock signal) inevitably induced by scattering and dephasing during the first measurement, an enhancement of the signal-to-noise ratio by 3.0(8) dB was demonstrated. The group of Eugene Polzik obtained similar results without the aid of a cavity by using a precision-stabilized Mach–Zehnder interferometer to detect the atomic index of refraction [81]. They have used this technique as the basis for a prototype squeezed clock [83].

A second squeezing method relies on repulsive interactions between atoms to deterministically entangle them and produce states with reduced uncertainty on the population difference between the two clock states [64, 65]. Unlike measurement-based squeezing, this method unconditionally prepares the same squeezed state on every experimental cycle, which has the practical benefit that the squeezing is independent of the performance of the detection apparatus: one can know the initial squeezed state without having to observe it. However, in the context of precision timekeeping, it has the salient drawback of relying upon the very same collisional energy shifts which are so detrimental to the clock's performance. Thus, while this technique has a bright future in the study of many-body entanglement and may be useful for atom interferometry, it is unlikely to be of much use in improving the performance of chip clocks.

A third method has been developed that combines the benefits of the previous two techniques: deterministic squeezing by cavity feedback [84, 85]. In this method, the atoms are placed inside an optical resonator, as in the resonator-enhanced readout of Section 8.6. However, instead of being projected into a squeezed state by measurement, the atomic ensemble evolves deterministically into a squeezed state due to the collective interaction of the atoms with the light field of the resonator, as follows. The atomic quantum noise tunes the resonator frequency just as a clock signal would, which for an incident light field tuned to the slope of the resonator line results in an intra-cavity intensity that depends on the atomic population difference. The light-shift-induced phase evolution in each individual atom thus depends on the population difference of all atoms in the ensemble, leading to quantum correlations between atoms. This approach generates spin dynamics similar to those of the one-axis twisting Hamiltonian in the original spin squeezing proposal of Kitagawa and Ueda [86], or to the repulsive interaction employed in the collisional squeezing experiments [64, 65]. However, since the incident light intensity can be switched to zero at will, the light-mediated interaction can be turned on only for the preparation of the initial squeezed state and then switched off to avoid perturbing the clock during the Ramsey precession time. The cavity feedback method has produced the largest spin squeezing at the time of writing, a 5.6(6) dB improvement in signal-to-noise ratio [70]. It has also been used to operate a proof-of-principle clock which, for integration times up to 50 s, achieved a stability 2.8(3) times better than the standard quantum limit of Eq. (8.1) [78]. The fractional stability of this clock was poor in concrete terms ($\sigma(\tau) = 1.1 \times 10^{-9} \mathrm{s}^{1/2}/\sqrt{\tau}$), primarily because a very short Ramsey precession time of 200 μs was used for the demonstration, but together with the experiment of reference [83] it shows that even the standard quantum limit need not be an insuperable obstacle to the improvement of stability in chip clocks.

References

1 Santarelli, G., Laurent, P., Lemonde, P., Clairon, A., Mann, A.G., Chang, S., Luiten, A.N., and Salomon, C. (1999) Quantum projection noise in an atomic

fountain: A high stability cesium frequency standard. *Phys. Rev. Lett.* **82**, 4619.
2. Wineland, D.J., Bollinger, J.J., Itano, W.M., Moore, F.L., and Heinzen, D.J. (1992) Spin squeezing and reduced quantum noise in spectroscopy. *Phys. Rev. A* **46**, R6797.
3. Wineland, D.J., Bollinger, J.J., Itano, W.M., and Heinzen, D.J. (1994) Squeezed atomic states and projection noise in spectroscopy. *Phys. Rev. A* **50**, R67.
4. Kasevich, M.A., Riis, E., Chu, S., and DeVoe, R.G. (1989) RF spectroscopy in an atomic fountain. *Phys. Rev. Lett.* **63**, 612.
5. Wynands, R. and Weyers, S. (2005) Atomic fountain clocks. *Metrologia* **42**, S64.
6. Dimopoulos, S., Graham, P.W., Hogan, J.M., and Kasevich, M.A. (2007) Testing general relativity with atom interferometry. *Phys. Rev. Lett.* **98**, 111102.
7. Laurent, P., Abgrall, M., Jentsch, C., Lemonde, P., Santarelli, G., Clairon, A., Maksimovic, I., Bize, S., Salomon, C., Blonde, D., Vega, J., Grosjean, O., Picard, F., Saccoccio, M., Chaubet, M., Ladiette, N., Guillet, L., Zenone, I., Delaroche, C., and Sirmain, C. (2006) Design of the cold atom PHARAO space clock and initial test results. *Appl. Phys. B* **84**, 683.
8. Esnault, F.X., Perrin, S., Tré, S.mine, Gué, S.randel, Holleville, D., Dimarcq, N., Hermann, V., and Delporte, J. (2007) Stability of the compact cold atom clock HORACE, in *Proceedings of the 2007 IEEE International Frequency Control Symposium, Jointly with the 21st European Frequency and Time Forum, Geneva, Switzerland*, pp. 1342–1345, IEEE.
9. Bollinger, J., Heinzen, D., Itano, W., Gilbert, S., and Wineland, D. (1991) A 303-MHz frequency standard based on trapped Be^+ ions. *IEEE Trans. Instrum. Meas.* **40**, 126.
10. Rosenband, T., Hume, D.B., Schmidt, P.O., Chou, C.W., Brusch, A., Lorini, L., Oskay, W.H., Drullinger, R.E., Fortier, T.M., Stalnaker, J.E., Diddams, S.A., Swann, W.C., Newbury, N.R., Itano, W.M., Wineland, D.J., and Bergquist, J.C. (2008) Frequency ratio of Al^+ and Hg^+ single-ion optical clocks; Metrology at the 17th decimal place. *Science* **319**, 1808.
11. Chou, C.W., Hume, D.B., Koelemeij, J.C.J., Wineland, D.J., and Rosenband, T. (2010) Frequency comparison of two high-accuracy Al+ optical clocks. *Phys. Rev. Lett.* **104**, 070802.
12. Migdall, A.L., Prodan, J.V., Phillips, W.D., Bergeman, T.H., and Metcalf, H.J. (1985) First observation of magnetically trapped neutral atoms. *Phys. Rev. Lett.* **54**, 2596.
13. Bagnato, V.S., Lafyatis, G.P., Martin, A.G., Raab, E.L., Ahmad-Bitar, R.N., and Pritchard, D.E. (1987) Continuous stopping and trapping of neutral atoms. *Phys. Rev. Lett.* **58**, 2194.
14. Hess, H.F., Kochanski, G.P., Doyle, J.M., Masuhara, N., Kleppner, D., and Greytak, T.J. (1987) Magnetic trapping of spin-polarized atomic hydrogen. *Phys. Rev. Lett.* **59**, 672.
15. Harber, D.M., Lewandowski, H.J., McGuirk, J.M., and Cornell, E.A. (2002) Effect of cold collisions on spin coherence and resonance shifts in a magnetically trapped ultracold gas. *Phys. Rev. A* **66**, 053616.
16. Treutlein, P., Hommelhoff, P., Steinmetz, T., Hänsch, T.W., and Reichel, J. (2004) Coherence in microchip traps. *Phys. Rev. Lett.* **92**, 203005.
17. Chu, S., Bjorkholm, J.E., Ashkin, A., and Cable, A. (1986) Experimental observation of optically trapped atoms. *Phys. Rev. Lett.* **57**, 314.
18. Grimm, R., Weidemuller, M., and Ovchinnikov, Y. (2000) Optical dipole traps for neutral atoms. *Adv. At. Mol. Opt. Phys.* **42**, 95.
19. Taïeb, R., Dum, R., Cirac, J.I., Marte, P., and Zoller, P. (1994) Cooling and localization of atoms in laser-induced potential wells. *Phys. Rev. A* **49**, 4876.
20. Kimble, H.J. et al., In R. Blatt, J. Eschner, D. Leibrfried, and F. Schmidt-Kaler (eds) (1999) *Proceedings of the XIV International Conference on Laser Spectroscopy*, vol. XIV, 80–89, World Scientific.

21 McKeever, J., Buck, J.R., Boozer, A.D., Kuzmich, A., Nägerl, H.-C., Stamper-Kurn, D., and Kimble, H.J. (2003) State-insensitive cooling and trapping of single atoms in an optical cavity. *Phys. Rev. Lett.* **90**, 133602.

22 Katori, H., Ido, T., and Kuwata-Gonokami, M. (1999) Optimal design of dipole potentials for efficient loading of Sr atoms. *J. Phys. Soc. Japan* **68**, 2479.

23 Ido, T., Isoya, Y., and Katori, H. (2000) Optical-dipole trapping of Sr atoms at a high phase-space density. *Phys. Rev. A* **61**, 061403.

24 Katori, H., Takamoto, M., Pal'chikov, V.G., and Ovsiannikov, V.D. (2003) Ultrastable optical clock with neutral atoms in an engineered light shift trap. *Phys. Rev. Lett.* **91**, 173005.

25 Takamoto, M. and Katori, H. (2003) Spectroscopy of the $^1S_0 - {}^3P_0$ clock transition of ^{87}Sr in an optical lattice. *Phys. Rev. Lett.* **91**, 223001.

26 Takamoto, M., Hong, F.-L., Higashi, R., and Katori, H. (2005) An optical lattice clock. *Nature* **435**, 321.

27 Santra, R., Arimondo, E., Ido, T., Greene, C.H., and Ye, J. (2005) High-accuracy optical clock via three-level coherence in neutral bosonic ^{88}Sr. *Phys. Rev. Lett.* **94**, 173002.

28 Ido, T., Loftus, T.H., Boyd, M.M., Ludlow, A.D., Holman, K.W., and Ye, J. (2005) Precision spectroscopy and density-dependent frequency shifts in ultracold Sr. *Phys. Rev. Lett.* **94**, 153001.

29 Boyd, M.M., Zelevinsky, T., Ludlow, A.D., Foreman, S.M., Blatt, S., Ido, T., and Ye, J. (2006) Optical atomic coherence at the 1-second time scale. *Science* **314**, 1430.

30 Boyd, M.M., Ludlow, A.D., Blatt, S., Foreman, S.M., Ido, T., Zelevinsky, T., and Ye, J. (2007) ^{87}Sr lattice clock with inaccuracy below 10^{-15}. *Phys. Rev. Lett.* **98**, 083002.

31 Blatt, S., Ludlow, A.D., Campbell, G.K., Thomsen, J.W., Zelevinsky, T., Boyd, M.M., Ye, J., Baillard, X., Fouché, M., Targat, R.L., Brusch, A., Lemonde, P., Takamoto, M., Hong, F.-L., Katori, H., and Flambaum, V.V. (2008) New limits on coupling of fundamental constants to gravity using ^{87}Sr optical lattice clocks. *Phys. Rev. Lett.* **100**, 140801.

32 Ye, J., Kimble, H.J., and Katori, H. (2008) Quantum state engineering and precision metrology using state-insensitive light traps. *Science* **320**, 1734.

33 Campbell, G.K., Boyd, M.M., Thomsen, J.W., Martin, M.J., Blatt, S., Swallows, M.D., Nicholson, T.L., Fortier, T., Oates, C.W., Diddams, S.A., Lemke, N.D., Naidon, P., Julienne, P., Ye, J., and Ludlow, A.D. (2009) Probing interactions between ultracold fermions. *Science* **324**, 360.

34 Lodewyck, J., Westergaard, P.G., and Lemonde, P. (2009) Nondestructive measurement of the transition probability in a Sr optical lattice clock. *Phys. Rev. A* **79**, 061401.

35 Ludlow, A.D., Zelevinsky, T., Campbell, G.K., Blatt, S., Boyd, M.M., H. G. de Miranda, M., Martin, M.J., Thomsen, J.W., Foreman, S.M., Ye, J., Fortier, T.M., Stalnaker, J.E., Diddams, S.A., Coq, Y.L., Barber, Z.W., Poli, N., Lemke, N.D., Beck, K.M., and Oates, C.W. (2008) Sr lattice clock at 1×10^{-16} fractional uncertainty by remote optical evaluation with a Ca clock. *Science* **319**, 1805.

36 Rosenbusch, P. (2009) Magnetically trapped atoms for compact atomic clocks. *Appl. Phys. B* **95**, 227.

37 Colombe, Y., Steinmetz, T., Dubois, G., Linke, F., Hunger, D., and Reichel, J. (2007) Strong atom-field coupling for Bose–Einstein condensates in an optical cavity on a chip. *Nature* **450**, 272.

38 Flambaum, V.V., Dzuba, V.A., and Derevianko, A. (2008) Magic frequencies for cesium primary-frequency standard. *Phys. Rev. Lett.* **101**, 220801.

39 Vanier, J. and Audoin, C. (1989) The Quantum Physics of Atomic Frequency Standards. Adam Hilger, Philadelphia.

40 Lewandowski, H.J., Harber, D.M., Whitaker, D.L., and Cornell, E.A. (2002) Observation of anomalous spin-state segregation in a trapped ultracold vapor. *Phys. Rev. Lett.* **88**, 070403.

41 Ramírez-Martínez, F., Lacroûte, C., Rosenbusch, P., Reinhard, F., Deutsch, C., Schneider, T., and Reichel, J. (2010) Compact frequency standard using

42 Leo, P.J., Julienne, P.S., Mies, F.H., and Williams, C.J. (2001) Collisional frequency shifts in ^{133}Cs fountain clocks. *Phys. Rev. Lett.* **86**, 3743.

43 Vuletić, V., Kerman, A.J., Chin, C., and Chu, S., Observation of low-field Feshbach resonances in collisions of cesium atoms. *Phys. Rev. Lett.* **82**, 1406 (1999).

44 Chin, C., Vuletić, V., Kerman, A.J., and Chu, S. (2000) High resolution Feshbach spectroscopy of cesium. *Phys. Rev. Lett.* **85**, 2717.

45 Leo, P.J., Williams, C.J., and Julienne, P.S. (2000) Collision properties of ultracold ^{133}Cs atoms. *Phys. Rev. Lett.* **85**, 2721.

46 Kerman, A., Chin, C., Vuletić, V., Chu, S., Leo, P., Williams, C., and Julienne, P. (2001) Determination of Cs–Cs interaction parameters using Feshbach spectroscopy. *CR Acad. Sci. Paris IV* **2**, 633.

47 Chin, C., Kerman, A.J., Vuletić, V., and Chu, S. (2003) Sensitive detection of cold cesium molecules formed on Feshbach resonances. *Phys. Rev. Lett.* **90**, 033201.

48 Chin, C., Vuletić, V., Kerman, A.J., Chu, S., Tiesinga, E., Leo, P.J., and Williams, C.J. (2004) Precision Feshbach spectroscopy of ultracold Cs$_2$. *Phys. Rev. A* **70**, 032701.

49 Gibble, K. (2009) Decoherence and collisional frequency shifts of trapped bosons and fermions. *Phys. Rev. Lett.* **103**, 113202.

50 Vogels, J.M., Freeland, R.S., Tsai, C.C., Verhaar, B.J., and Heinzen, D.J. (2000) Coupled singlet-triplet analysis of two-color cold-atom photoassociation spectra. *Phys. Rev. A* **61**, 043407.

51 Pethick, C.J. and Smith, H. (2002) *Bose Einstein Condensation in Dilute Gases*, Cambridge University Press, Cambridge, UK.

52 Deutsch, C., Ramirez-Martinez, R., Lacroûte, C., Reinhard, F., Schneider, T., Fuchs, J.N., Piéchon, F., Laloë, F., Reichel, J., and Rosenbusch, P. (2010) Spin self-rephasing and very long coherence times in a trapped atomic ensemble, *Phys. Rev. Lett.*, **105**, 020401.

53 Pasquini, T.A., Saba, M., Jo, G.-B., Shin, Y., Ketterle, W., Pritchard, D.E., Savas, T.A., and Mulders, N. (2006) Low velocity quantum reflection of Bose–Einstein condensates. *Phys. Rev. Lett.* **97**, 093201.

54 Gerritsma, R., Whitlock, S., Fernholz, T., Schlatter, H., Luijges, J.A., Thiele, J.-U., Goedkoop, J.B., and Spreeuw, R.J.C. (2007) Lattice of microtraps for ultracold atoms based on patterned magnetic films. *Phys. Rev. A* **76**, 033408.

55 Henkel, C., Pötting, S., and Wilkens, M. (1999) Loss and heating of particles in small and noisy traps. *Appl. Phys. B* **69**, 379.

56 Henkel, C. and Pötting, S. (2001) Coherent transport of matter waves. *Appl. Phys. B* **72**, 73.

57 Lin, Y., Teper, I., Chin, C., and Vuletić, V. (2003) Impact of the Casimir–Polder potential and Johnson noise on Bose–Einstein condensate stability near surfaces. *Phys. Rev. Lett.* **92**, 050404.

58 Killian, T.C., Fried, D.G., Willmann, L., Landhuis, D., Moss, S.C., Greytak, T.J., and Kleppner, D. (1998) Cold collision frequency shift of the 1S–2S transition in hydrogen. *Phys. Rev. Lett.* **81**, 3807.

59 Oktel, M.O. and Levitov, L.S. (2002) Collective dynamics of internal states in a Bose–Einstein gas. *Phys. Rev. A* **65**, 063604.

60 Zwierlein, M.W., Hadzibabic, Z., Gupta, S., and Ketterle, W. (2003) Spectroscopic insensitivity to cold collisions in a two-state mixture of fermions. *Phys. Rev. Lett.* **91**, 250404.

61 Hänsel, W., Hommelhoff, P., Hänsch, T., and Reichel, J. (2001) Bose–Einstein condensation on a microelectronic chip. *Nature* **413**, 498.

62 Ott, H., Fortágh, J., Schlotterbeck, G., Grossmann, A., and Zimmermann, C. (2001) Bose–Einstein condensation in a surface microtrap. *Phys. Rev. Lett.* **87**, 230401.

63 Schneider, S., Kasper, A., vom Hagen, C., Bartenstein, M., Engeser, B., Schumm, T., Bar-Joseph, I., Folman, R., Feenstra, L., and Schmiedmayer, J. (2003) Bose–Einstein condensation in a simple microtrap. *Phys. Rev. A* **67**, 023612.

64 Gross, C., Zibold, T., Nicklas, E., Estève, J., and Oberthaler, M.K. (2010) Non linear atom interferometer surpasses classical precision limit. *Nature* **464**, 1165.

65 Riedel, M.F., Böhi, P., Li, Y., Hänsch, T.W., Sinatra, A., and Treutlein, P. (2010) Atom-chip-based generation of entanglement for quantum metrology. *Nature* **464**, 1170.

66 Teper, I., Lin, Y.-J., and Vuletić, V. (2006) Resonator-aided single-atom detection on a microfabricated chip. *Phys. Rev. Lett.* **97**, 023002.

67 Brennecke, F., Donner, T., Ritter, S., Bourdel, T., Köhl, M., and Esslinger, T. (2007) Cavity QED with a Bose–Einstein condensate. *Nature* **450**, 268.

68 Heine, D., Wilzbach, M., Raub, T., Hessmo, B., and Schmiedmayer, J.(2009) Integrated atom detector: Single atoms and photon statistics. *Phys. Rev. A* **79**, 021804(R).

69 Schleier-Smith, M.H., Leroux, I.D., and Vuletić, V. (2010) States of an ensemble of two-level atoms with reduced quantum uncertainty. *Phys. Rev. Lett.* **104**, 073604.

70 Leroux, I.D., Schleier-Smith, M.H., and Vuletić, V. (2010) Implementation of cavity squeezing of a collective atomic spin. *Phys. Rev. Lett.* **104**, 073602.

71 Kuzmich, A., Bigelow, N.P., and Mandel, L. (1998) Atomic quantum nondemolition measurements and squeezing. *Europhys. Lett.* **42**, 481.

72 Kuzmich, A., Mandel, L., and Bigelow, N.P. (2000) Generation of spin squeezing via continuous quantum nondemolition measurement. *Phys. Rev. Lett.* **85**, 1594.

73 Takeuchi, M., Ichihara, S., Takano, T., Kumakura, M., and Takahashi, Y. (2007) Spin noise measurement with diamagnetic atoms. *Phys. Rev. A* **75**, 063827.

74 Takano, T., Fuyama, M., Namiki, R., and Takahashi, Y. (2009) Spin squeezing of a cold atomic ensemble with the nuclear spin of one-half. *Phys. Rev. Lett.* **102**, 033601.

75 Takano, T., Tanaka, S.-I.-R., Namiki, R., and Takahashi, Y. (2010) Manipulation of nonclassical atomic spin states. *Phys. Rev. Lett.* **104**, 013602.

76 Teper, I., Vrijsen, G., Lee, J., and Kasevich, M.A. (2008) Backaction noise produced via cavity-aided nondemolition measurement of an atomic clock state. *Phys. Rev. A* **78**, 051803.

77 Koschorreck, M., Napolitano, M., Dubost, B., and Mitchell, M.W. (2010) Sub-projection-noise sensitivity in broadband atomic magnetometry. *Phys. Rev. Lett.* **104**, 093602.

78 Leroux, I.D., Schleier-Smith, M.H., and Vuletić, V. (2010) Orientation-dependent entanglement lifetime in a squeezed atomic clock, *Phys. Rev. Lett.*, **105**, 250801.

79 Meyer, V., Rowe, M.A., Kielpinski, D., Sackett, C.A., Itano, W.M., Monroe, C., and Wineland, D.J. (2001) Experimental demonstration of entanglement-enhanced rotation angle estimation using trapped ions. *Phys. Rev. Lett.* **86**, 5870.

80 Estève, J., Gross, C., Weller, A., Giovanazzi, S., and Oberthaler, M.K. (2008) Squeezing and entanglement in a Bose–Einstein condensate. *Nature* **455**, 1216.

81 Appel, J., Windpassinger, P.J., Oblak, D., Hoff, U.B., Kjærgaard, N., and Polzik, E.S. (2009) Mesoscopic atomic entanglement for precision measurements beyond the standard quantum limit. *Proc. Natl. Acad. Sci. USA* **106**, 10960.

82 Bouchoule, I. and Mølmer, K. (2002) Preparation of spin-squeezed atomic states by optical-phase-shift measurement. *Phys. Rev. A* **66**, 043811.

83 Louchet-Chauvet, A., Appel, J., Renema, J.J., Oblak, D., and Polzik, E.S. (2009) Entanglement-assisted atomic clock beyond the projection noise limit, *New J. Phys.*, **12**, 065032.

84 Takeuchi, M., Ichihara, S., Takano, T., Kumakura, M., Yabuzaki, T., and Takahashi, Y. (2005) Spin squeezing via one-axis twisting with coherent light. *Phys. Rev. Lett.* **94**, 023003.

85 Schleier-Smith, M.H., Leroux, I.D., and Vuletić, V. (2010) Squeezing the collective spin of a dilute atomic ensemble by cavity feedback. *Phys. Rev. A* **81**, 021804.

86 Kitagawa, M. and Ueda, M. (1993) Squeezed spin states. *Phys. Rev. A* **47**, 5138.

9
Quantum Information Processing with Atom Chips

Philipp Treutlein, Antonio Negretti, and Tommaso Calarco

9.1
Introduction

Since the 1990s, when groundbreaking algorithms based on the laws of quantum mechanics for solving classically intractable computational problems were found, quantum information science has rapidly grown with the promise of building a quantum computer. Similar to present 'classical' computers, quantum hardware consists of memory and a processor. The former stores the information, the latter, with a set of gates, processes the information.

The concept of gate is fundamental in quantum computation [1]. Thus, let us first consider its classical analogue. A gate in a classical computer, which implements a Boolean function, is a device that accomplishes a well-defined operation on one or more bits. For instance, CMOS transistors realize the logical NOT operation. Instead, a quantum gate performs a unitary operation on the linear space of quantum bits (*qubits*). Thus, a quantum gate is the time evolution operator generated by a given Hamiltonian; control by external fields, according to the Hamiltonian structure, allows one to perform desired transformations on the qubit wave function. In the 1990s it was shown that a general N-qubit gate can be decomposed into $O(N^2)$ one- and two-qubit gates. Because of this crucial theoretical result, most of the schemes for quantum gates concern the implementation of one- and two-qubit operations, which will be the main topic of this chapter.

Atom chips combine many important features of a scalable architecture for quantum information processing (QIP): (1) the exquisite coherence properties of neutral atoms; (2) accurate control of the coherent evolution of the atoms in tailored micropotentials; (3) scalability of the technology through microfabrication, which allows the integration of many qubits in parallel on the same device while maintaining individual addressability; (4) interfaces to photons as flying qubits provided by on-chip optical cavities; and (5) the exciting perspective of interfacing quantum optical qubits with solid-state systems for QIP located on the chip surface. In this respect, the unique features of atom chips such as trap miniaturization and integration are powerful concepts from which many different systems can benefit, as

Atom Chips. Edited by Jakob Reichel and Vladan Vuletić
Copyright © 2011 WILEY-VCH Verlag GmbH & Co. KGaA, Weinheim
ISBN: 978-3-527-40755-2

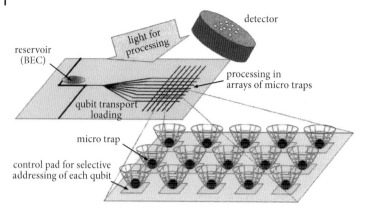

Figure 9.1 Schematic illustration of an atom chip quantum processor, adapted from [2].

the current efforts to miniaturize ion traps, initiated by the success of neutral atom chip traps, show.

Figure 9.1 illustrates the general idea of an atom-chip-based quantum information processor. It includes a reservoir of cold atoms, preferably in their motional ground state, and in a well-defined internal state. An ideal starting point for this is a BEC (Bose–Einstein condensate) in a chip trap. From there the atoms are transported using guides or moving potentials to a large array of processing sites. Either single atoms, or small ensembles of atoms, are then loaded into the qubit traps. Each qubit site can be addressed individually. Micro-fabricated wires and electrodes located close to the individual sites can be used for site-selective manipulations such as single-qubit gates. For two-qubit gates, interactions between adjacent sites are induced. For readout, micro-optics can be used to focus lasers onto each site separately, or the whole processor can be illuminated and single qubits are addressed by shifting them in and out of resonance using local electric or magnetic fields.

9.2
Ingredients for QIP with Atom Chips

A fully developed architecture for quantum information processing with atom chips comprises the following ingredients [2–4]:

1. **Qubit states with long coherence lifetime.** A qubit state pair has to be identified which can be manipulated with electromagnetic fields on the chip, but still allows for long coherence lifetimes in a realistic experimental situation. In particular, attention has to be paid to decoherence and loss mechanisms induced by the chip surface, which is typically at a distance of only a few micrometers from the atoms.

2. **Qubit rotations (single-qubit gates).** High-fidelity single-qubit rotations have to be implemented. This corresponds to high-contrast Rabi oscillations between the qubit states, with a period much shorter than the coherence lifetime.
3. **Single-qubit readout.** Individual qubits have to be read out with high efficiency. Since most scenarios for QIP with atom chips implement the qubit in single atoms, this requirement amounts to single-atom detection – ideally with a detector which is integrated on the chip.
4. **Single-qubit preparation.** A method has to be developed to deterministically prepare single qubits on the chip. In most proposals, this amounts to deterministic single-atom preparation. An exception are ensemble-based approaches to QIP. Furthermore, many proposals require the atoms to be prepared in the motional ground state of the trap, with very low occupation probability of excited states.
5. **Conditional quantum dynamics (two-qubit gates).** A universal two-qubit quantum gate is needed, which allows for high-fidelity gate operations under realistic conditions. A gate infidelity (error rate) below a certain threshold is required in order to allow for a fault-tolerant implementation of quantum computing. Depending on error models and recovery schemes, estimates of such a threshold vary from a few 10^{-3} for active error correcting codes [5] up to well above 10 % for error detection schemes [6].
6. **Interfaces to other systems.** Interfaces allow one to transfer quantum information from the atoms to other physical systems. One example is an interface between atoms for processing and storage and photons (optical or microwave) for transmission of information. Furthermore, quantum interfaces are the key ingredient of hybrid approaches to QIP, where the gate operations themselves rely on the coupling of atoms to ions, molecules, or solid-state quantum systems such as Cooper pair boxes and superconducting microwave cavities.

In the following sections, we discuss each point of the above list. We highlight the main challenges, review theoretical proposals which show how these ingredients can be implemented on atom chips, and report on the status of experimental realizations.

9.3
Qubit States with Long Coherence Lifetime

Two potentially conflicting requirements have to be met by the qubit states $\{|0\rangle, |1\rangle\}$ chosen for QIP on atom chips. On the one hand, both states have to couple to the electromagnetic fields which are used for trapping and manipulating the atoms. On the other hand, high-fidelity gate operations require a long coherence lifetime of superposition states $\alpha|0\rangle + \beta|1\rangle$, $(|\alpha|^2 + |\beta|^2 = 1)$, and thus the qubit has to be sufficiently robust against fluctuations of electromagnetic fields in realistic experimental situations. A peculiarity of chip-based traps is the presence of atom–surface interactions, which lead to additional loss and decoherence mechanisms which are not present in macroscopic traps (see Chapters 4 and 5). It is therefore important to investigate the coherence properties of the proposed qubit candidates close to the chip surface. In the following we discuss the different types of qubits which have been studied.

Hyperfine Qubits

An obvious qubit candidate are the long-lived ground-state hyperfine levels of the atoms. In most experiments, at least a part of the trapping potential is provided by static magnetic fields generated by wires or permanent-magnet structures on the atom chip. It is, therefore, desirable that both $|0\rangle$ and $|1\rangle$ are magnetically trappable. Long coherence lifetimes can be expected if states with equal magnetic moments are chosen, so that both states experience nearly identical trapping potentials in static magnetic traps, and the energy difference $h\nu_{10} = E_{|1\rangle} - E_{|0\rangle}$ between $|0\rangle$ and $|1\rangle$ is robust against magnetic field fluctuations. These requirements are satisfied by the $|F = 1, m_F = -1\rangle \equiv |0\rangle$ and $|F = 2, m_F = +1\rangle \equiv |1\rangle$ hyperfine levels of the $5S_{1/2}$ ground state of ^{87}Rb.

In [7], the coherence properties of this qubit were studied experimentally on an atom chip. Figure 9.2 shows measurements of the coherence lifetime close to the chip surface. The atoms were held in a magnetic trap, with a magnetic field of $B_0 \sim 3.23$ G in the trap center. At this field, both states experience the same first-order Zeeman shift and the remaining magnetic field dependence of the transition frequency ν_{10} is minimized. The coherence lifetime was measured at different atom–surface distances d using Ramsey spectroscopy. Coherence lifetimes exceeding 1 s were observed, comparable to those obtained in macroscopic traps. Within the experimental error, the Ramsey contrast does not show a dependence on atom–surface distance for $d = 5$–130 μm. These experiments confirm that this hyperfine qubit is very well suited for atom chip-based QIP. It is, therefore, considered in several schemes for atom chip quantum gates (see Section 9.7). Large arrays

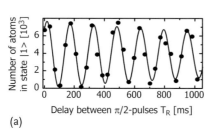
(a)

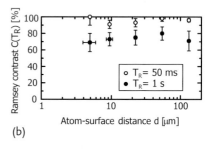
(b)

Figure 9.2 Coherence lifetime measurements of the hyperfine qubit $|F = 1, m_F = -1\rangle \equiv |0\rangle$ and $|F = 2, m_F = +1\rangle \equiv |1\rangle$ (adapted from [7]). (a) Ramsey spectroscopy of the $|0\rangle \leftrightarrow |1\rangle$ transition with atoms held at a distance $d = 9$ μm from the atom chip surface. An exponentially damped sine fit to the Ramsey fringes yields a $1/e$ coherence lifetime of $\tau_c = 2.8 \pm 1.6$ s. Each data point corresponds to a single shot of the experiment. (b) Contrast $C(T_R)$ of the Ramsey fringes as a function of d for two values of the time delay T_R between the $\pi/2$-pulses.

of qubits could be prepared in chip-based magnetic lattice potentials which are demonstrated in [8, 9].

Magnetic-field-insensitive hyperfine qubits were also experimentally studied in optical microtraps, created by an array of microlenses [10]. Using spin-echo techniques, coherence times of 68 ms were obtained, limited by spontaneous scattering of photons. Simultaneous Ramsey measurements in up to 16 microtraps were performed, demonstrating the scalability of this approach.

Vibrational Qubits

Qubits can also be encoded in the vibrational states of atoms in tight traps. This has been proposed both for optical [11, 12] and magnetic [13, 14] microtraps. The computational basis states can be two vibrational levels in a single trap, for example the ground and first excited vibrational level [11]. Alternatively, they can be defined by the presence of an atom in either the left or the right well of a double-well potential [12]. Initialization of the atoms in the lowest vibrational state of the trap with high fidelity is crucial in both of these schemes.

Vibrational states are usually more delicate to handle and to detect than hyperfine states. An advantage, on the other hand, is that an internal-state independent interaction is sufficient for two-qubit gates and collisional loss can be reduced as the two interacting qubits are in the same internal state. The proposals of [13, 14], therefore, consider a combination of hyperfine states for qubit storage and vibrational states for processing. Measurements of vibrational coherence near surfaces still have to be performed. The expected fundamental limits due to surface-induced decoherence, however, are comparable to those for hyperfine states (see Chapters 4 and 5).

Rydberg State Qubits

Rydberg states are attractive for QIP because of their strong electric dipole moment [15, 16]. The resulting dipole–dipole interaction between Rydberg atoms can be exploited for fast two-qubit quantum gates. Moreover, Rydberg qubits can be combined with long-lived ground state hyperfine qubits for information storage. For the $n \sim 50$ Rydberg states of Rb, typical lifetimes are ~ 100 µs for low angular momentum states up to ~ 30 ms for circular states. In [16] it is proposed to enhance the lifetime of circular Rydberg states into the range of seconds by using a micro-structured trap on a superconducting atom chip that simultaneously acts as a cavity with a microwave cut-off frequency high enough to inhibit spontaneous emission. Furthermore, it is shown that with the help of microwave state dressing, coherence lifetimes of similar magnitude could be achieved.

Ensemble-Based Qubits

A single qubit can be encoded in collective states of an ensemble of particles [17, 18]. State $|0\rangle$ corresponds to all particles in the ground state, while in state $|1\rangle$, a single excitation is shared collectively by the whole ensemble. To isolate this two-level system, a blockade mechanism is required that prevents the creation of two or more excitations. The necessary nonlinearity can be provided by the dipole–dipole interaction of Rydberg atoms [17–19], or by coupling the ensemble via a cavity to a single saturable two-level system such as a Cooper pair box [20, 21]. Ensemble qubits have the advantage that single-atom preparation is not required. Moreover, the Rabi frequency between the collective qubit states is enhanced by $\sqrt{N}$, where N is the number of atoms in the ensemble. The decay rates are the same as for a single particle if the decay is dominated by non-collective processes such as atom loss. Ensemble qubits in chip traps have been considered for ground-state atoms [21], Rydberg atoms [19], and polar molecules [20, 22].

Hybrid Systems Involving Solid-State Qubits
A particularly attractive feature of chip traps is the possibility to combine atomic or molecular qubits with solid-state qubits on the chip surface [20–24]. Such hybrid systems would combine fast processing in the solid state with long coherence times for information storage in the atomic system. An impressive degree of coherent control has been demonstrated for example for qubits based on superconducting circuits [25, 26]. Coherent dynamics as well as decoherence in these systems typically occur on a time scale of nanoseconds to microseconds, several orders of magnitude faster than in atomic gases. A very promising approach to couple atomic and superconducting qubits is the use of superconducting microwave resonators [20–22, 24]. To combine the necessary cryogenic technology with atom chips represents an experimental challenge which is currently pursued in several experiments.

9.4
Qubit Rotations (Single-Qubit Gates)

Single-qubit gates are unitary transformations in the Hilbert space of a single qubit. In the language of atomic physics, this corresponds to rotations on the Bloch sphere of the two-level system encoding the qubit. The necessary degree of control can be experimentally demonstrated by driving high-contrast Rabi oscillations between the qubit states. Atom-chip-based experiments have demonstrated such control with hyperfine qubit states. Coherent control of motional states has been demonstrated in the context of atom interferometry.

Hyperfine Qubits
Hyperfine qubit states can be coupled with oscillating microwave and/or radio-frequency (rf) magnetic fields. High-fidelity Rabi oscillations on the qubit transition $|F = 1, m_F = -1\rangle \leftrightarrow |F = 2, m_F = +1\rangle$ of ^{87}Rb discussed in Section 9.3 have already been demonstrated experimentally on an atom chip, see Figure 9.3 and [7, 27]. In this case, a two-photon transition with a microwave and an RF photon is involved. A two-photon Rabi frequency of a few kHz is easily achieved, so that single-qubit gates can be performed on a time scale of hundreds of microseconds, three to four orders of magnitude faster than the relevant coherence lifetimes. Al-

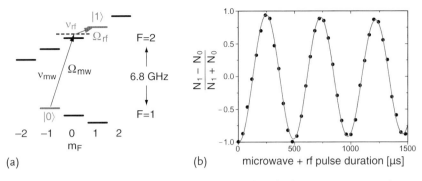

Figure 9.3 Qubit state rotations on the hyperfine transition $|F=1, m_F=-1\rangle \equiv |0\rangle$ and $|F=2, m_F=+1\rangle \equiv |1\rangle$ (adapted from [4]). (a) Ground-state hyperfine structure of ^{87}Rb in a weak magnetic field. The first-order Zeeman shift of the states $|0\rangle$ and $|1\rangle$ is approximately identical. The two-photon transition $|0\rangle \leftrightarrow |1\rangle$ is driven by a microwave ν_{mw} and a radio-frequency ν_{rf}. Ω_{mw} and Ω_{rf} are the single-photon Rabi frequencies of the microwave and RF transition, respectively. (b) Rabi oscillations on the two-photon transition recorded as a function of the microwave and RF pulse length. The two-photon Rabi frequency is $\Omega_{2ph}/2\pi = 2.1$ kHz, and the fidelity of a π-pulse is 96 %.

ternatively, hyperfine states can be coupled through a two-photon Raman transition driven by two laser beams [10, 28].

A nice feature of atom chips is that the qubit driving fields can be generated by chip-based waveguides (for microwaves) or just simple wires (for RF). This results in a stable, well-controlled coupling with tailored polarization. All elements for qubit manipulation can thus be integrated on chip. Moreover, chip-based driving fields can have strong near-field gradients. This is advantageous as it allows individual addressing of spatially separated qubits. On the other hand, care has to be taken to avoid dephasing due to strong gradients across a single qubit.

Vibrational Qubits

If the qubit is encoded in the ground and first excited vibrational levels of a single trap [11, 13, 14], qubit rotations can be induced by driving a two-photon Raman transition between the states with two lasers. Such transitions between vibrational levels are routinely employed in ion trap QIP [28] and have been demonstrated in optical dipole traps [29]. Similar experiments with neutral atoms in chip traps still have to be performed. As the vibrational levels have to be spectrally resolved, tight traps with large vibrational frequencies are required. On atom chips, sufficiently high vibrational frequencies of up to ~ 1 MHz are accessible.

If the qubit basis states are the left and right states of a double well [12], single-qubit gates can be performed by adiabatically lowering the barrier between the two wells and allowing tunneling to take place. This has strong connections to atom interferometry. Chip-based atom interferometers demonstrating versatile coherent control of the motional state of BECs have been realized, see for example [27, 30, 31] and Chapter 7.

Other Qubit Implementations

Several new atom chip experiments with relevance for QIP are currently being set up, involving among other things Rydberg atoms or superconducting structures on the chip (see Chapter 10). We expect chip-based experiments on coherent control in these systems in the near future.

9.5
Single-Qubit Readout (Single-Atom Detection)

Most schemes for QIP on atom chips consider individual atoms as the carriers of quantum information. While preparation, manipulation, and detection of single particles is a standard task in ion trap experiments, it is significantly harder to achieve with neutral atoms. This is because there is no equivalent to the strong Coulomb interaction that allows one to separate individual trapped ions, and that provides a tight, internal-state independent trapping potential in which the ions can be held during detection. The atom chip experiments on qubit state coherence and qubit rotations described in Sections 9.3 and 9.4, for example, were performed with atomic ensembles. While they indeed demonstrate the usefulness of the investigated states for QIP, eventually similar experiments will have to be performed with single atoms. An important achievement is therefore the integration of single-atom detectors with atom chip technology. In the following, we briefly highlight chip-based single-atom detectors which have been realized.

One approach to single-atom detection is photoionization and subsequent ion detection. This has been demonstrated on an atom chip in [32]. For atoms in Rydberg states, state-dependent electric field ionization is used as a standard single-atom detection method [16]. Detectors based on ionization are destructive, that is the atom is lost after detection. Fluorescence detection, which is also employed in ion trap experiments, can in principle provide non-destructive single-atom detection if the atom is held in a tight, state-independent trap such as an optical dipole trap. A simple and compact system for on-chip fluorescence detection are the fiber-based detectors demonstrated in [33].

High-finesse optical cavities are a particularly powerful system for single-atom detection. Non-destructive detection with negligible heating of atomic motion is possible. Moreover, optical cavities form the basis of certain quantum gate schemes, see Section 9.7.4, and of interfaces between storage qubits (atoms) and flying qubits (photons), see Section 9.8.2. An experiment demonstrating single-atom detectivity with a macroscopic cavity surrounding an atom chip is reported in [34]. Moreover, miniaturized optical cavities were developed which are directly integrated on the chip [35–38]. The single-atom strong coupling regime of cavity quantum electrodynamics was reached with such a cavity on an atom chip [37].

9.6
Single-Qubit Preparation (Single-Atom Preparation)

The availability of single-atom detectors on atom chips enables experiments on deterministic single-atom preparation. In many proposals for QIP, single atoms have to be prepared in the vibrational ground state of the trapping potential with very high probability – either because the qubit itself is encoded in vibrational states, or because atomic motion plays a crucial role in two-qubit gates, for example in the form of collisions between atoms.

A BEC can be seen as a large reservoir of atoms in the motional ground state. Proposals for the deterministic extraction of single atoms from a BEC have been made [39, 40]. They consider a tight microtrap into which condensate atoms can be transferred. Precise control of the atom number in the tight trap is provided by a collisional blockade mechanism. This is related to the Mott insulator transition, which could be used to prepare a large array of single atoms in an on-chip lattice. An alternative is heralded single-atom preparation, where atoms are probabilistically coupled from a BEC reservoir to an initially unoccupied internal state, which is continuously monitored with a non-destructive single-atom detector such as an optical cavity. Once an atom is detected, the coupling is turned off.

Single neutral atoms can already be prepared by loading an optical dipole trap from a magneto-optical trap, exploiting again collisional blockade [41]. This approach could be used for optical microtraps. Initially, the atoms are thermally distributed over many vibrational states. Raman sideband cooling could be used for ground-state cooling.

9.7
Conditional Dynamics (Two-Qubit Gates)

Two-qubit gates are the heart of a quantum processor, as they are required for the generation of entanglement between the qubits. In this section we present schemes for two-qubit gates that can be implemented with neutral atoms on an atom chip.

Let us consider the dynamics of an arbitrary number of atoms (no matter if charged or not) in a time- and state-dependent three-dimensional trapping potential $V_k(r, t)$, $r = (x, y, z)$ governed by the Hamiltonian operator [42, 43]

$$\hat{H}(t) = \sum_{k=0}^{1} \int dr\, \hat{\Psi}_k^\dagger(r) \left[-\frac{\hbar^2}{2m} \nabla^2 + V_k(r, t) \right] \hat{\Psi}_k(r)$$

$$+ \sum_{k,\ell=0}^{1} \frac{1}{2} \int drdr'\, \hat{\Psi}_k^\dagger(r) \hat{\Psi}_\ell^\dagger(r') U_{k\ell}(r, r') \hat{\Psi}_\ell(r') \hat{\Psi}_k(r) . \quad (9.1)$$

Here m is the atomic mass, $\hat{\Psi}_k^\dagger(r)$, $\hat{\Psi}_k(r)$ are atomic field creation and annihilation operators for the logic state $|k\rangle$, and $U_{k\ell}(r, r')$ is the two-atom interaction potential for the qubit states $|k\rangle$ and $|\ell\rangle$, with $k, \ell = 0, 1$. Our goal is the realization of a two-qubit gate with two atoms, each of them carrying a qubit of information usually encoded in an extra degree of freedom (e.g., a pair of hyperfine states) other than their motional state. In this specific case, the full many-body problem described by the Hamiltonian (9.1) can be reduced to a Schrödinger equation for two trapped particles and this will be assumed in the following.

The quantum gate we aim to implement is a phase gate having the following truth table: $|\epsilon_1\rangle|\epsilon_2\rangle \to e^{i\phi_{\epsilon_1\epsilon_2}}|\epsilon_1\rangle|\epsilon_2\rangle$, where $|\epsilon_1\rangle, |\epsilon_2\rangle$ are the logic qubit states with $\epsilon_{1,2} = 0, 1$. When the phase ϕ takes on the value of π, the combination of a phase gate with two Hadamard gates yields a controlled-NOT gate. In this respect it is an important quantum gate. Since it requires only to produce a phase shift for the state $|1\rangle|1\rangle$ such a gate has become of interest, because it requires a state-dependent interaction that is relatively straightforward to realize physically.

Let us explain the basic principle to obtain a conditional phase shift ϕ when two atoms are trapped in a microscopic potential. Initially, at $t = 0$, we assume that the two particles are in the respective ground states of the trapping potential and that their wave functions are well separated from each other so that their overlap is negligible. At times $0 < t < T_g$ the potential wells are changed in such a way that the atomic wave functions are displaced differently depending on their logical state $|k\rangle$ and a state-dependent wave function overlap results. The particles interact for a time T_g, the gate operation time, and at $t = T_g$ the initial situation is restored. With this approach we get state-dependent phase shifts of two kinds: a purely kinematic one, $\theta_k + \theta_\ell$, due to the single particle motion in the trapping potential; and an interaction phase, $\theta_{k\ell}$, due to the coherent interactions among the atoms. Thus, we can summarize the ideal phase gate with the mapping [43, 44]

$$|\epsilon_1\rangle|\epsilon_2\rangle|\psi_{\epsilon_1\epsilon_2}\rangle \to e^{i\phi_{\epsilon_1\epsilon_2}}|\epsilon_1\rangle|\epsilon_2\rangle|\psi_{\epsilon_1\epsilon_2}\rangle, \qquad (9.2)$$

where the motional state $|\psi_{\epsilon_1\epsilon_2}\rangle$ has to factor out at the beginning and at the end of the gate operation. In the ideal transformation (9.2) we grouped together the kinematic and global two-particle phases. Indeed, the application of single-qubit operations affords $\phi = \theta_{11} - \theta_{01} - \theta_{10} + \theta_{00}$ [45].

We conclude this section by introducing the concept of gate fidelity $F \in [0, 1]$, which will be a useful quantity later in the chapter to assess the gate performance. Basically, it is the projection of the physical state obtained by actually manipulating the system onto the logical state that the gate aims to attain, averaged over degrees of freedom (e.g., motion) that cannot be accurately controlled.

9.7.1
Internal-State Qubits and Collisional Interactions

In order to obtain conditional dynamics, as we discussed in the previous section, either the trapping potential or the interaction term should be state-dependent. In the case of ultra-cold neutral atoms, the interaction between atoms is mediated by two-body collisions, whose dominant contribution is s-wave scattering described by

$$U_{k\ell}(\mathbf{r}, \mathbf{r}') = \frac{4\pi\hbar^2 a_s^{k\ell}}{m} \delta^3(\mathbf{r} - \mathbf{r}'),\qquad(9.3)$$

where $a_s^{k\ell}$ is the s-wave scattering length for the internal states $|k\rangle$ and $|\ell\rangle$. Because of the short range of the pseudopotential (9.3), the wave functions of the atoms have to overlap in order to interact, and for identical atoms in the same logical state, s-wave scattering is only possible for bosons, and therefore in the following we will consider bosonic atomic species. As the potential given in Eq. (9.3) assumes elastic collisions, the states $|0\rangle$ and $|1\rangle$ have to be chosen such that they remain the same after the collision.

One of the most effective theoretical models for an atom chip phase gate has been proposed in [42]. In this scheme the control of the interaction between the atoms is realized by changing the shape of a microscopic potential depending on the internal state of the atoms. Three conditions are assumed: (i) the shape of the potential is harmonic; (ii) the atoms are initially cooled to the vibrational ground state of two potential wells centered at $\mathbf{r} = \mathbf{r}_0$ and $\mathbf{r} = -\mathbf{r}_0$; (iii) the change in the form of the trapping potential is instantaneous. The principle of the gate is the following: at time $t = 0$ the barrier between the atoms, say in the x-direction, is suddenly removed (selectively) for atoms in the logical state $|1\rangle$, whereas for atoms in state $|0\rangle$ the potential is not changed. An atom in state $|1\rangle$ finds itself in a new harmonic potential centered at $\mathbf{r} = 0$ with a frequency ω, smaller than the one of the separated wells, ω_0. The atoms in state $|1\rangle$ are allowed to perform an integer number of oscillations and at $t = T_g$ the initial wells are restored. In this process the particles acquire both a kinematic phase due to their oscillations in the traps and an interaction phase due to their collisions. In the tight transverse confinement regime, where the frequency ($\omega_\perp$) of the well in the y-z-directions is much larger than that (ω, ω_0) in the x-direction, the gate dynamics can be well approximated by a one-dimensional model with a contact potential $U_{k\ell}^{1D}(x, x') = 2\hbar\omega_\perp a_s^{k\ell} \delta(x - x')$. The kinematic phase $2n\pi\omega_\perp/\omega$ ($T_g = 2\pi n/\omega$, $n \in \mathbb{N}$) due to the radial confinement is common to all states, while the one due to the oscillation in the axial direction is state-dependent. Because of the harmonicity of the trapping potentials, almost perfect revivals of the wave packet occur. By choosing $\omega = j\omega_0$, with $j \in \mathbb{N}$, and in the limit where the interaction does not induce any relevant alteration in the shape of the two-particle wave function, the gate operation time T_g can be fixed by looking at the revival where the total accumulated phase ϕ assumes a defined value, for example, π. Because of the form of $U_{k\ell}^{1D}$ the frequency $\omega_\perp$ can be adjusted in order to fix the value of ϕ [42].

Atom chips can provide microscopic state-dependent potential landscapes in which atoms can be trapped and manipulated for the implementation of the above scheme. In [2] it was pointed out that a combination of static magnetic and static electric fields could be used for this purpose. However, several issues have to be addressed that could prevent a successful experimental realization of the scheme discussed above: (i) the trapping potentials are usually anharmonic; (ii) the fidelity is strongly reduced by wave packet distortion due to undesired collisions in some of the qubit basis states [42]; (iii) transverse excitations of the atoms can arise during the collisions if the ratio $\omega_\perp/\omega$ is not properly chosen for the 1D condition. Those processes would significantly reduce the gate fidelity.

An analysis of the limitations due to anharmonicity of the potentials is carried out in [46]. In that analysis a cubic and a quartic term is added to the harmonic potential in order to include the next leading order terms in the Taylor series expansion of an arbitrary potential. While a cubic anharmonicity is well tolerated, the quartic correction poses severe restrictions to the correct performance of the gate that are not easy to satisfy on atom chips. Thus, for a correct performance, the atoms have to be forced to a given dynamics.

The variant of [47] to the original proposal [42] can be regarded as the first attempt towards a realistic implementation of the collisional phase gate on an atom chip. It employs the hyperfine qubit states $|0\rangle \equiv |F = 1, m_F = -1\rangle$ and $|1\rangle \equiv |F = 2, m_F = 1\rangle$ of ^{87}Rb whose favorable coherence properties were already discussed in Section 9.3. Moreover, its key ingredient, the coherent manipulation of these states with a state-dependent trapping potential, was realized in a recent experiment [27]. Let us analyze the features of this scheme (see Figure 9.4). The state-dependent potential is split into

$$V_k(\mathbf{r}, t) = u_c(\mathbf{r}) + \lambda(t) u_k(\mathbf{r}), \quad (9.4)$$

where $u_c(\mathbf{r})$ is a common part and $u_k(\mathbf{r})$ a qubit-state-dependent part ($k = 0, 1$). The common part of the potential is a time-independent double-well potential along x that can be realized by a static magnetic potential, which is nearly identical for the chosen qubit states. As in [42], the dynamics can be reduced to 1D assuming a tight confinement in the transverse dimensions y, z. The state-dependent part can be realized by a microwave near-field potential (see below). It is modulated with a function $\lambda(t)$, with $0 \leq \lambda(t) \leq 1$. At times $t < 0$, when the gate is in its initial state, we have $\lambda(t) = 0$ and the atoms are subject to $u_c(\mathbf{r})$ only. Each atom is prepared

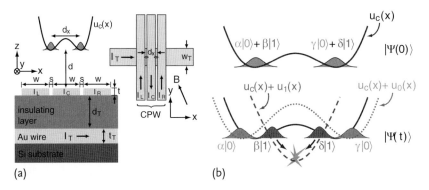

Figure 9.4 (a) Layout of the atom chip for the microwave collisional phase gate. (b) State-selective potential, atomic wave functions, and principle of the gate operation. Top: the state-independent potential $u_c(x)$ along the x direction. Bottom: the state-dependent potential $u_c(x) + u_k(x)$ (here $\lambda(t) = 1$). The atomic wave functions after half an oscillation period are shown.

in the motional ground of one of the wells of the double-well potential. During the time $0 \leq t \leq T_g$, $\lambda(t) \neq 0$ and the potential is state-dependent. The effect of $u_k(r)$ is twofold: $u_1(r)$ removes the barrier of the double well for state $|1\rangle$ and atoms in this state start to oscillate; the potential $u_0(r)$ shifts the minima of the double well for state $|0\rangle$ further apart in the x-direction (see Figure 9.4b), whereas in the original proposal those atoms do not experience any trap change. In this way, unwanted collisions (atoms in state $|01\rangle$), which are a major source of infidelity, are strongly reduced and the map (9.2) is implemented.

The state-dependent potential can be realized with microwave near-fields with a frequency near the hyperfine splitting of ^{87}Rb of 6.8 GHz. Unlike the optical potentials created by non-resonant laser beams, which can be tightly focussed due to their short wavelength, the centimeter wavelength λ_{mw} of microwave radiation poses severe limitations on far-field traps. On atom chips, however, the atoms are trapped at distances $d \ll \lambda_{mw}$ from the chip surface, and therefore they can be manipulated by microwave signals in on-chip transmission lines. In the near-field of the source currents and voltages, the microwave fields have the same position dependence as the static fields created by equivalent stationary sources. The maximum field gradients depend on the size of the transmission line conductors and on the distance d, not on λ_{mw}. Therefore, state-dependent microwave potentials varying on the micrometer scale can be realized. In a related way, radio-frequency fields can be used to generate near-field potentials (see Chapter 7).

When we consider the hyperfine levels $|F, m_F\rangle$ of the $5S_{1/2}$ ground state of a ^{87}Rb atom, the magnetic component of the microwave field $B_{mw}(r)\cos(\omega t)$ couples the $|1, m_1\rangle$ to the $|2, m_2\rangle$ sublevels, with Rabi frequencies

$$\Omega^{2,m_2}_{1,m_1}(r) = \frac{\langle 2, m_2|\hat{\mu} \cdot B_{mw}(r)|1, m_1\rangle}{\hbar}, \tag{9.5}$$

for the different transitions (in the rotating-wave approximation). In Eq. (9.5), $\hat{\mu} = \mu_B g_J \hat{J}$ is the operator of the electron magnetic moment ($g_J \simeq 2$). In a combined static magnetic and microwave trap, as considered here, both the static field $B_s(r)$ and the microwave field $B_{mw}(r)$ vary with position. This leads to a position-dependent microwave coupling with in general all polarization components present. The detuning of the microwave from the resonance of the transition $|1, m_1\rangle \rightarrow |2, m_2\rangle$ is:

$$\Delta^{2,m_2}_{1,m_1}(r) = \Delta_0 - \frac{\mu_B}{2\hbar}(m_2 + m_1)|B_s(r)|, \tag{9.6}$$

where $\Delta_0 = \omega - \omega_0$ is the detuning from the transition $|1, 0\rangle \rightarrow |2, 0\rangle$, and the different Zeeman shifts of the levels have been taken into account. The limit of large detuning $|\Delta^{2,m_2}_{1,m_1}|^2 \gg |\Omega^{2,m_2}_{1,m_1}|^2$ allows for long coherence lifetimes of the qubit states in the microwave potential. In this limit, the magnetic microwave potentials for the sublevels of $F = 1$ (left) and $F = 2$ (right) are given by

$$V^{1,m_1}_{mw}(r) = \frac{\hbar}{4} \sum_{m_2} \frac{|\Omega^{2,m_2}_{1,m_1}(r)|^2}{\Delta^{2,m_2}_{1,m_1}(r)}, \quad V^{2,m_2}_{mw}(r) = -\frac{\hbar}{4} \sum_{m_1} \frac{|\Omega^{2,m_2}_{1,m_1}(r)|^2}{\Delta^{2,m_2}_{1,m_1}(r)}. \tag{9.7}$$

As desired, the potentials for $F = 1$ and $F = 2$ have opposite signs, leading to a differential potential for the qubit states $|0\rangle \equiv |1, -1\rangle$ and $|1\rangle \equiv |2, 1\rangle$.

In addition to the magnetic microwave field, the electric field $\boldsymbol{E}_{\text{mw}}(\boldsymbol{r}) \cos(\omega t + \varphi)$ also leads to energy shifts. By averaging over the fast oscillation of the microwave at frequency ω, which is much faster than the atomic motion, the electric field leads to a time-averaged quadratic Stark shift. Hence, the total microwave potential for state $|0\rangle$, $u_0(\boldsymbol{r})$ in (9.4), is

$$u_0(\boldsymbol{r}) = -\frac{\alpha}{4}|\boldsymbol{E}_{\text{mw}}(\boldsymbol{r})|^2 + \frac{\hbar}{4}\sum_{m_2=-2}^{0}\frac{\left|\Omega_{1,-1}^{2,m_2}(\boldsymbol{r})\right|^2}{\Delta_{1,-1}^{2,m_2}(\boldsymbol{r})}, \tag{9.8}$$

while the microwave potential for state $|1\rangle$ is

$$u_1(\boldsymbol{r}) = -\frac{\alpha}{4}|\boldsymbol{E}_{\text{mw}}(\boldsymbol{r})|^2 - \frac{\hbar}{4}\sum_{m_1=0}^{+1}\frac{\left|\Omega_{1,m_1}^{2,+1}(\boldsymbol{r})\right|^2}{\Delta_{1,m_1}^{2,+1}(\boldsymbol{r})}. \tag{9.9}$$

The atom chip layout shown in Figure 9.4a allows one to realize the desired state-selective potential. It consists of two layers of gold metallization on a high resistivity silicon substrate, separated by a thin dielectric insulation layer. The wires carry stationary (DC) currents, which, when combined with appropriate stationary and homogeneous magnetic bias fields, create the state-independent potential $u_c(\boldsymbol{r})$. In addition to carrying DC currents, the three wires on the upper gold layer form a coplanar waveguide (CPW) for a microwave at frequency ω. The microwave fields guided by these conductors create the state-dependent potential $u_k(\boldsymbol{r})$. The combination of DC and microwave currents in the same wires is possible by the use of bias injection circuits.

Making use of quantum optimal control techniques [48], a gate operation time $T_g = 1.11$ ms with a fidelity $F = 0.996$ can be obtained, as shown in [47]. We emphasize that with this T_g and the long coherence lifetime of the qubit pair chosen (~ 1 s at a few μm distance from the chip [7], see Section 9.3), thousands of gate operations can be accomplished. The fidelity calculation includes the effect of several error sources: trap losses and decoherence due to the chip surface, undesired two-photon transitions induced by the microwave, mixing of the hyperfine levels due to the microwave coupling, and qubit dephasing due to technical noise. In the limit of large microwave detuning, the admixture of other states with different magnetic moments to the qubit states is strongly reduced. A last important point is related to the difficulty to prepare the atoms in the vibrational ground state with close to 100 % efficiency. This effect, modeled by a finite temperature, has been also included in the analysis. For temperatures $T \leq 20$ nK in the initial double-well trap, the fidelity is not reduced significantly.

A key ingredient of the quantum gate discussed here is the state-dependent microwave near-field potential. In a recent experiment [27], such a potential was realized on an atom chip, and it was used for the coherent manipulation of a two-component BEC in a superposition of the qubit states $|0\rangle$ and $|1\rangle$. The BEC was

9.7 Conditional Dynamics (Two-Qubit Gates)

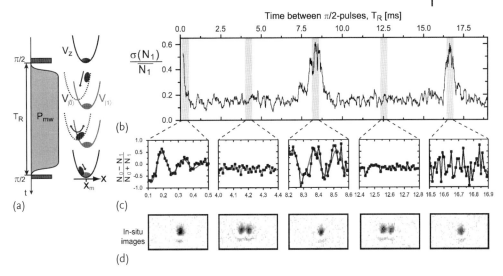

Figure 9.5 *Coherent state-dependent splitting and recombination of a BEC with microwave near-fields. The contrast of Ramsey fringes on the $|0\rangle \leftrightarrow |1\rangle$ transition is modulated due to the periodic splitting and recombination of the motional wave function of the BEC. (a) Experimental sequence. (b) As a measure of the wave function overlap, $\sigma(N_1)/\bar{N}_1$ is shown as a function of T_R, where $\sigma(N_1)$ is the standard deviation and $\bar{N}_1$ the mean of N_1 obtained from a running average over one period of the Ramsey fringes. (c) Corresponding Ramsey fringe data for selected values of T_R. (d) In situ images of the atomic density distribution of $|0\rangle$ and $|1\rangle$, for T_R corresponding to center of the windows in (c).*

state-selectively split and recombined, entangling atomic internal and motional states in a reversible way, as required for the atom chip quantum gate. Figure 9.5a illustrates the experimental sequence. It consists of a Ramsey $\pi/2 - \pi/2$ sequence on the $|0\rangle \leftrightarrow |1\rangle$ transition in combination with state-dependent splitting and recombination of the motional wave functions. After the first $\pi/2$-pulse, the microwave on the CPW is switched on within 50 µs, which corresponds to a sudden displacement of the potential minimum for state $|0\rangle$ by 4.3 µm. The wave function of state $|0\rangle$ is thus set into oscillation in the shifted potential. After a variable delay, the microwave is switched off within 50 µs, followed by the second $\pi/2$-pulse and state-selective detection to determine the number of atoms N_0 (N_1) in state $|0\rangle$ ($|1\rangle$). Figures 9.5b–d show the experimental results. The Ramsey interference contrast is modulated by the wave function overlap of the two states and thus periodically vanishes and reappears again due to the oscillation of state $|0\rangle$. Precisely at the time when state $|0\rangle$ has performed a full oscillation a sharp recurrence of the contrast is observed. This recurrence of high-contrast interference fringes proves that the combined evolution of internal and motional state is coherent.

We conclude this section with a last remark. The scheme proposed in [49] for atoms confined in optical lattices could also be implemented with state-dependent microwave potentials. Here at time $t = 0$ the atom α in the logical state $|k\rangle$ experiences the potential $V_\alpha^k(\mathbf{r}, t) = V(\bar{\mathbf{r}}^k + \delta \mathbf{r}_\alpha^k(t) - \mathbf{r})$, which is initially ($t < 0$) centered at position $\bar{\mathbf{r}}_k$. The centers of the potentials move according to the trajectories $\delta r_a^k(t)$ with the condition $\delta r_a^k(0) = \delta r_a^k(T_g) = 0$, and such that the first atom collides with the second one if and only if they are in the logic states $|0\rangle$ and $|1\rangle$. Such a scheme could be implemented with microwave potentials by using a microwave state dressing scheme where only state $|0\rangle$ effectively couples to the microwaves, as for example in the experiment of [27].

9.7.2
Motional-State Qubits and Collisional Interactions

Based again on the conditional phase shifts induced by the collision between cold atoms, a number of proposals rely on the manipulation of quantum information stored in motional degrees of freedom. The original proposals [11, 12, 50] dealt with optical lattices, but those schemes can also be implemented using microscopic potentials on atom chips. Besides magnetic, microwave, or radio-frequency traps, chip-based optical traps are of interest in this context. By illuminating a 2D array of refractive or diffractive microlenses with laser light, a 2D set of diffraction-limited laser foci can be formed. Atoms can be confined in the optical dipole potentials generated by the laser foci [51]. In a first experiment, arrays with more than 80 sites were loaded with ensembles of about 1000 trapped ^{85}Rb atoms in the center of the 2D configuration and about 100 at the edges [52]. Theoretical proposals for two-qubit gates in this system rely again on the spatial overlap of two qubits out of initially separated locations. This can be accomplished by illuminating the array of microlenses with two laser beams with a finite relative angle of propagation creating two interleaved sets of dipole traps [51]. The variation of the relative angle yields a variation in the mutual distance between the trap sets. An important feature of these optical micropotentials is the relatively large separation of neighboring sites (~ 125 μm) which enables individual addressing [52].

In the quantum gate scheme of [50], a double-well potential contains one atom per well. The logic states $|0\rangle$ and $|1\rangle$ are identified with the single particle ground and excited states of each well, respectively. Initially the barrier is sufficiently high that tunneling between the lowest four eigenstates of a single trapped atom is negligible. When the barrier is lowered in such a way that the single particle excited states (the qubit state $|1\rangle$) of the potential do overlap, tunneling takes place and the energy shift due to the atom–atom interaction increases exponentially. The interaction lasts for a time sufficient to accumulate the required phase shift for a phase gate and subsequently the initial trapping configuration is restored by increasing the barrier again. An accurate use of quantum interference between two-particle states yields an optimized gate duration of 38 ms with an infidelity $1 - F \approx 6.3 \times 10^{-6}$. In the proposal of [11] the scenario is very similar and it uses the same qubit set. While in [50] the two-qubit gate is physically realized by lowering and increasing the barrier of the double-well potential, in [11] the (initially) separated traps adiabatically approach (or separate from) each other. In that way it is possible to obtain $T_g \sim 20$ ms for a $\sqrt{\text{SWAP}}$ two-qubit gate.

The proposal of [11, 12] uses again motional states, but not strictly the vibrational states of the trap. In the scheme each qubit consists of two separated traps and a single atom. Now, the computational basis $\{|0\rangle, |1\rangle\}$ is formed in this way: the ground state of the left trap represents $|0\rangle \equiv |0\rangle_L$, whereas the ground state of the right trap represents $|1\rangle \equiv |0\rangle_R$. One- and two-qubit quantum gates are performed by adiabatically approaching the trapping potentials and allowing for tunneling to take place. We note that in such a scheme four wells are needed to implement a two-qubit gate, either arranged in a 1D configuration with the traps on a line or side-by-side in a 2D configuration. Taking into account the different error sources present in this scheme, like fluctuations of the trap positions, photon scattering, and heating, one obtains an error rate of about 0.02, with a single-qubit operation time of 4 ms, and $T_g \sim 10$ ms for a two-qubit operation such as a phase gate. Even though the error rate is rather large, the scheme offers several advantages: (1) decoherence due to spontaneous emission reduces the fidelity only marginally; (2) no momentum transfer is effected for single and two-qubit gates; (3) a state-dependent interaction is not required for the implementation of two-qubit gates; (4) the readout is done with a laser beam focused onto one trap minimum and detecting the fluorescence light; (5) since one- and two-qubit gates are realized using the same technique, that is, by approaching the traps adiabatically, the complexity of the experimental setup would be reduced.

Both of these schemes can be implemented on an atom chip by means of a combination of static and microwave fields, with the need of trapping only one hyperfine level. However, one can combine the nice coherence properties of the qubit states $|F = 2, m_F = 1\rangle$ and $|F = 1, m_F = -1\rangle$ of the ground state of ^{87}Rb, and the entanglement produced by cold collisions via the motional states ($|g\rangle, |e\rangle$). In [13] two different ways of realizing this concept have been proposed: (a) duplicate the logical state of the storage levels in the motional levels, where $|1g\rangle \leftrightarrow |1e\rangle$; (b) swap the logical states of the two degrees of freedom, $|1g\rangle \leftrightarrow |0e\rangle$. Here we consider only the swap scheme.

Given the initial state $|\varphi_0\rangle = (a|00\rangle + b|01\rangle + c|10\rangle + d|11\rangle)|gg\rangle$, the swap scheme takes place in three steps: (i) we selectively excite the operation state and de-excite the storage states $|\varphi_1'\rangle = |00\rangle (a|gg\rangle + b|ge\rangle + c|eg\rangle + d|ee\rangle)$, that is, we swap their logic states; then (ii) the operation states get a dynamical phase ϕ: $|\varphi_2'\rangle = |00\rangle (a|gg\rangle + b|ge\rangle + c|eg\rangle + e^{i\phi} d|ee\rangle)$ through collisions; finally (iii) we swap again the storage and operation states $|\varphi_3'\rangle = (a|00\rangle + b|01\rangle + c|10\rangle + e^{i\phi} d|11\rangle) |gg\rangle$. Such a swap gate scheme is not restricted to internal and external degrees of freedom of cold atoms, but it can be applied to any system with at least two degrees of freedom. A selective excitation of vibrational states is required when cold atoms are employed. In order to realize it, the use of two-photon Raman transitions has been suggested. The experimental implementation of such transitions with ^{87}Rb atoms is rather delicate and a careful analysis is given in [14]. In a static version of the swap scheme, where the barrier is fixed and it is designed in such a way that the left and right single particle excited states overlap, an operation time for a phase gate of 16.25 ms with a gate fidelity $F > 0.99$ has been predicted [13]. In an optimized version of the gate dynamics, where the barrier is lowered and increased in order to get faster operation times for a desired value of infidelity, it has been possible to achieve fidelities of 0.99 in 6.3 ms and of 0.999 in 10.3 ms [14].

We conclude by noting that the concept of spatially delocalized qubits of [11] could also be used to implement discrete-time quantum walks [53].

9.7.3
Alternative Chip-Specific Approaches to Entanglement Generation

High-fidelity entanglement of cold neutral atoms can be achieved by combining several already available techniques such as the creation or dissociation of neutral diatomic molecules with the manipulation of atoms by linear atom optics elements that can be integrated on atom chips, like atomic beam splitters, phase shifters, and interferometers.

Let us consider the free-space decay of a two-atom system (e.g., a diatomic molecule) with zero total momentum. The two atoms will freely propagate along correlated directions due to momentum conservation: If one of the two atoms leaves along a specific direction, say from the left side a_1 with momentum $\mathbf{k}_a$, the remaining atom will certainly leave along the corresponding direction a_2 (opposite to a_1) with momentum $-\mathbf{k}_a$. However, the decay in free space leads to freely propagating atoms along many pairs of correlated directions such that the probability for the two atoms moving along any specified pair is small. By means of integrated, miniaturized atom optical devices on atom chips this drawback can be overcome. By restricting the decay to a limited phase space given by the atom optical microstructure one can reduce the available decay modes significantly to only the few desired modes. If there are only two directions the two atoms will move along, one can deterministically obtain the path-entangled state $|\Phi\rangle \propto |a_1, a_2\rangle + |b_1, b_2\rangle$, where $|a\rangle$ and $|b\rangle$ are two orthonormal spatial states of atoms [54].

For instance, a diatomic molecule can be guided into a molecule beam splitter, which can split the molecule into either path a or path b. In each of the one-dimensional guides, the molecule can be dissociated into correlated atoms, which will propagate along the two pairs of correlated directions. If the released decay energy for each atom is smaller than the transverse level spacing in the guides, the decay can only occur in the lowest energy state of the transverse modes, and it is restricted to only one mode per path. In this case each two-atom correlated decay leads to an atom pair entangled in the specified paths (spatial modes).

9.7.4
Cavity-QED-Based Schemes

Recent experimental advances in cavity QED on a chip have yielded results that promise the full integration and scalability of such cavities. Microscopic Fabry–Pérot cavities whose open structure gives access to the central part of the cavity field have been developed. In such cavities strong coupling between a single atom and the cavity mode has been obtained [37]. In these experiments a BEC was employed, which can be located deterministically everywhere in the cavity and positioned entirely within a single antinode of the standing-wave cavity mode field. This gives rise to a controlled and tunable coupling rate.

On the theoretical side, proposals for a quantum computer based on a cavity QED model have been put forward. The scheme of [55] assumes N atoms coupled to a single quantized mode of a high-finesse cavity. Quantum operations (e.g., a controlled-NOT gate) are realized via the coupling of the atoms with individual lasers and their entanglement is mediated by the exchange of a single cavity photon. In a similar setup a controlled-NOT gate can be performed through a sequence of (destructive) measurements on the atoms and quantum non-demolition measurements of the atom number [56]. Because of the randomness of the measurement outcome in this case the gate operation is probabilistic. Nevertheless, this scheme is more robust against decoherence and cavity losses than the one of [55]: while in that proposal the infidelity scales as $1/\sqrt{2C_0}$, in [56] it scales as $\log(2C_0)/2C_0$, where C_0 is the cooperativity parameter.

Another interesting possibility of quantum information processing is given by distributed quantum networks [57], whose nodes are optical cavities. This would allow for applications in entangled state cryptography, teleportation, purification and distributed quantum computation. The combination of integrated micro-optical cavities and optical fibers as transmission lines would allow the implementation of such networks also on atom chips.

9.7.5
Quantum Gate Schemes that Can Be Adapted from Other Contexts

Quantum gate proposals initially meant for systems confined in, for example, optical lattices such as the scheme for moving potentials discussed briefly in Section 9.7.1 can be also realized in on-chip optical lattices [58] or magnetic microtrap lattices [8, 9]. In Section 9.7.1 we suggested to move the centers of the traps by means of microwave potentials. The results of [58] show that it is possible to realize 1D and 2D optical lattices on a chip, where the traps are the nodes of the evanescent wave field above an optical waveguide resulting from the interference of different waveguide modes. With a laser power of $\sim 1\,\text{mW}$ it is possible to produce tight traps, 150 nm above the on-chip waveguide surface, with trap frequencies on the order of 1 MHz, and with a spatial periodicity of about 1 μm. Moreover, the individual qubits are readily addressable, and it is possible to move 1D arrays of qubits by adjusting the phases of the waveguide modes. The drawback of such technology is that to get strong confinement with waveguides made from existing materials and using low laser powers, one needs to work extremely close to the waveguide surface implying a relevant impact on the qubit coherence. Alternatively one can use current-carrying wires and a perpendicularly magnetized grooved structure [8]. This solution allows for the trapping and cooling of ultra-cold atoms by means of the current-carrying wires, whereas the magnetic microstructure generates a 1D permanent magnetic lattice with a spacing between neighboring sites on the order of 10 μm and with trap frequencies of up to 90 kHz.

On the basis of such experimental achievements, schemes that exploit the interaction of atoms excited to low-lying Rydberg states [15] can be also used on atom chips [16]. An appropriate sequence of laser pulses above a waveguide can excite the qubits into Rydberg states and entangle them via electric dipole–dipole interactions [58]. While the phase gate model suggested in [15] relies only on the strong dipole–dipole interaction, in another recent proposal [59] one can combine the Rydberg blockade mechanism with the rapid laser pulse sequence of the well-known stimulated Raman adiabatic passage (STIRAP). This combination with the engineering of a time-dependent relative phase $\phi_R(t)$ between the Rabi frequencies of the two STIRAP laser pulses affords a higher degree of control of the phase of Eq. (9.2) through the manipulation of geometrical phases.

So far, we discussed only qubits that are represented by individual two-level systems. This kind of qubits requires high control and addressability of individual particles, which raise important challenges for experimentalists. Some theoretical investigations, however, use a symmetric collective state of a mesoscopic atomic ensemble [17], where only one excitation is present in the system. The dipole excitation blockade mechanism of Rydberg atoms works in that direction, because it prevents multiple excitations in an ensemble. The same mechanism used to implement a fast phase gate with two Rydberg atoms can be extended to qubits stored in few-µm-spaced atomic clouds, where each atomic ensemble is a qubit. Alternatively, collective states in ensembles of multi-level quantum systems can be used to store the information [18]. Quantum operations such as one- and two-bit gates are then implemented by collective internal state transitions taking place in the presence of an excitation blockade mechanism, such as the one provided by the Rydberg blockade.

Another interesting solution to the issue of single-atom addressability via a laser is quantum computation with neutral atoms, based on the concept of 'marker' atoms, that is, auxiliary atoms that can be efficiently transported in state-independent periodic external traps to operate quantum gates between physically distant qubits [48]. Here, again, qubits are represented by internal long-lived atomic states, and qubit atoms are stored in a regular array of microtraps. These qubit atoms remain frozen at their positions during the quantum computation. In addition to the atoms representing the qubits, an auxiliary 'marker atom' (or a set of marker atoms) is considered, which can be moved between the different lattice sites containing the qubits. The marker atoms can either be of a different atomic species or of the same type as the qubit atoms, but possibly employing different internal states. These movable atoms serve two purposes. First, they allow addressing of atomic qubits by 'marking' a single lattice site due to the marker qubit interactions: the corresponding molecular complex can be manipulated with a laser without the requirement of focusing on a particular site. Second, the movable atoms play the role of 'messenger' qubits which allow one to transport quantum information between different sites in the optical lattice, and thus to entangle distant atomic qubits. The required qubit manipulation and trapping techniques for such a scheme are essentially the same as the ones previously presented, and it is therefore well suited for atom chips.

9.8
Hybrid Approaches to QIP on a Chip

Atom chips are ideally suited as a platform for hybrid approaches to QIP, in which different systems such as neutral atoms, ions, photons, and solid-state systems are combined. In fact, one could argue that chip traps for atoms, ions and molecules are the enabling technology which makes such hybrid approaches possible.

9.8.1
Hybrid Approaches to Entanglement Generation

Atomic or molecular qubits trapped on an atom chip could be coupled to solid-state systems for QIP on the chip surface. The resulting hybrid quantum systems would combine fast processing (solid state) with long coherence times (atomic or molecular systems). Many interesting combinations of systems are conceivable. The proposals of [20–24, 60] suggest to use superconducting wires or superconducting microwave cavities to 'wire up' several atomic qubits or to couple atomic qubits to a Cooper pair box on the chip surface. The atomic systems considered in this context are Rydberg atoms [23], atomic ions [60], polar molecules [20, 22, 24], and ground-state neutral atoms [21].

In this context, the proposal of [24] suggests an extension of the atom chip concept to ultra-cold polar molecules. It proposes a scalable cavity-QED-type quantum computer architecture, where entanglement of distant qubits stored in long-lived rotational molecular states is achieved via exchange of microwave photons. Polar molecules have stable internal states that can be controlled by electrostatic fields. This controllability is due to their rotational degree of freedom in combination with the asymmetry of their structure (unlike atoms). By applying moderate laboratory electric fields, rotational states with transition frequencies in the microwave range can be mixed, and the molecules acquire large dipole moments, which are the key property that makes them effective qubits in a quantum processing system. Furthermore, the application of electric field gradients leads to large mechanical forces, allowing one to trap the molecules. For instance, the electrostatic Z-trap for polar molecules proposed in [24] creates a non-zero electric field minimum in close proximity to the surface, analogous to Ioffe–Pritchard-type magnetic traps for neutral atoms.

In [21], a related idea is put forward involving an ensemble of ultra-cold neutral atoms coupled magnetically to a superconducting microwave resonator. Strong coupling between a single microwave photon in the coplanar waveguide resonator and a collective hyperfine qubit state in the atomic ensemble could be achieved. Integrated on an atom chip, such a system could be used to interconnect solid-state and atomic qubits and to perform a range of interesting cavity quantum electrodynamics experiments.

The proposals of [20, 22] suggest a quantum computer model where ensembles of cold molecules are used as a stable quantum memory by means of collective spin states, whereas a Cooper pair box, connected to the molecular ensemble via a stripline cavity, is used to perform one- and two-qubit operations and readout. In [20] the ground state and a symmetric collective state with only one excitation present in the ensemble are the qubit states, and therefore an ensemble of molecules carries only two logic states. On the other hand, in [22] a 'holographic quantum register' is encoded in collective excitations with defined spatial phase variations, where each phase pattern is addressed by optical Raman processes with classical fields. This allows one to store hundreds of qubits in just one sample. This way of encoding does not pose special problems for an atom chip implementation, where usually single-atom addressing with laser beams is required.

9.8.2
Interfacing Atoms (Storage/Processing Qubits) with Photons (Flying Qubits)

Strong atom–photon coupling has been demonstrated with trapped atoms in an on-chip cavity [37]. Such a system could be used to interface storage/processing qubits and flying qubits. In a similar way, microwave cavities could be employed as a quantum bus on the chip, as described in the previous section, where a collective atomic quantum state can be mapped into a photonic state and vice versa. With such on-chip cavities quantum networks as described in Section 9.7.4 can be realized. Quite interesting is also the possibility offered by integrated cavities to perform quantum non-demolition measurements of the BEC atom number, therefore allowing the preparation of atomic quantum superposition states [61]. Additionally, by employing Raman transitions for transferring a small and exact known number N of BEC atoms into a different internal state, it is possible, by means of a transverse (to the cavity axis) laser beam, to convert such N condensed atoms into a N-photon Fock state, which would allow the deterministic production of photons [37].

9.8.3
Quantum Information Technology for Precision Measurement and Other Applications

Quantum information technology, in particular techniques for coherent manipulation and entanglement generation, already find applications in other areas of science. In the currently emerging field of quantum metrology [62], many-particle entanglement in the form of spin-squeezed states is investigated as a way to overcome the standard quantum limit of interferometric measurement, which limits today's best atomic clocks. Techniques originally developed for QIP on atom chips enable such experiments with chip-based atomic clocks and interferometers.

A class of hybrid systems which is interesting in the context of precision measurement are atoms coupled to micro- or nano-mechanical oscillators [63–65]. Mechanical oscillators have applications in ultra-sensitive force detection. A strongly coupled atom-oscillator system provides a quantum interface allowing the coherent transfer of quantum states between the mechanical oscillator and atoms, opening the door to coherent manipulation, preparation, and measurement of micromechanical objects via the well-developed tools of atomic physics.

9.9
Conclusion and Outlook

When the first atom chip experiments were performed, quantum information processing was already considered a promising direction for future applications. A significant number of theoretical proposals for implementing quantum gates on an atom chip have since then been put forward, and we have tried to review them in this chapter. Several of these proposals have been worked out in detail, including investigations of various kinds of imperfections such as decoherence due to atom–surface interactions. High-fidelity quantum gates compatible with the requirements for fault-tolerant QIP seem experimentally feasible on atom chips even in the presence of these imperfections. Moreover, ideas and techniques developed for QIP are now also being investigated with great success in the context of quantum metrology or quantum simulations with atom chips.

On the experimental side, impressive progress was made in the chip-based coherent control of ultra-cold atoms. Coherent manipulation of long-lived hyperfine qubit states was demonstrated close to a chip surface [7], and coherent control of motional states is routinely achieved in chip-based atom interferometers (see Chapter 7). Chip-based lattices were created [9] which could store a large register of qubits. The experimental achievement of strong coupling of atoms to a chip-based optical cavity [35, 37] paves the way for deterministic single-atom control. In a recent experiment, ultra-cold atoms were coherently manipulated with an internal-state-dependent potential on an atom chip [27], a key ingredient of the proposed quantum gate of [42, 47]. An important experimental milestone for the near future is the controlled generation of entanglement between atoms on a chip.

While the development of chip-based near-field traps was pioneered for ultra-cold neutral atoms, it has already triggered similar developments for other systems such as ions or molecules. As trapped ions are currently one of the frontrunners in the field of QIP, the recent demonstration of chip-based ion traps is particularly exciting (see Chapter 13). One of the most promising directions of future research is the development of hybrid architectures for QIP, where the advantageous features of neutral atoms, ions, photons, and solid-state systems are combined in a single chip-based device. The many beautiful experiments on atom chips reported in this book have been an essential inspiration for this development.

References

1 Nielsen, M.A. and Chuang, I.L. (2000) *Quantum Computation and Quantum Information*, Cambridge University Press, Cambridge.

2 Schmiedmayer, J., Folman, R., and Calarco, T. (2002) Quantum information processing with neutral atoms on an atom chip. *J. Mod. Opt.* **49**, 1375.

3 DiVincenzo, D. (2000) The Physical Implementation of Quantum Computation. *Fortschr. Phys.* **48**, 771.

4 Treutlein, P., Steinmetz, T., Colombe, Y., Lev, B., Hommelhoff, P., Reichel, J., Greiner, M., Mandel, O., Widera, A., Rom, T., Bloch, I., and Hänsch, T.W. (2006) Quantum information process-

ing in optical lattices and magnetic microtraps. *Fortschr. Phys.* **54**, 702.
5 Steane, A. (2003) Overhead and noise threshold of fault-tolerant quantum error correction. *Phys. Rev. A* **68**, 42322.
6 Knill, E. (2005) Scalable quantum computing in the presence of large detected-error rates. *Phys. Rev. A* **71**, 42322.
7 Treutlein, P., Hommelhoff, P., Steinmetz, T., Hänsch, T.W., and Reichel, J. (2004) Coherence in microchip traps. *Phys. Rev. Lett.* **92**, 203005.
8 Singh, M., Volk, M., Akulshin, A., Sidorov, A., McLean, R., and Hannaford, P. (2008) One-dimensional lattice of permanent magnetic microtraps for ultracold atoms on an atom chip. *J. Phys. B: At. Mol. Opt. Phys.* **41**, 065301.
9 Whitlock, S., Gerritsma, R., Fernholz, T., and Spreeuw, R.J.C. (2009) Two-dimensional array of microtraps with atomic shift register on a chip. *New. J. Phys.* **11**, 023021.
10 Lengwenus, A., Kruse, J., Volk, M., Ertmer, W., and Birkl, G. (2007) Coherent manipulation of atomic qubits in optical micropotentials. *Appl. Phys. B* **86**, 377.
11 Eckert, K., Mompart, J., Yi, X.X., Schliemann, J., Bruß, D., Birkl, G., and Lewenstein, M. (2002) Quantum computing in optical microtraps based on the motional states of neutral atoms. *Phys. Rev. A* **66**, 042317.
12 Mompart, J., Eckert, K., Ertmer, W., Birkl, G., and Lewenstein, M. (2003) Quantum computing with spatially delocalized qubits. *Phys. Rev. Lett.* **90**, 147901.
13 Cirone, M.A., Negretti, A., Calarco, T., Krüger, P., and Schmiedmayer, J. (2005) A simple quantum gate with atom chips. *Eur. Phys. J. D* **35**, 165.
14 Charron, E., Cirone, M.A., Negretti, A., Schmiedmayer, J., and Calarco, T. (2006) Theoretical analysis of a realistic atom-chip quantum gate. *Phys. Rev. A* **74**, 012308.
15 Jaksch, D., Cirac, J.I., Zoller, P., Rolston, S.L., Côté, R., and Lukin, M.D. (2000) Fast quantum gates for neutral atoms. *Phys. Rev. Lett.* **85**, 2208.
16 Mozley, J., Hyafil, P., Nogues, G., Brune, M., Raimond, J.-M., and Haroche, S. (2005) Trapping and coherent manipulation of a Rydberg atom on a microfabricated device: a proposal. *Eur. Phys. J. D* **35**, 43.
17 Lukin, M.D., Fleischhauer, M., Cote, R., Duan, L.M., Jaksch, D., Cirac, J.I., and Zoller, P. (2001) Dipole blockade and quantum information processing in mesoscopic atomic ensembles. *Phys. Rev. Lett.* **87**, 037901.
18 Brion, E., Mølmer, K., and Saffman, M. (2007) Quantum computing with collective ensembles of multilevel systems. *Phys. Rev. Lett.* **99**, 260501.
19 Yan, H., Yang, G., Shi, T., Wang, J., and Zhan, M. (2008) Quantum gates with atomic ensembles on an atom chip. *Phys. Rev. A* **78**, 034304.
20 Rabl, P., DeMille, D., M. Doyle, J., Lukin, M.D., Schoelkopf, R.J., and Zoller, P. (2006) Hybrid quantum processors: Molecular ensembles as quantum memory for solid state circuits. *Phys. Rev. Lett.* **97**, 033003.
21 Verdú, J., Zoubi, H., Koller, C., Majer, J., and Ritsch, H. (2009) Strong magnetic coupling of an ultracold gas to a superconducting waveguide cavity. *Phys. Rev. Lett.* **103**, 043603.
22 Tordrup, K., Negretti, A., and Mølmer, K. (2008) Holographic quantum computing. *Phys. Rev. Lett.* **101**, 40501.
23 Sørensen, A.S., van der Wal, H.C., Childress, L.I., and Lukin, M.D. (2004) Capacitive coupling of atomic systems to mesoscopic conductors. *Phys. Rev. Lett.* **92**, 063601.
24 Andre, A., DeMille, D., Doyle, J.M., Lukin, M.D., Maxwell, S.E., Rabl, P., Schoelkopf, R.J., and Zoller, P. (2006) A coherent all-electrical interface between polar molecules and mesoscopic superconducting resonators. *Nat. Phys.* **2**, 636.
25 Steffen, M., Ansmann, M., Bialczak, R.C., Katz, N., Lucero, E., McDermott, R., Neeley, M., Weig, E.M., Cleland, A.N., and Martinis, J.M. (2006) Measurement of the entanglement of two superconducting qubits via state tomography. *Science* **313**, 1423.
26 Majer, J., Chow, J.M., Gambetta, J.M., Koch, J., Johnson, B.R., Schreier, J.A.,

Frunzio, L., Schuster, D.I., Houck, A.A., Wallraff, A., Blais, A., Devoret, M.H., Girvin, S.M., and Schoelkopf, R.J. (2007) Coupling superconducting qubits via a cavity bus. *Nature* **449**, 443.

27 Böhi, P., Riedel, M.F., Hoffrogge, J., Reichel, J., Hänsch, T.W., and Treutlein, P. (2009) Coherent manipulation of Bose-Einstein condensates with state-dependent microwave potentials on an atom chip. *Nat. Phys.* **5**, 592.

28 Wineland, D.J., Monroe, C., Itano, W.M., Leibfried, D., King, B.E., and Meekhof, D.M. (1998) Experimental issues in coherent quantum-state manipulation of trapped atomic ions. *J. Res. Natl. Inst. Stand. Technol.* **103**, 259.

29 Morinaga, M., Bouchoule, I., Karam, J.C., and Salomon, C. (1999) Manipulation of motional quantum states of neutral atoms. *Phys. Rev. Lett.* **83**, 4037.

30 Wang, Y.-J., Anderson, D.Z., Bright, V.M., Cornell, E.A., Diot, Q., Kishimoto, T., Prentiss, M., Saravanan, R.A., Segal, S.R., and Wu, S. (2005) Atom Michelson interferometer on a chip using a Bose–Einstein condensate. *Phys. Rev. Lett.* **94**, 090405.

31 Hofferberth, S., Lesanovsky, I., Fischer, B., Verdu, J., and Schmiedmayer, J. (2006) Radiofrequency-dressed-state potentials for neutral atoms. *Nat. Phys.* **2**, 710.

32 Stibor, A., Kraft, S., Campey, T., Komma, D., Günther, A., Fortágh, J., Vale, C.J., Rubinsztein–Dunlop, H., and Zimmermann, C. (2007) Calibration of a single-atom detector for atomic microchips. *Phys. Rev. A* **76**, 033614.

33 Wilzbach, M., Heine, D., Groth, S., Liu, X., Hessmo, B., and Schmiedmayer, J. (2008) A simple integrated single-atom detector, preprint arXiv:0801.3255.

34 Teper, I., Lin, Y.-J., and Vuletić, V. (2006) Resonator-aided single-atom detection on a microfabricated chip. *Phys. Rev. Lett.* **97**, 023002.

35 Aoki, T., Dayan, B., Wilcut, E., Bowen, W.P., Parkins, A.S., Kippenberg, T.J., Vahala, K.J., and Kimble, H.J. (2006) Observation of strong coupling between one atom and a monolithic microresonator. *Nature* **443**, 671.

36 Steinmetz, T., Colombe, Y., Hunger, D., Hänsch, T.W., Balocchi, A., Warburton, R.J., and Reichel, J. (2006) Stable fiber-based Fabry–Pérot cavity. *Appl. Phys. Lett.* **89**, 111110.

37 Colombe, Y., Steinmetz, T., Dubois, G., Linke, F., Hunger, D., and Reichel, J. (2007) Strong atom-field coupling for Bose–Einstein condensates in an optical cavity on a chip. *Nature* **450**, 272.

38 Trupke, M., Goldwin, J., Darquie, B., Dutier, G., Eriksson, S., Ashmore, J., and Hinds, E.A. (2007) Atom detection and photon production in a scalable, open, optical microcavity. *Phys. Rev. Lett.* **99**, 063601.

39 Diener, R.B., Wu, B., G. Raizen, M., and Niu, Q. (2002) Quantum tweezer for atoms. *Phys. Rev. Lett.* **89**, 070401.

40 Mohring, B., Bienert, M., Haug, F., Morigi, G., Schleich, W.P., and Raizen, M.G. (2005) Extracting atoms on demand with lasers. *Phys. Rev. A* **71**, 053601.

41 Schlosser, N., Reymond, G., and Grangier, P. (2002) Collisional blockade in microscopic optical dipole traps. *Phys. Rev. Lett.* **89**, 023005.

42 Calarco, T., Hinds, E.A., Jaksch, D., Schmiedmayer, J., Cirac, J.I., and Zoller, P. (2000) Quantum gates with neutral atoms: Controlling collisional interactions in time-dependent traps. *Phys. Rev. A* **61**, 022304.

43 Calarco, T., Briegel, H.-J., Jaksch, D., Cirac, J., and Zoller, P. (2000) Quantum computing with trapped particles in microscopic potentials. *Fortschr. Phys.* **48**, 945.

44 Calarco, T., Briegel, H.-J., Jaksch, D., Cirac, J.I., and Zoller, P. (2000) Entangling neutral atoms for quantum information processing. *J. Mod. Opt.* **47**, 2137.

45 Calarco, T., Cirac, J.I., and Zoller, P. (2001) Entangling ions in arrays of microscopic traps. *Phys. Rev. A* **63**, 062304.

46 Negretti, A., Calarco, T., Cirone, M.A., and Recati, A. (2005) Performance of quantum phase gates with cold trapped atoms. *Eur. Phys. J. D* **32**, 119.

47 Treutlein, P., Hänsch, T.W., Reichel, J., Negretti, A., Cirone, M.A., and Calarco,

T. (2006) Microwave potentials and optimal control for robust quantum gates on an atom chip. *Phys. Rev. A* **74**, 022312.

48 Calarco, T., Dorner, U., Julienne, P.S., Williams, C.J., and Zoller, P. (2004) Quantum computations with atoms in optical lattices: Marker qubits and molecular interactions. *Phys. Rev. A* **70**, 012306.

49 Jaksch, D., Briegel, H.-J. Cirac, J.I., Gardiner, C.W., and Zoller, P. (1999) Entanglement of atoms via cold controlled collisions. *Phys. Rev. Lett.* **82**, 1975.

50 Charron, E., Tiesinga, E., Mies, F., and Williams, C. (2002) Optimizing a phase gate using quantum interference. *Phys. Rev. Lett.* **88**, 077901.

51 Birkl, G. and Fortágh, J. (2007) Micro traps for quantum information processing and precision force sensing. *Laser Photon. Rev.* **1**, 12.

52 Dumke, R., Volk, M., Müther, T., Buchkremer, F.B.J., Birkl, G., and Ertmer, W. (2002) Micro-optical realization of arrays of selectively addressable dipole traps: A scalable configuration for quantum computation with atomic qubits. *Phys. Rev. Lett.* **89**, 097903.

53 Eckert, K., Mompart, J., Birkl, G., and Lewenstein, M. (2005) One- and two-dimensional quantum walks in arrays of optical traps. *Phys. Rev. A* **72**, 012327.

54 Zhao, B., Chen, Z.-B., Pan, J.-W., Schmiedmayer, J., Recati, A., Astrakharchik, G.E., and Calarco, T. (2007) High-fidelity entanglement via molecular dissociation in integrated atom optics. *Phys. Rev. A* **75**, 042312.

55 Pellizzari, T., Gardiner, S.A., Cirac, J.I., and Zoller, P. (1995) Decoherence, continuous observation, and quantum computing: A cavity QED model. *Phys. Rev. Lett.* **75**, 3788.

56 Sørensen, A.S. and Mølmer, K. (2003) Measurement induced entanglement and quantum computation with atoms in optical cavities. *Phys. Rev. Lett.* **91**, 097905.

57 Cirac, J.I., Zoller, P., Kimble, H.J., and Mabuchi, H. (1997) Quantum state transfer and entanglement distribution among distant nodes in a quantum network. *Phys. Rev. Lett.* **78**, 3221.

58 Christandl, K., Lafyatis, G.P., Lee, S.-C., and Lee, J.-F. (2004) One- and two-dimensional optical lattices on a chip for quantum computing. *Phys. Rev. A* **70**, 032302.

59 Møller, D., Madsen, L.B., and Mølmer, K. (2008) Quantum gates and multiparticle entanglement by Rydberg excitation blockade and adiabatic passage. *Phys. Rev. Lett.* **100**, 170504.

60 Tian, L., Rabl, P., Blatt, R., and Zoller, P. (2004) Interfacing quantum-optical and solid-state qubits. *Phys. Rev. Lett.* **92**, 247902.

61 Nielsen, A.E.B., Poulsen, U.V., Negretti, A., and Mølmer, K. (2009) Atomic quantum superposition state generation via optical probing. *Phys. Rev. A* **79**, 023841.

62 Giovannetti, V., Lloyd, S., and Maccone, L. (2004) Quantum-enhanced measurements: Beating the standard quantum limit. *Science* **306**, 1330.

63 Tian, L. and Zoller, P. (2004) Coupled ion-nanomechanical systems. *Phys. Rev. Lett.* **93**, 266403.

64 Treutlein, P., Hunger, D., Camerer, S., Hänsch, T.W., and Reichel, J. (2007) Bose–Einstein condensate coupled to a nanomechanical resonator on an atom chip. *Phys. Rev. Lett.* **99**, 140403.

65 Hammerer, K., Wallquist, M., Genes, C., Ludwig, M., Marquardt, F., Treutlein, P., Zoller, P., Ye, J., and Kimble, H.J. (2009) Strong coupling of a mechanical oscillator and a single atom. *Phys. Rev. Lett.* **103**, 063005.

Part Four New Directions

10
Cryogenic Atom Chips

Gilles Nogues, Adrian Lupaşcu, Andreas Emmert, Michel Brune, Jean-Michel Raimond, and Serge Haroche

10.1
Introduction

It is quite remarkable to observe that in 1995, when Weinstein and Libbrecht wrote their seminal article on microscopic atomic traps [1], the very same research group published experimental results on long-lived neutral atom traps in a cryogenic environment [2]. They reported trapping lifetimes of nearly 10 minutes for magnetostatic traps created by centimeter-size superconducting coils. Moreover they claimed that "with a cryogenic system one can use superconducting magnets and SQUID detectors to trap and non-destructively sense spin-polarized atoms". This sentence foresees some of the possibilities of cryogenic atom chips using superconducting materials. Low temperatures and superconductivity do not only offer ultra-high vacuum conditions and high currents without dissipation. They also open the way to a new class of experiments where ultra-cold atoms will be integrated on a chip together with solid-state devices whose coherent manipulation is only possible at low temperatures. A good example of such a device is a Superconducting Quantum Interference Device (SQUID) coupled to the magnetic moment of the trapped atomic cloud [3]. Other interesting candidates like coplanar waveguide resonators or the nano-electro-mechanical system (NEMS) are also worth mentioning.

It took some time after the initial proposal of [1] to develop and operate the first atom chips (see Chapter 1) at room temperature and an even longer time to operate them in cryogenic conditions. The first results on superconducting atom chips were reported at ENS Paris in 2006 [4] and at NTT Basic research labs in 2007 [5]. Many groups in the world are now building experiments along this line of research and exciting developments are expected. The first part of this chapter (Section 10.2) will present a review of the existing experiments on cryogenic chips. We will try to show by which aspects they may be similar to or differ from equivalent room-temperature setups. In Section 10.3 we will give a few examples of new experiments that could be carried out with cryogenic chips.

Atom Chips. Edited by Jakob Reichel and Vladan Vuletić
Copyright © 2011 WILEY-VCH Verlag GmbH & Co. KGaA, Weinheim
ISBN: 978-3-527-40755-2

10.2
Superconducting Atom Chip Setup: Similarities and Differences with Conventional Atom Chips

10.2.1
Experimental Considerations

Superconducting atom chips differ in many aspects from the usual atom chip experiments at room temperature. We focus in this section on the technical issues associated with building a setup that can be properly operated at cryogenic temperatures. Although they may appear technical, such differences are often a direct illustration of the physics of superconducting materials.

10.2.1.1 Chip Fabrication and Wiring

Cryogenic atom chips use superconducting films to carry the required trapping currents. Indeed, the typical power dissipated by the Joule effect in normal metal chips ranges from 1 to 10 W. This is not compatible with cryogenic systems. Among all the possible superconducting materials, present experimental efforts are focusing on niobium (Nb, $T_c = 9.2$ K) or niobium nitride (NbN, $T_c = 15$ K) [4]. More complicated alloys such as magnesium diboride (MgB$_2$, $T_c = 39$ K) [5], or even high-T_c oxides like YBCO (YBa$_2$Cu$_3$O$_{7-x}$, $T_c = 87$ K) [6] are also being investigated. It is important to note that the critical temperatures of Nb and MgB$_2$ given here correspond to the bulk material properties. For thin films, especially if the thickness is below a few 10 nm, the presence of defects can significantly reduce T_c. The maximum critical current in the smallest Nb wire (thickness 1 µm, width 40 µm) in [4] is 1.8 A at 4.2 K. It corresponds to a critical current density of 4.5 MA cm^{-2}. Current densities as high as 10 MA cm^{-2} are reported for MgB$_2$ at 4.2 K [5] and YBCO at 50 K [6]. Those values are of the same order of magnitude as the current densities typically observed in normal metal atom chips at room temperature [7] and hence superconductors will not offer the possibility to reduce the size of the conductors at similar currents. They will essentially differ in the fact that absolutely no heat is dissipated in the chip during the trapping.

Niobium-based chips are relatively easy to fabricate. They use standard techniques already applied to the case of normal metal chips (see Chapter 3). Superconducting Nb films, with thicknesses ranging from a few nm to a few µm, can be deposited on a Silicon substrate using cathodic sputtering. Standard optical or electronic lithography techniques can then be used to define the trapping wire and contact pads. Reactive ion etching (RIE) with fluor ions is used to remove undesired Niobium surfaces. Figure 10.1 shows the process for the fabrication of the chip of [4]. The superconducting current-carrying wires are complete after step 7. The complete process includes two additional steps for depositing a layer of gold on top of the wires in order to reflect the laser beams necessary for the operation of an on-chip mirror-MOT.

The technology involved in the fabrication of the MgB$_2$ chip of [5] is more demanding. A 1.6-µm-thick layer of superconductor is grown by molecular beam

10.2 Superconducting Atom Chip Setup: Similarities and Differences with Conventional Atom Chips

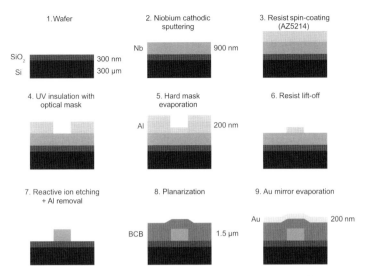

Figure 10.1 Complete fabrication process for the niobium chip of [4]. Starting from a silicon wafer, the superconducting Nb film is deposited by cathodic sputtering. The steps 3 to 6 correspond to the optical lithography used for creating a structured hard mask of aluminum on top of the Nb layer. The hard mask protects the wires during the reactive ion etching (step 7). Steps 8 and 9 allow one to deposit a gold mirror on top of the chip in order to reflect the lasers arriving on it.

epitaxy (MBE) on a sapphire substrate. For MgB_2 the patterning of the circuit cannot rely on lithography and etching techniques. In the case of [5] for example, it was performed by removing unnecessary parts of the film by ion milling.

Similar film deposition methods as in [5] are used for the fabrication of the high-T_c chip described in [6]. A 600–800-nm film of YBCO is grown by epitaxy on a yttria-stabilized zirconia (YSZ) single-crystal substrate in order to match the lattice parameters. It is important to control carefully the substrate temperature during the process. In the case of YBCO, it is even important to match the temperature expansion coefficients of the film and the substrate. A difference between the two expansion coefficients could be the source of additional surface tensions during the cooling of the chip that can significantly degrade the superconducting properties of the film. The patterning of the YBCO layer can be done by combining standard lithography techniques together with wet chemical etching or laser ablation.

A very important issue for all setups is the connection of the wire to the outside. It is relatively difficult to pass the required current in the structure and to avoid at the same time local heating due to the Joule effect in the connection wires or the contact pad. Because of the very small heat capacity of the chip materials at low temperature, even a moderate dissipation can rapidly lead to a local temperature increase above T_c, in which case superconductivity is lost. A possible solution to this problem consists in using only superconducting materials both for the lead wires and their contact on the film [8]. Another radical solution consist in suppressing completely the connecting wires and to induce permanent currents in a superconducting loop as shown in [5]. If a bias magnetic field is applied perpendicularly to a loop of superconductor before its transition to the superconducting state, the circuit will then react in order to keep constant the magnetic flux through the loop. As a consequence persistent currents will exist in the loop after switching off the external field. It is possible to switch off or even tune the current by performing a local transition to the dissipative normal state. This is done in [5] by shining a laser on a portion of the loop.

10.2.1.2 The Cryogenic Cell

The first experimental question that arises when one starts a low-temperature experiment is the choice of the cryogenic system. It will greatly determine the kind of studies one is able to carry out. In the case of atom chips the cryostat should be compatible with standard conditions for ultra-cold atom manipulation, like ultra-high vacuum (UHV), good optical access, and large magnetic field gradients. As was pointed out in [2], trapping ultra-cold atoms in a cryostat can lead to a very long lifetime because of the huge adsorption on the cold surface surrounding the cloud. UHV conditions can be reached without baking out the cold cell.

The problem of the optical access to the cloud is much more difficult to solve. It seems reasonable to keep the imaging system at room temperature. However, a large numerical aperture for the collection lens outside the cryostat implies that the superconducting atom chip will be directly exposed to blackbody radiation at 300 K with a large solid angle. This is usually not admissible. The coldest parts of the experiment are protected from the room-temperature radiation by thermal shields at intermediate temperatures. It is also necessary to bring some lasers inside the setup in order to trap the atoms and to probe them.

In many experiments, the trapping of the ultra-cold cloud is achieved by combining magnetic fields created by on-chip wires together with external bias coils. The typical size of the latter is of the order of a few cm and they must be placed at similar distances from the trapped cloud. Having the coils outside the cryostat limits the volume of the cryostat and might generate potentially harmful eddy currents in the cold setup while changing the external fields. On the other hand it might be technically difficult to arrange and thermalize bias coils inside the cryostat.

The first two experiments that have reported the operation of superconducting chips so far use radically different approaches:

- In [4] (see Figure 10.2a) the cold cell is in direct thermal contact with a 4.2-K liquid He reservoir. The volume of the cell is large enough (a few dm^3) to accommodate the chip holder as well as the external bias coils. The latter need, therefore, to be superconducting. The first imaging lens is 20 cm away from the trapped cloud. An intermediate thermal shield at liquid nitrogen temperature (77 K) screens most of the blackbody radiation. Cold viewports are mounted at 4.2 and 77 K. A good thermalization of all the wires connected to the cold part of the setup allows one to run the experiment for hours without difficulties. As opposed to room-temperature experiments, one cannot rely on the atomic

10.2 Superconducting Atom Chip Setup: Similarities and Differences with Conventional Atom Chips | 315

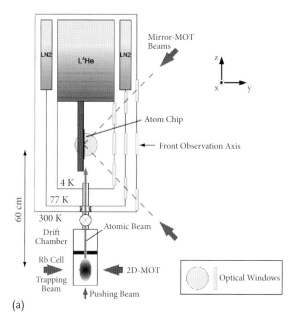

(a)

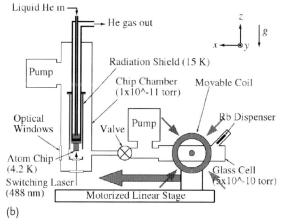

(b)

Figure 10.2 Scheme of the experimental setup for [4] (a) and [5] (b).

background vapor pressure to load the trap with atoms because it is dramatically low at cryogenic temperature. The problem can be circumvented by using a low-velocity beam of atoms which is recaptured in a magneto-optical trap (MOT) in the cryostat [4]. An additional UHV chamber at room temperature is therefore required in order to prepare the cold atoms.

- In [5] the experimental cell is much more compact and mounted at the end of a cold finger (see Figure 10.2b). The chip can be cooled down to 3 K but its temperature is significantly higher during normal operation. It is protected by a shield at 15 K. A better optical access is provided but thermalization of the lead wires might be a problem. Atoms are magnetically trapped outside the cryostat in an external UHV chamber and conveyed inside the cryostat with movable coils [5].

10.2.2
Trapping and Cooling: First Results

10.2.2.1 Magnetic Trap

As we have seen previously in Section 10.2.1.1, typical current densities inside superconducting slabs are similar to what is generally used in normal metal atom chips. As a consequence the trapping parameters such as the trapping frequencies, the trap volume, the distance to the chip, and so on ... do not differ significantly from the values observed for room-temperature chips (at least as long as the cloud is not too close to the surface, see Section 10.3.1). A typical experimental sequence for the ENS experiment is described in [4]. It was later refined in [9] where the first observation of a BEC (Bose–Einstein condensate) on a superconducting atom chip was reported. All the usual steps of normal metal chip experiments [10], see Chapter 2) are present. About 5×10^7 atoms are loaded in 5 s in a mirror-MOT whose magnetic field is created by the lower part of a rectangular coil placed 1.5 mm behind the chip. It mimics a centimeter-size U-wire whose magnetic field, combined with the homogeneous field created by the bias coils inside the cryostat, produces a quadrupole field a few mm away from the chip surface. Atoms are then transferred in 20 ms with an efficiency of 86 % to an on-chip mirror-MOT that uses a U-shaped superconducting wire on the chip. The currents in the U-wire and the external bias fields are then changed in 20 ms in order to compress the MOT and bring the atomic cloud to a distance of about 500 μm from the surface. The lasers are detuned during the compression step as well as during the 2-ms-long optical molasses that follows it. About 10^7 are present at the end of the molasses, at a temperature of 20 μK. Finally, an optical pumping to a low-field seeking state is performed before the transfer to the magnetic trap. The values of the experimental parameters at each step (current, bias field, laser power, and detuning) are close to those typical of room-temperature atom chips. Between 1 and 3×10^6 atoms are trapped 400 μm away from the surface at a temperature between 60 and 20 μK. Those numbers were measured by observation of the atomic cloud using standard absorption imaging 100 ms after the transfer to the magnetic trap.

The situation is slightly different for the permanent current magnetic trap presented in [5]. As explained in Section 10.1, the trapping current is induced in a superconducting loop long before the atoms are brought into the system. Moreover, there is no intermediate mirror-MOT close to the chip surface and the atoms are magnetically trapped and transferred inside the cryostat with the help of external movable coils (see Section 10.2.1.2 and Figure 10.2b). The center of the quadrupole magnetic trap created by those coils is simply superposed with the location of the permanent trap, then the coils are rapidly switched off. The number of atoms and the cloud temperature 100 ms after this transfer are 3×10^5 and 200 μK, respectively.

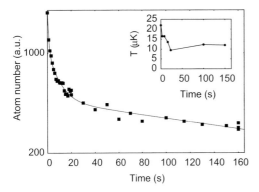

Figure 10.3 Number of atoms (log scale) as a function of time measured in [11]. The solid line is a double exponential fit of the data. During the first part the trapping atoms are rapidly lost because of evaporation. The atomic cloud temperature decreases during this phase (see inset). In the last part of the curve atomic loss is significantly reduced (trapping lifetime $\tau = 350$ s)

For both experiments corresponding to [4] and [5], relatively long trapping lifetimes have been observed. Figure 10.3 shows the number of atoms as a function of time in the ENS setup at a distance of 250 μm from the surface [11]. It clearly displays a double exponential behavior. Atoms are rapidly lost at the beginning of the sequence. During this time interval the cloud temperature decreases with the same characteristic time. The loss of atoms during that time interval is attributed to natural evaporation: because of the finite depth of the trap the hottest atoms produced by collision can escape while the remaining cloud cools down after rethermalization. After 20 s, the temperature remains constant and the trapping lifetime dramatically increases to 350(70) s. This value, in the minute range, is comparable to the one observed in [2]. If one assumes that the lifetime is limited by collisions with a residual background gas of He at 4 K, the corresponding pressure would be $\sim 1 \times 10^{-11}$ mbar. The actual pressure is probably much smaller than this value. From a systematic study of the trap lifetime as a function of the distance to the trapping wire, one concludes that the main source of loss here remains the technical current noise in the wire which induces magnetic Zeeman transitions towards untrapped states. This phenomenon is removed with the use of permanent current. In [5], atoms were observed for a time as long as 15 s. Surface evaporation takes place during the first second, followed by another regime where the atom number remains constant within an accuracy of 10% (corresponding lifetime ≥ 80 s). A detailed study of lifetime as a function of distance was recently published by the NTT group [12], that suggests that phenomena related to the superconducting nature of the surface, like magnetic field distortion resulting from flux penetration, can have a strong impact on the trap decay.

10.2.2.2 Forced Evaporation and Quantum Degeneracy

Further cooling of the atomic sample has been realized at ENS by using the standard technique of forced evaporative cooling [9]. In order to speed up the evaporation process, the elastic collision rate is first increased by an adiabatic compression of the cloud. After the compression, the Zeeman transition frequency towards untrapped states at the bottom of the trap potential is approximately 9 MHz. This value minimizes the influence of technical noise, present at lower frequencies, at the expense of a reduced confinement of the cloud. With these settings, the atoms are located 85 μm away from the surface. Calculations of the magnetic field give axial and longitudinal trapping frequencies equal to $2\pi \times 6\,\text{kHz}$ and $2\pi \times 100\,\text{Hz}$, respectively. A RF knife whose initial frequency corresponds to the energy of the hottest atoms in the trap is slowly ramped down. It is produced with the help of a current that is fed in an extra superconducting wire, located on the chip about 3.3 mm away from the atoms. Under these conditions, elastic collisions produce hot atoms which are expelled, while the mean temperature of the remaining sam-

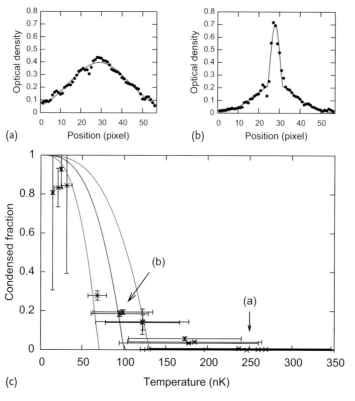

Figure 10.4 (a) Horizontal cross-section of the absorption image of the atomic cloud after 17 ms of expansion, for a final evaporation ramp frequency $f = 9.871\,\text{MHz}$. The solid line is a Gaussian fit. (b) Same as (a) with $f = 9.841\,\text{MHz}$. It shows the emergence of a condensed fraction, fitted by a Thomas–Fermi distribution superimposed on a thermal background (solid line). (c) Condensed fraction of atoms as a function of the cloud temperature T. The latter is varied by adjusting the final frequency of the evaporation ramp. The solid lines are theoretical fractions in the absence of interactions for the transition temperatures $T_c = 70$, 100, and 130 nK.

ple decreases. The frequency of the knife is ramped down in order to match the temperature decrease, and thus force the process.

Figure 10.4a,b show two cross-sections in the absorption image of the atomic cloud for the final RF frequencies $f = 9.871$ and 9.841 MHz, respectively. The magnetic trap is first adiabatically decompressed, then suddenly switched off. Both images are taken 17 ms after releasing the cloud. A sharp peak clearly appears on Figure 10.4b on top of a broader thermal Gaussian profile. This bimodal distribution is strong evidence for the presence of a BEC. The condensed part of the sample can be fitted by a 2D Thomas–Fermi distribution, whose amplitude is used to determine the fraction of condensed atoms for Figure 10.4c. The cloud temperature T on the abscissa of the graph has been obtained by fitting the width of the thermal fraction with a Gaussian. The solid lines in Figure 10.4c correspond to the fit of the condensed fraction in the absence of interaction between atoms for three different critical temperatures ($T_c = 70, 100$, and 130 nK), which are qualitatively compatible with the experimental points. The corresponding critical temperature and number of atoms at the onset of BEC are $T_c = 100(30)$ nK and $N = 1.0(5) \times 10^4$, respectively. Those results also prove that it is possible in present experiments to produce and observe almost pure condensates with about 1000 atoms.

10.3
Perspectives for Cryogenic Atom Chips: A New Realm of Investigations

We present now some experiments that could be undertaken in a near future with cryogenic chips. In Section 10.3.1 we show that many properties of the superconducting film can affect the magnetic trap. Hence atom chips offer a new toolbox for testing the different theoretical models of current flow and electrodynamic response in superconductors. Section 10.3.2 will be devoted to the integration of ultra-cold trapped atoms with standard superconducting elements like SQUIDs or coplanar resonators. Finally Section 10.3.3 will discuss the possibility to trap on a chip a single circular Rydberg atom excited from the ultra-cold cloud. Indeed those highly sensitive quantum states require cryogenic conditions for coherent operations over long times.

10.3.1
Probing the Superconducting Film Current Distribution

The results of Section 10.2 demonstrate that it is possible to produce an ultra-cold atomic cloud close to superconducting films at low temperature. The properties of such a surface are dramatically different from those of a normal metal. For example it is well known that a superconducting material has the ability to generate permanent currents in order to screen an applied magnetic field (Meissner effect). In the case of an atom chip, those currents are likely to affect the trapping potential [13] and reduce its depth when the atoms are brought very close to the surface. In the experiment of [13] atoms are trapped close to a relatively large cylindrical superconducting wire which is well described by the Meissner state. But most of the materials used in present chips (see Section 10.2.1.1 are thin films of superconductor in the mixed state, with a magnetic flux which partially penetrates through the material. This case is treated in [14] where the Bean Model of the superconducting mixed state is used to describe the current distribution. One of the most important predictions of this model is that the current flow maintains a memory of its previous physical states and hence exhibits a hysteretic behavior. This is very different from the Meissner case, where the current distribution corresponds to a thermal equilibrium and shows no memory effect. Evidence of this phenomenon has been recently observed with the ENS setup: permanent currents in the superconducting film strongly distort the trapping potential seen by the atoms as in [13]. However, the current distribution depends dramatically on the magnetic field applied to the chip at the time of its transition to the superconducting state. As a consequence it is possible to set completely different trap configurations by controlling the value of this applied field [15]. The agreement with the Bean Model is very good.

The presence of vortices could also introduce a corrugation of the trapping potential at distances of the order of the inter-vortex separation. The latter depends on the local applied magnetic field and ranges typically from 0.1 to a few μm [16]. It would be fascinating to access the distribution and the dynamics of those structures. Present microscopic magnetic-field imaging experiments with trapped BEC [17] on atom chips (see Chapter 7) could provide a way to make those observations.

The response of the superconducting film to an AC electromagnetic field is also of importance. Chapters 4 and 5 explain in great detail the interaction of an ultracold cloud with a surface at micrometer distance. Thermal current fluctuations in the film result in a random magnetic field at the cloud position. The latter can be relatively strong because the radiation takes place in the near-field regime. Some spectral components at RF frequency for the random field can induce atomic transitions towards untrapped high-field-seeking Zeeman sublevels, in which case the atom is expelled from the trap. The fluctuation-dissipation theorem lets us guess that the loss rate from the trap is directly related to the dissipation on the surface at the Zeeman transition frequency. The latter is in the kHz to MHz range, that is, much smaller than typical superconducting bandgap frequencies. Hence one expects to observe very low dissipation and significantly longer trapping times for superconductors compared to normal metals (for a detailed review, see Section 4.3.2). Moreover, measuring the trapping time close to superconducting bodies will offer a way to test the different theoretical models for the electrodynamics of superconductors [18, 19]. For example, the presence of vortices in the superconducting films used so far could possibly affect the spin-flip rate of the atoms. As long as the film is superconducting the vortices are pinned. However, the random hopping of the vortex lines from one pinning center to another at finite temperature generates an additional flux noise [16]. Moreover, the response of the vortex lattice to an electromagnetic wave at the Zeeman transition frequency adds an additional term of dissipation. Recent calculations suggest that it could be the dominant source of loss from the trap [20].

10.3.2
Integration of Atom Chips with Superconducting Circuit Elements

10.3.2.1 Coupling with a Superconducting Qubit

We have just seen that the presence of vortices in the superconducting film will probably play a crucial role in the atom–surface interaction. A clear imaging of the vortices as in [17] remains, however, very challenging. A possible way to overcome this difficulty could be to design a larger piece of superconductor (with dimensions around 10 µm) through which magnetic flux is quantized. This can be very easily achieved by making a simple loop of superconducting material. The most important point here is that, in the particular case where the loop is interrupted by a Josephson-junction (a very thin isolating barrier), it becomes possible to create quantum superpositions of different flux-states corresponding to different supercurrents in the loop. The characteristic energy scale for this phenomenon is the Josephson energy J. It can be precisely controlled by replacing the single junction by two split junctions, in which case the device is called a DC-SQUID. Such systems are very powerful quantum circuit elements. They can be operated in a fully coherent way if the temperature T is lower than J/k_B. They have proved to be very good candidates for the physical realization of a quantum bit (qubit) [21].

Reference [22] studies the interaction of a BEC close to a DC-SQUID. The experimental situation considered is presented in Figure 10.5. The diameter of the SQUID is 10 µm. A field B_0 of 100 mG perpendicular to the loop is enough to apply a half quantum of flux $\Phi_0/2 = h/4e$ through it. It is, therefore, easy to prepare it in a superposition of two flux-states. One can consider without loss of generality that the flux inside the loop is either 0 or Φ_0 (states $|0\rangle$ and $|1\rangle$). Hence the SQUID states correspond to two different values of supercurrent in the loop with equal amplitude and opposite directions. Let's assume now that atoms in a BEC are trapped 10 µm away from the loop. The magnetic field at the bottom of the trap can be adjusted with an external bias field along the x-direction. But the field created by the loop adds a contribution along the x-direction and distorts the potential. The displacement of the trap center depends on the supercurrent and one observes that the distance between the minima of the two resulting perturbed configurations associated to $|0\rangle$ and $|1\rangle$ is of the order of the diameter of the superconducting loop. Moreover, the amplitude of the field perturbation is 5 mG which is much larger than the typical chemical potential for the BEC presented in Section 10.2.2.2.

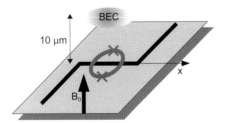

Figure 10.5 Experimental conditions of [22].

This means that the atoms of the BEC can be in either one of two distinguishable modes corresponding to the fundamental level for two well-separated traps. If one adiabatically brings the cloud close to the chip, the final quantum state for the SQUID+BEC system is entangled and can be written as [22]:

$$|\Psi\rangle = \frac{1}{\sqrt{2}}(|0\rangle \otimes |0, N\rangle_{01} + |1\rangle \otimes |N, 0\rangle_{01}),\qquad(10.1)$$

where $|0, N\rangle_{01}$ is the BEC state representing zero atoms in one perturbed configuration of the trap and N atoms in the other configuration. The final SQUID-BEC state is an entangled state of both systems. Strong correlations between the internal state of the SQUID and the position of the condensate are expected. However, if one tries to observe the atomic cloud directly, this operation corresponds to a local operation on one of the two subsystems. One needs to trace the different SQUID states and the BEC is then a mere statistical mixture of the two modes. It is, however, possible to go beyond that limitation. For example, manipulating the SQUID state and measuring it makes it possible to perform a projective measurement for the SQUID in the $|0\rangle \pm |1\rangle$ state basis [21]. Because of the strong correlations between the two parts of the entangled state, the BEC is then projected onto:

$$|\Psi\rangle_{BEC} = \frac{1}{\sqrt{2}}(|0, N\rangle_{01} \pm |N, 0\rangle_{01}),\qquad(10.2)$$

which is a coherent superposition of two macroscopically distinguishable states, or in other words a "Schrödinger-cat state". Such states are very important from the fundamental point of view as they provide a tool to explore the frontier between the microscopic and macroscopic worlds. For example, the loss of a single atom from the trap is enough to destroy the quantum superposition. Very long trapping lifetimes are therefore necessary. Another problem is that the magnetic potential seen by the atoms is changed as soon as the SQUID state is determined. As a consequence, the two modes of the BEC in Eq. (10.2) are no longer stationary and will evolve and interfere in a complex manner. Moreover, the coherent operation of the superconducting qubit requires that the performances of the cryogenic system are pushed to temperatures no larger than 100 mK. This is a very challenging and stimulating proposal as it combines two emerging fields of physics: atom chips and superconducting qubit manipulation.

10.3.2.2 Coupling with a Superconducting Resonator: On-Chip CQED

Many recent experiments on superconducting qubits have been realized by coupling them to an on-chip microwave resonator [23] (see Figure 10.6). The latter is made of a coplanar waveguide: a slab of superconducting material between two lateral ground planes which is interrupted at each end in order to create two "mirrors." It offers ideal conditions for observing the strong coupling regime: the mode volume is very small as the photon energy is "squeezed" in a 1D structure. The Q factor at low temperature ranges from 10^4 to 10^6. It now becomes possible to reproduce with solid-state devices all the experiments which had been pioneered with atomic systems in the field of cavity quantum electrodynamics (CQED).

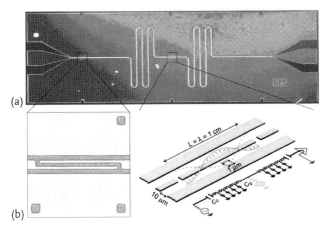

Figure 10.6 The circuit QED chip of [23] (a). A scheme of the principle of the chip is displayed at the bottom right. It couples a coplanar waveguide resonator with a superconducting qubit. The resonator is made of a film of superconducting niobium. A conductor is placed centrally between two lateral ground planes. In order to define a cavity, a "mirror" on both ends is made by cutting the line (b). The resonant frequency is 6 GHz (wavelength 5 cm, which explains the large dimensions of the chip). The Q factor is adjusted by tuning the capacitance of each mirror. It is of the order of 10^4 in that case.

On the other hand, those circuit QED cavities could become very powerful tools if one brings atomic systems into their vicinity. It appears quite natural to try to integrate them with atom chips. As an example, it was recently proposed to use a coplanar waveguide as a means to couple two separate polar molecules [24]. The latter would be electrically trapped on the chip.

In order to evaluate the achievable coupling for this kind of experiment, let us consider the case of an atom trapped at a distance z from the coplanar resonator of Figure 10.7. The atom could be magnetically trapped with the help of an additional on-chip wire (not shown in the figure) through which a DC current is flowing. Both electric and magnetic field amplitudes of the cavity mode decrease rapidly with distance from the surface [24]. The characteristic length of this decay is of the order of the gap w between the central conductor and the ground planes. As a consequence a significant coupling is possible only if $z \sim w$. We assume that this is the case and limit ourselves to gap distances ranging from 1 to 10 µm. In order to account for the decay of the field we will assume that the electric field at the cloud position E (resp. the magnetic field H) is only a fraction $\eta = 0.1$ of the maximum electric field in the chip plane E_0 (resp. H_0).

We will first study the hyperfine transition of ^{87}Rb. This particular transition has been used for demonstrating the coherent manipulation of trapped atoms on an atom chip [25]. The cavity mode frequency $\omega_0/2\pi$ is about 6.8 GHz. It is in the frequency range in which circuit QED cavities have been realized so far, with Q factors as high as 10^6. We present here a simplified calculation of the atom-cavity coupling. A more detailed study can be found in [26]. One can evaluate the vacuum Rabi coupling g following the same procedure as in [24, 27]. The zero-point energy

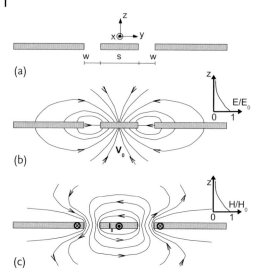

Figure 10.7 Cross-section of the coplanar resonator displaying the geometrical parameters of the transmission line (a). We consider the case where the central conductor width s is of the order of the gap w. The corresponding electric and magnetic field lines are shown in (b) and (c), respectively. Both field amplitudes decrease rapidly with a characteristic length of the order of w.

of the resonator can be written as:

$$\mathcal{E}_0 = \frac{1}{2}\hbar\omega_0 = \frac{1}{2}CV_0^2 + \frac{1}{2}LI_0^2, \qquad (10.3)$$

where V_0 and I_0 are the resonator voltage and current quantum fluctuations. C and L are the resonator characteristic capacitance and inductance. If the resonator is matched with standard microwave devices, its impedance is $Z = 50\,\Omega$ and $C = \pi/2\omega_0 Z$. For the considered frequency one gets $V_0 = \sqrt{\hbar\omega_0/2C} = 1.75\,\mu\text{V}$ and $I_0 = V_0/Z = 35\,\text{nA}$. One can then calculate the magnetic induction at the cloud position $B = \eta\mu_0 H_0 = \eta\mu_0 I_0/w = 4.4\,\mu\text{G}$ for $w = 10\,\mu\text{m}$. Because the hyperfine transition is only coupled through the magnetic dipole Hamiltonian the resulting vacuum Rabi frequency is $g = \mu_B B = 2\pi \times 6\,\text{Hz}$ (μ_B is the Bohr magneton). We are not in the strong coupling regime, g being 3 orders of magnitude smaller than the cavity linewidth $\kappa = \omega_0/Q = 2\pi \times 6.8\,\text{kHz}$ for $Q = 10^6$.

On the other hand the use of a hyperfine transition allows us to obtain very long atomic lifetime γ. In the case of present experiments, one can assume that γ is limited by transitions towards untrapped Zeeman levels or any other loss mechanism from the magnetic trap. In order to be conservative, we consider $\gamma = 0.1\,\text{s}^{-1}$ at 10 μm. The corresponding cooperativity factor $C_0 = Ng^2/(2\kappa\gamma)$, where N is the atom number coupled to the cavity, is equal to 0.03 for $N = 1$. This means that the onset of quantum effects is observable with about 30 atoms coupled to the cavity. A BEC with $N = 10^4$ atoms would clearly modify the cavity transmission spectrum. By reciprocity the atoms could act on the cavity field and shift its frequency, the cloud playing in a way the role of a dielectric slab whose index of refraction depends on the atomic internal state. The measurement of the resulting phase-shift by standard microwave homodyning techniques could then provide information on the internal state of the BEC [26, 28].

Turning now to the case of an electric dipole transition, one expects the coupling to be much larger. In the case of a single dipolar molecule (transition frequency around 10 GHz), [24] evaluates a vacuum Rabi frequency of 40 kHz at 1 μm from the surface, which fulfills the conditions for strong coupling. It could be possible also to replace the molecules by highly excited atoms in Rydberg states [27]. Experiments on circular Rydberg atoms, highly excited quantum levels with maximum angular momentum, have proven them extremely sensitive systems to electric and magnetic fields, both static and dynamic. As an illustration, the electric dipole on the transition between two adjacent circular levels with principal quantum number $n = 50$ and 51 ($\omega_0/2\pi = 50$ GHz, for more details see Section 10.3.3) is $d = 1776 \times ea_0$, where e is the charge of the electron and a_0 is the Bohr radius. Its coupling frequency to the cavity described above is then $g = \eta d E_0/\hbar = \eta d V_0/\hbar w = 2\pi \times 375$ kHz for $w = 10$ μm. A strong coupling regime could be observed if the Q factor does not degrade significantly at the transition frequency, which is still an open question.

10.3.3
Atom Chips for Circular Rydberg States

Circular Rydberg states are ideal tools for the investigation of numerous quantum phenomena [29]. Its remarkable sensitivity, in conjunction with its relative ease of detection, makes a single circular Rydberg atom an excellent probe of microwave field photon number ranging from one to a few tens [29] and a very good candidate for implementation of simple quantum information algorithms. A circular state with principal quantum number n can only emit a photon on a σ-polarized transition towards the lower circular state $n - 1$. For $n = 51$ the corresponding free-space radiative lifetime is 30 ms. However, it has been shown that cavity QED effects can dramatically inhibit the spontaneous emission and hence increase the state lifetime.

A possible trapping geometry was proposed in [30]. It is presented in Figure 10.8. Trapping is achieved with the help of the strong quadratic Stark polarizability α of levels e and g [$\alpha \equiv 200$ Hz/(V/m)^2 for weak fields] in a field $\mathbf{E}$ approximately aligned with Oz. Since circular states are high-field seekers, there can be no static field trap. Therefore, we turn to a dynamic scheme, reminiscent of the Paul ion traps (see Chapter 13), combining static and oscillating fields. Our design is comparable to that discussed in [31] for ground-state atoms. We note that atom chips have already demonstrated the ability to trap ground-state atoms with electric fields [32]. In both articles however, the required trapping fields are orders of magnitude larger than that necessary for trapping highly polarizable Rydberg states.

The trap is composed of two chips facing each other made up of concentric electrodes. Different voltages can be applied to the electrodes. In Figure 10.8 the static

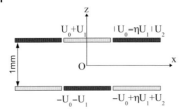

Figure 10.8 Section of the proposed circular Rydberg trap in a vertical xOz plane, with applied potentials. The trap has a cylindrical symmetry around Oz. The diameter of the inner electrode is 1 mm. The plate spacing, 1 mm, is appropriate for spontaneous emission inhibition. The electrodes are shaded according to the phase of the oscillating potential U_1.

potential U_0 creates the homogeneous directing field $E_0 \boldsymbol{u}_z$. The potential $U_1 < U_0$, oscillating at frequency ω_1, creates a smaller, AC, approximately hexapolar field $\boldsymbol{E}_1$. In order to cancel the amplitude of E_1 at the origin O for all times, we apply the potential $\pm \eta U_1$ to the outer electrodes. The factor η is determined by the electrode geometry. Calculations of [31] show that the resulting atomic Stark energy, $-\alpha (E_0 \boldsymbol{u}_z + \boldsymbol{E}_1)^2$, has a roughly quadratic spatial dependence around the origin O. It looks, however, like a saddle-point potential with trapping and anti-trapping directions depending on the sign of U_1. Stable trapping configurations are obtained only if one modulates U_1. Finally, the yet smaller static potential U_2 creates an approximately quadrupolar field. This provides a force, nearly constant in the trap region, compensating gravity (antiparallel to Oz).

The main advantage of the above geometry is that radiative lifetime for both levels can be greatly increased by inhibiting their spontaneous emission. Let us assume that the atom is placed between two parallel infinite conducting plates in a geometry similar to the one of Figure 10.8. We apply a voltage across the plates (voltage $\pm U_0$) in order to fix a quantization axis for the quantum levels. Hence both levels e and g only have a single decay channel: the emission of a millimeter-wave photon circularly polarized with respect to the quantization axis. This emission only couples to the transverse electric (TE) modes of the cavity made by the two plates. If the latter are separated by a distance $d < \lambda/2$ ($\lambda \sim 6$ mm being the emitted photon wavelength), there exists no TE mode at the transition frequency and the decay is inhibited. Levels e and g can simultaneously be made long-lived.

A detailed assessment of the trap performances is given in [33]. It relies on numerical simulations of the electric field created by the various voltages in the setup. Then we numerically integrate a set of atomic trajectories with randomly chosen initial conditions compatible with an ultra-cold cloud in a standard magnetic microtrap at an adjustable temperature T_0. For $U_0 = 0.2$ V, $U_2 = -0.003$ V and a hexapolar potential $U_1 = 0.155$ V oscillating at $\omega_1 = 2\pi \times 430$ Hz, the resulting motion is the combination of a fast micromotion, at frequency $\omega_1/2\pi$, with a slow oscillation whose longitudinal (along Oz) and transverse (orthogonal to Oz) frequencies are 175 and 64 Hz respectively. For such low frequencies the quantization axis for the atomic levels follows adiabatically the space- and time-dependent direction of the local electric field. At $T_0 = 90\,\mu K$, the motion has an extension of about 100 µm. We have checked that for $T_0 > 100\,nK$, the atomic excursion in the Rydberg trap is much larger than the de Broglie wavelength, allowing a classical treatment of the trajectory. The trap depth, T_d, which we define as the T_0 value for which half of the atoms remain within 400 µm of the origin, is $T_d = 180\,\mu K$, well within the reach of standard ultra-cold atom experiments. Deeper traps, with trapping frequencies in the kHz range, are achievable with higher but still moderate voltages (U_0 around 10 V).

In order to take full advantage of a Rydberg atom microtrap for quantum information or high-precision spectroscopy, one needs to keep a superposition of states e and g coherent over the longest possible time. The inhibition of spontaneous emission limits the effect of radiative decoherence for both states. Taking into account many possible imperfections to the inhibition, [33] estimates a radiative decoherence time of about $\tau_r = 3\,s$ in the particular case of a superposition of states e and g. The most important cause of decoherence is actually the absorption of blackbody radiation photons. The value of τ_r is reached only if the trap is cooled down to 1 K, corresponding to 0.1 mean thermal photon at the Rydberg transition frequency. Very low temperatures are therefore required.

However, the main source of dephasing in the microtrap is due to the slightly different Stark polarizabilities of levels e and g. Figure 10.9 shows the energy levels as a function of the electric-field amplitude. The frequency ν_{eg} of the transition has a strong dependence on the electric-field amplitude E, $\delta\nu_{eg} = \nu_{eg}(E) - \nu_{eg}(0) = -25.5\,Hz/(V/m)^2 \times E^2$. If one considers the case of an atom prepared from a cloud

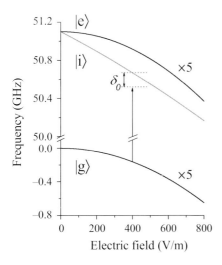

Figure 10.9 Energies of levels g, e, and i as a function of the electric field. Circular states g and e have a small, approximately quadratic Stark effect, magnified in this figure by a factor 5. Level i has a much larger, linear Stark effect. An additional microwave field (vertical arrow) predominantly mixes g and i and reduces the difference in Stark polarizability between e and the resulting dressed level $\tilde{g}$.

at 0.3 µK under the trapping conditions previously presented, it experiences over its trajectory a mean electric-field amplitude $E_a = 400$ V/m, with an excursion of $\Delta E = \pm 1$ V/m. The corresponding frequency broadening $\Delta \nu_{eg} = 2 \delta \nu_{eg} E_a \Delta E$ is of the order of 20 kHz. This broadening is inhomogeneous as it differs from one trajectory to another and could only be controlled by perfect control of the initial conditions of the atomic motion. The dephasing time associated to this motional broadening is of the order of a few tens of µs. In addition to this dephasing, the trapping frequencies (ω_ρ, ω_z) for states e and g differ by about 10%. The trajectories for the two states are, therefore, rapidly separated ('Stern–Gerlach' effect). Coherence would be lost should this separation exceed the wave-packet coherence length, of the order of the de Broglie wavelength (about 0.5 µm for the conditions considered).

Similar broadenings are observed in many traps for ground-state atoms, where the potential energy is also, in general, level-dependent. However, a proper choice of external parameters like bias field or laser frequency can reduce significantly this effect. This possibility was clearly demonstrated in [25] in the context of magnetic atom chips, paving the way for future atom chip clock devices (see Chapter 8). We proposed the artificial re-creation of a similar situation by shining an additional microwave field, linearly polarized, on the trapped atom [30]. To first approximation, this field couples g to the Rydberg state i ($n = 51$, $m = m_g = 49$), which experiences a large linear Stark effect [polarizability 1 MHz/(V/m)]. The Stark polarizability of the 'dressed' $\tilde{g}$ level is thus significantly modified. The dressing field has a smaller effect on e which is far off-resonance. The Rabi frequency Ω_0 and the detuning δ_0 of the dressing microwave can therefore be tailored to cancel the first- and to minimize the second-order terms in the expansion of $\nu_{eg}(E)$ around E_a. The calculation of the energy shift for levels $\tilde{g}$ and e is given in [33]. The optimal parameters Ω_0 and δ_0 have reasonable values and limit the variation of the 'dressed' transition frequency to ~ 10 Hz over the ± 1 V/m electric field range explored by an atom for $T_0 = 0.3$ µK. The level dressing reduces the transition broadening by over 3 orders of magnitude. Combined with standard refocusing techniques, like spin-echoes, it allows one to preserve the coherence of a state superposition for times as long as one second.

10.4
Conclusion

The operation of atom chips in a cryogenic environment has become a reality. The present experimental observations are very similar to the results usually found for room-temperature systems, although very long trapping times have already been observed. However, the understanding and control of permanent currents for trapping the cloud has already provided initial evidence that new phenomena, that only exist at low temperature, significantly affect the atomic ensemble. The coupling of superconducting micro- or nanostructures like SQUIDs or coplanar waveguides to the ultra-cold cloud will certainly lead to very interesting results in the coming years. We also think that cryogenic conditions offer an ideal environment around the atoms for preparation in exotic states like Rydberg levels. A new realm of investigations, with important results for both atomic and solid-state physics, is opening.

References

1 Weinstein, J.D. and Libbrecht, K.G. (1995) Microscopic magnetic traps for neutral atoms. *Phys. Rev. A* **52**, 4004.
2 Willems, P.A. and Libbrecht, K.G. (1995) Creating long-lived neutral-atom traps in a cryogenic environment. *Phys. Rev. A* **51**, 1403.
3 Walraven, J.T.M. and Silvera, I.F. (1981) Measurement of the static magnetization of spin-polarized atomic hydrogen with a SQUID magnetometer. *Physica B* **107**, 517.
4 Nirrengarten, T., Qarry, A., Roux, C., Emmert, A., Nogues, G., Brune, M., Raimond, J.-M., and Haroche, S. (2006) Realization of a superconducting atom chip. *Phys. Rev. Lett.* **97**, 200405.
5 Mukai, T., Hufnagel, C., Kasper, A., Meno, T., Tsukada, A., Semba, K., and Shimizu, F., Persistent supercurrent atom chip, *Phys. Rev. Lett.* **98**, 260407 (2007).
6 Müller, T., Wu, X., Mohan, A., Eyvazov, A., Wu, Y., and Dumke, R. (2008) Towards a guided atom interferometer based on a superconducting atom chip. *New J. Phys.* **10**, 073006.
7 Groth, S., Krüger, P., Wildermuth, S., Folman, R., Fernholz, T., Schmiedmayer, J., Mahalu, D., and Bar-Joseph, I. (2004) Atom chips: Fabrication and thermal properties. *Appl. Phys. Lett.* **85**, 2980.
8 Garfield, J.R.B. (1997) Superconducting contacts for use in niobium thin film applications. *Rev. Sci. Instrum.* **68**, 1906.
9 Roux, C., Emmert, A., Lupaşcu, A., Nirrengarten, T., Nogues, G., Brune, M., Raimond, J.-M. and Haroche, S. (2008) Bose–Einstein condensation on a superconducting atom chip. *Eur. Phys. Lett.* **81**, 56004.
10 Hänsel, W., Hommelhoff, P., Hänsch, T.W. and Reichel, J. (2001) Bose–Einstein condensation on a microelectronic chip. *Nature* (London) **413**, 498.
11 Emmert, A., Lupaşcu, A., Nogues, G., Brune, M., Raimond, J.-M., and Haroche, S. (2009) Measurement of the trapping lifetime close to a cold metallic surface on a cryogenic atom-chip. *Eur. Phys. J D* **51**, 173.
12 Hufnagel, C., Mukai, T. and Shimizu, F. (2009) Stability of a superconductive atom chip with persistent current. *Phys. Rev. A* **79**, 053641.
13 Cano, D., Kash, B., Hattermann, H., Koelle, D., Kleiner, R., Zimmermann, C., and Fortágh, J. (2008) Impact of the Meissner effect on magnetic microtraps for neutral atoms near superconducting thin films. *Phys. Rev. A* **97**, 063408.
14 Dikovsky, V., Sokolovsky, V., Zhang, B., Henkel, C. and Folman, R. (2009) Superconducting atom chips: Advantages and challenges. *Eur. Phys. J. D* **51**, 247.
15 Emmert, A., Lupaşcu, A., Brune, M., Raimond, J.-M., Haroche, S. and Nogues, G. (2009) Microtraps for neutral atoms using superconducting structures in the critical state. *Phys. Rev. A* **80**, 061604.
16 Scheel, S., Fermani, R. and Hinds, E.A. (2007) Feasibility of studying vortex noise in two-dimensional superconductors with cold atoms. *Phys. Rev. A* **75**, 064901.
17 Wildermuth, S., Hofferberth, S., Lesanovsky, I., Haller, E., Andersson, M.L., Groth, S., Bar-Joseph, I., Krüger, P., and Schmiedmayer, J. (2005) Bose–Einstein condensates: Microscopic magnetic field imaging. *Nature* **435**, 440.
18 Hohenester, U., Eiguren, A., Scheel, S., and Hinds, E.A. (2007) Spin-flip lifetimes in superconducting atom chips: Bardeen–Cooper–Schrieffer versus Eliashberg theory. *Phys. Rev. A* **76**, 033618.

19 Skagerstam, B.K. and Rekdal, P.K. (2007) Photon emission near superconducting bodies. *Phys. Rev. A* **76**, 052901.

20 Nogues, G., Roux, C., Nirrengarten, T., Lupaşcu, A., Emmert, A., Brune, M., Raimond, J.-M., Haroche, S., Plaçais, B., and Greffet, J.-J. (2009) Effect of vortices on the spin-flip lifetime of atoms in superconducting atom-chips. *EPL* **87**, 13002.

21 Clarke, J. and Wilhelm, F.K. (2008) Superconducting quantum bits. *Nature* **453**, 1031.

22 Singh, M. (2009) Macroscopic entanglement of a Bose–Einstein condensate on a superconducting atom chip. *Opt. Express* **17**, 2600.

23 Wallraff, A., Schuster, I.D., Blais, Frunzio, A.L., Huang, R.-S., Majer, J., Kumar, S., Girvin, S.M., and Schoelkopf, R.J. (2004) Circuit quantum electrodynamics: Coherent coupling of a single photon to a Cooper pair box. *Nature* (London) **431**, 162.

24 André, A., DeMille, D., Doyle, J.M., Lukin, M.D., Maxwell, S.E., Rabl, P., Schoelkopf, R.J., and Zoller, P. (2006) A coherent all-electrical interface between polar molecules and mesoscopic superconducting resonators. *Nat. phys.* **2**, 636.

25 Treutlein, P., Hommelhoff, P., Steinmetz, T., Hänsch, T.W., and Reichel, J. (2004) Coherence in microchip traps. *Phys. Rev. Lett.* **92**, 203005.

26 Verdú, J., Zoubi, H., Koller, C., Majer, J., Ritsch, H., and Schmiedmayer, J. (2009) Strong magnetic coupling of an ultracold gas to a superconducting waveguide cavity. *Phys. Rev. Lett.* **103**, 043603

27 Sorensen, S.A., van der Wal, C.H., Childress, L.I., and Lukin, M.D. (2004) Capacitive coupling of atomic systems to mesoscopic conductors. *Phys. Rev. Lett.* **92**, 063601.

28 Wallraff, A., Schuster, D.I., Blais, A., Frunzio, L., Majer, J., Devoret, M.H., Girvin, S.M., and Schoelkopf, R.J. (2005) Approaching unit visibility for control of a superconducting qubit with dispersive readout. *Phys. Rev. Lett.* **95**, 060501.

29 Raimond, J.-M., Brune, M., and Haroche, S. (2001) Manipulating quantum entanglement with atoms and photons in a cavity. *Rev. Mod. Phys.* **73**, 565.

30 Hyafil, P., Mozley, J., Perrin, A., Tailleur, J., Nogues, G., Brune, M., Raimond, J.-M., and Haroche, S. (2004) Coherence preserving trap architecture for long-term control of giant Rydberg atoms. *Phys. Rev. Lett.* **93**, 103001.

31 Peik, E. (1999) Electrodynamic trap for neutral atoms. *Eur. Phys. J. D* **6**, 179.

32 Krüger, P., Luo, X., Klein, M.W., Brugger, K., Haase, A., Wildermuth, S., Groth, S., Bar-Joseph, I., Folman, R., and Schmiedmayer, J. (2003) Trapping and manipulating neutral atoms with electrostatic fields. *Phys. Rev. Lett.* **91**, 233201.

33 Mozley, J., Hyafil, P., Nogues, G., Brune, M., Raimond, J.-M., and Haroche, S. (2005) Trapping and coherent manipulation of a Rydberg atom on a microfabricated device: a proposal. *Eur. Phys. J. D* **35**, 43.

34 Hulet, R.G., Hilfer, E.S., and Kleppner, D. (1985) Inhibited spontaneous emission by a Rydberg atom. *Phys. Rev. Lett.* **55**, 2137.

11
Atom Chips and One-Dimensional Bose Gases

I. Bouchoule, N.J. van Druten, and C.I. Westbrook

11.1
Introduction

As this book indicates, the technology of atom chips is currently enjoying great success in a large variety of experiments on degenerate quantum gases. Because of their geometry and their ability to create highly confining potentials, they are particularly well adapted to realizing one-dimensional situations [1–9]. This characteristic has contributed to a revival of interest in the study of 1D Bose gases with repulsive interactions, a system which provides a vivid example of an exactly solvable quantum many-body system [10–12]. The quantum many-body eigenstates [10, 11] and thermodynamics [12] can be calculated without resorting to approximations. In addition, the 1D Bose gas shows a remarkably rich variety of physical regimes that are very different both from those found in 2D and in 3D. One dramatic example of the difference is the tendency for a 1D Bose gas to become more *strongly* interacting as its density *decreases* [10]. Finally, and in a more practical vein, a good understanding of its behavior is relevant for guided-wave atom lasers [13] and trapped-atom interferometry [14]. Because of the effects of interactions, the analogy to the manipulation of light in single-mode fibers needs to be examined carefully.

An atom chip is not the only means of producing a 1D Bose gas. Optical trapping has been used to generate similarly elongated trap geometries. In particular, a 2D optical lattice can be used to generate a 2D array of 1D tubes [15–19]. Because of the massively parallel nature of this system, it is possible to work with only a few atoms per tube, and still get a sizeable signal per experimental cycle. Thus, the strongly interacting regime alluded to above can be reached. This regime has yet to be reached with an atom chip. But as we will show here, a key feature of atom chips is that they produce individual samples in which one does not intrinsically average over many realizations. Fluctuation phenomena are, therefore, readily accessible, an aspect which we will treat later in this chapter.

In the following we first give an introduction to the various regimes of the homogeneous 1D Bose gas, with particular emphasis on the behavior of the density profiles and the density fluctuations in the context of approximate models. Then we

Atom Chips. Edited by Jakob Reichel and Vladan Vuletić
Copyright © 2011 WILEY-VCH Verlag GmbH & Co. KGaA, Weinheim
ISBN: 978-3-527-40755-2

will discuss the exact solution and how it differs from the approximations. Next, we discuss some of the important issues involved in realizing 1D gases in a 3D trap. Finally, we describe a series of experiments performed in Orsay and Amsterdam using atom chips to explore and illustrate features of the 1D Bose gas.

11.2
Regimes of One-Dimensional Gases

First, we review some theoretical results concerning the one-dimensional Bose gas with repulsive interactions. Most of these results are derived in [10, 12, 20–24]. Here we will concentrate on intuitive arguments, and the reader is referred to the above references for more careful demonstrations. The system is described by the Hamiltonian

$$H = -\frac{\hbar^2}{2m}\int dz\, \psi^+ \frac{\partial^2}{\partial z^2} \psi + \frac{g}{2}\int dz\, \psi^+ \psi^+ \psi \psi , \tag{11.1}$$

where ψ is the field operator in second quantization, and g is the coupling constant characterizing the interactions between particles. From this coupling constant, one can deduce an intrinsic length scale related to the interactions,

$$l_g = \frac{\hbar^2}{mg}, \tag{11.2}$$

as well as an energy scale:

$$E_g = \frac{mg^2}{2\hbar^2} = \frac{\hbar^2}{2ml_g^2}. \tag{11.3}$$

In thermal equilibrium, the gas is described by the temperature T and the linear atomic density n. Rescaling these two quantities by the intrinsic scales introduced above, and setting Boltzmann's constant equal to unity (i.e., measuring temperature in units of energy) we find that the properties of the gas are functions of the dimensionless quantities

$$t = \frac{T}{E_g}, \tag{11.4}$$

and

$$\gamma = \frac{mg}{\hbar^2 n} = \frac{1}{nl_g}, \tag{11.5}$$

the latter being the famous Lieb–Liniger parameter [10].

It is useful to also introduce two other relevant scales, namely the thermal de Broglie wavelength,

$$\lambda_{\mathrm{dB}} = \hbar\sqrt{\frac{2\pi}{mT}} = l_g\sqrt{\frac{4\pi}{t}}, \tag{11.6}$$

and the quantum degeneracy temperature

$$T_d = \frac{\hbar^2 n^2}{2m} = \frac{E_g}{\gamma^2}.$$ (11.7)

In the above (t, γ) parametrization, quantum degeneracy ($T \approx T_d$, or equivalently $n\lambda_{dB} \approx 1$) is reached around

$$t \approx \frac{1}{\gamma^2}.$$ (11.8)

The thermal equilibrium for the Hamiltonian of Eq. (11.1) has been extensively studied theoretically [12, 22]. Without going into great detail however, we can present some important features of this system. Several regimes may be identified in the parameter space (γ, t), as sketched in Figure 11.1. We begin by noting that the region $\gamma \gg 1$, $t \ll 1$ (dark gray area) defines a strongly interacting regime that occurs at low density and low temperature, often referred to as the Tonks–Girardeau gas [20, 25, 26].

In the weakly interacting regime, $\gamma < 1$, several subregimes are identified. These are the regimes which to date have been accessible in atom chip experiments, and we shall elaborate further on their nature in the discussion below. The two main regimes are the nearly ideal gas regime (white area) and the quasi-condensate regime (light gray area). Each one permits an approximate description that we

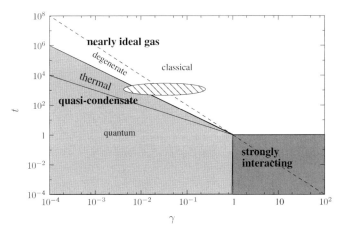

Figure 11.1 Physical regimes of a 1D Bose gas with repulsive contact interactions in the parameter space (γ, t), adapted from [22]. The dashed diagonal line separates the degenerate and non-degenerate gases. The strongly interacting regime is shown in dark gray. The weakly interacting regime is divided into the nearly ideal gas regime (also called decoherent regime) shown in white and the quasicondensate regime shown in light gray. Note that the nearly ideal gas can be degenerate. The quasi-condensate regime is divided into the thermal and quantum regimes. The lines represent smooth (and often wide) crossovers rather than phase transitions. The crossovers are given in Eqs. (11.10), (11.11), (11.8), (11.25), and (11.41). The dashed area shows the parameter space investigated in the experiments presented in this chapter.

present later in this section and which allows the identification of subregimes. For the moment we simply wish to emphasize that no phase transition occurs in the 1D Bose gas and that all the boundaries represent smooth (and often broad) crossovers in behavior.

11.2.1
Strongly versus Weakly Interacting Regimes

We first comment on the distinction between strong and weak interactions. Following the approach of [20], we study the scattering wave function of two atoms interacting via the potential $g\delta(z_1 - z_2)$, where z_1 and z_2 are the position of the two atoms. For this, we consider the wave function ψ in the center-of-mass frame, with reduced mass $m/2$ and subject to the potential $g\delta(z)$. The effect of the potential is described by the continuity condition

$$\frac{\partial}{\partial z}\psi(0_+) - \frac{\partial}{\partial z}\psi(0_-) = \frac{mg}{2\hbar^2}\psi(0), \qquad (11.9)$$

where 0_+ (0_-) denotes the limit when z goes to zero through positive (negative) values. Let us consider the scattering solution for an energy $E = \hbar^2 k^2/m$. Since we consider bosons, we look for even wave functions of the form $\cos(k|z| + \phi)$. The continuity conditions give ϕ and thus the value $\psi(0)$. We find then that the energy E_g given by Eq. (11.3) is the relevant energy scale and that for $E \ll E_g$, $\psi(0)$ is close to zero, while, for $E \gg E_g$, $\psi(0)$ is close to one, as illustrated in Figure 11.2.

The above results hold for a gas of particles since the continuity relation (11.9) holds for the many-body wave function when two atoms are close to the same place. Thus, as long as the typical energy of the particles is much lower than E_g, the many-body wave function vanishes when two particles are at the same position: the gas is then in the strongly interacting, or Tonks–Girardeau regime. The vanishing of the wave function when two particles are at the same place mimics the Pauli exclusion principle and the gas acquires some similarities with a gas of non-interacting fermions. More precisely, in this strong interaction regime, the available wave functions of the many-body problem are, up to a symmetrization factor, the wave functions of an ideal Fermi gas [26]. Since the wave function vanishes when two atoms

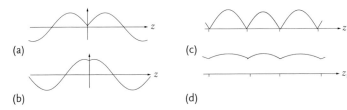

Figure 11.2 Strong interaction vs. weak interaction regime. We show the wave function in the center-of-mass frame of two atoms for (a) strong interactions, scattering energy E much smaller than $E_g = mg^2/2\hbar^2$ and (b) weak interactions, E much larger than E_g. We also plot the wave function $\psi(z_1, z_2, z_3, \ldots)$ for given positions of $z_2, z_3, \ldots$ in (c) the strongly interacting regime and (d) the weakly interacting regime.

are at the same place, the energy of the system is purely kinetic energy and the eigenenergies are those of the Fermi system. Thus, the 1D strongly interacting Bose gas and the ideal 1D Fermi gas share the same energy spectrum. This implies in particular that all thermodynamic quantities are identical for both systems.

To identify the parameter space of the strongly interacting regime, we suppose the gas to be strongly interacting and then require that the typical energy of the atoms be smaller than E_g. To estimate the typical energy per atom, we use the Bose–Fermi mapping presented above. If the gas is degenerate, the temperature is smaller than the degeneracy temperature T_d, Eq. (11.7), and T_d corresponds to the "Fermi" energy of the atoms. The typical atom energy is therefore T_d and it is of order E_g if

$$\gamma \simeq 1. \tag{11.10}$$

The strongly interacting regime thus requires $\gamma \gg 1$. If the gas is nondegenerate, the typical energy of the equivalent Fermi gas is T and interactions become strong when $T = E_g$ or

$$t \simeq 1. \tag{11.11}$$

We then find that the gas is strongly interacting for $t \ll 1$.

The condition (11.10) is often derived using the following alternative argument, valid at zero temperature. At zero temperature, there are two extremes for the possible solutions for the wave function $\psi(z_1, z_2, \ldots)$. As seen in Figure 11.2, either the wave function vanishes when two atoms are at the same place, or the wave function is almost uniform, corresponding to the strongly and weakly interacting configurations, respectively. In the weakly interacting configuration, the kinetic energy is negligible and the interaction energy per particle, of the order of gn, determines the total energy. In the strongly interacting configuration, on the other hand, the interaction energy vanishes while the typical kinetic energy per particle is $\hbar^2 n^2/m$. Comparing these two energies, we find that the strongly interacting configuration is favorable only for $\gamma > 1$.

11.2.2
Nearly Ideal Gas Regime

At sufficiently high temperatures, interactions between atoms have little effect and the gas is well described by an ideal Bose gas. In [22], this regime was referred to as the "decoherent regime"; we will call it the (nearly) ideal Bose gas regime. A 1D ideal Bose gas at thermal equilibrium is well described using the grand canonical ensemble, introducing the chemical potential μ. All properties of the gas are calculated using the Boltzmann law which states that, for a given one-particle state of momentum $\hbar k$, the probability to find N atoms in this state is proportional to $e^{-(\hbar^2 k^2/(2m)-\mu)N/T}$; note that $\mu < 0$ in this description. In the following, we use a quantization box of size L (tending to infinity in the thermodynamic limit) and periodic boundary conditions so that the available states are the momentum states with momentum $k = 2\pi j/L$ where j is an integer.

Let us first consider the linear gas density. From the Boltzmann law, we find that the mean population is the Bose distribution

$$\langle n_k \rangle = \frac{1}{e^{\left(\frac{\hbar^2 k^2}{2m} - \mu\right)/T} - 1} . \tag{11.12}$$

The atom number, and thus the linear density, is obtained by summing the population over the states and one finds

$$n = \frac{1}{\lambda_{dB}} g_{1/2}(e^{\frac{\mu}{T}}) , \tag{11.13}$$

where $g_{1/2}(x)$ is one of the Bose functions

$$g_n(x) = \sum_{l=1}^{\infty} \frac{x^l}{l^n} , \tag{11.14}$$

also known as the polylogarithmic functions [27, 28]. Unlike in 3D systems, where the excited-state density is given by $\rho_e = g_{3/2}(e^{\mu/T})/\lambda_{dB}^3$ in this approach [27], no saturation of the excited states occurs (the function $g_{1/2}$ diverges as $\sqrt{-\pi T/\mu}$ as $\mu \to 0$ from below, whereas $g_{3/2}(1) = 2.612$ is finite): in the thermodynamic limit no Bose–Einstein condensation is expected and the gas is well described by a thermal gas at any density.

Two asymptotic regimes may be identified: the non-degenerate regime for which $-\mu \gg T$ and $\hbar^2 n^2/m \ll T$ and the degenerate regime for which $-\mu \ll T$ and $\hbar^2 n^2/m \gg T$. In the non-degenerate regime, the linear density is well approximated by the Maxwell–Boltzmann formula

$$n = \frac{1}{\lambda_{dB}} e^{\frac{\mu}{T}} . \tag{11.15}$$

In this regime $n\lambda_{dB}$ is much smaller than unity. In the degenerate regime, the states of energy much smaller than T are highly occupied and the linear density is given by

$$n = \frac{T}{\hbar} \sqrt{\frac{m}{-2\mu}} . \tag{11.16}$$

This density is much larger than $1/\lambda_{dB}$, that is $n\lambda_{dB} \gg 1$.

As we will discuss in the experimental section, fluctuations are also very important for characterizing the gas. It is thus instructive to consider the correlation functions. The normalized one body correlation function is $g^{(1)}(z) = \langle \psi^+(0)\psi(z)\rangle/n$, where ψ is the field operator in the second quantization picture. Using the expansion of the field operator in the plane wave basis $\psi(z) = \sum_k a_k e^{-ikz}/\sqrt{L}$ where a_k is the annihilation operator for the mode k, we find $g^{(1)}(z) = \sum_k \langle n_k \rangle e^{-ikz}/(Ln)$. Here $n_k = a_k^+ a_k$ is the atom number operator for the mode k. Simple analytical expressions are found in the nondegenerate and highly degenerate limits. In the non-degenerate limit ($-\mu \gg T$ or, equivalently $n \ll 1/\lambda_{dB}$), we find

$$g^{(1)}(z) \simeq e^{-\pi \frac{z^2}{\lambda_{dB}^2}} . \tag{11.17}$$

As the gas becomes more degenerate, the correlation length increases and, in the degenerate regime ($-\mu \ll T$ or, equivalently $n \gg 1/\lambda_{dB}$), we find

$$g^{(1)}(z) \simeq e^{-\frac{mTz}{n\hbar^2}} = e^{-\frac{2\pi z}{n\lambda_{dB}^2}}. \tag{11.18}$$

In this regime the correlation length, about $n\lambda_{dB}^2$, is much larger than the de Broglie wavelength (and the mean inter-particle distance $1/n$) since $\lambda_{dB} \gg 1/n$.

Next we consider the normalized density-density or two-body correlation function

$$g^{(2)}(z) = \langle \psi^+(z)\psi^+(0)\psi(0)\psi(z) \rangle / n^2. \tag{11.19}$$

This function is proportional to the probability of finding an atom at position z and another atom at position $z = 0$. It is given by

$$n^2 g^{(2)}(z) = \sum_{k_1 k_2 k_3 k_4} \langle a_{k_1}^+ a_{k_2}^+ a_{k_3} a_{k_4} \rangle e^{ik_1 z} e^{-ik_4 z} / L^2. \tag{11.20}$$

Using Bose commutation relations and the fact that, since atoms do not interact, different momentum state populations are uncorrelated, the sum simplifies to:

$$n^2 g^{(2)}(z) = \sum_{k_1 \neq k_2} \langle n_{k_1} \rangle \langle n_{k_2} \rangle \left(1 + e^{i(k_1-k_2)z}\right) / L^2 + \sum_{k} \langle a_k^+ a_k^+ a_k a_k \rangle / L^2. \tag{11.21}$$

In the last term, the commutation relations give: $\langle a_k^+ a_k^+ a_k a_k \rangle = \langle n_k^2 \rangle - \langle n_k \rangle$, and in thermal equilibrium one has:

$$\langle n_k^2 \rangle = \langle n_k \rangle + 2\langle n_k \rangle^2. \tag{11.22}$$

Therefore we find:

$$g^{(2)}(z) = 1 + \left|g^{(1)}(z)\right|^2, \tag{11.23}$$

a result which one can also obtain directly from Wick's theorem [29]. Equation (11.23) means that the probability of finding atoms within less than a correlation length in a thermal Bose gas is twice that of finding two atoms far apart. This phenomenon is often referred to as "bunching" and has been observed in cold atoms in several experiments [30–32]. Bunching is closely related to density fluctuations. As one can see from Eq. (11.22), in a thermal gas, fluctuations in the occupation of a single quantum state, $\delta n_k^2 = \langle n_k^2 \rangle - \langle n_k \rangle^2$, show a "shot noise" term, $\langle n_k \rangle$ and an "excess noise" term, $\langle n_k \rangle^2$. The density fluctuation experiment described later in this chapter has demonstrated this behavior.

Validity of the ideal gas treatment. The two-body correlation function has been used to characterize the crossover between the ideal gas and quasi-condensate regimes [22]. When interactions become important, they impose an energy cost on density fluctuations and the latter tend to smooth out. This amounts to a reduction in the value of $g^{(2)}(0)$. In the quasi-condensate regime which we discuss in the next section, the bunching effect is absent and $g^{(2)}(0)$ is close to unity. The ideal Bose gas description fails when the typical interaction energy per particle gn is not negligible compared to $-\mu$. Using Eq. (11.16) one finds that the ideal Bose gas description fails when the temperature is no longer much smaller than the crossover temperature, which we define as

$$T_{co} = T_d \sqrt{\gamma} . \tag{11.24}$$

Using the reduced dimensionless temperature $t = T/E_g$, this can be written as

$$t_{co} = \frac{1}{\gamma^{3/2}} . \tag{11.25}$$

This line separates the nearly ideal gas regime from the quasi-condensate regime in Figure 11.1. Note that, in terms of chemical potential, the domain of validity of the ideal gas model is $-\mu \gg \mu_{co}$ where we define the crossover chemical potential as

$$\mu_{co} = \frac{T}{t^{1/3}} . \tag{11.26}$$

In making this estimate, we have assumed that the gas is degenerate at the crossover. From Eq. (11.24), one can see that if one is in the weakly interacting regime ($\gamma \ll 1$) this assumption is indeed true. The experiments described below confirm that one can observe the effects of degeneracy before the onset of the reduction of density fluctuations.

A precursor of the reduction of density fluctuations is shown by a perturbative calculation valid in the nearly ideal gas regime which gives, to lowest order in g [22],

$$g^{(2)}(0) \simeq 2 - 4(T_{co}/T)^2 . \tag{11.27}$$

To accurately treat the crossover regime however, it is necessary to make use of the exact solution to the 1D Bose gas model. The exact solution in the crossover regime is discussed in Section 11.2.4.

The correlation lengths of the gas are important parameters of the gas that will be used in the following to estimate the validity criteria of the local density approximation. In the degenerate regime, the correlation length is $l_c \simeq n\lambda_{dB}^2$ (see Eq. (11.18)). Using Eq. (11.24), we find that, close to the crossover, the correlation length of the gas is close to the healing length

$$\xi = \frac{\hbar}{\sqrt{mgn}} . \tag{11.28}$$

11.2.3
Quasi-Condensate Regime

On the other side of the crossover, that is for $T \ll T_{co}$, the bunching effect is entirely suppressed and the $g^{(2)}$ function is close to unity for any z. This regime is the quasi-condensate regime.[26] In this section, we present a description of the gas, valid in the quasi-condensate regime. This description permits a simple estimate of the density fluctuations. We thus verify *a posteriori* that the quasi-condensate regime is obtained for $T \ll T_{co}$. We also give a simple calculation of phase fluctuations in the quasi-condensate regime.

In the quasi-condensate regime density fluctuations are strongly reduced compared to their value in an ideal Bose gas where the bunching effect is responsible for density fluctuations of the order of n^2. In other words:

$$\delta n^2 \ll n^2. \tag{11.29}$$

In this regime, a suitable description is realized by writing the field operator as $\psi = e^{i\theta}\sqrt{n + \delta n}$ where the real number n is the mean density and the operator δn and the phase operator θ are conjugate: $[\delta n(z), \theta(z')] = i\delta(z - z')$. Note that the definition of a local phase operator is subtle and the condition Eq. (11.29) is not well defined since, because of shot noise, δn^2 is expected to diverge in a small volume. A rigorous and simple approach consists in discretizing the space so that in each cell a large number of atoms is present while the discretization step is much smaller than the correlation length of density and phase fluctuations [33].

Following this prescription, one first minimizes the grand canonical Hamiltonian $H - \mu N$ with respect to n to obtain the equation of state

$$\mu = gn. \tag{11.30}$$

To second order in δn, this is the correct expression of the chemical potential. This equality ensures that the Hamiltonian has no linear terms in δn and $\nabla \theta$. Linearizing the Heisenberg equations of motion in δn and $\nabla \theta$, we obtain [33]

$$\begin{cases} \hbar \frac{\partial \theta}{\partial t} = -\frac{1}{2\sqrt{n}} \left(-\frac{\hbar^2}{2m} \Delta + 2gn \right) \frac{\delta n}{\sqrt{n}} \\ \hbar \frac{\partial \delta n}{\partial t} = 2\sqrt{n} \left(-\frac{\hbar^2}{2m} \Delta \right) \theta \sqrt{n} \end{cases} \tag{11.31}$$

These equations are the so-called hydrodynamic equations. They are derived from a Hamiltonian quadratic in δn and $\nabla \theta$, that can be diagonalized using the Bogoliubov procedure [33]. It is not the purpose of this chapter to provide detail on this calculation and to give exact results within this theory. We will simply give arguments that enable an estimate of the density fluctuations and of their correlation length. This estimate will then be used to check that $\delta n^2 \ll n^2$, as assumed in Eq. (11.29). We will show that this condition is the same as the condition $T \ll T_{co}$ where T_{co} is given in Eq. (11.24). After that, we will give similar arguments to estimate the phase fluctuations. Since in the following we will study the gas properties versus the chemical potential, it is instructive to rewrite the condition $T \ll T_{co}$ in terms of chemical potential. Using Eq. (11.30), we find that the quasi-condensate regime is valid as long as $\mu \gg \mu_{co}$ where μ_{co} is given by Eq. (11.26).

26) It is also called the coherent regime since the $g^{(2)}$ function is close to unity, as in a coherent state. On the other hand, the first-order correlation function still decays and so the gas is not strictly coherent in this sense. Within this terminology, the ideal Bose gas regime is called the decoherent regime [22].

11.2.3.1 Density Fluctuations

To estimate the density fluctuations introduced by the excitations, it is convenient to divide the excitations into two groups: the excitations of low wave vector for which the phase representation is most appropriate and the excitations of high wave vector for which a particle point of view is most convenient.

In the following, we use the expansions on sinusoidal modes $\theta = \sum_{k>0} \sqrt{2} (\theta_{ck} \cos(kz) + \theta_{sk} \sin(kz))$ and $\delta n = \sum_{k>0} \sqrt{2} (\delta n_{ck} \cos(kz) + \delta n_{sk} \sin(kz))$. Here δn_{jk} and θ_{jk} are conjugate variables ($[\delta n_{jk}, \theta_{j'k'}] = (i/L)\delta_{jj'}\delta_{kk'}$) where j stands for c or s. For modes of small wave vector k, the excitations are phonons, or density waves, for which the relative density modulation amplitude $\delta n_{jk}/n$ is much smaller than the phase modulation amplitude θ_{jk}. In this case, the local velocity of the gas is given by $\hbar \nabla \theta / m$ and the kinetic energy term is simply $Ln\hbar^2 k^2 \theta_{jk}^2/(2m)$. The Hamiltonian for this mode then reduces to

$$H_{jk} = L \left(\frac{g \delta n_{jk}^2}{2} + n\hbar^2 k^2 \frac{\theta_{jk}^2}{(2m)} \right). \tag{11.32}$$

This Hamiltonian could also have been derived from the equations of motion given in Eq. (11.31), provided that the quantum pressure term $\hbar^2/(2m)\Delta\delta n/n$ is neglected: indeed, for a given wave vector k, the Laplacians in Eq. (11.31) give a factor k^2 and Eq. (11.31) are simply the equations of motion derived from the Hamiltonian Eq. (11.32). For temperatures much larger than ng, the thermal population of these phonon modes is large and classical statistics apply. Thus, the mean energy per quadratic degree of freedom is $T/2$ and we obtain

$$\langle \delta n_{jk}^2 \rangle = \frac{T}{(Lg)}. \tag{11.33}$$

and

$$\langle \theta_{jk}^2 \rangle = \frac{mT}{(Lnk^2\hbar^2)}. \tag{11.34}$$

We can now check the validity of the assumption $\delta n_k/n \ll \theta_k$: it is valid as long as $k \ll \sqrt{mgn}/\hbar$. Since k values are spaced by $2\pi/L$, there are about $L\sqrt{mgn}/(\pi\hbar)$ modes that satisfy this condition. Since the contribution of each of these modes to the relative density fluctuations is given in Eq. (11.33), we find that the contribution of these low momentum excitations to the relative density fluctuations is of the order of

$$\frac{\langle \delta n^2 \rangle_{\text{phonons}}}{n^2} \simeq \frac{T}{n\hbar\sqrt{\frac{gn}{m}}} \simeq \frac{T}{T_d\sqrt{\gamma}} \simeq \frac{T}{T_{\text{co}}}. \tag{11.35}$$

For wave vectors much larger than $\sqrt{mgn}/\hbar$, the phase-density representation is not the most appropriate. An excitation of wave vector $k \gg \sqrt{mgn}/\hbar$ corresponds to the presence of an atom of momentum k, whose wave function is $e^{ikz}/\sqrt{L}$ and whose energy is $\hbar^2 k^2/(2m)$. The annihilation operator for this mode is a_k

as introduced in Section 11.2.2. For temperatures much larger than $\hbar^2 k^2/m$, the thermal population of this mode is large and classical field theory, in which a_k is treated as a c-number, is adequate. We then find that a_k has a Gaussian distribution which satisfies $\langle |a_k|^2 \rangle = 2mT/(\hbar^2 k^2)$. The density fluctuations caused by the presence of such high momentum atoms result mainly from the interference between the atomic field $a_k e^{ikz}/\sqrt{L}$ and the atomic field of long wavelength spatial variations, whose amplitude is close to $\sqrt{n}$. The density fluctuations are thus $\delta n = \sqrt{n}(a_k e^{ikz} + a_k^* e^{-ikz})/\sqrt{L}$. We then find that the contribution of the mode of wave vector k to density fluctuations is $\delta n_k^2 = 4nmT/(L\hbar^2 k^2)$. Summing the contributions of the modes for all $k > \sqrt{mgn}/\hbar$, we obtain an estimate of the density fluctuations $\langle \delta n^2 \rangle_{\text{atoms}}$ caused by high momentum excitations:

$$\frac{\langle \delta n^2 \rangle_{\text{atoms}}}{n^2} \simeq \frac{T}{T_d \sqrt{\gamma}} \simeq \frac{T}{T_{\text{co}}}. \tag{11.36}$$

One also sees from the above argument that the density fluctuations fall off as $1/k^2$ above $k = \sqrt{mgn}/\hbar$. The inverse of this scale gives the length scale of density fluctuations and we find that this correlation length is the healing length ξ defined in Eq. (11.28).

From Eq. (11.35) and Eq. (11.36), we find that $\delta n^2/n^2 \simeq T/T_{\text{co}}$. Thus, the quasi-condensate treatment is valid as long as $T \ll T_{\text{co}}$. In conclusion, we have shown that T_{co} gives the limit of both the ideal gas regime, valid as long as $T \gg T_{\text{co}}$, and the limit of the quasi-condensate regime, valid for $T \ll T_{\text{co}}$. Equivalently, in terms of chemical potential, as long as the chemical potential is positive and much larger than μ_{co} of Eq. (11.26), the gas is in the quasi-condensate regime whereas for negative chemical potential of absolute value much larger than μ_{co} the gas is in the ideal gas regime. This is illustrated in Figure 11.3. The two regimes differ by the fact that the $g^{(2)}(z)$ function is modified: it is close to one for any z in the quasi-condensate regime while $g^{(2)}(0) = 2$ in the ideal gas regime.

11.2.3.2 Phase Fluctuations

In the quasi-condensate regime, although the gas is coherent with respect to the $g^{(2)}$ function, it is not coherent with respect to the $g^{(1)}$ function. This is why the gas is called a *quasi*-condensate. The phase fluctuations have been measured experimentally in various experiments where the quasi-condensate presented a one-dimensional character [34–38]. The description of the quasi-condensate given above permits a simple calculation of those phase fluctuations as we now show. Phase fluctuations are given by

$$\langle (\theta(z) - \theta(0))^2 \rangle = \sum_{k>0} 2 \langle \theta_{ck}^2 \rangle (\cos(kz) - 1)^2 + \sum_{k>0} 2 \langle \theta_{sk}^2 \rangle \sin^2(kz). \tag{11.37}$$

Using Eq. (11.34) and $(\cos(kz) - 1)^2 + \sin^2(kz) = 2(1 - \cos(kz))$ this gives

$$\langle (\theta(z) - \theta(0))^2 \rangle = 4 \left(\frac{mT}{Ln\hbar^2} \right) \sum_{k>0} \frac{1 - \cos(kz)}{k^2}. \tag{11.38}$$

Transforming $\sum_k$ into $L/(2\pi)\int_0^\infty dk$ and using $\int_0^\infty(1-\cos(kz))/k^2 dk = \pi z/2$, we obtain

$$\langle(\theta(z)-\theta(0))^2\rangle = \frac{mTz}{n\hbar^2} = \frac{2\pi z}{n\lambda_{dB}^2}. \tag{11.39}$$

Since density fluctuations are very small, the $g^{(1)}$ function is about $g^{(1)}(z) = n\langle e^{i(\theta(z)-\theta(0))}\rangle$. Since the Hamiltonian is quadratic, we can use the Wick theorem to compute $\langle e^{i(\theta(z)-\theta(0))}\rangle$, which gives $\langle e^{i(\theta(z)-\theta(0))}\rangle = e^{-\langle(\theta(z)-\theta(0))^2\rangle/2}$. We find

$$g^{(1)}(z) \simeq e^{-mTz/(2n\hbar^2)}. \tag{11.40}$$

Comparing this to Eq. (11.18), we observe that the behavior of $g^{(1)}$ is close to that in the ideal gas regime. The factor of 2 difference in the correlation length formulae is because for the ideal gas regime, both density and phase fluctuations contribute to $g^{(1)}$ whereas only phase fluctuations remain in the quasi-condensate regime. The crossover from the ideal gas regime to the quasi-condensate regime, at a temperature T_{co}, Eq. (11.25), corresponds to the situation where the correlation length of phase fluctuations, given by Eq. (11.40), equals the correlation length of density fluctuations given by Eq. (11.28).

In both this section and the previous one, we assumed that the temperature is high enough that the population of the relevant modes (whose wavelengths are of the order of ξ) is much greater than unity. This is no longer the case when T reaches values of the order or smaller than gn. For lower temperatures, quantum fluctuations are expected to be dominant. This is the so-called quantum quasi-condensate and the boundary between the thermal quasi-condensate regime and the quantum quasi-condensate regime is at $T \approx ng$, corresponding to

$$t \approx \frac{1}{\gamma} \tag{11.41}$$

and is shown as a line in Figure 11.1. A recent experiment using an atom chip reported these quantum phase fluctuations [39]. In the experiments we describe here however, the temperature is high enough that thermal fluctuations dominate.

11.2.4
Exact Thermodynamics

In Sections 11.2.2 and 11.2.3, we discussed models that apply independently in the asymptotic limits of the nearly ideal gas regime ($T \gg T_{co}$ or equivalently $-\mu \gg \mu_{co}$) and the quasi-condensate regime ($T \ll T_{co}$ or equivalently $\mu \gg \mu_{co}$), respectively. While the above classification gives very useful insight, it should be emphasized that the boundary between these two regimes is a smooth crossover, not a sharp transition and that neither of the two theories presented above account for the physics in the vicinity of the crossover. Since in many cases we are interested in the precise behavior near the crossover from the ideal gas to the quasi-condensate regime, it is not sufficient to use the asymptotic results.

As already mentioned in the Introduction, the 1D Bose gas with repulsive delta-function interactions is an example of an exactly solvable model [40, 41]. This allows us to quantitatively compare predictions of the two approximate descriptions to the exact results, and verify the regions of validity of the approximations. Furthermore, the exact results will turn out to be important for an accurate description of the experiments.

Exactly solvable models typically occur in lower dimensions (1D quantum systems [40, 41] and 2D classical systems [42]) and allow one to obtain exact solutions for the quantum many-body eigenstates through a method known as the "Bethe Ansatz" (due to Hans Bethe [43]), for *any* value of the interaction strength. For the repulsive delta-interacting 1D Bose gas (with periodic boundary conditions), these solutions were first obtained by Eliot Lieb and Werner Liniger [10, 11]. Furthermore, the method based on the Bethe Ansatz can be extended to also obtain the thermodynamics exactly (for *any* temperature), via a method due to C.N. Yang and C.P. Yang [12].

For a concise and lucid description of the Yang–Yang method to obtain the exact thermodynamics of the 1D Bose gas and the related equations, we refer the reader to the original literature [12]. In brief, each exact quantum many-body eigenstate of the Lieb–Liniger Hamiltonian Eq. (11.1) is characterized by a set of distinct integer quantum numbers and a corresponding set of distinct quasi-momenta k, obtained through the Bethe Ansatz. For a large system, one can consider the distribution of these quasi-momenta $\rho(k)$ and also of the "holes" $\rho_h(k)$, the latter corresponding to the "missing" values in the set of integers characterizing the individual quantum states. By considering the entropy for given distributions $\rho(k)$ and $\rho_h(k)$, Yang and Yang showed that the condition of thermal equilibrium leads to a set of non-linear integral equations that can be solved by iteration. Subsequently, from the resulting distributions thermodynamic quantities such as pressure and free energy can be obtained. Once these quantities have been found, further thermodynamic quantities can be calculated using the standard thermodynamic relations.

Although numerical solutions to the Yang–Yang equations were already obtained at an early stage by C.P. Yang [44], important further insight into the Yang–Yang thermodynamics was gained much more recently by Kheruntsyan, Gangardt, Drummond, and Shlyapnikov [22, 23]. They calculated both density and the normalized local density-density correlation function $g^{(2)}(0)$, and compared these results to approximate results in the various regimes discussed above. The former, $n(\mu, T)$, is obtained as part of the equation of state. The latter is obtained from the derivative of free energy with respect to the coupling constant g, using the Hellmann–Feynman theorem.

As an important example, a comparison to the approximate results of the previous sections is shown in Figure 11.3, for a fixed scaled temperature of $t = 1000$. This value is in the relevant range for the experiments to be described below. Such curves as a function of chemical potential μ are particularly useful to describe the behavior in a trap, since in this case one has a well-defined global temperature, while the density varies (within the local density approximation) with the local

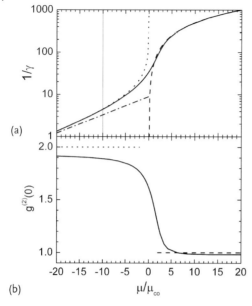

Figure 11.3 Normalized density ($1/\gamma$) and local pair correlation $g^{(2)}(0)$ as a function of chemical potential scaled to the crossover chemical potential μ_{co} given in Eq. (11.26) for fixed temperature corresponding to $t = 1000$. Numerical results from the Yang–Yang equations (solid lines, courtesy K. Kheruntsyan) are compared to the ideal Bose gas model (dotted line in (a), Eq. (11.13)) and the quasi-condensate model (dashed line in (a), Eq. (11.30)). The vertical line in (a) indicates the degeneracy chemical potential $-\mu = T$. The classical Maxwell–Boltzmann prediction Eq. (11.15) is shown as a dashed-dotted line. In (b) the asymptotic values of $g^{(2)}(0)$ are indicated for both the ideal gas regime ($g^{(2)}(0) = 2$ for $\mu \ll -\mu_{co}$, dotted line), and the quasi-condensate regime ($g^{(2)}(0) = 1$ for $\mu \gg \mu_{co}$, dashed line). Adapted from [38, 45].

chemical potential $\mu(z)$ according to $\mu(z) = \mu - V(z)$, where $V(z)$ is the trapping potential. This will be discussed in more detail in the following section.

Figure 11.3a shows that the exact density ($n \propto 1/\gamma$) indeed approaches the ideal gas behavior as μ/μ_{co} becomes sufficiently negative, while for large positive μ/μ_{co} it approaches the quasi-condensate result. There is a large range in density (more than a factor 4) over which neither asymptotic description gives correct predictions. In the same vein, the local density-density correlation function $g^{(2)}(0)$ (Figure 11.3b) smoothly crosses over from 2, the value for an ideal gas, to about 1, as expected for a quasi-condensate. This smoothness is characteristic of crossover behavior, and is drastically different from the step-like behavior typical for a 3D gas.

Looking more closely at Figure 11.3, one sees that the ideal gas description begins to fail for a gas that is only moderately degenerate: already at $\mu/T = -0.5$ ($\mu/\mu_{co} = -5$ for the considered t parameter), a chemical potential for which $n\lambda_{dB} \approx 10$ and the population in the $k = 0$ mode according to Eq. (11.12) is ≈ 1.5, the ideal Bose gas prediction is off by about 10%. This is because the interaction-induced crossover is sufficiently wide that for the used value of t (1000), the chemical potential at degeneracy ($\mu/T \approx -1$) is not very far removed. The narrowness of the degenerate ideal gas regime is also seen in Figure 11.1. To achieve well-separated regimes, one would need to work at much higher t and much smaller γ. For $t = 1000$, the effect of degeneracy is nevertheless visible before the quasi-condensate crossover. This is shown by comparing the density with both the true ideal gas model and the Maxwell–Boltzmann model in Figure 11.3: at $\mu/\mu_{co} \simeq -5$ the ideal gas model gives a prediction for the density accurate within 10 % as mentioned above (and the nearly ideal gas description can thus be expected to be applicable) while the Maxwell–Boltzmann prediction is off by a factor of about 2.

Concerning the local pair correlation function $g^{(2)}(0)$, it deviates from the ideal gas value of 2 for the entire range plotted in the figure. The experiments presented in this chapter however (see Section 11.4), are not precise enough to detect this deviation. Finally, the value of $g^{(2)}(0)$ can take values below unity in the quasi-condensate regime. We will briefly return to this point in Section 11.4.3.3.

Despite its power, the Yang–Yang theory does not permit calculation of any non-thermodynamic quantities. For example, only the *local* value of the density correlation function $g^{(2)}(0)$ has been obtained from thermodynamics, while the full behavior of $g^{(2)}(z)$ has been obtained from the exact solution only at zero temperature [46]. At finite temperature, the behavior of $g^{(2)}(z)$ has been obtained only by perturbative calculations valid in each asymptotic regime [47], but they do not describe the crossover itself. An alternative approach uses the fact that the crossover appears in a highly degenerate gas. In this case, the modes are highly populated and a classical field approach is possible [48, 49].

11.3
1D Gases in the Real World

In real experimental situations, the atomic gas is neither homogeneous nor purely one dimensional. As usual, in our experiments the trapping is to a good approximation harmonic. The trap has cylindrical symmetry and is characterized by a tight radial trapping frequency $\omega_\perp$ and a much lower axial trapping frequency ω. Here, we briefly summarize the main issues related to realizing a 1D system in this trapping geometry. We first discuss the link between transverse effects related to $\omega_\perp$ and we present a model based on the Yang–Yang thermodynamics, valid at low enough linear densities, that takes into account these transverse degrees of freedom. We then discuss the effect of the longitudinal trapping potential. We finish by discussing the link with 3D physics, in particular with regard to the usual Bose–Einstein condensation in 3D.

11.3.1
Transverse Trapping and Nearly 1D Bose Gases

Strictly speaking the conditions to be 1D in a transversely trapped gas are that both temperature and chemical potential are much smaller than the radial vibration quantum, $T, \mu \ll \hbar\omega_\perp$. If this is the case, the gas is frozen in the transverse direction both thermally and in terms of chemical potential, and the (many-body) wave functions can be factorized into the product of a transverse part (the Gaussian ground-state wave function of the radial trap) and an axial part. The system is then kinematically one dimensional. Studying the scattering properties, [20] has shown that the interactions can be modeled by an effective 1D coupling constant g and, as long as the 3D scattering length a is much smaller than the typical size of the transverse oscillator wave function, $l_\perp = \sqrt{\hbar/m\omega_\perp}$,

$$g = 2a\hbar\omega_\perp . \tag{11.42}$$

In most experiments on atom chips, neither of the above conditions on temperature and chemical potential are well fulfilled, and it is necessary to also take into account the transverse degrees of freedom.

It is useful to consider the linear density n_1, obtained from the actual 3D density $\rho(x, y, z)$ through integration

$$n_1(z) = \iint dx dy \rho(x, y, z) . \tag{11.43}$$

When the gas is strictly 1D, one can identify n_1 with the 1D density n. We will present our main experimental results in terms of this linear density, because it turns out that n_1 is often the key parameter, in particular when considering the crossover to 3D at low temperatures.

This is in particular true for the quasi-condensate regime, at temperatures $T \ll \hbar\omega_\perp$. In this regime, the chemical potential is close to its value at zero temperature, which is given by the solution of the radial Gross–Pitaevskii equation[50, 51]. It was found from comparison to numerical integration of the radial Gross–Pitaevskii equation [50, 51] that, in the quasi-condensate regime, to good approximation the chemical potential can be expressed as[27]

$$\mu = \hbar\omega_\perp \left(\sqrt{1 + 4n_1 a} - 1\right) . \tag{11.44}$$

For linear density $n_1 \ll 1/4a$, we find that $\mu \simeq 2\hbar\omega_\perp a n$. We recover here the chemical potential gn of the 1D case. At larger linear density, the chemical potential is reduced compared to the formula $2\hbar\omega_\perp a n_1$. This reflects the fact that, for large densities, the transverse cloud size is increased with respect to the transverse vibrational ground state. As another example of how n_1 is the relevant quantity for low enough temperatures, we note that the expression Eq. (11.40) for the phase coherence length remains correct also on the 3D side of the crossover, if we replace the 1D density n by the linear density n_1 [52].

---

27) An additional factor -1 has been introduced in brackets in Eq. (11.44) compared to [51]. This subtracts the radial zero-point energy $\hbar\omega_\perp$, so that $\mu = 0$ corresponds to the energy of the lowest energy ($k = 0$) state, as in the treatment in Section 11.2.

11.3.2
Applying 1D Thermodynamics to a 3D Trapped Gas

Another case that one can consider is when the interaction energy is in the 1D regime, $ng \ll \hbar\omega_\perp$, while temperature is in the 1D–3D crossover, $T \simeq \hbar\omega_\perp$. A model for this regime was introduced in [8], and we describe it here. The key step is to separately consider the radial states. Under the above conditions only the radial ground state is significantly affected by the interactions, while the radially excited states can still be treated as an ideal gas. Thus, for the radial ground state, the solution $n_{YY}(\mu, T)$ to the Yang–Yang equations must be used. Each radially excited state with radial quantum number $j \geq 1$ is now considered as an independent ideal 1D gas, in thermal equilibrium with the rest of the cloud. Each of the radially excited states is thus taken to have a density (cf. Eq. (11.13))

$$n_e(\mu_j, T) = \frac{1}{\lambda_{dB}} g_{1/2}\left(\exp\left(\frac{\mu_j}{T}\right)\right), \tag{11.45}$$

where an effective chemical potential μ_j has been introduced that takes into account the radial excitation energy,

$$\mu_j = \mu - j\hbar\omega_\perp. \tag{11.46}$$

Taking into account the degeneracy factor $j + 1$ of the radially excited states, the total linear density in this model thus becomes

$$n_l(\mu, T) = n_{YY}(\mu, T) + \sum_{j=1}^{\infty}(j + 1)n_e(\mu_j, T). \tag{11.47}$$

As long as $\mu < \hbar\omega_\perp$, we have $\mu_j < 0$ which is necessary to avoid divergence of $g_{1/2}$ in Eq. (11.45). In fact, from the previous discussion in Section 11.2.4, for our parameters ($t \approx 1000$), we can expect an ideal gas treatment of the radially excited density to begin to break down for $\mu_j/T > -0.5$ since this is where interactions will become important. In practical cases where $T \approx \hbar\omega_\perp$, the model should thus be accurate as long as $\mu < 0.5\hbar\omega_\perp$, while for $\mu > 0.5\hbar\omega_\perp$ the model will start to become inaccurate.

11.3.3
Longitudinal Trapping

Experimentally, cold gases are axially confined in a confining potential $V(z)$ and the cloud is not infinite and homogeneous as assumed in the previous section. However, as seen below, for weak enough axial confinement, the results for homogeneous gases can be applied using a local density approximation. In the first following subsection, we present the local density approximation and discuss its predictions. We then evaluate the condition of validity of this approximation.

11.3.3.1 Local Density Approximation

If the confinement is weak enough that the correlation length of the gas is, at each position, much smaller than the length of the mean density variations, then the gas may be divided into small slices in which the thermodynamics of uniform systems applies. A slice located at position z is in equilibrium with the rest of the gas. It is thus described by the grand canonical ensemble at temperature T and at a chemical potential μ_0. The energy of the gas contained in this slice is shifted by the quantity $V(z)$. It is equivalent to assuming that the chemical potential is $\mu_0 - V(z)$, while the energy of the gas is unshifted. Thus, the local properties of the gas are that of a homogeneous infinite gas at temperature T and local chemical potential $\mu(z) = \mu_0 - V(z)$. This is the so-called local density approximation.

Within the local density approximation, all the results presented in the previous section hold. Thus, performing local analysis, one can observe all the features of homogeneous 1D gases: the presence of the ideal gas regime, which includes the degenerate regime, the crossover towards a quasi-condensate and the quasi-condensate regime. In particular, a quasi-condensate appears in the center of the trap, when the peak density exceeds the crossover density n_{co} given by Eq. (11.24).

It is often interesting to investigate the behavior of the gas using the extensive variable N, where N is the total atom number. As long as the peak density is much smaller than n_{co}, the density profile is well described using the equation of state $n(\mu, T)$ of an ideal Bose gas. Then the total atom number is easily computed and, for gases that are degenerate at the trap center, we obtain [45]

$$N = \frac{T}{\hbar \omega} \ln\left(\frac{T}{|\mu_0|}\right). \tag{11.48}$$

The atom number at the crossover towards a quasi-condensate is obtained when the peak density reaches n_{co}. Inserting Eqs. (11.16) and (11.24) into Eq. (11.48) we find that the atom number at the crossover is approximately

$$N_{co} = \frac{T}{(\hbar \omega)} \ln\left(\left(\frac{\hbar^2 T}{mg^2}\right)^{\frac{1}{3}}\right) = \frac{T}{3\hbar \omega} \ln\left(\frac{t}{2}\right). \tag{11.49}$$

Since $t^{1/3} \gg 1$ (see text below Eq. (11.24)), this equation can be inverted to give a crossover temperature

$$T_{co} = \frac{N\hbar\omega}{\ln\left(\left(N\frac{\hbar^3\omega}{mg^2}\right)^{\frac{1}{3}}\right)}. \tag{11.50}$$

A comparison of this formula with a numerical calculation using Yang–Yang thermodynamics shows very good agreement [45]. Since $t \gg 1$ at the crossover, Eq. (11.49) shows that the ratio N_{co}/N_d, where $N_d = \hbar\omega/T$ is the atom number at degeneracy, is larger than one at the crossover. Thus, even considering the extensive variable N, the degenerate ideal gas regime is in principle identifiable. However, the ratio N_{co}/N_d only grows as a logarithm of t and it is in practice difficult to have N_{co}/N_d very large.

11.3.3.2 Validity of the Local Density Approximation

All the previous results use the local density approximation, which requires that the correlation length l_c of the gas be much smaller than the scale L of variation of the density. At the crossover, the correlation length of the gas is about $l_c \simeq \xi = \hbar/\sqrt{mgn}$, as seen in Section 11.2. To estimate L, let us approach the crossover from the ideal gas regime. The density profile of the central part of the cloud, obtained using Eq. (11.16) and the local chemical potential $\mu(z) = \mu_0 - m\omega^2 z^2/2$, turns out to be a Lorentzian of width $\sqrt{|\mu_0|/m\omega^2}$. Thus, $L \simeq \sqrt{|\mu_0|/m\omega^2} \simeq (gT_{co}/m\hbar\omega^3)^{1/3}$ at the crossover. We thus find that the condition of validity of the local density approximation, $l_c \ll L$, can be rewritten as

$$\omega \ll \omega_{co} = \frac{(E_g T^2)^{\frac{1}{3}}}{\hbar} = \frac{\mu_{co}}{\hbar}, \tag{11.51}$$

a result which has been derived in [45].

If the local density approximation (11.51) is not satisfied, the discrete structure of the trap energy levels has to be taken into account. In the opposite limit, $\omega \gg \omega_{co}$, the quantization of energy levels plays a role while the gas is still described by an ideal Bose gas. Then, it has been shown in [53] that one expects a condensation phenomenon to occur at a temperature

$$T_C = \frac{N\hbar\omega}{\ln(2N)}. \tag{11.52}$$

In contrast to the crossover described in the previous section (referred to now as the interaction-induced crossover), this is a finite size phenomenon since T goes to zero when the trap confinement ω goes to 0, $N\omega$ being fixed. This condensation phenomenon will dominate the interaction-induced crossover when $T_C > T_{co}$. This condition is equivalent to $\omega \gg \omega_{co}$, which shows the consistency of our analysis.

Experimentally, the condition (11.51) to observe the interaction-induced crossover is very easily satisfied: using Eq. (11.42), the condition (11.51) reduces to

$$\omega \ll \omega_\perp \left(\frac{T}{\hbar\omega_\perp}\right)^{\frac{2}{3}} \left(\frac{a}{l_\perp}\right)^{\frac{2}{3}}. \tag{11.53}$$

One can check that, for most alkali atoms, in trapping potentials with $\omega_\perp$ ranging from 1 to several tens of kilohertz and for temperatures between $0.1\hbar\omega_\perp$ and $\hbar\omega_\perp$, this condition is easily fulfilled, unless a is extremely small ($a < 0.1$ nm). Thus, one expects that a trapped 1D gas undergoes the interaction-induced crossover towards a quasi-condensate and that the local density approximation is valid to describe the gas.

11.3.4
3D Physics versus 1D Physics

Experimentally, one expects a crossover from a one-dimensional behavior to a three-dimensional behavior as the temperature of the gas increases and, at large enough temperature, one expects to recover the physics of a three-dimensional gas. The physics of a 3D gas is very different from that of a 1D gas. The most striking difference is that, even in the absence of interactions, a 3D Bose gas undergoes a phase transition towards a Bose–Einstein condensate due to saturation of the population of the excited states. This is in contrast to 1D gases where, in the absence of interactions between atoms, the gas behaves, for any density, as a thermal gas in which bosonic bunching is present. For weakly interacting gases, in both 1D and 3D gases, a transition towards a (quasi-)condensate is expected. However, these transitions are different in nature and this difference can be captured by studying the validity of mean-field theories in both cases.

In 3D weakly interacting gases ($\rho a^3 \ll 1$), the effect of interactions between atoms at the onset of Bose–Einstein condensation is very small. This is why 3D Bose gases with weak interactions are well described by mean-field theories. For instance, the thermodynamics is given with a very good approximation by the Hartree–Fock–Bogoliubov self-consistent theory [54, 55]. In such a theory, at temperatures larger than the critical condensation temperature, the gas is described by the Hartree–Fock approach, in which correlations between atoms introduced by interactions are neglected. Condensation is then expected, as for an ideal Bose gas, when the density reaches $2.612\ldots/\lambda_{dB}$. For higher densities, a non-zero condensate wave function appears, which is the order parameter of this second-order phase transition. The experimental value of the critical temperature in weakly interacting ultra-cold Bose gases is in good agreement with this theory [56].

However, even for weakly interacting gases, such a mean-field theory is expected to fail very close to the critical point of temperature T_c. This is due to the large long wavelength fluctuations that develop in the vicinity of the transition. On the condensate side, that is for $T < T_c$, the Hartree–Fock–Bogoliubov self-consistent theory is valid only if the fluctuations of the condensate wave function, averaged over a volume of the order of the correlation length, are smaller than its mean-field value. This is the so-called Ginzburg criteria and it gives [55]

$$\frac{T_c - T}{T_c} \gg a\rho^{\frac{1}{3}} . \tag{11.54}$$

The same criterion (up to an absolute value) is true above T_c. The region around the transition where $|T_c - T|/T_c$ is of the order or smaller than $a\rho^{1/3}$ is not expected to be described by a mean field. The beyond mean-field effect includes a modification of the transition temperature. Since interactions tend to decrease long wavelength density fluctuations, they favor the appearance of a condensate and, for small $a\rho^{1/3}$, an increase of the critical temperature is expected [57–59]. Such a modification is very small in cold-atom experiments and has never been observed. A second non-mean-field effect is the modification of the critical exponent that describes the divergence of the correlation length in the vicinity of the critical point. The critical exponent was measured recently in dilute atomic gases by measuring beat notes between the atomic field extracted at different places in the atomic cloud, in agreement with beyond mean-field theories [60].

The physics is very different in 1D systems, since long wavelength fluctuations play an enhanced role compared to 3D systems. The crossover towards a quasi-condensate is, in 1D gases, a phenomenon driven by interactions. More precisely, the crossover towards a quasi-condensate is produced by the correlations between atoms introduced by the interactions. It cannot be captured by the Hartree–Fock theory because Hartree–Fock theory neglects correlations between atoms introduced by interactions. Thus, in real systems the failure of the Hartree–Fock theory to describe the appearance of a (quasi-)condensate is a signature of the 1D nature of the physics involved.

11.4 Experiments

In this section we will discuss several experiments that have been carried out in both Orsay and Amsterdam using atom chips which probe the ideas discussed in the previous sections. Atom chip setups are very well suited to the study of one-dimensional geometry since very tight atom guides are easily realized by going close to a current-carrying microwire, as shown in Chapter 2. The atom chips which were used in the experiments presented below are sufficiently similar that we will attempt to describe both at once. We will refer the reader to the individual experiments for more detailed information. A picture of each experiment is shown in Figure 11.4. The atom chips we used employed current-carrying wires to create magnetic trapping fields for ^{87}Rb atoms in the $F = 2, m_F = 2$ state. Magneto-optical traps, laser cooling, and evaporative cooling were used to load atoms into the chip-based traps, which tended to be highly confining but rather shallow. Typical currents were on the order of a few amperes and the atoms were at a distance of several tens of micrometers from the wire surface. Typical transverse confinement frequencies ($\omega_\perp/2\pi$) were about 3 kHz, while longitudinal frequencies were on the order of 10 Hz. This transverse frequency corresponds to a temperature $\hbar\omega_\perp/k_B$ of 144 nK, and evaporative cooling was able to reach a temperature equal to or slightly above this value. For Rb atoms, with 3D scattering length $a = 5.24$ nm, the energy scale E_g corresponds to 0.20 nk, and 144 nK in reduced temperature units corresponds to $t = 720$. Since the longitudinal trapping potential is roughly harmonic, the linear atom density varied in space, and thus a single sample permits one to probe a large range in density at constant temperature. A single density profile thus corresponded to a horizontal line in Figure 11.1. The value of the parameter γ was typically between 10^{-1} to 10^{-3}. The data consisted of absorption images of the cloud, taken either *in situ* or after a very short expansion time. Temperature measurements were made by fitting the wings of the cloud, or by fitting to the Yang–Yang model (see description below).

The first set of measurements we describe are simple observations of the density profiles of nearly one-dimensional gases on an atom chip. The measurements were carried out with two purposes in mind. In the first measurements, carried out in Orsay, emphasis was placed on proving that in the region of the crossover between

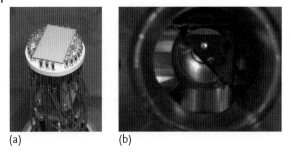

Figure 11.4 (a) Chip used in the experiment in Amsterdam (photograph taken from [38]). (b) Picture of the experiment in Orsay: the chip is facing down at 45° in the small stainless steel vacuum chamber.

the ideal gas and quasi-condensate regimes, no theoretical approach which neglected interaction-induced correlations between particles, in particular the Hartree–Fock approach, could explain the profiles. In the second set, carried out in Amsterdam, it was shown that the exact thermodynamic treatment accounted very well for the entire observed profile, notably when the gas was in the crossover regime. After examining the profiles we move to another type of measurement in which the absorption images were analyzed to give information about density fluctuations. Although these measurements where chronologically the first, we will treat them last.

11.4.1
Failure of the Hartree–Fock Model

A typical density profile is shown in Figure 11.5. Superimposed on the data are three different theoretical predictions. The dashed line shows the profile as predicted by the ideal gas model. In the wings of the profile this model should be valid, and indeed the fit to the wings of the distribution was used to deduce the temperature and the chemical potential of the gas. Clearly, however the ideal gas prediction begins to rapidly deviate from the data, because, without interactions, a 1D Bose gas can accommodate arbitrarily high densities at a given temperature. The dashed-dotted line shows the prediction of the quasi-condensate model Eq. (11.44), at the same chemical potential as was found by fitting the wings. This model accurately reproduces the high density part of the distribution, but not the presence of so many atoms in the wings of the distribution.

The Hartree–Fock theory is a variational method in which the atoms are described by a gas of non-interacting bosons subject to an effective potential V_{HF} due to the mean field of the other atoms. Minimizing the free energy of the gas, one finds

$$V_{\mathrm{HF}}(r) = 2g_{3\mathrm{D}}\rho(r) , \qquad (11.55)$$

where $g_{3\mathrm{D}} = 4\pi\hbar^2 a/m$ is the 3D coupling constant and ρ is the 3D gas density. This theory is thus self-consistent, since for a given chemical potential and

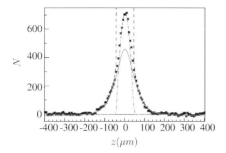

Figure 11.5 Failure of Hartree–Fock theory in a quasi-1D gas. The experimental profile (crosses) is compared with the profiles expected for a quasi-condensate (dotted-dashed), for an ideal Bose gas (dashed), and to the profile predicted by the Hartree–Fock theory (continuous line) for the same temperature and chemical potential. The vertical axis is the number of detected atoms per 6 μm longitudinal pixel. The temperature of the gas was $T = 360 \, \text{nK} = 2.75 \hbar \omega_\perp$. Adapted from [6].

temperature, $\rho(r)$ depends on V_{HF}. The factor 2 reflects the bunching, which is present in the Hartree–Fock approximation since the gas is described by a gas of non-interacting bosons.

Using minimization techniques, the Hartree–Fock density profile was calculated in [6] for the experimental three-dimensional trapping potential and for the temperature and chemical potential found by fitting the wings of the distribution. One sees that the Hartree–Fock density profile, shown as a solid line in Figure 11.5, reproduces the wings of the density profile, and does not diverge as does the ideal gas profile. It does not however, reproduce the high density part of the profile. Moreover, the Hartree–Fock calculation shows that the Hartree–Fock gas is far from being saturated: the population of the ground state is very small, and no condensation is expected according to this mean-field model. The excess of atoms in the center is the onset of a quasi-condensate, although the cloud is not deep into the quasi-condensate regime. This peak is formed by the effects of interactions altering the two-body correlation function so as to lower the interaction energy relative to a Hartree–Fock gas at the same density.

11.4.2
Yang–Yang Analysis

Two more examples of axial density profiles measured [8] at two different temperatures and a peak linear density of $\approx 50 \, \mu\text{m}^{-1}$ are shown in Figure 11.6. These profiles were fitted to the model based on the exact Yang–Yang solutions described in Section 11.3.2. The fits are shown in the Figure as continuous curves, and the resulting temperature T and chemical potential μ are also indicated. The chemical potential μ and the temperature T are the only free parameters in the model, and it was found that the full set of *in situ* measurements could be explained by the Yang–Yang-based model [8]. For comparison, the ideal gas prediction and the quasi-condensate prediction are also shown. Clearly, the Yang–Yang-based model

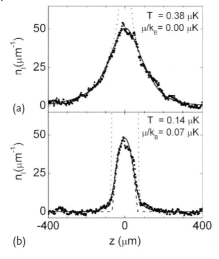

Figure 11.6 (a),(b) Comparison of experiment to Yang–Yang thermodynamics. The model described in Section 11.3.2 is fitted to two examples of measured linear densities n_l (dots) in the Amsterdam experiment [8, 38]. The resulting fits (continuous curves) yield chemical potential μ and temperature T as indicated. Dotted curves: ideal gas profile at the same temperature and chemical potential exhibiting divergence for $\mu(z) = 0$. Dashed curve in (b): quasi-condensate profile with the same peak density as the experimental data. In these experiments $\hbar\omega_\perp = 158$ nK. Adapted from [8].

describes the entire profiles well, while the approximate models fail, in particular the smooth crossover between the two approximate models in the region where $\mu(z) \approx 0$ is captured very well by the model.

This analysis was further corroborated by measurements of the axial momentum distribution [8], obtained using Bose gas focusing [37]. The tails of the momentum distribution were used to extract temperatures, and these were found to agree very well with the temperatures derived from the Yang–Yang fit to the *in situ* data. The full momentum distribution is not a thermodynamic quantity, and can thus not be obtained directly from the Yang–Yang analysis.

The similarity of the measured density profiles of Figures 11.5 and 11.6b clearly suggests that the same physics of an interaction-induced crossover applies to both experiments. Although it is tempting to apply the Yang–Yang-based analysis of Section 11.3.2 also to the data of Figure 11.5, this has not been done. It is likely that the result would not be quantitatively accurate, because at the higher linear densities and temperatures of Figure 11.5, the validity limits of the model of Section 11.3.2 are reached near the peak of the profile (since both $\mu \approx \hbar\omega_\perp$ and $T \approx \hbar\omega_\perp$). Interactions are then expected to also play a role in the radially excited states, and also interactions among the different radial states will be significant.

11.4.3
Measurements of Density Fluctuations

As we have emphasized in Section 11.2, the transition towards a quasi-condensate in 1D gases is characterized by the inhibition of atom bunching, the large density fluctuations characteristic of a thermal Bose gas. A direct measurement of the density fluctuations through the crossover thus captures an essential characteristic of the crossover.

The measurement of density fluctuations proceeds similarly to the density profile measurements. The difference is that many (about 300) profiles are acquired and, roughly speaking, for each observation pixel, we compute the variance of the density measurements as well as the mean. We can relate this variance to the density fluctuations predicted by various theoretical approaches as described below.

The measurements involve several subtleties requiring careful normalizations and corrections of the data. These are described in detail in [7, 61]. The measurement requires a high degree of reproducibility in the data. The atom chip geometry permits the construction of a very compact apparatus with low sensitivity to vibration. The images contain not only noise due to atom fluctuations, but also photon shot noise. The photon noise must be carefully characterized and subtracted. Examples of the data are shown in Figures 11.7 and 11.8.

11.4.3.1 A Local Density Analysis

The pixel size in the experiment is $\Delta = 6\,\mu\mathrm{m}$. The pixel size is much larger than the correlation length of the gas which is always smaller than a micrometer in these experiments, but much smaller than the longitudinal length scale of mean density variation. Thus, the data should reproduce number fluctuations predicted in a

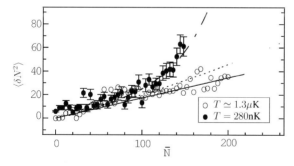

Figure 11.7 Density fluctuations of a gas on an atom chip. The atom number variance in an observation pixel (6 μm) is plotted as a function of the mean number. The open circles are the fluctuations measured for a hot cloud ($T = 1.3\,\mu\mathrm{K}$ corresponds to $10\hbar\omega_\perp$) for which bunching is unobservable because of the large number of transverse states involved. The variance is due to atom shot noise. Full circles correspond to a colder cloud, at a temperature $T = 2.1\hbar\omega_\perp$. The increase in fluctuations is due to bunching. The theoretical prediction for an ideal Bose gas at the same temperature is given by the dashed curve. The prediction for a non-degenerate cloud, Eq. (11.61), is shown as the dotted curve. The degeneracy of the gas is evident. Adapted from [7].

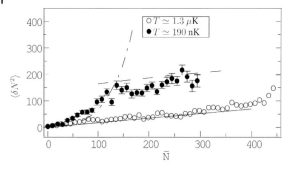

Figure 11.8 Density fluctuations in the quasi-condensate regime. The dashed-dotted curve is the prediction for an ideal Bose gas at the same temperature as in Figure 11.7. The dashed curve is the prediction for a quasi-condensate. In units of transverse energy, the temperature is $T = 1.4\hbar\omega_\perp$. Adapted from [7].

longitudinal local density treatment. More precisely, the gas contained in the pixel located at position z can be described as a gas, confined transversely by the transverse potential of frequency $\omega_\perp$ and confined longitudinally by a box-like potential of size Δ. The properties of this slice, which can exchange energy and particles with the rest of the gas, are well described within the grand-canonical ensemble. The energy shift $V(z)$ of this slice can be converted to a shift $-V(z)$ of the chemical potential. This is the local density approximation, already discussed in Section 11.3. Since Δ is large compared to correlation length of the gas, the boundary conditions used to compute thermodynamic quantities are all equivalent and we use the periodic boundary conditions in the following.

Within the local density approximation, the confinement potential $V(z)$ is irrelevant to analyze the atom number fluctuations. The atom number fluctuation δN^2 in each pixel depends only on the temperature T and on the local chemical potential. Equivalently, δN^2 is a function of T and $\langle N \rangle$, since the linear density is a monotonically increasing function of the chemical potential. We thus choose, for each cloud temperature, to represent the measured atom number fluctuation as a function of the mean atom number in the pixel. Experimental results are shown in Figures 11.7 and 11.8.

11.4.3.2 Ideal Gas Regime: Observation of Bunching

If the gas within a pixel can be considered ideal, we can use the results of Section 11.2.2. The fluctuations of atom number n_i in each one-atom quantum state $|i\rangle$ are

$$\langle n_i^2 \rangle - \langle n_i \rangle^2 = \langle n_i \rangle + \langle n_i \rangle^2 \,. \tag{11.56}$$

The fluctuations of the total atom number N are thus

$$\langle N^2 \rangle - \langle N \rangle^2 = \langle N \rangle + \sum_i \langle n_i \rangle^2 \,, \tag{11.57}$$

where the sum is performed over all the quantum states. The mean values $\langle n_i \rangle$ are given by the Bose distribution and the fluctuations of N are easily computed.

A rough calculation is as follows: if M quantum states are populated with similar populations, Eq. (11.57) simplifies to

$$\langle N^2 \rangle - \langle N \rangle^2 = \langle N \rangle + \langle N \rangle \frac{\langle N \rangle}{M} . \tag{11.58}$$

The first term of the right-hand side is the shot noise term, expected for uncorrelated, statistically independent atoms. The second term on the right-hand side is the effect of the bunching. We see from this expression that as long as $\langle N \rangle / M$ is much smaller than 1, the bunching term is negligible compared to the shot noise term. The ratio $\langle N \rangle / M$ is approximately the phase space density of the gas and is much smaller than 1 if the gas is nondegenerate. Thus, one expects the measured atom number fluctuations to be dominated by the shot noise term for gases at high temperature. This is observed experimentally for non-degenerate clouds, as shown by the open circles in Figure 11.7. The linearity of the measured value of $\langle N^2 \rangle - \langle N \rangle^2$ versus $\langle N \rangle$ shows that the fluctuations are given by the shot noise. The fact that the slope is smaller than the expected slope of 1 is due to the fact that the optical resolution (about 10 µm) is larger than the pixel size [7].

One can also give a more precise calculation of the fluctuations. For this purpose, we index the quantum states by the integer n_x and n_y, which label the transverse vibrational levels, and the longitudinal wave vector k_z, which takes values in multiples of $2\pi/\Delta$. For a highly non-degenerate gas, $|\mu| \gg T$, the population of each state is given by the Boltzmann law

$$\langle n_{n_x, n_y, k_z} \rangle = A e^{-\left(\frac{\hbar^2 k_z^2}{2m} + \hbar \omega_\perp (n_x + n_y)\right)/T} , \tag{11.59}$$

where the normalization factor A is

$$A = \frac{N \lambda_{dB}}{\Delta} \left(1 - e^{-\frac{\hbar \omega_\perp}{T}}\right)^2 . \tag{11.60}$$

Inserting this into Eq. (11.57), we obtain

$$\langle N^2 \rangle - \langle N \rangle^2 = \langle N \rangle + \langle N \rangle^2 \frac{\lambda_{dB}}{\sqrt{2}\Delta} \tanh^2 \left(\frac{\hbar \omega_\perp}{2T}\right) . \tag{11.61}$$

We thus recover an expression similar to Eq. (11.58), with

$$M = \frac{\sqrt{2}\Delta}{\left(\lambda_{dB} \tanh^2 \left(\frac{\hbar \omega_\perp}{2T}\right)\right)} .$$

The tanh term accounts for the number of populated transverse states. The term $\sqrt{2}\Delta/\lambda_{dB}$, which accounts for the longitudinal states, may be recovered by a semi-classical analysis: the volume of the occupied phase space is $\Omega \simeq \Delta \sqrt{mT}$ and the number of quantum states contained in this volume is of the order of $\Omega/\hbar$. Equation (11.61) is valid if the gas is nondegenerate. When the gas becomes degenerate, the distribution of the mean occupation number $\langle n \rangle$ versus the state energy becomes more peaked around zero. This amounts to a reduction of the effective number of occupied states M and the effect of bunching is larger than the prediction of Eq. (11.61). For highly degenerate gases, Eq. (11.61) underestimates the true fluctuations, which become large compared to the shot noise level.

The bunching effect is quite clear for a cold enough cloud as shown in Figure 11.7. In this experiment, the bunching term is even larger than the shot noise term, indicating that the gas is degenerate. The degeneracy is also shown by a comparison of the data with Eq. (11.61) shown as a dotted line. This equation, valid for a non-degenerate gas, underestimates the measured fluctuations. On the other hand, a calculation of Eq. (11.57) using the true Bose occupation factor is in much better agreement with the data. This comparison shows that, at least as concerns fluctuations, the gas is well described by an ideal, degenerate Bose gas.

11.4.3.3 Quasi-Condensate Regime: Saturation of Atom Number Fluctuations

At sufficiently high density and low temperature, repulsive interactions between atoms are no longer negligible. As described in Section 11.2, one expects the interactions to reduce the density fluctuations to lower the interaction energy. The gas then enters the quasi-condensate regime. For the temperature $T = 2.1\hbar\omega_\perp$ of the data in Figure 11.7, using Eq. (11.42) and assuming a purely 1D gas, Eq. (11.24) gives a density at the crossover of about 130 atoms per pixel. Although Eq. (11.24) does not apply since the gas is not purely 1D, this rough estimate shows that the crossover to a quasi-condensate is achievable at slightly higher atom number and/or lower temperature. Measurements of atom number fluctuations in a regime where the cloud center is in the quasi-condensate regime are shown in Figure 11.8. Whereas at low atomic density the measured fluctuations are in agreement with the ideal Bose gas prediction, one sees a saturation of the fluctuations at higher densities.

To calculate the fluctuations, we first suppose the gas to be purely one dimensional, with a coupling constant g given by Eq. (11.42). In a local density approximation, we consider atom number fluctuations in a longitudinal box of length Δ in equilibrium with a reservoir of energy at temperature T and a reservoir of particles at chemical potential μ. As explained in Section 11.2.3, the Hamiltonian is quadratic in δn in the quasi-condensate approximation and δn can be expanded as a sum of independent modes indexed by the wave vector k. The atom number fluctuations $N - \langle N \rangle$ are obtained by integrating δn over the pixel size. Thus, the only excitation that leads to atom number fluctuation is the zero momentum mode. Its energy, derived from Eq. (11.32), is

$$H_{k=0} = \Delta \frac{g}{2} \delta n_0^2. \tag{11.62}$$

Using the equipartition theorem, we find $\delta n_0^2 = T/(g\Delta)$. The atom number fluctuations, which are $(\Delta \delta n_0)^2$, are thus given by:

$$\langle N^2 \rangle - \langle N \rangle^2 = \frac{\Delta T}{g}. \tag{11.63}$$

Thus, we expect the atom number fluctuations to be independent of $\langle N \rangle$. The shot noise term is not present in this quasi-condensate regime: interactions between atoms prevent even the shot noise fluctuations and at temperatures smaller than gn, one expects to observe sub-shotnoise fluctuations. This feature is also seen in Figure 11.3 where $g^{(2)}(0)$ goes below unity in the exact solution.

In the experimental results shown in Figure 11.8, the typical interaction energy is about $0.7\hbar\omega_\perp$. Under these conditions, the transverse degrees of freedom cannot be neglected and the result of Eq. (11.63) must be corrected. More precisely, the phonons, which are longitudinal density waves, are associated with a breathing of the transverse shape of the cloud. For phonons of frequency much smaller than the transverse frequency, the transverse shape of the cloud follows adiabatically the ground-state equilibrium state for the local linear density. We denote by $E_{eq}(n)$ the energy of the gas per unit length for a linear density n. The phonon Hamiltonian of Eq. (11.32) is the term of the Hamiltonian of order two in δn and $\nabla \theta$. Thus, the interaction term of the phonons is $E_{int} = \frac{1}{2} L(\partial^2 E_{eq}/\partial n^2)\delta n_k^2$. Since $\partial E_{eq}/\partial n$ is the chemical potential of the gas (to zero order in δn), we can rewrite the former expression as $E_{int} = \frac{1}{2} L(\partial \mu/\partial n)\delta n_k^2$. In particular, the zero momentum term is

$$H_{k=0} = \frac{1}{2} L \frac{\partial \mu}{\partial n} \delta n_0^2 . \tag{11.64}$$

Then, the atom number fluctuations are

$$\langle N^2 \rangle - \langle N \rangle^2 = \frac{\Delta T}{\frac{\partial \mu}{\partial n}} . \tag{11.65}$$

Although we derived this expression in the approximate quasi-condensate theory using expansion of the Hamiltonian to second order in δn, we recover here a well-known result of statistical physics. More precisely, as shown in [62], Eq. (11.65) holds for any system in equilibrium with a particle reservoir at chemical potential μ and with an energy reservoir of temperature T.

To apply Eq. (11.65) to the experiment, we need the equation of state $\mu(n)$. Using Eqs. (11.44) and (11.65), one can compute the expected fluctuations. Figure 11.8 shows that the results are in fairly good agreement with the measured atom number fluctuations.

11.5 Conclusion

We hope that we have given the reader a useful overview of the physics of 1D gases in the weakly interacting regime. These systems are rich and manifest several different regimes separated by smooth crossovers. It is often necessary to appeal to many different physical models to understand them. The existence of exact solutions allows us to test the models and to explore their validity in the face of highly non-trivial many-body correlations. Although the experimental and theoretical work we have described is quite extensive, we believe that much work remains to be done. The study of fluctuation phenomena is still at an early stage. For example, the exact thermodynamics should permit a careful comparison with data such as that in Figure 11.8, and improved experiments should probe larger parameter ranges and possibly even permit measurements of the correlation length. It may also be possible in the near future to enter the strongly interacting regime using an atom chip. Experiments are also capable of measuring momentum distributions [8], but so far no quantitative theoretical comparison has been made. Finally, measurements of fluctuations and correlations in momentum space are also experimentally feasible [32].

Acknowledgments

We thank K. Kheruntsyan for a critical reading of the manuscript. NJvD acknowledges stimulating discussions with J.T.M. Walraven, G.V. Shlyapnikov, J.-S. Caux, and A.H. van Amerongen. The Amsterdam work was supported by FOM (Stichting Fundamenteel Onderzoek der Materie) and NWO (Nederlandse Organisatie voor Wetenschappelijk Onderzoek). The work in Orsay was supported by the Institut Francilien pour la Recherche en Atomes Froids, and by the SCALA program of the EU.

References

1 Reichel, J. (2002) Microchip traps and Bose–Einstein condensation. *Appl. Phys. B* **74**, 469.
2 Folman, R., Krüger, P., Schmiedmayer, J., Denschlag, J., and Henkel, C. (2002) Microscopic atom optics: From wires to an atom chip. *Adv. At. Mol. Opt. Phys.* **48**, 263.
3 Fortágh, J. and Zimmermann, C. (2007) Magnetic microtraps for ultracold atoms. *Rev. Mod. Phys* **79**, 235.
4 Wildermuth, S., Hofferberth, S., Lesanovsky, I., Groth, S., Krüger, P., Schmiedmayer, J., and Bar-Joseph, I. (2006) Sensing electric and magnetic fields with Bose–Einstein condensates. *Appl. Phys. Lett.* **88**, 264103.
5 Schumm, T., Hofferberth, S., Andersson, L.M., Wildermuth, S., Groth, S., Bar-Joseph, I., Schmiedmayer, J., and Krüger, P. (2005) Matter-wave interferometry in a double well on an atom chip. *Nat. Phys.* **1**, 57.
6 Trebbia, J.-B., Estève, J., Westbrook, C.I., and Bouchoule, I. (2006) Experimental evidence for the breakdown of a Hartree–Fock approach in a weakly interacting Bose gas. *Phys. Rev. Lett.* **97**, 250403 (pages 4).
7 Estève, J., Trebbia, J.-B., Schumm, T., Aspect, A., Westbrook, C.I., and Bouchoule, I. (2006) Observations of density fluctuations in an elongated Bose gas: Ideal gas and quasicondensate regimes. *Phys. Rev. Lett.* **96**, 130403 (pages 4).
8 van Amerongen, H.A., van Es, J.P., Wicke, J.P., Kheruntsyan, K.V., and van Druten, J.N. (2008) Yang–Yang thermodynamics on an atom chip. *Phys. Rev. Lett.* **100**, 090402.
9 (2005) Special Issue – Atom chips: manipulating atoms and molecules with microfabricated structures. *Eur. J. Phys. D.* **35**.
10 Lieb, E.H. and Liniger, W. (1963) Exact analysis of an interacting Bose gas. I.

The general solution and the ground state. *Phys. Rev.* **130**, 1605.

11 Lieb, E.H. (1963) Exact analysis of an interacting Bose gas. II. The excitation spectrum. *Phys. Rev.* **130**, 1616.

12 Yang, C.N. and Yang, C.P. (1969) Thermodynamics of a one-dimensional system of bosons with repulsive delta-function interaction. *J. Math. Phys.* **10**, 1115.

13 Guerin, W., Riou, J.-F., Gaebler, J.P., Josse, V., Bouyer, P., and Aspect, A. (2006) Guided quasicontinuous atom laser. *Phys. Rev. Lett* **97**, 200402 (pages 4).

14 Hofferberth, S., Lesanovsky, I., Fischer, B., Schumm, T., and Schmiedmayer, J. (2007) Non-equilibrium coherence dynamics in one-dimensional Bose gases. *Nature* **449**, 324.

15 Kinoshita, T., Wenger, T., and Weiss, D.S. (2004) Observation of a one-dimensional Tonks–Girardeau gas. *Science* **305**, 1125.

16 Kinoshita, T., Wenger, T., and Weiss, D.S. (2005) Local pair correlations in one-dimensional Bose gases. *Phys. Rev. Lett.* **95**, 190406.

17 Laburthe Tolra, B., O'Hara, K.M., Huckans, J.H., Phillips, W.D., Rolston, S.L., and Porto, J.V. (2004) Observation of reduced three-body recombination in a correlated 1D degenerate Bose gas. *Phys. Rev. Lett.* **92**, 190401.

18 Paredes, B., Widera, A., Murg, V., Mandel, O., Fölling, S., Cirac, I., Shlyapnikov, G.V., Hänsch, T.W., and Bloch, I. (2004) Tonks–Girardeau gas of ultracold atoms in an optical lattice. *Nature* **429**, 277.

19 Moritz, H., Stöferle, T., Köhl, M., and Esslinger, T. (2003) Exiting collective oscillations in a trapped 1D gas. *Phys. Rev. Lett.* **91**, 250402.

20 Olshanii, M. (1998) Atomic scattering in the presence of an external confinement and a gas of impenetrable bosons. *Phys. Rev. Lett.* **81**, 938.

21 Petrov, D., Shlyapnikov, G., and Walraven, J. (2000) Regimes of quantum degeneracy in trapped 1D gases. *Phys. Rev. Lett.* **85**, 3745.

22 Kheruntsyan, K.V., Gangard, D.M., Drummond, P.D., and Shlyapnikov, G.V. (2003) Pair correlations in a finite-temperature 1D Bose gas. *Phys. Rev. Lett.* **91**, 040403.

23 Kheruntsyan, K.V., Gangardt, D.M., Drummond, P.D., and Shlyapnikov, G.V. (2005) Finite-temperature correlations and density profiles of an inhomogeneous interacting one-dimensional Bose gas. *Phys. Rev. A* **71**, 053615 (pages 17).

24 Popov, V.N. (1983) *Functional Integrals in Quantum Field Theory and Statistical Physics*, Reidel, Dordrecht, The Netherlands.

25 Tonks, L. (1936) The complete equation of state of one, two and three-dimensional gases of hard elastic spheres. *Phys. Rev.* **50**, 955.

26 Girardeau, M. (1960) Relationship between systems of impenetrable bosons and fermions in one dimension. *J. Math. Phys.* **1**, 516.

27 Huang, K. (1987) *Statistical Mechanics*, John Wiley & sons, New York, 2nd edn.

28 Lewin, L. (1981) *Polylogarithms and Associated Functions*, Elsevier North Holland Inc., New York, NY, USA.

29 Wick, G.C. (1950) The evaluation of the collision matrix. *Phys. Rev.* **80**, 268.

30 Yasuda, M. and Shimizu, F. (1996) Observation of two-atom correlation of an ultracold neon atomic beam. *Phys. Rev. Lett.* **77**, 3090.

31 Fölling, S., Gerbier, F., Widera, A., Mandel, O., Gericke, T., and Bloch, I. (2005) Spatial quantum noise interferometry in expanding ultracold atom clouds. *Nature* **434**, 481.

32 Schellekens, M., Hoppeler, R., Perrin, A., Gomes, J.V., Boiron, D., Aspect, A., and Westbrook, C.I. (2005) Hanbury Brown–Twiss effect for ultracold quantum gases. *Science* **310**, 648.

33 Mora, C. and Castin, Y. (2003) Extension of Bogoliubov theory to quasi-condensates. *Phys. Rev. A* **67**, 053615.

34 Dettmer, S., Hellweg, D., Ryytty, P., Arlt, J.J., Ertmer, W., Sengstock, K., Petrov, D.S., Shlyapnikov, G.V., Kreutzmann, H., Santos, L., and Lewenstein, M. (2001) Observation of phase fluctuations in elongated Bose–Einstein condensates. *Phys. Rev. Lett.* **87**, 160406.

35 Hellweg, D., Cacciapuoti, L., Kottke, M., Schulte, T., Sengstock, K., Ertmer, W.,

and Arlt, J.J. (2003) Measurement of the spatial correlation function of phase fluctuating Bose–Einstein condensates. *Phys. Rev. Lett.* **91**, 010406.

36 Richard, S., Gerbier, F., Thywissen, J.H., Hugbart, M., Bouyer, P., and Aspect, A. (2003) Momentum spectroscopy of 1D phase fluctuations in Bose–Einstein condensates. *Phys. Rev. Lett.* **91**, 010405.

37 Shvarchuck, I., Buggle, C., Petrov, D.S., Dieckmann, K., Zielonkowski, M., Kemmann, M., Tiecke, T.G., von Klitzing, W., Shlyapnikov, G.V., and Walraven, J.T.M. (2002) Bose–Einstein condensation into nonequilibrium states studied by condensate focusing. *Phys. Rev. Lett.* **89**, 270404.

38 van Amerongen, A.H. (2008) *One-dimensional Bose gas on an atom chip*, Ph.D. thesis, Universiteit van Amsterdam.

39 Hofferberth, S., Lesanovsky, I., Schumm, T., Imambekov, A., Gritsev, V., Demler, E., and Schmiedmayer, J. (2008) Probing quantum and thermal noise in an interacting many-body system. *Nat. Phys.* **4**, 489.

40 Korepin, V.E., Bogoliubov, N.M., and Izergin, A.G. (1993) *Quantum Inverse Scattering Method and Correlation Functions*, Cambridge University Press, Cambridge, England.

41 Takahashi, M. (1999) *Thermodynamics of One-Dimensional Solvable Models*, Cambridge University Press, Cambridge, England.

42 Baxter, R.J. (1982) *Exactly Solved Models in Statistical Mechanics*, Academic Press, New York.

43 Bethe, H. (1931) Zur Theorie der Metalle. *Z. Phys.* **71**, 205.

44 Yang, C.P. (1970) One-dimensional system of bosons with repulsive δ-function interactions at a finite temperature T. *Phys. Rev. A* **2**, 154.

45 Bouchoule, I., Kheruntsyan, K.V., and Shlyapnikov, G.V. (2007) Interaction-induced crossover versus finite-size condensation in a weakly interacting trapped one-dimensional Bose gas. *Phys. Rev. A* **75**, 031606 (pages 4).

46 Caux, J.-S. and Calabrese, P. (2006) Dynamical density-density correlations in the one-dimensional Bose gas. *Phys. Rev. A* **74**, 031605 (pages 4).

47 Sykes, A.G., Gangardt, D.M., Davis, M.J., Viering, K., Raizen, M.G., and Kheruntsyan, K.V. (2008) Spatial nonlocal pair correlations in a repulsive 1D Bose gas. *Phys. Rev. Lett.* **100**, 160406 (pages 4).

48 Castin, Y., Dum, R., Mandonnet, E., Minguzzi, A., and Carusotto, I. (2000) Coherence properties of a continuous atom laser. *J. Mod. Opt.* **47**, 2671.

49 Blakie, P.B. and Davis, M.J. (2005) Projected Gross–Pitaevskii equation for harmonically confined Bose gases at finite temperature. *Phys. Rev. A* **72**, 063608 (pages 12).

50 Menotti, C. and Stringari, S. (2002) Collective oscillations of a one-dimensional trapped Bose–Einstein gas. *Phys. Rev. A* **66**, 043610.

51 Gerbier, F. (2004) Quasi-1D Bose–Einstein condensates in the dimensional crossover regime. *Europhys. Lett.* **66**, 771.

52 Petrov, D.S., Shlyapnikov, G.V., and Walraven, J.T.M. (2001) Phase-fluctuating 3D Bose–Einstein condensates in elongated traps. *Phys. Rev. Lett.* **87**, 050404.

53 Ketterle, W. and J. van Druten, N. (1996) Bose–Einstein condensation of a finite number of particles trapped in one or three dimensions. *Phys. Rev. A* **54**, 656.

54 Griffin, A. (1996) Conserving and gapless approximations for an inhomogeneous Bose gas at finite temperatures. *Phys. Rev. B* **53**, 9341.

55 Giorgini, S., Pitaevskii, L.P., and Stringari, S. (1996) Condensate fraction and critical temperature of a trapped interacting Bose gas. *Phys. Rev. A* **54**, R4633.

56 Gerbier, F., Thywissen, J.H., Richard, S., Hugbart, M., Bouyer, P., and Aspect, A. (2004) Critical temperature of a trapped, weakly interacting Bose gas. *Phys. Rev. Lett.* **92**, 030405.

57 Holzmann, M. and Krauth, W. (1999) Transition temperature of the homogeneous, weakly interacting Bose gas. *Phys. Rev. Lett.* **83**, 2687.

58 Arnold, P. and Moore, G. (2001) BEC transition temperature of a dilute homo-

geneous imperfect Bose gas. *Phys. Rev. Lett.* **87**, 120401.

59 Kashurnikov, V.A., Prokof'ev, N.V., and Svistunov, B.V. (2001) Critical temperature shift in weakly interacting Bose gas. *Phys. Rev. Lett.* **87**, 120402.

60 Donner, T., Ritter, S., Bourdel, T., Öttl, A., Köhl, M., and Esslinger, T. (2007) Critical behavior of a trapped interacting Bose gas. *Science* **315**, 1556.

61 Trebbia, J.-B. (2007) *Etude de gaz quantiques dégénés quasi-unidimensionnels confinés par une micro-structure*, Ph.D. thesis, Université Paris Sud – Paris XI.

62 Kittel, C. and Kroemer, H. (1980) *Thermal Physics*, Freeman and Company, New York, NY, USA, 2nd edn.

12
Fermions on Atom Chips

Marcius H.T. Extavour, Lindsay J. LeBlanc, Jason McKeever, Alma B. Bardon, Seth Aubin, Stefan Myrskog, Thorsten Schumm, and Joseph H. Thywissen

12.1
Introduction

Degenerate Fermi gases (DFGs) earned their place at the leading edge of degenerate quantum gas research with the first demonstration of an atomic ^{40}K DFG in 1999 [1]. Since then, DFGs have been central to many important advances in the field, including molecular Bose–Einstein condensates (BECs), resonant superfluidity, fermion lattice physics, Mott insulator phase, and boson-fermion mixtures (see [2] and references therein).

Not long after the first DFG was produced, efficient loading of cold atoms into atom chip microtraps enabled the first demonstrations of Bose–Einstein condensation on an atom chip [3, 4]. In subsequent years, research efforts in DFGs and atom chips progressed independently, but were combined with the 2006 demonstration in Toronto of a DFG of ^{40}K on an atom chip [5].

In this chapter we provide the background to our work, including ideal fermion statistics and the challenges particular to evaporative cooling in a chip trap. We then describe our observations of Fermi degeneracy in ^{40}K, and of the Ramsauer–Townsend effect. Finally, we discuss several works in progress: species selective manipulation, interaction tuning.

This chapter begins with a review of statistical and thermodynamic properties of the ideal, non-interacting Fermi gas. We discuss our experimental approach to producing a ^{40}K DFG and a ^{40}K-^{87}Rb DFG-BEC mixture, and our observation of the Ramsauer–Townsend effect. We also include a discussion of how trap depth constrains the possible wire geometries for chip traps. Finally, we discuss several works in progress: species selective manipulation, and interaction tuning using a Feshbach resonance.

Atom chips offer a wide array of techniques for trapping and manipulating ultra-cold atoms using a single micro-fabricated device (see Chapter 3). These techniques, discussed extensively in this book, include magnetostatic potentials (see Chapters 1 and 2), electrostatic potentials (see Chapter 13), dynamic radio-

Atom Chips. Edited by Jakob Reichel and Vladan Vuletić
Copyright © 2011 WILEY-VCH Verlag GmbH & Co. KGaA, Weinheim
ISBN: 978-3-527-40755-2

frequency (RF) and microwave dressed potentials (see Chapters 7 and 9), and integrated optical potentials (discussed here and in Chapter 3).

With these tools atom chips are well poised to advance active research in fermionic many-body systems. The "Z" trap configuration often used for chip traps makes an elongated potential that, when smooth enough, can produce one-dimensional gases. One-dimensional fermion physics might include Luttinger liquids [6–8], confinement-induced molecule formation [9], and spin-charge separation [10].

RF-dressed double-well potentials on atom chips [11–14] have only recently been applied to ultra-cold fermions [15]. This combination is an exciting step in the direction of fermion quantum atom optics, including interferometry [16], mesoscopic quantum pumping circuit simulations [17], antibunching [18], and number statistics [19–21]. (See also Chapter 7.)

Spin-independent optical potentials can trap the spin mixtures necessary for strong inter-particle interactions in fermions. Optical traps near the surface of an atom chip [22, 23] provide opportunities to combine strongly interacting DFGs with near-field RF and microwave probes. We discuss a first step in this direction in Section 12.7.

Finally, there are several attractive technical advantages in using atom chips for fermions. Large collision rates, made possible by the strong confinement of microtraps relative to conventional, centimeter-scale traps, permit rapid sympathetic evaporative cooling to quantum degeneracy. This obviates the need for minute-scale vacuum lifetimes, multi-chamber vacuum systems, and Zeeman slowers. Dramatically shorter experimental cycle times from atomic vapor to DFG also become possible, a point of practical value in day-to-day laboratory research.

12.2
Theory of Ideal Fermi Gases

Ultra-cold Fermi gases differ from ultra-cold Bose gases in their simplest theoretical description in two important ways: first, there is no macroscopic occupation of the single-particle ground state; second, spin-polarized Fermi gases are completely noninteracting at ultra-cold temperatures [24, 25]. Ideal thermodynamic functions are thus excellent descriptors of cold spin-polarized fermions, even as $T \to 0$. In this section we review fermion thermodynamics, calculate trapped density distributions, and discuss observable signatures of Fermi degeneracy.

12.2.1
Thermodynamics

In the grand canonical ensemble description of an ideal Fermi gas, the mean occupation number of the single-particle energy state ϵ is

$$\langle n_\epsilon \rangle = \frac{1}{e^{\beta(\epsilon-\mu)} + 1} = \frac{1}{\mathcal{Z}^{-1} e^{\beta\epsilon} + 1}, \tag{12.1}$$

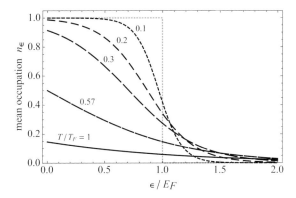

Figure 12.1 Universal curves of the mean Fermi–Dirac occupation of single particle states ϵ vs. ϵ/E_F, shown at select values of T/T_F. The gray, dotted line indicates the $T = 0$ filled Fermi sea (see text).

where $\beta \equiv 1/k_B T$, k_B is the Boltzmann constant, μ is the chemical potential of the gas, and $\mathcal{Z} \equiv e^{\beta \mu}$ is the fugacity. The occupation number is bounded $0 \leq n_\epsilon \leq 1$ as a result of the Pauli exclusion principle. Figure 12.1 shows the occupation number as a function of single-particle energy for various temperatures. The $T = 0$ ideal Fermi gas is characterized by a filled "Fermi sea": all energy levels ϵ for which $\epsilon \leq E_F$ are fully occupied ($n_\epsilon = 1$), while those for which $\epsilon > E_F$ are empty ($n_\epsilon = 0$). The Fermi energy E_F is equal to the chemical potential μ at $T = 0$. At high temperatures the gas is described by a Boltzmann-like distribution [26].

Since a trapped gas is not in contact with number or energy reservoirs in experiments, we ignore fluctuations in the total number and energy predicted by the grand canonical ensemble description, taking N and E to be the *average* total number and total energy, respectively. These can be calculated using the discrete sums

$$N = \sum_\epsilon \langle n_\epsilon \rangle \quad \text{and} \quad E = \sum_\epsilon \epsilon \langle n_\epsilon \rangle , \qquad (12.2)$$

where the sums run over all discrete states. In the limit of a large number of occupied states we can take the continuum limit, writing

$$N = \int_{\epsilon=0}^\infty g(\epsilon)\langle n_\epsilon \rangle d\epsilon \quad \text{and} \quad E = \int_{\epsilon=0}^\infty g(\epsilon)\epsilon\langle n_\epsilon \rangle d\epsilon , \qquad (12.3)$$

where $g(\epsilon) = \epsilon^2/2(\hbar\bar{\omega})^3$ is the energy density of states for a harmonically trapped gas in three dimensions, and $\bar{\omega} \equiv (\omega_x \omega_y \omega_z)^{1/3}$ is the geometric mean harmonic trap frequency.

Integrals of this type can be evaluated using the Fermi–Dirac integrals [26]

$$f_n(C) \equiv \frac{1}{\Gamma(C)} \int_0^\infty \frac{a^{n-1} da}{C^{-1} e^a + 1} = -\text{Li}_n(-C) \quad (0 \leq C < \infty, \ n > 0) , \qquad (12.4)$$

where $\text{Li}_n(C) = \sum_{j=1}^{\infty} C^j/j^n$ is a polylogarithmic function and $\Gamma(C)$ is the gamma function. For $n = 1$, $f_1 = \ln(1 + C)$. Using Eq. (12.4) we find that the average total number and energy are

$$N = (\beta\hbar\bar{\omega})^{-3} f_3(\mathcal{Z}) \quad \text{and} \quad E = 3k_B T(\beta\hbar\bar{\omega})^{-3} f_4(\mathcal{Z}) . \tag{12.5}$$

The number and energy at zero temperature can be found using the zero temperature limit of the Fermi function,

$$\lim_{T \to 0} f_n(\mathcal{Z}) = \frac{(\beta\mu)^n}{\Gamma(n+1)} . \tag{12.6}$$

As mentioned above, the Fermi energy E_F is defined as the zero temperature limit of μ. For convenience, we will also refer to the "Fermi temperature" $T_F \equiv E_F/k_B$, even though this temperature does not correspond to a phase transition, as is the case for T_c of Bose gases. Re-writing Eq. (12.5) in terms of E_F, we find

$$N = \frac{1}{6}\left(\frac{E_F}{\hbar\bar{\omega}}\right)^3 \quad \text{and} \quad E = \frac{3}{4} N E_F . \tag{12.7}$$

The chemical potential and fugacity at finite temperature can be found numerically by solving

$$6 f_3(\mathcal{Z}) = (\beta E_F)^3 . \tag{12.8}$$

Using the Sommerfeld expansion of the polylogarithms, one obtains low- and high-temperature approximations to the chemical potential in a three-dimensional harmonic trap [28]:

$$\mu \approx \begin{cases} E_F\left[1 - \frac{\pi^2}{3}\left(\frac{k_B T}{E_F}\right)^2\right] & \text{for } k_B T \ll E_F, \text{ and} \\ -k_B T \ln\left[6\left(\frac{k_B T}{E_F}\right)^3\right] & \text{for } k_B T \gg E_F. \end{cases} \tag{12.9}$$

Low Dimensionality Under certain conditions, a $T = 0$ Fermi gas in an anisotropic magnetic trap having $\omega_\perp \gg \omega_\|$ may become effectively one dimensional. If the atom number and temperature are such that $E_F < \hbar\omega_\perp$, the transverse degrees of freedom are "frozen out" and fermions occupy only the longitudinal energy levels of the trap. The maximum number of fermions N_{1D} that can populate such a one-dimensional configuration at $T = 0$ is equal to the aspect ratio of the trap: $N_{1D} = \omega_\perp/\omega_\|$. This scenario is especially relevant to atom chip micro-magnetic traps, whose aspect ratios can be on the order of 10^2 to 10^4 [29].

12.2.2
Density Distribution

Apart from the choice of $g(\epsilon)$, many of the expressions derived in Section 12.2.1 resemble the textbook treatment of a uniform Fermi gas. In this section, we calculate the non-uniform position and momentum distributions of *trapped* fermions. The position distribution is observable *in situ* (with sufficient spatial resolution), while the momentum distribution is observable in time-of-flight. We calculate these distributions by two different conceptual starting points: first, using semi-classical integrals; and second, using the local density approximation.

Semi-Classical Approximation Taking the energies $\epsilon \equiv \epsilon(r, p) = p^2/2M + U(r)$ to be those of a classical free particle at position r, where $U(r)$ is the trapping potential and M is the atomic mass, we can integrate the occupation function of Eq. (12.1) over the momentum degrees of freedom to find the semi-classical position distribution:

$$n(r) = \int \frac{d^3p}{(2\pi\hbar)^3} \left[\mathcal{Z}^{-1} e^{\beta \epsilon(r,p)} + 1 \right]^{-1}, \tag{12.10}$$

in which we have used the semi-classical phase-space volume of one quantum state, $(2\pi\hbar)^3$. Integration using Eq. (12.4) yields

$$n(r) = \Lambda_T^{-3} f_{3/2}\left(\mathcal{Z} e^{-\beta U(r)} \right), \tag{12.11}$$

where $\Lambda_T = \sqrt{2\pi\hbar^2 \beta/M}$ is the thermal de Broglie wavelength. Unlike the corresponding expression for ideal bosons, Eq. (12.11) is valid *at all temperatures*. The difference lies in the fact that we have ignored the occupation of the single-particle ground state in taking the continuum limit (see Eq. (12.3)), evidenced by the vanishing density of states $g(\epsilon)$ for $\epsilon = 0$. This does not pose a problem for fermions, for which the occupation of the ground state $\mathcal{Z}/(1 + \mathcal{Z}) \leq 1$. For bosons, however, the corresponding expression $\mathcal{Z}/(1 - \mathcal{Z})$ diverges; the continuum limit completely ignores the condensed fraction, whose contribution to the thermodynamics must be accounted for separately [26].

Local Density Approximation An alternate conceptual approach to the calculation of inhomogeneous distributions is the local density approximation. We start with the expression for the density of a *uniform* Fermi gas $n_{\text{uniform}} = \Lambda_T^{-3} f_{3/2}(\mathcal{Z})$ [26] and assume that local properties can be described by a local chemical potential $\mu - U(r)$ and hence a local fugacity $\mathcal{Z} e^{-\beta U(r)}$. We immediately recover Eq. (12.11), and obtain its zero-temperature using Eq. (12.6)

$$n_{\text{uniform}} \xrightarrow{T=0} \frac{1}{6\pi^2} \left[\frac{2M}{\hbar^2} E_F \right]^{3/2}. \tag{12.12}$$

This implies that long-range properties of the Fermi gas may be ignored, unlike in a BEC, which exhibits long-range phase coherence. In fact, the local density approximation and the semi-classical approach generally yield identical results for non-interacting fermions [24].

Specializing to the case of a three-dimensional harmonic oscillator potential $U(r) = \frac{1}{2} M(\omega_x^2 x^2 + \omega_y^2 y^2 + \omega_z^2 z^2)$ we obtain

$$n(r) = \Lambda_T^{-3} f_{3/2}\left(\mathcal{Z} \exp\left[-\frac{\beta M}{2} \left(\omega_x^2 x^2 + \omega_y^2 y^2 + \omega_z^2 z^2 \right) \right] \right). \tag{12.13}$$

At zero temperature,

$$n(r) = \frac{8N}{\pi^2 \bar{R}_{TF}^3}\left[1 - \frac{x^2}{X_{TF}^2} - \frac{y^2}{Y_{TF}^2} - \frac{z^2}{Z_{TF}^2}\right]^{3/2} \Theta\left(1 - \frac{x^2}{X_{TF}^2} - \frac{y^2}{Y_{TF}^2} - \frac{z^2}{Z_{TF}^2}\right) \quad (12.14)$$

where $\bar{R}_{TF} = \sqrt{2E_F/M\bar{\omega}^2}$ is the mean Thomas–Fermi radius of the cloud, $X_{TF} = \sqrt{2E_F/M\omega_x^2}$ and so on are the Thomas–Fermi lengths along each trapping axis, and $\Theta(\cdots)$ is the Heaviside step function.

The momentum distribution can also be calculated using either the local density or semi-classical approach. Experimentally, we observe the momentum distribution in a time-of-flight density image, for which the distribution is obtained by rescaling all spatial coordinates

$$\frac{x_i \to x_i}{\sqrt{1 + \omega_i^2 t^2}}$$

in Eq. (12.13) along the direction $i \in \{x, y, z\}$, and renormalizing to conserve particle number [25].

12.2.3
Crossover to Fermi Degeneracy

The $T = 0$ filled Fermi sea is quantum degenerate in the sense that it represents the absolute many-particle ground state of this non-interacting system. The meaning of the term "degenerate" here should not be confused with its more conventional meaning for a gas of bosons, for which degeneracy implies multiple or macroscopic occupation of the single-particle ground state (see Chapter 2). Multiple occupancy is forbidden for fermions, and so they are always "nondegenerate" in the more conventional sense of the word.

What, then, is the nature of the transition to quantum degeneracy in fermions? In contrast to the boson case, there is no phase transition into or out of the filled Fermi sea. As is the case with bosons, however, high-temperature expansions for thermodynamic quantities fail around $\mathcal{Z} = 1$. At lower temperatures the behavior differs dramatically from the predictions of classical, Boltzmann statistics. As $T \to 0$, $\mathcal{Z} \to 1^-$ for ideal bosons, whereas $\mathcal{Z} \to \infty$ for ideal fermions with the scaling $\mathcal{Z} \approx e^{\beta E_F}$, as implied by Eq. (12.9). It is interesting to note the quantitative relationship between fugacity and degeneracy for fermions:

$$n_0 \Lambda_T^3 = f_{3/2}(\mathcal{Z}), \quad (12.15)$$

where $n_0 \equiv n(0)$ is the central density of the cloud. Thus, $n_0 \Lambda_T^3 \simeq 0.77$ when $\mathcal{Z} = 1$ for fermions, which occurs at $T \simeq 0.57 T_F$. By comparison, $n_0 \Lambda_T^3 \simeq 2.61$ when $\mathcal{Z} = 1$ for bosons, at $T = T_c$.

The lack of a marked phase transition raises the question of what an experimental signature of Fermi degeneracy might be. Unlike a BEC, the non-interacting

DFG has an isotropic momentum distribution in time-of-flight, even when released from an anisotropic trap [28]. Thus, the aspect ratio of the cold cloud cannot be a signature of degeneracy. Instead, observations of Fermi degeneracy rely on two signatures: the average energy per particle, and the shape of the time-of-flight density distribution.

Using Eq. (12.5) we may write the average energy per particle as

$$\frac{E}{N} = 3k_B T \frac{f_4(\mathcal{Z})}{f_3(\mathcal{Z})}. \tag{12.16}$$

The finite zero-temperature limit of Eq. (12.16) is $3E_F/4$, corresponding to Fermi pressure [26, 27]. By comparison, the corresponding expression for the Boltzmann gas is $E/N = 3k_B T$, which tends toward zero at zero temperature.

The second signature of Fermi degeneracy is evident when the observed fermion time-of-flight distribution is compared to the predictions of the Boltzmann and Fermi–Dirac models. The latter is obtained by integration of Eq. (12.13) along the imaging line of sight.[28]

Taking z as the imaging direction, we find

$$\tilde{n}(x, y, t) = \frac{N}{2\pi r_x(t) r_y(t) f_3(\mathcal{Z})} f_2\left(\mathcal{Z} \exp\left[-\frac{x^2}{2r_x^2(t)} - \frac{y^2}{2r_y^2(t)}\right]\right), \tag{12.17}$$

where $r_i^2(t) = (\omega_i^{-2} + t^2)/\beta M$ is the cloud size in the $i \in \{x, y, z\}$ direction after a time t of free expansion. By comparison, the spatial distribution for an expanding cloud of classical particles is

$$\tilde{n}_{cl}(x, y, t) = \frac{N}{2\pi r_x(t) r_y(t)} \exp\left[-\frac{x^2}{2r_x^2(t)} - \frac{y^2}{2r_y^2(t)}\right], \tag{12.18}$$

using the same definitions for $r_i(t)$. Both of these signatures have been observed, as discussed further in Section 12.5.5.

12.3
The Atom Chip

The experiments described in this chapter were carried out at the University of Toronto with an atom chip designed and fabricated in the atom chip group at the Laboratoire Charles Fabry de l'Institut d'Optique [30, 31]. This section details the conductor layout, material composition, and supporting electrical and mechanical infrastructure for the chip. See Chapter 3 for a more thorough review of atom chip fabrication.

---

28) When integrating Fermi functions over Gaussian degrees of freedom, it is useful to note that $\int_{-\infty}^{\infty} dx\, f_n(Ce^{-x^2}) = \sqrt{\pi} f_{n+1/2}(C)$.

12.3.1
Chip Construction and Wire Pattern

The atom chip consists of gold wires electroplated onto a cleaved 16 mm × 28 mm × 600 µm silicon substrate. The chip was patterned using photolithography, with the metal wires deposited by evaporation and electroplating. A 20-nm titanium adhesion layer and 200-nm gold seed layer were evaporated onto the SiO_2 surface oxide, followed by solution electroplating of gold to a final wire height of 6 µm. The final result is a mostly bare silicon substrate hosting eight gold electrical contact pads connecting five separate conductors on its surface [30]. The wires can sustain an operating DC current of 5 A, though we never exceed 2 A in the central Z-wire.

We use two of the five chip wires in the work presented here, highlighted in dark gray in Figure 12.2a. The central Z-wire forms the basis of our static micromagnetic trap. An adjacent, thinner wire is used as a near-field antenna for delivering RF and microwave fields to the trapped atoms.

12.3.2
Electrical and Mechanical Connections

The chip is fastened to a copper support "stack" without glue or screws, but by mechanical pressure only. Custom-made MACOR ceramic C-clamps press flattened strips of beryllium-copper foil onto the gold contact pads for DC and RF electrical connections, while at the same time pressing the chip onto the stack (see

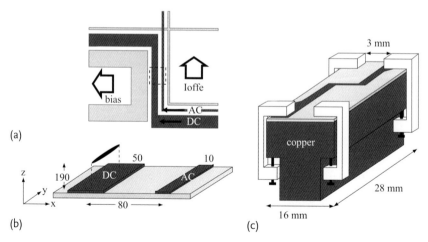

Figure 12.2 Atom chip conductor pattern and mounting schematic diagram. (a) Central atom chip conductor pattern, highlighting the Z-wire ("DC") and adjacent, thinner antenna wire ("AC") in dark gray. Thick arrows indicate the directions of external, uniform magnetic fields. (b) Close-up view of dashed region in (a), showing wire widths, center-to-center separation, and location of the trapped atomic cloud (black ellipsoid). All dimensions are in micrometers. (c) The atom chip is pressed onto a fly-cut copper block using ceramic MACOR C-clamps (white). Electrical connections are not shown.

Figure 12.2c). The beryllium-copper foil strips are in turn connected to ceramic-insulated copper wires which deliver current from the air side into the UHV chamber and onto the chip. The entire assembly is mounted vertically in the vacuum chamber with the atom chip at the bottom, face down [32, 33]. The stack also acts as a heat-sink for the atom chip, and was machined from oxygen-free high-conductivity (OFHC) copper.

12.3.3
The Z-Wire Magnetic Trap

We use an anisotropic, Ioffe–Pritchard-type Z-wire microtrap [34] throughout (see also Chapters 1 and 2). The external DC bias fields used for magnetic confinement are supplied by three pairs of magnetic field coils mounted outside of the vacuum chamber. With this setup we have achieved atom–surface distances between 80 µm and 300 µm, harmonic oscillation frequencies $\omega_{x,z} \sim 2\pi \times 200$ Hz to $2\pi \times 2.5$ kHz, $\omega_y \sim 2\pi \times 32$ to 50 Hz, and trap depths as large as $k_B \times 1.4$ mK.

The effects of wire surface roughness on the magnetic potential have been studied extensively for atom chips of this type [30, 35, 36]. Though sub-micrometer-scale wire rugosity is not a problem at our 190-µm working distance, we do observe three large and unexpected minima in the longitudinal magnetic potential. Though these "defects" prevent us from reaching the largest trap anisotropies available in Z-traps, we use them to our advantage during evaporative cooling. Magnetic gradients applied along y tilt the longitudinal potential so that atoms preferentially fill the largest and deepest of these local potential minima. In this local trap, the longitudinal oscillation frequency is two to three times larger than it would be in the absence of the defect. The increased mean oscillation frequency $\bar{\omega}$ (see Section 12.2.1) and accompanying gains in collision rate allow us to evaporate to quantum degeneracy more rapidly and efficiently than would otherwise be possible.

12.4
Loading the Microtrap

Our experimental approach to creating a ^{40}K DFG on an atom chip is motivated in large part by the scarcity of ^{40}K, whose natural isotopic abundance is only 0.012 %. Even when using a potassium dispenser with ^{40}K enriched to 5 %, we find that large MOT beams are essential to capture enough ^{40}K in a background UHV pressure compatible with magnetic trapping. Rather than using a reflected surface MOT [37], which would require a mirror-coated chip of length ~ 6 cm to accommodate our 4-cm-diameter MOT beams, we first load a large conventional MOT several centimeters beneath the atom chip, then magnetically trap and transport the atoms to the chip. Once loaded into the Z-trap, we sympathetically cool ^{40}K to quantum degeneracy with a ^{87}Rb reservoir to minimize ^{40}K atom number loss. This approach also obviates the need for a mirror-coated atom chip.

In this section and the next, we review our laser cooling, magnetic trapping and transport, chip loading, and evaporative cooling steps, noting that further detail is available in [5, 15, 32]. We also add to our previous work with a discussion of the roles of trap depth and trap volume in micro-trap experiments, with particular emphasis on chip loading in our setup.

12.4.1
Laser Cooling and Magnetic Transport to the Chip

Both ^{40}K and ^{87}Rb are initially trapped and cooled in a dual-species MOT formed by six counter-propagating 4-cm-diameter beams centered 5 cm below the atom chip [15, 32]. We use a single-chamber vacuum system, and load the MOT directly from atomic vapor created using a combination of dispensers and light-induced atom desorption (LIAD). During each MOT loading cycle, commercial high-power 405-nm LEDs irradiate our 75 mm × 75 mm × 160 mm Pyrex vacuum cell for several seconds to generate the atomic vapor. We use commercial Rb and home-made K dispensers to replenish the ^{40}K and ^{87}Rb coatings on the interior walls of the UHV chamber as needed – typically every few days or weeks [15]. By this method we achieve ^{40}K and ^{87}Rb atom numbers of roughly 6×10^6 and 6×10^8 in the MOT.

After optical molasses cooling of ^{87}Rb, both ^{87}Rb and ^{40}K are optically pumped into the stretched internal magnetic hyperfine states, $|F = 2, m_F = 2\rangle$ and $|F = 9/2, m_F = 9/2\rangle$, respectively. With all lasers extinguished, the mixture is magnetically trapped in a quadrupole magnetic trap and transported vertically to the surface of the atom chip using external coils. From there atoms are smoothly transferred from the quadrupole magnetic trap into the Z-trap located 190 μm from the chip surface, which has harmonic ^{40}K oscillation frequencies $\omega_{x,z} = 2\pi \times 823 \pm 7$ Hz and $\omega_y = 2\pi \times 46 \pm 1$ Hz, and a trap depth of $\sim k_B \times 1.05$ mK.

12.4.2
Loading Bosons and Fermions onto the Atom Chip

Along with the advantages of high compression and fast collision rate characteristic of atom chip microtraps comes a disadvantage: small trap volume. The volume occupied by a trapped gas depends on its temperature, unlike in the uniform "box" potential we are accustomed to from thermodynamics. The trap volume is not typically discussed when creating "macroscopic" magnetic traps with large coils, since their trap depths can be orders of magnitude larger than is required to confine laser-cooled atoms. For microtraps, however, the trap volume may limit the number of atoms that can "fit" into the trap, in a way that we will quantify in the following subsections. Our discussion points out a clear route to larger atom number, when it would be desirable.

An atom is trapped when its energy is less than the trap depth U_{td}. A good model of a thermalized gas in a trap-depth-limited trap is a truncated Boltzmann distribution [38], where truncation occurs at η times the temperature, that is

$U_{td} = \eta k_B T$. For the collision rates typical of atom chip traps, free evaporation resulting from the limited trap depth occurs when $\eta \lesssim 3$; by contrast, *efficient* evaporation occurs when $\eta \gtrsim 5$.

The laser cooling discussed in Section 12.4.1 allows atoms to be delivered to the chip at temperatures less than U_{td}. However, our loading efficiency is typically 10% or less, while phase-space density is roughly preserved: we load roughly 2×10^{7} ^{87}Rb atoms and 2×10^{5} ^{40}K into the Z-trap. We will attempt to explain the factors limiting this efficiency in the following subsections.

The initial loaded number of fermions places an upper bound on the number of ultra-cold fermions we can produce. Furthermore, though we load many more bosons than fermions, bosons are continually lost during the evaporative cooling process, as described in Section 12.5.1. The limited amount of "refrigerant" eventually limits the number of fermions that can be cooled, or their final temperature. It is, therefore, important to understand our loading process and ways in which it can be improved.

12.4.3
Effective Trap Volume

Our discussion here concerns the number, temperature, and density of a gas at the start of evaporative cooling, having already been loaded into the microtrap. We will assume an initial phase-space density ρ_0, which is typically $\lesssim 10^{-5}$ for laser-cooled atoms. At the densities typical of this point in the experimental cycle, the density distribution of the gas is well approximated as that of an ideal non-degenerate gas.

From Eq. (12.11), the density distribution of an ideal Fermi gas in three dimensions is $n(r) = \Lambda_T^{-3} f_{3/2}(\mathcal{Z} e^{-\beta U(r)})$, where $U(0) \equiv 0$. The corresponding expression for ideal bosons is obtained by using the thermodynamic Bose–Einstein function $g_{3/2}(\cdots)$ in place of $f_{3/2}(\cdots)$ in this expression. In either case, at low fugacity

$$n(r) \xrightarrow{\mathcal{Z} \ll 1} \Lambda_T^{-3} \mathcal{Z} e^{-\beta U(r)} . \tag{12.19}$$

Integrating both sides of the equation, we recover the total atom number

$$N = n_0 \int e^{-\beta U(r)} dr , \tag{12.20}$$

where the volume of integration is defined by $U < U_{td}$, and $n_0 = n(0)$.

In analogy with a uniform gas, we define the effective volume

$$V_{\text{eff}} \equiv \int e^{-\beta U(r)} dr , \tag{12.21}$$

such that $n_0 = N/V_{\text{eff}}$ [38]. For a simple three-dimensional box of side L, $V_{\text{eff}} = V = L^3$.

In the limit of $\eta \gg 1$, we can integrate the full Boltzmann distribution to find the trap volume in several typical cases. For a three-dimensional simple harmonic oscillator potential,

$$V_{\text{eff}}^{\text{SHO}} = \left(\frac{2\pi}{M\bar{\omega}^2}\right)^{\frac{3}{2}} (k_B T)^{\frac{3}{2}}, \tag{12.22}$$

where $\bar{\omega}$ is the geometric mean oscillation frequency. For the three-dimensional quadrupole (linear) trap,

$$V_{\text{eff}}^{\text{QT}} = 8\pi \bar{F}^{-3} (k_B T)^3, \tag{12.23}$$

where $U \equiv |\mathbf{F} \cdot \mathbf{r}|$ and $\bar{F}$ is the geometric mean gradient. Finally, a hybrid two-dimensional quadrupole and one-dimensional box model gives

$$V_{\text{eff}}^{\text{2QB}} = 2\pi L \bar{F}^{-2} (k_B T)^2. \tag{12.24}$$

In all of the above cases, the effective volume has a power-law dependence $V_{\text{eff}} = C_\delta T^\delta$.

12.4.4
A Full Tank of Atoms: Maximum Trapped Atom Number

Using the effective volume, we can now relate the trapped atom number to the initial phase-space density ρ_0, which is equivalent to the degeneracy parameter $n_0 \Lambda_T^3$ for $\mathcal{Z} \ll 1$:

$$N = \rho_0 \Lambda_T^{-3} V_{\text{eff}}(T). \tag{12.25}$$

Since $k_B T = U_{\text{td}}/\eta$ by definition, we can write out the explicit temperature dependence in Eq. (12.25) to find

$$N_{\max} = \rho_0 \left(\frac{M}{2\pi\hbar^2}\right)^{\frac{3}{2}} C_\delta \left(\frac{U_{\text{td}}}{\eta}\right)^{\delta + \frac{3}{2}}, \tag{12.26}$$

for a trap with a δ power law effective volume.

This equation shows us why the loaded atom number is typically smaller in microtraps than macrotraps. First, the trap depth U_{td} is typically smaller, which reduces atom number with a power law as fast as $U_{\text{td}}^{9/2}$ for a three-dimensional quadrupole. Second, even for comparable trap depths, the stronger trapping strength of a microtrap reduces C_δ: $C_3 \propto \bar{\omega}^{-3}$ in the case of a three-dimensional harmonic oscillator, for instance.

Equation (12.26) also demonstrates the importance of large currents in trapping wires. Consider the case in which a three-dimensional harmonic microtrap is formed above a single long wire, for which $\omega_\perp \propto B_{0\perp}/I$ where $B_{0\perp}$ is the perpendicular bias field and I is the wire current. Since the trap depth increases linearly with $B_{0\perp}$, Eq. (12.26) suggests that the maximum atom number at a fixed η and phase-space density is $N_{\max} \propto I^2 B_{0\perp}/\omega_z$. Assuming that the distance from the trapped atoms to the chip surface is fixed, and that $\omega_z \propto \sqrt{I}$, $N_{\max} \propto I^{5/2}$. Thus, we see that larger wire currents allow an increase in the number of trapped atoms.

12.4.5
Effect of Geometry on Loaded Atom Number

We now evaluate the trap volume and the expected maximum atom number loaded into several well-studied chip traps. We will start with the earliest proposed traps, described in [39] by Libbrecht, and assume they are loaded with ^{87}Rb in the $|2, 2\rangle$ state. For a single-loop quadrupole trap of radius 10 μm and using a 1-A current, the gradient is 5.4×10^5 G/cm and the trap depth $k_B \times 21$ mK. Assuming the trap can be loaded with $\eta \geq 4$ (corresponding to an initial temperature of ≤ 5 mK), the trap volume would be $V_{\text{eff}} \leq 310$ μm^3. At an initial phase-space density of 10^{-6}, 2×10^4 atoms could be loaded into the trap. However, a quadrupole trap has a magnetic field zero at its center, and is thus unsuitable for trapping ultra-cold atoms.

The Ioffe "(c)" configuration of [39] consists of concentric half-loops with a 10 μ minimum diameter. Using a 1-A current, the trap has a depth of $k_B \times 1.3$ mK and a curvature that gives $\bar{\omega}/2\pi \approx 94$ kHz. Although the trap is impressively strong, its effective volume is only 0.4 μm^3; less than one atom would be trapped at a phase-space density of 10^{-6}. For this reason, larger trap volumes than those of Libbrecht's pioneering geometries were required to achieve quantum degeneracy in an atom chip microtrap.

Finally, let us consider the Reichel Z-trap [37] (see also Chapter 2). The potential at the center of the trap is harmonic, with a typical geometric mean frequency of $\bar{\omega}/2\pi \approx 300$ Hz in our setup. The trap depth is limited by the transverse applied field, for which a typical value of 20 G gives $U_{\text{td}} \approx k_B \times 1.3$ mK. Assuming the trap is loaded at $\eta = 4$, we find that the effective volume is 1.3×10^7 μm^3, and the maximum trapped atom number 1.2×10^7 at an initial phase-space density of 10^{-6}. Although approximate, our calculation shows that the Z-trap geometry is capable of loading six to seven orders of magnitude more atoms than the Libbrecht geometry for the same initial phase-space density.

Furthermore, the calculation suggests that the loaded atom number in our experiment is limited by trap depth and volume. For our geometric mean field curvature of 3×10^4 G/cm^2 and initial temperature of 300 μK, the effective trap volume is 3×10^7 μm^3. One would expect 3×10^7 ^{87}Rb atoms at $\rho_0 \approx 10^{-6}$, and 3×10^5 ^{40}K atoms at $\rho_0 \approx 4 \times 10^{-8}$. This is consistent with our observations to within an order-of-magnitude (2×10^7 for ^{87}Rb and 2×10^5 for ^{40}K), and demonstrates that we are close to, if not at, the maximum possible number of loaded atoms, given the phase-space density and temperature after magnetic transport to the chip.

12.5
Rapid Sympathetic Cooling of a K-Rb Mixture

In this section we describe the sympathetic evaporative cooling of ^{40}K with ^{87}Rb in a microtrap, by which we produce a pure ^{40}K DFG or a dually degenerate BEC–DFG mixture. Particular emphasis is placed on the temperature dependence of the K-Rb scattering length, and its effect on K-Rb rethermalization and evaporation efficiency. The section concludes with a discussion of experimental signatures of quantum degeneracy in DFGs.

12.5.1
Forced Sympathetic RF Evaporation

We reach dual quantum degeneracy in ^{40}K and ^{87}Rb via sympathetic RF evaporative cooling of ^{87}Rb in a Z-trap having $B_0 \simeq 2.6$ G, ^{40}K harmonic oscillator frequencies $\omega_{x,z} = 2\pi \times 823 \pm 7$ Hz and $\omega_y = 2\pi \times 46 \pm 1$ Hz, and a trap depth of $k_B \times 1.05$ mK [5, 40–43]. In our case, ^{40}K is cooled indirectly by thermalizing elastic collisions with ^{87}Rb. By sweeping the RF evaporation frequency from 28.6 to 3.65 MHz in as little as 6 s, we reach $T/T_F \simeq 0.1 - 0.2$ with $\epsilon_F \simeq k_B \times 1.1$ μK and as many as 4×10^4 ^{40}K atoms, faster than has been possible in conventional magnetic traps [5, 15].[29] This rapid evaporation to degeneracy is made possible by the strong atom chip confinement.

As is evident in Figure 12.3, the ^{40}K is cooled to quantum degeneracy ($\mathcal{Z} \geq 1$) with only a five-fold loss in atom number, while the ^{87}Rb is evaporated with log-slope efficiency $-\partial[\log(\rho_0)]/\partial[\log(N)] = 2.9 \pm 0.4$, where ρ_0 is the peak phase-space density. When evaporating ^{87}Rb alone to BEC, the evaporation efficiency can be as high as 4.0 ± 0.1; sacrificing some evaporation efficiency, a more rapid evaporation can produce a BEC in just 2 s. By contrast, for ^{40}K-^{87}Rb mixtures, we observe that RF sweep times faster than 6 s are not successful in achieving dual degeneracy. The reason is that ^{87}Rb and ^{40}K rethermalize more slowly than ^{87}Rb alone, particularly

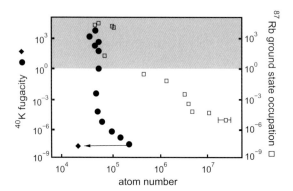

Figure 12.3 Sympathetic cooling to dual ^{40}K-^{87}Rb quantum degeneracy. Spin-polarized fermions without a bosonic bath cannot be successfully evaporatively cooled (diamond). However, if bosonic ^{87}Rb (squares) is evaporatively cooled, the fermionic ^{40}K is sympathetically cooled (circles) by thermalizing elastic collisions with ^{87}Rb. The vertical axes indicate the evolution of phase-space density en route to dual quantum degeneracy ($\mathcal{Z} \geq 1$) during evaporation: ^{40}K fugacity $\mathcal{Z}$ on the left, and ^{87}Rb ground-state occupation on the right. A typical run-to-run spread in atom number is indicated on the right-most point; all vertical error bars are smaller than the marker size.

29) A ^{6}Li DFG has been produced in an all-optical setup in as little as 3.5 s [44].

during the initial high-temperature stages of evaporation. Direct measurements of the ^{40}K and ^{87}Rb temperatures indicate that ^{40}K thermalization lags that of ^{87}Rb despite an experimentally optimized RF frequency sweep that is slower at higher temperatures and accelerates at lower temperatures [5].

12.5.2
K-Rb Cross-Thermalization

We have studied the ^{40}K-^{87}Rb inter-species scattering cross-section σ_{KRb} by measuring cross-thermalization rates at temperatures between 10 and 200 µK [5]. We compare the results, shown in Figure 12.4, to the σ_{KRb}-vs.-temperature behavior predicted by two scattering models. The simpler model, assuming only s-wave contact-interaction scattering, predicts $\sigma_{KRb} = 4\pi a^2/(1 + a^2 k^2) > \sigma_{RbRb}$ throughout the stated temperature range, where $a = a_{KRb}$ is the ^{40}K-^{87}Rb s-wave scattering length, and k the relative wavevector in the center-of-mass frame. The second more detailed model is based on an effective-range atom–atom scattering theory which includes the next-order correction to the s-wave scattering phase shift [45, 46] and is in good agreement with our measurements.

We attribute the observed reduction in scattering cross-section to the onset of the Ramsauer–Townsend effect, in which the s-wave scattering phase and cross-section approach zero for a particular value of relative energies between particles [5, 47].

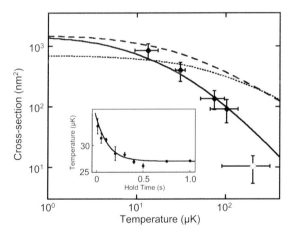

Figure 12.4 K-Rb cross-species thermalization. Measurements of σ_{KRb} (diamonds) are compared with the s-wave-only (dashed) and effective-range (solid) scattering models (see text). For reference, the s-wave σ_{RbRb} is also shown (dotted). Inset: We measure the cross-thermalization by abruptly reducing the temperature of ^{87}Rb and watching the temperature of ^{40}K relax with time. The data shown has an asymptotic ^{40}K temperature of 27 µK. The highest temperature point (open diamond) did not completely thermalize and lies off of the effective range prediction. A more sophisticated analysis may be required for this point, owing to severe trap anharmonicity at this high temperature. The vertical error bars are statistical (one standard deviation); the horizontal error bars show the spread in initial and final ^{40}K temperature during rethermalization.

Despite the high-temperature reduction in cross-section, however, ^{40}K and ^{87}Rb are relatively good sympathetic cooling partners. By comparison, ^{6}Li-^{87}Rb sympathetic cooling measurements [48, 49] suggest a zero-temperature cross-section approximately 100 times smaller than σ_{KRb}; in other words, a maximum ^{6}Li-^{87}Rb cross-section roughly equal to the lowest ^{40}K-^{87}Rb value we measure.

12.5.3
Density-Dependent Loss

The large and negative inter-species scattering length a_{KRb} results in a strong ^{40}K-^{87}Rb attractive interaction. At low temperatures and high atomic densities, this interaction creates an additional mean-field confinement that can lead to massive and sudden losses during sympathetic cooling [40, 41, 43]. These studies of density-dependent interaction-driven losses point to boson-boson-fermion[30] 3-body decay as the underlying mechanism for the collapse of the mixture. This effect ultimately limits the atom numbers in ^{40}K-^{87}Rb mixtures.

These effects manifest themselves in our experiment as ^{40}K and ^{87}Rb number losses near the end of evaporation. While we are able to produce BECs of up to 3×10^5 atoms when we work with ^{87}Rb alone, the simultaneous ^{87}Rb and ^{40}K atom numbers are restricted to at most $\sim 10^5$ and 4×10^4 respectively (see Figure 12.3).

12.5.4
Required Temperature

The reduction in the K-Rb elastic scattering cross-section discussed in Section 12.5.3 may lead the reader to wonder why we started evaporative cooling at such a high temperature. The reason is that the ^{87}Rb-^{87}Rb elastic collision rate required for efficient evaporative cooling within our magnetic trap lifetime imposes a lower bound on the initial temperature. In this section we will shift our focus to *single-species* collision rates in order to understand this constraint.

Evaporative cooling requires a trap lifetime that is some multiple (typically 10^3) of γ_{coll}^{-1}, where $\gamma_{coll} = n_0 \sigma v_r$ is the collision rate at the center of the trap; n_0 is the central density, σ is the elastic scattering cross-section, and $v_r = \sqrt{8 k_B T / \pi M}$ is the relative velocity of collision partners [50].

Since $\rho_0 \approx n_0 \Lambda_T^3$ at the start of evaporation (see Section 12.4.4), we can express the central density in terms of the phase-space density:

$$\gamma_{coll} = \frac{\sigma \rho_0 M}{\pi^2 \hbar^3} (k_B T)^2, \qquad (12.27)$$

which is independent of atom number.

This relation can be used to give a lower bound on temperature in a broad range of traps. Defining γ_{coll}^{min} as the minimum scattering rate,

$$(k_B T)^2 \geq \gamma_{coll}^{min} \frac{\pi^2 \hbar^3}{M \sigma \rho_0}. \qquad (12.28)$$

---

30) Fermion-fermion-boson 3-body decay is precluded by the Pauli exclusion principle.

In the case of ^{87}Rb, approximating σ by its low temperature limit $8\pi a_s^2$, where a_s is the s-wave scattering length, we find that the minimum temperature at the beginning of evaporative cooling is

$$T_0^{\min} = 300\,\mu\text{K} \times \left(\frac{10^{-6}}{\rho_0}\right)^{1/2} \left(\frac{\gamma_{\text{coll}}^{\min}}{150\,\text{s}^{-1}}\right)^{1/2}, \qquad (12.29)$$

for *any* loaded atom number or trap geometry. In our case, the Rb-Rb elastic collision rate at the start of RF evaporation is roughly $150\,\text{s}^{-1}$, and the phase-space density 10^{-6} or slightly higher. Equation (12.29) predicts $T_0^{\min} = 300\,\mu\text{K}$, which exactly coincides with the temperature we measure at the start of evaporation. (The excellent agreement should be taken with a grain of salt, since the model is only approximate.) Can we gain anything by decompressing or compressing the trap? Our trap lifetime $\tau_t \sim 5\,\text{s}$ is only 750 times larger than the collision time $\gamma_{\text{coll}}^{-1} = 1/150 \simeq 6.7\,\text{ms}$ – less than optimal, assuming that we require $\tau_t \gtrsim 1000 \gamma_{\text{coll}}^{-1}$ for efficient evaporation. Any adiabatic decompression would decrease our collision rate below $\gamma_{\text{coll}}^{\min}$ and result in significant loss of evaporation efficiency. If the trap lifetime were longer, decompression *would* reduce the temperature without sacrificing evaporation efficiency; $\tau_t > 1000\gamma_{\text{coll}}^{-1}$ would be much easier to satisfy even as $\gamma_{\text{coll}}^{-1}$ decreased in the decompression. Compressing the trap near the end of evaporation would increase in K-Rb 3-body loss, which would exacerbate the losses we already observe due to attractive ^{40}K-^{87}Rb interactions near the end of evaporation (see Section 12.5.3). Compressing near the beginning of evaporation should not induce much loss since densities are relatively small at that point, but compression causes dramatic loss in our case due to the limited trap depth. For these reasons, we have little choice but to start our evaporation in the regime where the Ramsauer–Townsend effect is significant.

12.5.5
Experimental Signatures of Fermi Degeneracy

Following the discussion in Section 12.2.3, we assess the degree of Fermi degeneracy in ^{40}K by fitting ideal Boltzmann and Fermi gas theory to time-of-flight absorption data. Unlike the Boltzmann gas, whose spatial width tends toward zero as $T \to 0$, Pauli exclusion results in a finite-sized Fermi gas with a finite average momentum, even at $T = 0$ [27]. This Fermi pressure (see Section 12.2.3) is evident in Figure 12.5a, in which the in-trap cloud width deviates from the Boltzmann prediction for $T/T_F \lesssim 0.5$. Absorption images taken at $T/T_F = 0.95$ and 0.35 are overlayed with a circle indicating E_F in Figure 12.5b,c, demonstrating that the average momentum of the Fermi gas plateaus at low temperature.

In addition to the expansion energy, a second measure of the degree of Fermi degeneracy is the deviation of the time-of-flight spatial profile from the Gaussian envelope predicted by Boltzmann statistics. The Gaussian fit is excellent at high temperatures, but fails at low temperatures [1]. Figure 12.5d shows the residuals of a Gaussian fit using Eq. (12.18) to a gas at $T/T_F = 0.1$, compared to those of a fit

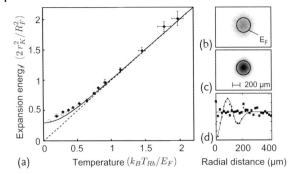

Figure 12.5 (a) The apparent *in situ* fermion temperature measured by Gaussian fits to ^{40}K time-of-flight absorption data, plotted versus temperature of both thermal (diamonds) and Bose-condensed (circles) ^{87}Rb. The data follows a curve derived from Gaussian fits to artificial, perfect Fermi distributions (solid line), deviating from the corresponding Boltzmann prediction (dashed) at temperatures below $T/T_F \approx 0.5$. Quantum degeneracy of ^{87}Rb has little effect on the ^{40}K expansion energy in this data (circles vs. diamonds). (b, c) Absorption images for $T/T_F = 0.95$ and 0.35 respectively, overlayed with black circles of radius R_F rescaled after time-of-flight (see Section 12.2.2). (d) Residuals from radial fits of Gaussian (dots) and Fermi–Dirac (squares) envelope functions to absorption data at $T/T_F = 0.1$. The good Fermi–Dirac fit and poor Gaussian fit are evidence of a Fermi degenerate sample.

using Eq. (12.17), which assumes Fermi–Dirac statistics. Though the Gaussian and Fermi–Dirac profiles are very similar, careful analysis shows that the Fermi–Dirac function is a better fit, with a reduced χ^2 three times lower than the Gaussian fit.

12.6
Species-Selective RF Manipulation

For atoms confined to a magnetic trap, an applied field oscillating at radio frequencies (RF) can resonantly couple adjacent m_F states. The atomic system in the combined static and time-varying magnetic fields can be described by a Hamiltonian with new, uncoupled eigenstates – the so-called "adiabatic" or "RF-dressed" states [51]. The spatial dependence of these states, and thus the spatial character of the trapping potential, can be manipulated by varying the RF amplitude B_{RF} and frequency ω_{RF} (see Chapter 7 for a thorough review of RF-dressed potentials). In this section we discuss two useful regimes for manipulating ultra-cold boson–fermion mixtures on atom chips using these effects.

The first type of RF manipulation is species-selective evaporative cooling. As implied in Section 12.5, RF manipulation can drive ^{87}Rb spin-flip transitions to untrapped spin states at the edge of the ^{87}Rb dressed potential without inducing loss in ^{40}K, as a consequence of the explicit m_F and g_F dependence of the magnetic trapping potentials. In the second type of RF manipulation, B_{RF} is large enough that Landau–Zener tunneling between dressed states is suppressed and the usual adiabatic condition is satisfied [50]. Atoms adiabatically follow the RF-dressed magnetic eigenstates and remain trapped in a double-well potential [52–54]. This effect was first demonstrated on both thermal [51, 55] and quantum degenerate Bose gases [11, 56] and is now a well-established method for dynamically "splitting" an ultra-cold Bose gas [11–14, 57]. We have added to this work by demonstrating the simultaneous creation of a ^{87}Rb double well and a ^{40}K single well [15]. This result demonstrates that the adiabatic potentials experienced by each species can be dramatically different in an applied RF field, owing to the different values of g_F for ^{87}Rb and ^{40}K.

12.6.1
Sympathetic RF Evaporation

In this section we consider a simple sympathetic cooling scenario, typical in our experiments; an RF "knife" is used to evaporatively cool ^{87}Rb, which then acts as a refrigerant for the ^{40}K.[31] Ideally, this scheme would causes no loss of ^{40}K atoms; here we explore to what extent this is true.

A critical parameter in evaporative cooling is $\eta = U_{td}/k_B T$ (see Section 12.4.2). The larger the choice of η, the fewer atoms are removed from the trap, since a smaller fraction of the kinetic energy distribution lies above U_{td} [50]. In sympathetic cooling of ^{40}K by ^{87}Rb, we would ideally like the RF field to have no effect on the ^{40}K trap depth. However, we shall see that for our scheme the RF knife *does* limit the trap depth for ^{40}K, though the effect is small. This effect may be quantified by evaluating η for ^{40}K in our cooling scheme.

We begin by considering atoms in spin states $|F, m_F\rangle$ in a static ("DC") magnetic trap, whose potential is $U(\mathbf{r}) = m_F g_F \mu_B B_{DC}(\mathbf{r})$. The minimum magnetic field amplitude is $B_0 \equiv B_{DC}(0)$ using the notation $B_{DC}(\mathbf{r}) \equiv |\mathbf{B}_{DC}(\mathbf{r})|$. We apply a weak, sinusoidally time-varying magnetic field of amplitude B_{RF} and frequency ω_{RF}. If $\hbar \omega_{RF} > g_F \mu_B B_0$, then atoms in state m_F at positions $\mathbf{r}_1$ for which

$$g_F \mu_B B_{DC}(\mathbf{r}_1) = \hbar \omega_{RF} \qquad (12.30)$$

can undergo spin-flips to adjacent m_F states, including untrapped ones. This imposes an effective trap depth $U_{td} = g_F m_F \mu_B (B_{DC}(\mathbf{r}_1) - B_0)$ on these atoms, since any atom with energy above U_{td} is ejected from the trap. Therefore, we can write

$$U_{td} \equiv \eta k_B T = m_F(\hbar \omega_{RF} - g_F \mu_B B_0) \qquad (12.31)$$

for a gas of atoms in state $|F, m_F\rangle$ in thermal equilibrium at temperature T.

For two species in thermal equilibrium in the same magnetic trap, we can write down simultaneous equations like (12.31) for both species. From these, we infer the relationship

$$\eta_K = \left(\frac{m_F^{(K)}}{m_F^{(Rb)}}\right) \eta_{Rb} + m_F^{(K)} \left[g_F^{(Rb)} - g_F^{(K)}\right] \frac{\mu_B B_0}{k_B T}. \qquad (12.32)$$

---

31) Microwave magnetic fields at 6.8 GHz may also be used to species-selectively evaporate ^{87}Rb via hyperfine transitions [48].

between the two η parameters in the case of ^{40}K and ^{87}Rb, using the fact that ω_{RF} and B_0 are common to both species. This convenient form highlights the role of m_F and g_F in the RF evaporation of a mixture of atomic species.

Moving now to a more specific case, we note that typical sympathetic cooling ramps are done with ^{87}Rb in the $|2, 2\rangle$ state, and the ^{40}K in $|9/2, 9/2\rangle$, since this mixture is stable with respect to many inelastic collisions [40]. Equation (12.32) yields

$$\eta_K = \frac{9}{4}\eta_{Rb} + \frac{5}{4}\frac{\mu_B B_0}{k_B T}, \tag{12.33}$$

from which we immediately conclude that $\eta_K > 9\eta_{Rb}/4$ for all values of temperature. Having $\eta_K > \eta_{Rb}$ ensures that ^{87}Rb can be evaporated without inducing significant loss in the ^{40}K population, as required for efficient sympathetic evaporative cooling.

For a typical experimental value of $\eta_{Rb} \approx 8$, which is roughly constant throughout the evaporation, Figure 12.6 demonstrates that η_K rises sharply for all trappable $m_F^{(K)}$ states as sympathetic evaporation proceeds and the ^{40}K temperature decreases. ^{40}K spin-flip losses are most likely to occur when η_K is smallest. In our experiments, this occurs at the beginning of evaporation, when $B_0 = 5.7$ G, $T \sim 300$ µK [5, 15], and $\eta_K = 8.1$.

We can also use Eq. (12.33) to compare η_K and η_{Rb} in the scenario in which all the ^{87}Rb is evaporated away to leave a pure ^{40}K DFG. Taking $\eta_{Rb} \to 0$ and working at a typical pure-DFG temperature $T = 220$ nK [5], we find $\eta_K > 220$ for all trappable ^{40}K sublevels. This confirms that ^{87}Rb can be fully and selectively ejected from the trap without causing any significant ^{40}K loss.

Finally, we note that it is possible in principle to evaporate ^{40}K without inducing ^{87}Rb loss. For RF frequencies $\hbar\omega_{RF} \leq g_F^{(Rb)}\mu_B B_0$, only ^{40}K atoms are ejected from the trap, at positions r_2 such that $\hbar\omega_{RF} = g_F^{(K)}\mu_B B(r_2)$. Following the derivation of Eq. (12.31), we can express the ^{40}K trap depth in this scenario

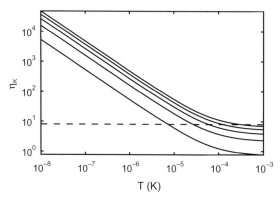

Figure 12.6 Evaporation parameter η_K for ^{40}K atoms in a typical sympathetic cooling ramp with $B_0 = 5.7$ G and $\eta_{Rb} \approx 8$ (dashed line, see text). The curves represent different magnetically trappable values of $m_F^{(K)}$ (1/2 to 9/2, from bottom to top). η_K rises sharply as evaporation proceeds and the temperature decreases, while η_{Rb} remains constant.

as $U_{\text{td}}^{(K)} = m_F^{(K)}[g_F^{(Rb)} - g_F^{(K)}]\mu_B B_0$. In our experiment, using $B_0 = 5.7\,\text{G}$ and the stretched states of ^{40}K and ^{87}Rb, $U_{\text{td}} = 5\mu_B B_0/4 \simeq k_B \times 480\,\mu\text{K}$. Thus, we expect to be able to evaporate $T \lesssim 480\,\mu\text{K}$ ^{40}K clouds without affecting the ^{87}Rb population.

12.6.2
Species-Selective Double Wells

In this section we discuss the species-selective nature of RF-dressed double wells. In a two-species mixture, this effect permits the simultaneous formation of a double-well potential for ^{87}Rb and single-well potential for ^{40}K [15], and vice versa. We use an RF amplitude $B_{\text{RF}} \ll B_0$, for which the effective adiabatic potentials[32] may be written [11–13]

$$U_{\text{eff}}(r) = m_F' \sqrt{\delta(r)^2 + \Omega(r)^2} \tag{12.34}$$

where

$$\delta(r) = \mu_B g_F B_{\text{DC}}(r) - \hbar \omega_{\text{RF}}$$
$$\Omega(r) = \mu_B g_F B_{\text{RF}\perp}(r)/2, \tag{12.35}$$

$B_{\text{DC}}(r) \equiv |\mathbf{B}_{\text{DC}}(r)|$ is the static magnetic field amplitude, $B_{\text{RF}\perp}(r)$ is the amplitude of the B_{RF} component which is perpendicular to $\mathbf{B}_{\text{DC}}(r)$ at point r, and m_F' is the new, effective magnetic quantum number. The parameters $\delta(r)$ and $\Omega(r)$ can be identified as the local detuning and Rabi frequency of the RF field, respectively. Using Eq. (12.34), we can calculate effective adiabatic potentials for ^{87}Rb and ^{40}K and use them to illustrate the simultaneous creation of single- and double-well potentials in a ^{40}K-^{87}Rb mixture.

For clarity, we consider the formation of the ^{87}Rb double well and ^{40}K single well separately. Our starting point is a $|9/2, 9/2\rangle$–$|2, 2\rangle$ ^{40}K-^{87}Rb mixture confined to a static, anisotropic harmonic Z-trap directly above the RF wire (see Figure 12.7a) with $B_0 = 1.214\,\text{G}$, $\omega_{x,z} = 2\pi \times 1.23\,\text{kHz}$, and $\omega_y = 2\pi \times 13.7\,\text{Hz}$.

Rb Double Well An RF field with initial frequency $\omega_{\text{RF}} = 2\pi \times 800\,\text{kHz}$ and detuning $\delta^{(Rb)} = -50\,\text{kHz}$ is applied by ramping up its amplitude from zero to the final value $B_{\text{RF}} = 200\,\text{mG}$. A potential barrier is formed at $r = 0$ by sweeping the RF frequency through the resonant point $\delta^{(Rb)}(0) = 0$. As the RF field is applied,

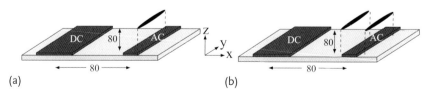

Figure 12.7 Atoms confined to single-well (a) and double-well (b) potentials induced by RF manipulation 80 μm above the chip surface.

---

32) For $B_{\text{RF}} \approx B_0$, Eq. (12.34), which relies on the rotating-wave approximation (RWA), is no longer valid [53, 54, 58].

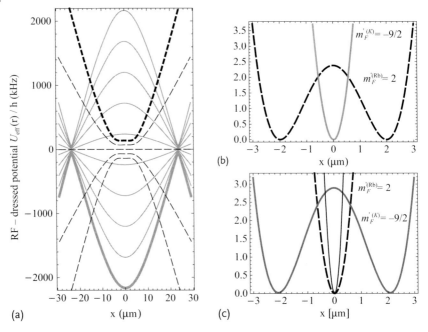

Figure 12.8 Simultaneous adiabatic dressed potentials for ^{87}Rb and ^{40}K. (a) The ^{87}Rb-double-well case. ^{87}Rb (black) and ^{40}K (gray) effective dressed potentials are plotted as a function of the spatial coordinate x (see Figure 12.7). Each curve corresponds to a single value of m'_F for ^{87}Rb and ^{40}K, with the uppermost curves corresponding to $m'_F = 2$ and $m'_F = 9/2$, respectively. ^{87}Rb atoms populate their upper-most $m'_F = 2$ dressed state (thick, dashed), while ^{40}K atoms populate their lower-most $m'_F = -9/2$ single-well dressed state (thick, gray). (b) A closer view of the ^{87}Rb double well and ^{40}K single well, plotted together on a single vertical U_{eff}/h axis in units of kHz. Both curves have been shifted vertically to align their potential minima at zero kHz. (c) A closer view of the ^{40}K double well and ^{87}Rb single well, with vertical axis similar to (b). The dressed ^{87}Rb single well (black, dashed) deviates slightly from the undressed single well (solid, thin black), illustrating the slight loss of radial trap curvature.

each undressed state is adiabatically connected to one dressed state; here $m_F^{(\text{Rb})} = 2$ is connected to $m_F'^{(\text{Rb})} = 2$, shown as the upper-most black curve in Figure 12.8a. After sweeping to a final RF frequency $\omega_{\text{RF}} = 2\pi \times 860\,\text{kHz}$, the barrier height is $h \times 2.4\,\text{kHz}$ and the x-direction double-well separation is $4\,\mu$ (see Figure 12.8b). The $m_F'^{(\text{Rb})}$ level repulsion at the double-well minima is $70\,\text{kHz}$, sufficient to prevent Landau–Zener spin-flips at our working temperatures $T \lesssim 1\,\mu\text{K}$. The ^{87}Rb population thus remains trapped in the $m_F'^{(\text{Rb})} = 2$ dressed level.

K Single Well The trapping potential for the ^{40}K atoms is affected in a very different way for the same magnetic field configuration. In our current example (Figure 12.8a,b), the detuning for ^{40}K is positive; $\delta^{(K)}(0) = 2\pi \times 482\,\text{kHz}$ at the trap minimum. Near $r = 0$, the RF dressing adiabatically connects $m_{F(K)} = 9/2$ to $m_{F'}^{(K)} = -9/2$. Since $[\delta^{(K)}(0)]^2 \gg [\Omega^{(K)}(0)]^2$, the potential curvature near $r = 0$

is largely unaffected by the RF coupling (see Eq. (12.34)). More quantitatively, the dressed states are most deformed where $\delta^{(K)}(r) \approx 0$, which corresponds to $x \simeq \pm 23\,\mu m$, and a potential energy of roughly 110 µK above the local minimum at $r = 0$. Since our experiments are typically conducted with $T \lesssim 1\,\mu K$, we can be satisfied that the ^{40}K potential retains its original form near $r = 0$ without inducing any ^{40}K loss.

One important feature of this single- and double-well arrangement is that the ^{87}Rb double-well separation and barrier height may be tuned over a wide range by adjusting ω_{RF} and B_{RF} without affecting the shape of the ^{40}K potential.

An obvious extension of the work described here and in [15] would be to reverse the roles of boson and fermion, creating a double well for fermions overlapped with a single well for bosons. The magnetic hyperfine structure of ^{40}K and ^{87}Rb makes this possible in a ^{40}K-^{87}Rb mixture, but with slightly different results than in the ^{87}Rb-double-well case.

Following the ^{87}Rb-double-well example of the preceding section, here we sweep the RF frequency from $\omega_{RF} = 2\pi \times 338\,kHz$ to $\omega_{RF} = 2\pi \times 383\,kHz$. In the same static trap with $B_0 = 1.214\,G$, $\delta^{(K)}$ changes sign from -50 to $+5\,kHz$, while $\delta^{(Rb)}$ remains negative throughout. This creates a ^{40}K double well in the $m_F'^{(K)} = 9/2$ state with x-direction well separation $\sim 4\,\mu m$, barrier height $h \times 2.9\,kHz$ at $r = 0$, and $m_F'^{(K)}$ level repulsion $\sim 70\,kHz$ at $x = \pm 2.1\,\mu m$. In contrast to the ^{87}Rb-double-well scenario, here both ^{40}K and ^{87}Rb adiabatically follow their respective uppermost dressed levels, which exhibit their strongest spatial deformation near the trap center. While ^{40}K experiences a double-well potential, the ^{87}Rb $m_F'^{(Rb)} = 2$ potential is a single well with slightly reduced radial curvature from the initial, undressed $m_F = 2$ potential, as shown in Figure 12.8c.

In addition to the species-selectivity of this process, it should be emphasized that atom chips are particularly well-suited to creating adiabatic dressed-state potentials due to the proximity of the atoms to chip wire RF antennae. The double wells described in this section were created using RF Rabi frequencies $\Omega \sim 100$–$200\,kHz$, though we can achieve values as large as 1 MHz with tens of milliamperes rms in the chip wire antenna. By comparison, achieving $\Omega \approx 1\,MHz$ with an air-side RF antenna would require a circular coil of radius 3 cm and 3 turns bearing 10 A rms of AC current.

12.7
Fermions in an Optical Dipole Trap near an Atom Chip

In our discussion of fermions on atom chips thus far, we have focused on spin-polarized ^{40}K in magnetic traps. In this section we describe an extension of our atom chip capabilities with the incorporation of an external crossed-beam optical dipole trap skimming the surface of the atom chip. (See Chapter 3 for a discussion of other on-chip optical potentials.) Optical trapping enables the use of any and all hyperfine and Zeeman spin states in our experiments, as well as more flexible control of the magnetic field environment. These added features allow us to work

388 12 Fermions on Atom Chips

with strongly interacting ^{40}K spin mixtures, while retaining the atom chip benefits of near-field RF and microwave manipulation and rapid evaporation to Fermi degeneracy.

12.7.1
Optical Trap Setup

Two optical trapping beams are generated using the output of a single 500-mW Nd:YAG laser operating at $\lambda = 1064$ nm, and directed along the x- and y-directions through the vacuum cell and beneath the atom chip using air-side optics (see Figure 12.9). Ideally, the foci would be near enough to the chip surface to allow sufficient RF and microwave coupling, but not so near that the optical potential is degraded by light scattering off the edges of the chip as the beams are focussed. The beams should also be aligned to the existing Z-trap position for efficient transfer from the Z-trap into the dipole trap. To satisfy these constraints we focus the two beams at roughly 190 µ beneath the chip surface, with $1/e^2$ waists $w_0 \sim 18$ and 30 µm, and Rayleigh ranges $z_0 = \pi w_0^2/\lambda \sim 1.0$ and 2.7 mm, respectively, as shown in Figures 12.9b,c.

12.7.2
Loading the Optical Trap

We load the optical trap by ramping off the Z-trap magnetic fields and ramping on the optical beams one at a time. First, the y-direction peak beam intensity is

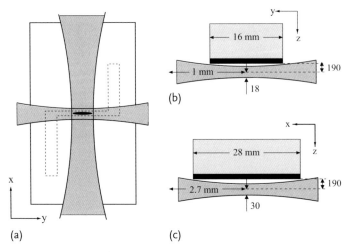

Figure 12.9 Alignment of the dipole beams to the atom chip. (a) Bottom view, showing the alignment with the Z-wire and the trapped atoms (black ellipse). (b, c) Side-views of beams skimming 190 µm below the atom chip surface. Beam waists and Rayleigh ranges are indicated, along with the outer dimensions of the atom chip (black) and copper mounting block (light gray). Dimensions are in units of micrometers unless otherwise indicated.

increased linearly from 0 to $\sim 2.9 \times 10^7$ mW/cm^2 in 100 ms. The Z-trap is then ramped off in 50 ms. Finally, the x-direction peak beam intensity is linearly ramped from 0 to $\sim 8.5 \times 10^6$ mW/cm^2 in 100 ms.

Unlike when loading the Z-trap from the macroscopic magnetic trap (see Section 12.4.1), the mode matching between the Z-trap and the initial single-beam optical trap is excellent. This allows us to load the optical trap with no observable loss in atom number, but not without introducing some heating: 6×10^3 atoms in the Z-trap at $T/T_F \simeq 0.56$ are all transferred into the optical trap, after which $T/T_F \simeq 1$.

12.7.3
Microwave and RF Manipulation

With the atoms now trapped solely by the optical trap, we can use the antenna wire (see Section 12.3.1) to apply microwave radiation with amplitude B_μ and frequency $\omega_\mu \sim 2\pi \times 1.3$ GHz to drive transitions between ^{40}K hyperfine ground states. To test and calibrate our microwave system, however, we first observed the effects of microwave radiation on magnetically trapped ^{40}K. By scanning the microwave frequency near 1.3 GHz in a Z-trap with magnetic minimum $B_0 = 5.22$ G, we observe a loss feature in the trapped atom number, corresponding to hyperfine transitions between the initial trapped state $|9/2, 9/2\rangle$ and the final, untrapped $|7/2, 7/2\rangle$ state (see Figure 12.10a).

Microwave and RF manipulation also allow us to create an interacting Fermi gas in the optical trap with a $|9/2, -9/2\rangle - |9/2, -7/2\rangle$ spin mixture after loading a spin-polarized sample from the Z-trap in state $|9/2, 9/2\rangle$. Atoms are transferred from $|9/2, 9/2\rangle$ to $|9/2, -7/2\rangle$ by rapid adiabatic transfer. In a 20-G external magnetic bias field, we sweep the microwave frequency from below the $|9/2, 9/2\rangle \rightarrow |7/2, 7/2\rangle$ transition ($\omega_\mu = 2\pi \times 1235.8$ MHz) to above the $|7/2, -7/2\rangle \rightarrow |9/2, -7/2\rangle$ transition ($\omega_\mu = 2\pi \times 1336.2$ MHz) in 500 ms. This is followed by a 5-ms RF frequency

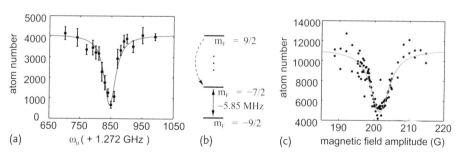

Figure 12.10 (a) Microwave-induced atom number ^{40}K loss from a magnetic Z-trap. (b) Schematic Zeeman level diagram of the ^{40}K $F = 9/2$ hyperfine ground state, indicating the microwave adiabatic rapid population transfer (dashed arrow) and RF pulse (solid arrow) used to create a $|9/2, -7/2\rangle - |9/2, -9/2\rangle$ spin mixture (see text). (c) 3-body loss of spin-mixed ^{40}K from an optical trap is due to an enhancement of the scattering cross-section induced by the Feshbach resonance near 202 G.

sweep from $\omega_{RF} = 2\pi \times 5.80$ MHz to $2\pi \times 5.90$ MHz (see Figure 12.10b), creating an equal spin mixture of $|9/2, -9/2\rangle$ and $|9/2, -7/2\rangle$.

The ^{40}K-^{40}K interaction strength may be dramatically increased using the well-known Feshbach resonance in a DC magnetic field of roughly 202 G [59]. Using external coils, we apply a magnetic bias field and monitor the ^{40}K population as a function of this field. We have observed the Feshbach resonance in our atom chip setup as a strong loss feature near 202 G, as shown in Figure 12.10b. The atom number loss is attributed to 3-body decay induced by the strong interactions near the Feshbach resonance.

12.8
Discussion and Future Outlook

The union of Fermi gases and atom chips is an important step forward in degenerate quantum gas research. In this chapter, we have described our work producing the first DFG on an atom chip. Using a large dual-species MOT, ^{40}K and ^{87}Rb are captured and loaded onto an atom chip in numbers large enough to enable sympathetic evaporative cooling to DFG and dual DFG-BEC quantum degeneracy. The strong confinement and large inter-species collision rate afforded by the micro-magnetic trap permits an evaporation to quantum degeneracy in as little as 6 s, faster than had been previously possible in magnetic traps. We have also demonstrated species-selective double-well potentials in a ^{40}K-^{87}Rb mixture, as well as the creation of a strongly interacting ^{40}K DFG in a crossed-beam optical dipole trap skimming the surface of the atom chip.

We also note several disadvantages associated with using micro-magnetic traps for fermions. Magnetic traps are not ideal for working with the spin-mixtures required for an interacting DFG, since spin-changing collisions can populate magnetically untrapped spin states, causing loss. We address this limitation by incorporating a crossed-beam optical dipole trap into our atom chip setup. The relatively small trap volumes of microtraps also limits the number of target ^{40}K and refrigerant ^{87}Rb atoms that can be loaded into the chip trap. Even at our maximum achievable trap depth, these limits on the number of refrigerant atoms ultimately limit the DFG to 4×10^4 atoms at $T/T_F \gtrsim 0.1$ in our system. By comparison, larger, colder, interacting DFGs are routinely produced in spin-independent optical potentials [2].

Despite these drawbacks, atom chips remain a useful and versatile tool for ultracold Fermi gas research. We close this chapter with descriptions of some of our ongoing and future research directions with a new atom chip: double-well potentials; interacting DFGs and DFG-BEC mixtures; and optical probes and traps.

New Atom Chip We have recently adopted a richer atom chip design, which improves on its predecessor (described in Section 12.3) in three important areas. First, the new conductor layout allows a wider selection of micro-magnetic traps with greater tunability of the oscillation frequencies. We are also now able to generate

RF double-well potentials at arbitrary angles and distances from the chip surface, as in [12]. Second, new near-field antenna wire geometries and high-frequency electrical feedthroughs improve microwave and RF impedance matching, reducing coupling losses between the air-side and the chip. Finally, all electrical connections to the chip are made on the back side of the substrate, increasing the optical access for probe and trapping beams compared to its predecessor. Designed and fabricated at the University of Toronto [60], the chip consists primarily of silver conductors that were patterned by photolithography and evaporatively deposited onto an aluminum nitride substrate.

Bose–Fermi Mixtures and Double-Well Potentials The species-selectivity of RF-dressed double-well potentials may be useful in studying boson–fermion interactions in ultra-cold atomic mixtures. The strong attractive interaction between ^{40}K and ^{87}Rb, known to impede sympathetic cooling in ^{40}K-^{87}Rb mixtures, depends on the inter-species collision rate and related peak ^{40}K and ^{87}Rb number densities [5, 40, 41, 43]. Adiabatic RF manipulation could be used to reduce the ^{87}Rb peak density by decompressing the $m_F^{(Rb)} = 2$ effective potential at the center of the trap during sympathetic RF evaporation. Ideally, the RF-dressed ^{40}K-^{87}Rb collision rate would be small enough to avoid a density-driven collapse, but still large enough to maintain good inter-species rethermalization for sympathetic cooling.

These potentials are also amenable to the study of phase coherence in the ^{40}K-^{87}Rb mixture as the potential barrier is raised. Recent studies of phase-coherent RF splitting of ^{87}Rb BECs on atom chips [11, 12, 14] and in optical traps [61, 62] have focused on interactions and tunneling in BEC-in-double-well systems. An interesting extension of this work would be to assess the effects of a background, unsplit fermion "bath" on the tunneling dynamics and coherence properties of this system.

Light Scattering by a DFG An inhibition of optical scattering is predicted to occur in DFGs when $E_F \gtrsim E_R$, where E_R is the recoil energy [63, 64]. The optical scattering rate depends on the availability of atomic recoil states ("final states"), which is constrained due to Pauli blocking in the filled or nearly filled Fermi sea of a trapped DFG. The large oscillation frequencies and Fermi energies available in anisotropic, harmonic atom chip microtraps ($E_F \propto \bar{\omega}$, see Section 12.2.1) offer a promising route toward measurements of this effect (see [65] and references therein).

References

1. DeMarco, B. and Jin, D.S. (1999) Onset of Fermi degeneracy in a trapped atomic gas. *Science* **285**, 1703.
2. Bloch, I., Dalibard, J., and Zwerger, W. (2008) Many-body physics with ultracold gases. *Rev. Mod. Phys.* **80**, 885.
3. Ott, H., Fortágh, J., Schlotterbeck, G., Grossmann, A., and Zimmermann, C. (2001) Bose–Einstein condensation in a surface microtrap. *Phys. Rev. Lett.* **87**, 230401.
4. Hänsel, W., Hommelhoff, P., Hänsch, T.W., and Reichel, J. (2001) Bose–

Einstein condensation on a microelectronic chip. *Nature* **413**, 498.

5. Aubin, S., Myrskog, S., Extavour, M.H.T., LeBlanc, L.J., McKay, D., Stummer, A., and Thywissen, J.H. (2006) Rapid sympathetic cooling to Fermi degeneracy on a chip. *Nat. Phys.* **2**, 384.

6. Mathey, L., Wang, D.W., Hofstetter, W., Lukin, M.D., and Demler, E. (2004) Luttinger liquid of polarons in one-dimensional boson-fermion mixtures. *Phys. Rev. Lett.* **93**, 120404.

7. Rauf, H., Pichler, T., Knupfer, M., Fink, J., and Kataura, H. (2004) Transition from a Tomonaga-Luttinger liquid to a Fermi liquid in potassium-intercalated bundles of single-wall carbon nanotubes. *Phys. Rev. Lett.* **93**, 096805.

8. Lieb, E.H. (1966) *Mathematical Physics in One Dimension: Exactly Soluble Models of Interacting Particles*, Perspectives in physics, Academic Press, New York, NY.

9. Moritz, H., Stöferle, T., Günter, K., Köhl, M., and Esslinger, T. (2005) Confinement induced molecules in a 1D Fermi gas. *Phys. Rev. Lett.* **94**, 210401.

10. Recati, A., Fedichev, P.O., Zwerger, W., and Zoller, P. (2003) Fermi one-dimensional quantum gas: Luttinger liquid approach and spin-charge separation. *J. Opt.B: Quantum Semiclass. Opt.* **5**, S55.

11. Schumm, T., Hofferberth, S., Wildermuth, L.M., Groth, S., Bar-Joseph, I., Schmiedmayer, J., and Krüger, P. (2005) Matter–wave interferometry in a double well on an atom chip. *Nat. Phys.* **1**, 57.

12. Hofferberth, S., Lesanovsky, I., Fischer, B., Verdu, J., Schmiedmayer, J., and Krüger, P. (2006) Radio-frequency dressed state potentials for neutral atoms. *Nat. Phys.* **2**, 710.

13. Lesanovsky, I., Schumm, T., Hofferberth, S., Andersson, L.M., Krüger, P., and Schmiedmayer, J. (2006) Adiabatic radio-frequency potentials for the coherent manipulation of matter waves. *Phys. Rev. A* **73**, 033619.

14. Jo, G.B., Will, S., A. Pasquini, T., Saba, M., Ketterle, W., and Pritchard, D.E. (2007) Long phase coherence time and number squeezing of two Bose–Einstein condensates on an atom chip. *Phys. Rev. Lett.* **98**, 030407.

15. Extavour, M.H.T., LeBlanc, L.J., Schumm, T., Cieslak, B., Myrskog, S., Stummer, A., Aubin, S., and Thywissen, J.H. (2006) *Dual-species quantum degeneracy of ^{40}K and ^{87}Rb on an atom chip*, in Atomic Physics, Proceedings of the 20th International Conference on Atomic Physics (eds C. Roos, H. Häffner, and R. Blatt), vol. 20, 241–249, American Institute of Physics.

16. Yurke, B. (1986) Input states for enhancement of fermion interferometer sensitivity. *Phys. Rev. Lett.* **56**, 1515.

17. Das, K. and Aubin, S. (2008) *Mesoscopic quantum pumping circuit simulations*, Manuscript in preparation.

18. Jeltes, T., McNamara, J.M., Hogervorst, W., Vassen, W., Krachmalnicoff, V., Schellekens, M., Perrin, A., Chang, H., Boiron, D., Aspect, A., and Westbrook, C.I. (2007) Comparison of the Hanbury Brown–Twiss effect for bosons and fermions. *Nature* **445**, 402.

19. Tran, M.N., Murthy, M.V.N., and Bhaduri, R.K. (2001) Ground-state fluctuations in finite Fermi systems. *Phys. Rev. E* **63**, 031105.

20. Tran, M.N. (2003) Exact ground-state number fluctuations of trapped ideal and interacting fermions. *J. Phys. A: Math. Gen.* **36**, 961.

21. Budde, M. and Mølmer, K. (2004) Number distributions for fermions and fermionized bosons in periodic potentials. *Phys. Rev. A* **70**, 053618.

22. Wang, Y.J., Anderson, D.Z., Bright, V.M., Cornell, E.A., Diot, Q., Kishimoto, T., Prentiss, M., Saravanan, R.A., Segal, R., and Wu, S. (2005) Atom Michelson interferometer on a chip using a Bose–Einstein condensate. *Phys. Rev. Lett.* **94**, 090405.

23. Colombe, Y., Steinmetz, T., Dubois, G., Linke, F., Hunger, D., and Reichel, J. (2007) Strong atom-field coupling for Bose–Einstein condensates in an optical cavity on a chip. *Nature* **450**, 272.

24. Castin, Y. (2008) *Trapped Fermi Gases*, in Proceedings of the International School of Physics Enrico Fermi (eds M. Inguscio,

W. Ketterle, and C. Salomon), vol. 164, 289–349, IOS Press.

25 Bruun, G.M. and Clark, C.W. (2000) Ideal gases in time-dependent traps. *Phys. Rev. A* **61**, 061601.

26 Pathria, R.K. (1996) *Statistical Mechanics*, 2nd ed., Butterworth Heineman, Woburn, MA.

27 Truscott, A.G., Strecker, K.E., McAlexander, W.I., Partridge, G.B., and Hulet, R.G. (2001) Observation of Fermi Pressure in a Gas of Trapped Atoms. *Science* **291**, 2570.

28 Butts, D.A. and Rokhsar, D.S. (1997) Trapped Fermi gases. *Phys. Rev. A* **55**, 4346.

29 Reichel, J. and Thywissen, J. (2004) Using magnetic chip traps to study Tonks–Girardeau quantum gases. *J. Phys. IV France* **116**, 265.

30 Estève, J. (2004) *Du miroir au guide d'onde atomique: effets de rugosité*, Ph.D. thesis, Laboratoire Charles Fabry de Institut d'Optique, Université Paris IV.

31 Aussibal, C. (2003) *Réalisation d'un condensat de Bose–Einstein sur une microstructure*, Ph.D. thesis, Laboratoire Charles Fabry, L'Université Paris IV Université Paris IV.

32 Aubin, S., Extavour, M.H.T., Myrskog, S., LeBlanc, L.J., Estève, J., Singh, S., Scrutton, P., McKay, D., McKenzie, R., Leroux, I.D., Stummer, A., and Thywissen, J.H. (2005) Trapping fermionic ^{40}K and bosonic ^{87}Rb in a chip trap. *J. Low Temp. Phys.* **140**, 377.

33 Extavour, M.H.T. (2004) *Design and construction of magnetic trapping and transport elements for neutral atoms*, Master's thesis, Department of Physics, University of Toronto.

34 Reichel, J. (2002) Microchip traps and Bose–Einstein condensates. *Appl. Phys. B* **75**, 469.

35 Estève, J., Aussibal, C., Schumm, T., Figl, C., Mailly, D., Bouchoule, I., Westbrook, C.I., and Aspect, A. (2004) Role of wire imperfections in micromagnetic traps for atoms. *Phys. Rev. A* **70**, 043629.

36 Schumm, T., Estève, J., Figl, C., Trebbia, J.B., Aussibal, C., Nguyen, H., Mailly, D., Bouchoule, I., Westbrook, C.I., and Aspect, A. (2005) Atom chips in the real world: the effects of wire corrugation. *Eur. Phys. J. D* **32**, 171.

37 Reichel, J., Hänsel, W., and Hänsch, T.W. (1999) Atomic micromanipulation with magnetic surface traps. *Phys. Rev. Lett.* **83**, 3398.

38 Luiten, O.J., Reynolds, M.W., and Walraven, J.T.M. (1996) Kinetic theory of the evaporative cooling of a trapped gas. *Phys. Rev. A* **53**, 381.

39 Weinstein, J.D. and Libbrecht, K.G. (1995) Microscopic magnetic traps for neutral atoms. *Phys. Rev. A* **52**, 4004.

40 Roati, G., Riboli, F., Modugno, G., and Inguscio, M. (2002) Fermi–Bose quantum degenerate ^{40}K-^{87}Rb mixture with attractive interaction. *Phys. Rev. Lett.* **89**, 150403.

41 Modugno, G., Roati, G., Riboli, F., Ferlaino, F., Brecha, R., and Inguscio, M. (2002) Collapse of a degenerate fermi gas. *Science* **297**, 2240.

42 Goldwin, J., Inouye, S., Olsen, M.L., Newman, B., DePaola, B.D., and Jin, D.S. (2004) Measurement of the interaction strength in a Bose–Fermi mixture with ^{87}Rb and ^{40}K. *Phys. Rev. A* **70**, 021601.

43 Ospelkaus, C., Ospelkaus, S., Sengstock, K., and Bongs, K. (2006) Interaction-driven dynamics of ^{40}K/^{87}Rb Fermi–Bose gas mixtures in the large-particle-number limit. *Phys. Rev. Lett.* **96**, 020401.

44 O'Hara, K.M., Hemmer, S.L., Gehm, M.E., Granade, S.R., and Thomas, J.E. (2002) Observation of a strongly interacting degenerate fermi gas of atoms. *Science* **298**, 2179.

45 Flambaum, V.V., Fribakin, G.F., and Harabati, C. (1999) Analytical calculation of cold-atom scattering. *Phys. Rev. A* **59**, 1998.

46 Anderlini, M., Courtade, E., Cristiani, M., Cossart, D., Ciampini, D., Sias, C., Morsch, O., and Arimondo, E. (2005) Sympathetic cooling and collisional properties of a Rb-Cs mixture. *Phys. Rev. A* **71**, 061401.

47 Townsend, J.S. (1992) *A Modern Approach to Quantum Mechanics*, McGraw-Hill, New York, NY.

48 Silber, C., Günther, G., Marzok, C., Deh, B., W. Courteille, P., and Zimmermann,

C. (2005) Quantum-degenerate mixture of fermionic lithium and bosonic rubidium gases. *Phys. Rev. Lett.* **95**, 170408.

49 Marzok, C., Deh, B., Courteille, W.P., and Zimmermann, C. (2007) Ultracold thermalization of ^{7}Li and ^{87}Rb. *Phys. Rev. A* **76**, 052704.

50 Ketterle, W. and J. van Druten, N. (1996) Evaporative cooling of trapped atoms. *Adv. At. Mol. Opt. Phys.* **37**, 181.

51 Zobay, O. and Garraway, B.M. (2001) Two-dimensional atom trapping in field-induced adiabatic potentials. *Phys. Rev. Lett.* **86**, 1195.

52 Colombe, Y. (2004) *Condensat de Bose–Einstein, champs évanescents et piégeage bidimensionnel*, Ph.D. thesis, Université Paris 13.

53 Lesanovsky, I., Schumm, T., Hofferberth, S., Andersson, L.M., Krüger, P., and Schmiedmayer, J. (2006) Adiabatic radio-frequency potentials for the coherent manipulation of matter waves. *Phys. Rev. A* **73**, 033619.

54 Lesanovsky, I., Hofferberth, S., Schmiedmayer, J., and Schmelcher, P. (2006) Manipulation of ultracold atoms in dressed adiabatic radio-frequency potentials. *Phys. Rev. A* **74**, 033619.

55 Colombe, Y., Knyazchyan, E., Morizot, O., Mercier, B., Lorent, V., and Perrin, H. (2004) Ultracold atoms confined in RF-induced two-dimensional trapping potentials. *Europhys. Lett.* **67**, 593.

56 White, M., Gao, H., Pasienski, M., and DeMarco, B. (2006) Bose–Einstein condensates in RF-dressed adiabatic potentials. *Phys. Rev. A* **74**, 023616.

57 van Es, J.P.J., Whitlock, S., Fernholz, T., van Amerongen, H.A., and van Druten, J.N. (2008) Longitudinal character of atom-chip-based RF-dressed potentials. *Phys. Rev. A* **77**, 063623.

58 Hofferberth, S., Fischer, B., Schumm, T., Schmiedmayer, J., and Lesanovsky, I. (2007) Ultracold atoms in radio-frequency dressed potentials beyond the rotating-wave approximation. *Phys. Rev. A* **76**, 013401.

59 Loftus, T., Regal, C.A., Ticknor, C., Bohn, J.L., and Jin, D.S. (2002) Resonant control of elastic collisions in an optically trapped Fermi gas of atoms. *Phys. Rev. Lett.* **88**, 173201.

60 Jervis, D. (2007) *Fabrication of an atom chip*, Master's thesis, Department of Physics, University of Toronto.

61 Gati, R., Albiez, M., Foelling, J., Hemmerling, B., and Oberthaler, M.K. (2006) Realization of a single Josephson junction for Bose–Einstein condensates. *Appl. Phys. B* **82**, 207.

62 Estève, J., Gross, C., Weller, A., Giovanazzi, S., and Oberthaler, M.K. (2008) Squeezing and entanglement in a Bose–Einstein condensate. *Nature advanced online publication*.

63 DeMarco, B. and Jin, D.S. (1998) Exploring a quantum degenerate gas of fermionic atoms. *Phys. Rev. A* **58**, R4267.

64 Busch, T., Anglin, J.R., Cirac, J.I., and Zoller, P. (1998) Inhibition of spontaneous emission in Fermi gases. *Europhys. Lett.* **44**, 1.

65 Shuve, B. and Thywissen, J.H. (2010) Enhanced Pauli blocking of light scattering in a trapped Fermi gas. *J. Phys. B: At. Mol. Opt. Phys.* **43**, 015301.

13
Micro-Fabricated Chip Traps for Ions

J.M. Amini, J. Britton, D. Leibfried, and D.J. Wineland

13.1
Introduction

Most chapters of this monograph focus on trapping and manipulating neutral atoms with magnetic and optical fields. In this chapter, we discuss the trapping of atomic ions. This is of current high interest because individual ions can be the physical representations of qubits for quantum information processing [1]. For recent reviews see [2, 3]. The goals are similar to those of neutral atom traps in that we wish to create micro-fabricated structures to trap, transport, and arrange ions in an array. Microfabrication holds the promise of forming large arrays of traps that would allow the scaling of current quantum information processing capabilities to the level needed to implement useful algorithms [4–7].

There are two primary types of ion traps used in low-energy atomic physics: Penning traps and Paul traps. In a Penning trap, charged particles are trapped by a combination of static electric and magnetic fields [8, 9]. In a Paul trap, a spatially varying sinusoidally oscillating electric field, typically in the radio-frequency (rf) domain, confines atomic or molecular ions in space [10]. In this review only the Paul type will be considered.

Neutral atom traps operate by a coupling between external trapping fields and atoms' electric or magnetic moments. Trap depths of a few kelvins are common. In ion traps, an ion is trapped by a coupling between the applied electric trapping fields and the atom's net (or overall) charge. Typical ion trap depths are 1 eV. This coupling does not depend on the ion's internal electronic state, leaving it largely unperturbed.

We begin this chapter with an introduction to the dynamics of ions confined in Paul traps based on the pseudopotential approximation. Subsequent topics include numeric and analytic models for various Paul trap geometries, a list of considerations for practical trap design and finally an overview of micro-fabricated trapping structures. A discussion of future directions concludes this chapter.

13.2
Radio-Frequency Ion Traps

In this section we discuss the equations of motion of a charged particle in a spatially inhomogeneous RF field based on the pseudopotential approximation model. We then present examples of suitable electrode geometries.

13.2.1
Motion of Ions in a Spatially Inhomogeneous RF Field

Most schemes for quantum information processing with trapped ions are based on a linear RF trap shown schematically in Figure 13.1a. This trap is essentially a linear quadrupole mass filter [10] with its ends plugged by static potentials [11]. The radial confinement (the x–y plane in Figure 13.1a) is provided by an RF potential applied to two of the electrodes with the other electrodes held at RF ground. In this linear geometry, the RF potential cannot generate full 3D confinement, so static potentials V_1 and V_2 applied to control electrodes provide axial (z-axis) confinement. We will assume the axial trapping fields are relatively weak so that the accompanying static radial fields do not significantly perturb the radial trapping.

Applying a potential of $V_0 \cos(\Omega_{rf} t)$ to the RF electrodes while grounding the other electrodes ($V_1 = V_2 = 0$), the RF potential near the geometric center of the

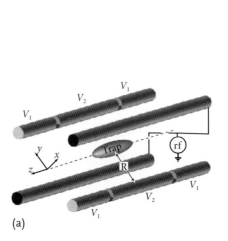
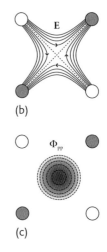
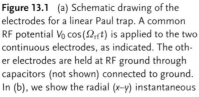

Figure 13.1 (a) Schematic drawing of the electrodes for a linear Paul trap. A common RF potential $V_0 \cos(\Omega_{rf} t)$ is applied to the two continuous electrodes, as indicated. The other electrodes are held at RF ground through capacitors (not shown) connected to ground. In (b), we show the radial (x–y) instantaneous electric fields from the applied RF potential. Contours of the pseudopotential due to this RF field are shown in (c). A static trapping potential is created along the z-axis by applying a positive potential $V_1 > V_2$ (for positive ions) to the outer segments relative to the center segments.

four rods takes the form

$$\Phi \approx \frac{1}{2} V_0 \cos(\Omega_{rf} t) \left(1 + \frac{x^2 - y^2}{R^2}\right), \tag{13.1}$$

where R is a distance scale that is approximately the distance from the trap axis to the nearest surface of the electrodes [4, 10, 11]. The resulting electric field is shown in Figure 13.1b. There is a field null at the trap center; the field magnitude increases linearly with distance from the center.

We can think of the RF electric field as analogous to the electric field from the trapping laser in an optical dipole trap [4, 12]. For a neutral atom, the laser's electric field induces a dipole moment. If the electric field is inhomogeneous, the force on the dipole, averaged over one cycle of the radiation, can give a trapping force. For detunings red of the atom's resonant frequency ω_0, the resulting potential is a minimum at high fields, while for detunings blue of ω_0 it is a minimum at low fields. An ion, however, is a free particle in the absence of a trapping field and its eigenfrequency is zero. The RF trapping potential is therefore analogous to a blue-detuned light field and the ion seeks the position of lowest intensity. In the case of Eq. (13.1), that corresponds to $x = y = 0$.

The motion for an ion placed in this field is commonly treated in one of two ways: as an exact solution of the Mathieu differential equation or as an approximate solution of a static effective potential called the 'pseudopotential'. The Mathieu solutions provide insights into trap stability and high frequency motion; the pseudopotential approximation is more straightforward and is convenient for the analysis of trap designs.

We define the pseudopotential that governs the secular motion as follows [13]. The motion of an ion in the RF field is a combination of fast 'micromotion' at the RF frequency on top of a slower 'secular' motion. For a particle of charge q and mass m in a uniform electric field $E = E_0 \cos(\Omega_{rf} t)$, the ion motion (neglecting a drift term) takes the form

$$x(t) = -x_{\mu m} \cos(\Omega_{rf} t), \tag{13.2}$$

where $x_{\mu m} = q E_0/(m \Omega_{rf}^2)$ is the amplitude of what we will call micromotion. If the RF field amplitude has a spatial dependence $E_0(x)$ along the x direction, there is a non-zero net force on the ion when we average over an RF cycle:

$$F_{net} = \langle q E(x) \rangle \approx -\frac{1}{2} q \left.\frac{d E_0(x)}{dx}\right|_{x \to x_s} x_{\mu m} = -\frac{q^2}{4 m \Omega_{rf}^2} \left.\frac{d E_0^2(x)}{dx}\right|_{x \to x_s}$$

$$= -\frac{d}{dx}(q \Phi_{pp}), \tag{13.3}$$

where x is evaluated at what we designate as the secular position x_s, and the pseudopotential Φ_{pp} is defined by

$$\Phi_{pp}(x_s) \equiv \frac{1}{4} \frac{q E_0^2(x_s)}{m \Omega_{rf}^2}. \tag{13.4}$$

We have made the approximation that the solution in Eq. (13.2) holds over an RF cycle and have dropped terms of higher order in the Taylor expansion of $E_0(x)$ around x_s. For regions near the center of the trapping potential, these approximations hold. In three dimensions, we make the substitution $E_0^2 \to |E|^2 = E_{0,x}^2 + E_{0,y}^2 + E_{0,z}^2$. Note that the pseudopotential depends on the magnitude of the electric field, not its direction.

For the quadrupole field given in Eq. (13.1), the pseudopotential is that of a 2D harmonic potential (see Figure 13.1c):

$$q\Phi_{pp} = \frac{1}{2} m \omega_r^2 (x^2 + y^2) , \tag{13.5}$$

where $\omega_r \simeq q V_0/(\sqrt{2} m \Omega R^2)$ is the resonant frequency. As an example, for ^{24}Mg$^+$ in a Paul trap with $V_0 = 50$ V, $\Omega_{rf}/2\pi = 100$ MHz and $R = 50$ μm, which are typical parameters for a micro-fabricated trap, the radial oscillation frequency is $\omega_r/2\pi = 14$ MHz.

The RF pseudopotential provides confinement of the ion in the radial (x–y) plane. Axial trapping is obtained by the addition of the static control potentials V_1 and V_2, as shown in Figure 13.1a.

For the axial frequency $\omega_z \ll \omega_r$, multiple ions trapped in the same potential well will form a linear 'crystal' along the trap axis due to a balance between the axial trapping potential and the ions' mutual Coulomb repulsion. The inter-ion spacing is determined by ω_z and the number of ions. The characteristic length scale of ion–ion spacing is

$$s = \left(\frac{q^2}{4\pi\epsilon_0 m \omega_z^2} \right)^{\frac{1}{3}} . \tag{13.6}$$

For a three-ion crystal the adjacent separation of the ions is $s_3 = (5/4)^{1/3} s$ [4]. For example, $s_3 = 5.3$ μm for ^{24}Mg$^+$ and $\omega_z/2\pi = 1.0$ MHz. For multiple ions in a linear Paul trap, ω_z is the frequency of the lowest vibrational mode (the center of mass mode) along the trap axis.

A single ion's radial motion in the potential given by Eq. (13.5) can be decomposed into uncoupled harmonic motion in the x and y directions, both with the same trap frequency ω_r. Because the potential is cylindrically symmetric about z, we could choose the decomposition about any two orthogonal directions, called the principle axes. We will see in Section 13.3.1 when discussing Doppler cooling that we need to break this cylindrical symmetry by the application of static electric fields. In that case, the choice of the principle axes becomes fixed with corresponding radial trapping frequencies ω_1 and ω_2, one for each principle axis.

13.2.2
Electrode Geometries for Linear Quadrupole Traps

Designs for miniaturized ion traps conserve the basic features of the Paul trap shown in Figure 13.1. Figure 13.2 shows a few geometries that have been experimentally realized. All these geometries generate a radial quadratic potential near

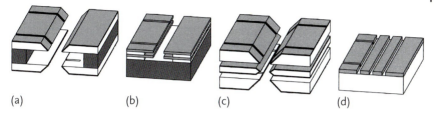

(a)   (b)   (c)   (d)

Figure 13.2 Examples of micro-fabricated trap structures: (a) two wafers mechanically clamped over a spacer [21–25], (b) two layers of electrodes fabricated onto a single wafer [26], (c) three wafers clamped with spacers (not shown) [27], and (d) surface electrode construction [14–20].

the trap axis, though the extent of deviations from the ideal quadrupole potential away from the axis will depend on the design.

In one particular geometry, the electrodes all lie in a single plane, as shown in Figure 13.2d with the ion suspended above the plane [14–20]. Trapping in such surface-electrode (SE) traps is possible over a wide range of geometries, albeit with 1/6 to 1/3 the motional frequencies and 1/30 to 1/200 the trap depth of more conventional quadrupolar geometries at comparable RF potentials and ion–electrode distances [14].

Advantages of the SE trap geometry over the other geometries shown in Figure 13.2 include easier fabrication and the possibility of integrating control electronics on the same trap wafer [6]. A SE trap at cryogenic temperature was demonstrated at MIT in 2008 [19].

Research on SE trap designs is ongoing and holds promise to yield complex geometries that would be difficult to realize in non-SE designs.

13.3
Design Considerations for Paul Traps

In this section, we will discuss the requirements that need to be addressed when designing a practical ion trap.

13.3.1
Doppler Cooling

For Doppler laser cooling of an ion in a trap, only a single laser beam is needed; trap strengths far exceed the laser beam radiation pressure. The cooling is offset by heating from photon recoil. Therefore, to cool in all directions, the Doppler cooling beam k-vector must have a component along all three principal axes of the trap [28]. This also implies that the trap frequencies are not degenerate, otherwise one principal axis could be chosen normal to the laser beam's k-vector.

Meeting the first condition is usually straightforward for non-SE-type traps, where access for the laser beam is fairly open (see Figure 13.3). For SE traps, where laser beams are typically constrained to run parallel to the chip surface, care has to

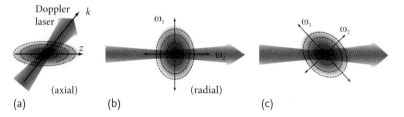

Figure 13.3 Doppler cooling with a single laser beam. The dashed lines are equipotential curves for the pseudopotential. The overlap with the axial direction (z) is fairly straightforward, as in (a), but care has to be taken that the orientation of the two radial modes ω_1 and ω_2 does not place one of the mode axes perpendicular to the laser beam, as shown in (b). For efficient cooling, the axes must be at an angle with respect to the laser beam k-vector (c).

be taken in designing the trap so that neither radial principle axis is perpendicular to the trap surface. Alternately, for SE traps, we could bring the Doppler laser beam at an angle to the surface but the beam would have to strike the surface. This can cause problems with scattered light affecting fluorescence detection of the ion and with charging of exposed dielectrics (see Section 13.3.3).

If any two trap frequencies are degenerate, then the trap axes in the plane containing those modes are not well defined and the motion in a direction perpendicular to the Doppler laser beam k-vector will not be cooled and will be heated due to photon recoil. The axial trap frequency can be set independent of the radial frequencies and can be chosen to prevent a degeneracy with either of the radial modes. However, the two radial modes could still be degenerate. There are several ways to break this degeneracy, but usually the axial trapping potential is sufficient. When we apply an axial trapping potential, Laplace's equation forces us to have a radial component to the electric field. In general, this radial field is not cylindrically symmetric about z and will distort the net trapping potential, as shown in Figure 13.4, thereby lifting the degeneracy of the radial frequencies. If this is not sufficient, offsetting *all* the control electrodes by a common potential with respect to the RF electrodes will result in a static field that has the same spatial dependence (that is the same function of x and y) as the field generated by the RF electrodes.

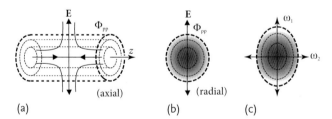

Figure 13.4 The degeneracy in the radial trap modes can be lifted by the radial component of the static axial confinement field. In (a), the quadrupole field is shown overlaid on the cylindrically symmetric pseudopotential. The radial component of the electric field (b) deforms the net potential seen by the ion (c), breaking the cylindrical symmetry.

This field, shown in Figure 13.1b, can be used to split the radial frequencies. We will refer to the axes' orientation resulting from the offset of all control electrodes as the 'intrinsic' trap axes since it does not depend on the segmentation of the control electrodes, but only on the overall geometry of the RF and control electrodes. The static axial potential might or might not define trap axes aligned with the intrinsic axes, but, overall, the control electrodes and axial potential can be configured to prevent either radial modes from being normal to the surface in an SE trap. Furthermore, in some cases additional control electrodes are designed into the trap to lift the degeneracy independent of both the axial potential and the intrinsic axes.

13.3.2
Micromotion

If the pseudopotential at the equilibrium position of a trapped ion is nonzero, then the ion motion will include a persistent micromotion component at frequency Ω_{rf}. There are two mechanisms that can generate a non-zero equilibrium pseudopotential. As the trapping structures become more complicated and the symmetry of the simple Paul trap in Figure 13.1 is broken, there can be a component of the RF field in the axial direction at the pseudopotential minimum; that is, the pseudopotential minimum need not be a pseudopotential zero. Since this effect is caused by the geometry of the trap, we refer to the resulting micromotion as 'intrinsic' micromotion [29]. Secondly, if there is a static electric field at the pseudopotential zero, the equilibrium position of an ion will be shifted away from the pseudopotential minimum. Because shim potentials can be applied to the control electrodes to null these fields [29], the micromotion due to this mechanism is called 'excess' micromotion.

Both intrinsic and excess micromotion can cause problems with the laser–ion interactions, such as Doppler cooling, ion fluorescence, and Raman transitions [4, 29]. An ion with micromotion experiences a frequency-modulated laser field due to the Doppler shift. In the rest frame of the ion, this modulation introduces sidebands to the laser frequency (as seen by the ion) at integer multiples of Ω_{rf} and reduces the laser beam's intensity at the carrier frequency, as shown in Figure 13.5. The strength of these sidebands is parametrized by the modulation index β, given by

$$\beta = \frac{2\pi x_{\mu m}}{\lambda} \cos\theta , \qquad (13.7)$$

where $x_{\mu m}$ is the micromotion amplitude, λ is the laser wavelength, and θ is the angle the laser beam k-vector makes with the micromotion. For laser beams tuned near resonance, ion fluorescence becomes weaker and can disappear entirely. As another example, when $\beta = 1.43$, the carrier and first micromotion sideband have equal strength. For $\beta < 1$, the fractional loss of on-resonance fluorescence is approximately $\beta^2/2$. As a rule of thumb, we aim for $\beta < 0.25$, which corresponds to a drop of less than five percent in on-resonant fluorescence.

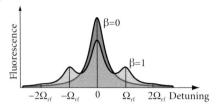

Figure 13.5 In the rest frame of the ion, micromotion induces sidebands of a probing laser. Here, the laser spectrum as viewed by the ions has been convoluted with the atomic linewidth.

For a given static electric field E_{dc} in the radial plane, an ion's radial displacement x_d from the trap center and the resulting excess micromotion amplitude $x_{\mu m}$ are

$$x_d = \frac{q E_{dc}}{m \omega_r^2}, \quad x_{\mu m} \simeq \sqrt{2} \frac{\omega_r}{\Omega_{rf}} x_d, \tag{13.8}$$

where ω_r is the radial trapping frequency.

Assume $^{24}\text{Mg}^+$, $\Omega_{rf}/2\pi = 100\,\text{MHz}$ and $\omega_r/2\pi = 10\,\text{MHz}$. A typical SE trap with $R \sim 50\,\mu\text{m}$ and an excess potential of 1 V on a control electrode will produce a radial electric field at the ion of $\sim 500\,\text{V/m}$. The resulting displacement is $x_d = 500\,\text{nm}$ and the corresponding micromotion amplitude is $x_{\mu m} = 70\,\text{nm}$. This results in a laser modulation index of $\beta = 1.14$.

Stray electric fields can be nulled if the control electrode geometry permits application of independent compensation fields along each radial principal axis. For the Paul trap in Figure 13.1, a common potential applied to one bank of the control electrodes can only generate a field at the trap center that is along the diagonal connecting the electrodes. We can compensate for other directions by applying, for example, a static potential offset to one of the RF electrodes or by adding extra compensation electrodes.

There are several experimental approaches to detecting and minimizing excess micromotion [29]. One technique uses the dependence of the fluorescence from a cooling laser beam on the micromotion modulation index. The micromotion can be minimized by maximizing the fluorescence when the laser is near resonance and minimizing the fluorescence when tuned to the RF sidebands.

Intrinsic micromotion can also be caused by an RF phase difference ϕ_{rf} between different RF electrodes. A phase difference can arise due to a path length difference or a differential capacitive coupling to ground for the leads supplying the electrodes with RF potential [20, 29]. We aim for $\beta < 0.25$ (see Section 13.3.2) for typical parameters, which requires $\phi_{rf} < 0.5°$.

13.3.3
Exposed Dielectric

Exposed dielectric surfaces near the trapping region can pose a problem due to charging of these surfaces and resulting stray electric fields. Charging can be caused by photo-emission by the probe laser or from electron sources such as

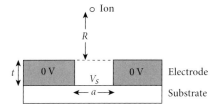

Figure 13.6 Model used for estimating the effect of stray charging. We assume that $R \gg a$ and $t \geq a/\pi$.

those used for loading ions into the traps. Depending on the resistivity of the dielectric, these charges can remain on the surfaces for minutes or longer, requiring time-dependent micromotion nulling or waiting a sufficient time for the charge to dissipate.

SE traps can be particularly prone to this problem. The metallic trapping electrodes are often supported by an insulating substrate and the spaces between the electrodes expose the substrate. The effect of charging these regions can be mitigated by increasing the ratio of electrode conductor thickness to the inter-electrode spacing.

Figure 13.6 illustrates a model for estimating how thick electrodes can suppress the field from a strip of exposed substrate charged to a potential V_s. The sidewalls are assumed to be conducting and grounded. Along the midpoint of the trench the potential drops exponentially with height [30]. Using this solution to relate V_s to the potential at the top of the trench, and employing the techniques described in Section 13.6.2 to relate the surface potential to a field at the ions, we obtain an approximate expression for the field seen by the ion:

$$|E| \simeq \frac{aV}{\pi R^2} \times \begin{cases} 1, & t = 0 \\ \frac{4}{\pi} e^{-\frac{\pi t}{a}}, & t \geq \frac{a}{\pi}, \end{cases} \quad (13.9)$$

where a is the width of the exposed strip of substrate, t is the electrode thickness, R is the distance from the trap surface to the ion, and we have assumed $R \gg a$. Thus, the effect of the stray charges drops off rapidly with the ratio of electrode thickness to gap spacing.

13.3.4
Loading Ions

Ions are loaded into traps by ionizing neutral atoms as they pass through the trapping region. The neutral atoms are usually supplied by a heated oven but can also come from background vapor in the vacuum or laser ablation of a sample.

It is necessary that the neutral atom flux reach the trapping region but not deposit on insulating spacers, which might cause shorting between adjacent trap electrodes. In practice, this is accomplished by careful shielding and, in some SE traps, undercutting of electrodes to form a shadow mask (see Figure 13.18). Alternately,

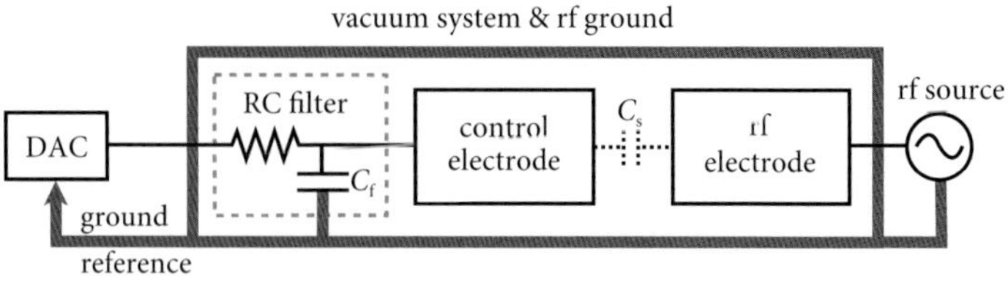

Figure 13.7 Figure showing typical filtering and grounding of a trap control electrode. Inside the vacuum system are low-pass RC filters which reduce noise from the control potential source and provide low impedance shorts to ground for the RF coupled to the control electrodes by stray capacitances $C_s \ll C_f$. The RC filters typically lie inside the vacuum system, within 2 cm of the trap electrodes. The control potential is referenced to the trap RF ground and is supplied over a properly shielded wire.

for SE traps, a hole machined through the substrate can be used to direct neutral flux from an oven on the back side of the wafer to a small region of the trap, preventing coating of the surface. This is called backside loading and has been demonstrated in several traps (see Section 13.7).

13.3.5
Electrical Connections

The control potentials and RF trapping potentials are delivered to the trap electrodes by wiring that includes conducting traces on the trap substrate. Care is needed to avoid several pitfalls.

The high-voltage RF potential is typically produced with resonant RF transformers [31–33]. RF losses in a micro-trap's electrodes or insulating substrate can degrade the resonator (loaded) quality factor (Q_L) and can cause ohmic heating of the microtrap itself. This can be mitigated by use of low-loss insulators (for example, quartz or alumina) and decreasing the capacitive coupling of the RF electrodes to ground through the insulators. Typical RF parameters are $\Omega_{rf}/2\pi = 10\text{--}100 \text{ MHz}$, $V_{rf} \simeq 100 \text{ V}$ and $Q_L \simeq 200$.

The RF electrodes have a small capacitive coupling C_s to each control electrode (typically less than 0.1 pF), which can result in RF potential on the control electrodes. This RF potential needs to be shunted to ground by a capacitor C_f as shown in Figure 13.7. A low-pass RC filter (typically $R = 1 \text{ k}\Omega$ and $C_f = 1 \text{ nF}$) on each control electrode is used to filter noise introduced by the externally applied control electrode potentials. The impedance of the lines between the control electrodes and C_f should be low or the RF shunting to ground will be compromised. Proper grounding, shielding, and filtering of the electronics supplying the control electrode potentials are also important to suppress pickup and ground loops (which can cause motional heating; see Section 13.3.6).

13.3.6
Motional Heating

Doppler and Raman cooling can place a trapped ion's harmonic motion into the ground state with high probability [4, 34–36]. If we are to use the internal states of an ion to store information, we must turn off the cooling laser beams during that period. Unfortunately, the ions do not remain in the motional ground state and this heating can reduce the fidelity of operations performed with the ions. One source of heating comes from laser interactions used to manipulate the electronic states [37]. Another source is ambient electric fields that have a frequency component at the ion's motional frequencies. We expect such fields from the Johnson noise on the electrodes [4, 38–40], but the heating rates observed experimentally are typically several orders of magnitude larger than the Johnson noise can account for. Currently, the source of this anomalous heating is not explained, but recent experiments [19, 39] indicate it is thermally activated and consistent with patches of fluctuating potentials with a size scale smaller than the ion–electrode spacing [38].

The spectral density of electric-field fluctuations S_E at the ion's position inferred from ion heating measurements in a number of traps is plotted versus the minimum ion–electrode separation R in Figure 13.8. The dependence of S_E on R and on the trap frequency ω follows a roughly $R^{-\alpha}\omega^{-\beta}$ scaling, where $\alpha \approx 3.5$ [38, 39] and $\beta \approx 0.8-1.4$ [16, 19, 38, 39]. In addition to being too small to account for these measured heating rates, Johnson noise scales as R^{-2} [4, 38]. One candidate mechanism that does scale as R^{-4} is noise caused by small fluctuating patch potentials on the electrode surfaces [38]. The potentials on these patches fluctuate at megahertz frequencies and generate a corresponding fluctuating electric field at the ion's equilibrium position. This field can lead to heating of the ion [38–40, 47–49].

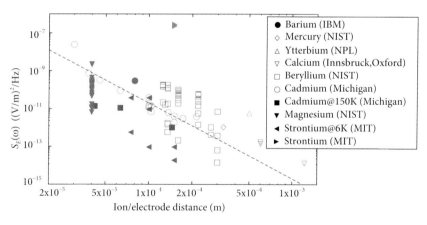

Figure 13.8 Spectral density of electric field fluctuations inferred from observed ion motional heating rates. Data points show heating measurements in ion traps observed in different ion species by several research groups [19–21, 25, 26, 34, 35, 38, 39, 41–46]. Unless specified, the data was taken with the trap at room temperature. The dashed line shows a R^{-4} trend for ion heating vs. ion–electrode separation R.

In the context of ion quantum information processing, microtraps are advantageous because quantum logic gate speeds and ion packing densities increase as the trap size decreases [4–6, 50]. However, these gains are at odds with the highly unfavorable dependence of motional heating on ion–electrode distance. For example, extrapolating from the room-temperature heating results of [16], a $R = 10\ \mu\mathrm{m}$ trap might exceed 10^6 quanta per second. Heating between gate operations can also be problematic because hot ions require more time to recool to the motional ground state.

13.4
Measuring Heating Rates

Heating rates have often been measured by observing an ion's energy increase after cooling to the motional ground state, a relatively complicated and technically challenging undertaking [34, 35]. This section outlines a method to measure ion motional heating with a single low-power laser beam [23, 46, 51]. Near resonance, an atom's fluorescence rate is influenced by its motion due to the Doppler effect. This can be exploited in the following way:

1. Cool a trapped ion to its Doppler limit.
2. Let it remain in the dark for some time. Ambient electric fields couple to the ion's motion and heat it.
3. Turn on the Doppler cooling laser and measure the ion's time-resolved fluorescence, as shown in Figure 13.9.
4. A fit to a theoretical model of the ion fluorescence rate versus time (during recooling) [51] gives an estimate of the ion's temperature at the end of step 2.

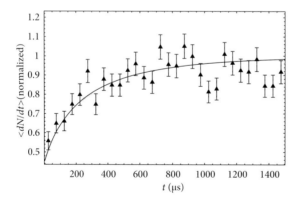

Figure 13.9 Plot showing normalized fluorescence rate dN/dt during Doppler cooling of a hot ion vs. the time the cooling laser is turned off (dark time). The experimental data averaged over many experiments is fitted to the 1D model [51] briefly discussed in the text. The fit has a single free parameter: the ion's temperature at the outset of cooling. The error bars are based on counting statistics. Data taken on $^{25}\mathrm{Mg}^+$ with a dark time of 25 s [51].

The theoretical model in [51] explored cooling of hot ions where the average modulus of the Doppler shift is on the order of, or greater than, the cooling transition line width Γ. The model is a one-dimensional semi-classical theory of Doppler cooling in the weak binding limit where $\omega_z \ll \Gamma$. It is assumed that hot ions undergo harmonic oscillations with amplitudes corresponding to the Maxwell–Boltzman energy distribution when averaged over many experiments.

As a one-dimensional model, only a single motional mode is assumed to be hot. Since the electric-field spectral density S_E at the ion is observed to scale approximately as $S_E \propto \omega^{-\beta}$, where $\beta \approx 0.8$–1.4 [16, 19, 38, 39], the heating is effectively 1D if $\omega_z \ll \omega_x, \omega_y$. This is also important experimentally because efficient Doppler cooling requires laser beam overlap with all modes simultaneously: a change in ion fluorescence can arise from heating of any mode.

Heating rates measured with the recooling technique were found to be in reasonable agreement with rates measured starting from the ground state and allowing heating to only a few average motional quanta [34, 35]. In these comparisons, heating seems to be approximately linear from the ground state to at least 10 000 motional quanta. The disadvantage of the recooling technique is that for small heating rates, the duration of step 2 can become quite long.

13.5
Multiple Trapping Zones

Much of the emphasis in the recent generation of ion traps is towards traps that can store ions in multiple trapping zones and can transport ions between the zones.

We can modify the basic Paul trap in Figure 13.1 to support multiple zones and ion transport by dividing the control electrodes into a series of segments as shown in Figure 13.10a. By applying appropriate potentials [21–23, 52–56] to these segments, an axial harmonic well can be moved along the length of the trap carrying ions along with it (Figure 13.10b). In the adiabatic limit (with respect to ω_z^{-1}), ions have been transported a distance of 1.2 mm in 50 μs with undetectable heating or internal-state decoherence [21].

As an example relevant to quantum information processing, we need to be able to take pairs of ions in a single zone (for example, zone 2 in Figure 13.10c) and separate them into independent zones (one ion in zone 1 and a second in zone 3) without excessive heating. Likewise, we need to reverse this process and combine the ions into a single well. Separating and recombining are more difficult tasks than ion transport; the theory is discussed in [53] and experimentally demonstrated in [21, 52]. The basis for these potentials is the quadratic and quartic terms of the axial potential. Proper design of the trap electrodes can increase the strength of the quartic term and facilitate faster ion separation and merging with less heating. Groups of two and three ions have been separated while heating the center of mass mode to less than 10 quanta and the higher order modes to less than 2 quanta [52].

The segmented Paul trap in Figure 13.10 forms a linear series of trapping zones, but other geometries are desirable. Of particular interest are junctions with linear

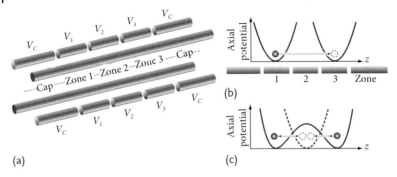

Figure 13.10 (a) Example of a multi-zone trap. By applying appropriate waveforms to the segmented control electrodes, ions can be (b) shuttled from zone to zone or (c) pairs of ions can be merged into a single zone or split into separate zones.

trapping regions extending from each leg. Specific junction geometries are discussed in Section 13.7. The broad goal is to create large interconnected trapping structures that can store, transport, and reorder ions so that any two ions can be brought together in a common zone [4, 5].

13.6
Trap Modeling

Calculation of trap depth, secular frequencies, and transport and separation waveforms requires detailed knowledge of the potential and electric fields near the trap axis. In the pseudopotential approximation, the general time-dependent problem is simplified to a slowly varying electrostatic one. For simple four-rod-type traps, good trap design is not difficult using numerical simulation owing to their symmetry. However, SE trap design is more complicated since the potential may have large anharmonic terms and highly asymmetric designs are common. Fortunately, for certain SE trap geometries, analytic solutions exist. These closed-form expressions permit efficient parametric optimization of electrode geometries not practical by numerical methods. In this section, we will first discuss the full 3D calculations and then introduce the analytic solutions.

13.6.1
Modeling 3D Geometries

There are several numerical methods for solving the general electrostatic problem. In the NIST simulations, the boundary element method implemented in a commercial software package was used. In contrast to the finite element method, the solutions from the boundary element method are in principle differentiable to all orders. A simulation consists of calculating the potential due to each control electrode when that electrode is set to a fixed non-zero potential and all others are grounded. The solution for an arbitrary set of potentials on the control electrodes is then a linear combination of these particular solutions. Similarly, the pseudopotential is obtained by scaling the field calculated for a finite potential on the RF electrodes and ground on the control electrodes and then squaring the field according to Eq. (13.4).

13.6.2
Analytic Solutions for Surface-Electrode Traps

Numerical calculations work for any electrode geometry, but they are slow and not well suited to automatic optimization of SE trap electrode shapes. For the special case of SE traps, an analytic solution exists subject to a few realistic geometric constraints. Electrodes are modeled as a collection of separately biased regions embedded in an infinite ground plane (see Figure 13.11) without gaps between the electrodes. The electric field that would be observed from a biased region is proportional to the magnetic field produced by a current flowing along its perimeter [57]. The problem is then reduced from solving Laplace's equation to integrating a Biot–Savart-type integral around the patch boundary. Furthermore, for patches that have boundaries composed of straight line segments, the integrals have analytic solutions. The application of this technique to SE traps is given in [58].

The main shortcoming of this method is the requirement that there be no gaps between the electrodes. Typical SE trap fabrication techniques produce 1–5 μm gaps which can only be accounted for at the level important to ion dynamics by full numerical simulations.

Fields for arbitrarily shaped patches can be calculated using this Biot–Savart technique, but for simplicity we restrict ourselves to strips that extend to infinity in the z-direction of Figure 13.11. For this particular case, we can also derive potentials from the calculated fields. A strip extending from $x = a$ to $x = b$ with $a < b$ held at potential U_s leads to a spatial potential

$$\Phi_s(a,b) = \frac{U_s}{\pi} \times \begin{cases} \tan^{-1}\left(\dfrac{x-a}{y}\right) - \tan^{-1}\left(\dfrac{x-b}{y}\right), & -\infty < a < b < \infty \\ \dfrac{\pi}{2} - \tan^{-1}\left(\dfrac{x-b}{y}\right), & a = -\infty \\ \dfrac{\pi}{2} + \tan^{-1}\left(\dfrac{x-a}{y}\right). & b = \infty \end{cases}$$

(13.10)

The potentials of multiple, non-overlapping strips can then be summed for more complex structures.

Two basic SE trap geometries are the 'four-wire' trap and the 'five-wire' trap. An example four-wire trap consists of an RF electrode from $x = -d$ to $x = 0$ and another semi-infinite RF electrode from $x = d$ to $x = \infty$ (see Figures 13.11a and 13.16). An example five-wire trap consists of two symmetric RF electrodes from $x = -3/2d$ to $x = -1/2d$ and $x = 1/2d$ to $x = 3/2d$. Their respective potentials

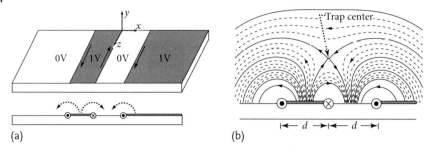

(a) (b)

Figure 13.11 Surface-electrode trap composed of two RF electrodes embedded in a ground plane (four-wire trap) (a). The field lines from the Biot–Savart-type integral are shown in (b).

are given by

$$\Phi_{4w} = \Phi_s(-d, 0) + \Phi_s(d, \infty); \quad \Phi_{5w} = \Phi_s\left(-\frac{3d}{2}, -\frac{d}{2}\right) + \Phi_s\left(\frac{d}{2}, \frac{3d}{2}\right).$$
(13.11)

From the electric fields and Eq. (13.4) we can derive the pseudopotential. Note that the potential minima coincide with the points of zero electric field that lie in the line of symmetry around $x = 0$ at $y_{4w} = d$ and $y_{5w} = \sqrt{3}d/2$, respectively. For an ion of mass m and charge q, the trap frequencies along the two degenerate radial directions are

$$\omega_{4w} = \frac{qU_s}{\sqrt{2}m\pi\Omega_{rf}d^2}; \quad \omega_{5w} = \sqrt{\frac{2}{3}}\frac{qU_s}{m\pi\Omega_{rf}d^2},$$
(13.12)

where Ω_{rf} is the RF-drive frequency. Figure 13.12 shows the general shape of the pseudopotential well along the y-axis at $x = 0$ for the four-wire trap (for the five-wire trap the potential looks very similar). The potential is zero at $y = d$ where the ion is trapped, then rises to a maximum and finally asymptotically drops towards zero for $y \to \infty$. The positions of the maxima are at

$$s_{4w} = d\sqrt{2 + \sqrt{5}}; \quad s_{5w} = d\sqrt{3/4 + \sqrt{3}},$$
(13.13)

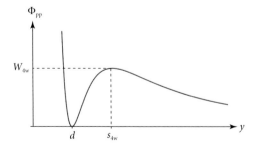

Figure 13.12 Analytic pseudopotential of the four-wire trap along y at $x = 0$. The trapping zero is at $y = d$; the maximum defining the well depth is at $s_{4w} = d\sqrt{2 + \sqrt{5}}$.

and the pseudopotential well depth (in eV) is

$$W_{4w} = \left(\frac{qU_s^2}{4m\Omega_{rf}^2}\right)\frac{2}{\pi^2 d^2(11+5\sqrt{5})} \; ; \; W_{5w} = \left(\frac{qU_s^2}{4m\Omega_{rf}^2}\right)\frac{1}{\pi^2 d^2(7+4\sqrt{3})}.$$
(13.14)

To get an idea of practical parameters, we can calculate the radial frequency and pseudopotential well depth of a four-wire trap with a geometry similar to the trap described in [16]. For $\Omega_{rf}/2\pi = 87$ MHz, $U_s = 103.2$ V, $d = 40$ µm, and m the mass of a ^{24}Mg$^+$ ion, we get $\omega_{4w}/2\pi = 16.9$ MHz and $W_{4w} = 203$ meV.

13.7
Trap Examples

Having covered the general principles for Paul trap designs, we now give specific examples of micro-fabricated ion traps. A number of fabrication techniques have been used for microtraps, starting with assembling multiple wafers to form a traditional Paul trap type design [20–25, 27]. Recently, trap fabrication has been extended to monolithic designs using substrate materials such as Si, GaAs, quartz, and printed circuit board [15–20, 26, 59]. The fabrication process includes such micro-fabrication standards as photolithography, metalization, and chemical vapor deposition as well as other less used techniques such as laser machining.

The micro-fabricated equivalent to the prototypical four-rod Paul trap can use two insulating substrates patterned with electrodes that are then clamped or bonded together with an insulating spacer. This approach has been implemented in a number of traps [20–22, 24, 25, 52] using two substrates, as shown in Figure 13.13a. Alternatively, it is possible to build this structure into a single monolithic device [26], as indicated schematically in Figure 13.14a.

Reference [27] describes a three-wafer trap design like that shown in Figure 13.13b incorporating a 'T'-shaped junction. At NIST, a two-layer trap with an 'X' junction has recently been implemented [25] and is shown in Figure 13.15. Such two-dimensional geometries will be important in order to combine arbitrarily selected qubits from an array together in the same trap zone.

Another approach demonstrated recently used two patterned substrates, without slots, that are mounted with the conducting layers facing each other [60] (see Figure 13.14b). The array of conducting gold electrode strips is driven with RF that alternates between a phase of 0 and 180° from one strip to the next. This creates a pseudopotential that is near zero for much of the space between the wafers but which rises sharply near the substrates. When combined with static potentials at the edges of the wafers, this trap generates a near field-free region bounded by 'hard' potential walls (Figure 13.14b). Arrays of cylindrical Paul-type traps have been microfabricated on silicon for use as mass spectrometers [59].

Surface electrode traps have the benefit of using standard micro-fabrication methods where layers of metal and insulator are deposited on the surface of the

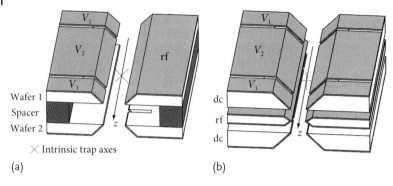

Figure 13.13 Multi-wafer traps can be formed by mechanically clamping or bonding multiple substrates to form (a) a four-rod quadrupolar Paul trap type structure or (b) a modified Paul trap using a three-layer structure [27]. The segmentation of the control electrodes on the bottom substrate is similar to that of the top substrate.

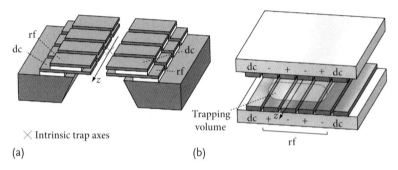

Figure 13.14 (a) Four-rod Paul trap realized by successively deposited layers of GaAs and AlGaAs on a GaAs wafer [26]. In (b), conducting gold strips deposited on two glass substrates and alternately driven at opposite phases of an RF source (phases denoted by '+' and '−') generates a trapping volume between the substrates [60]. Static potentials at the edges of the trap along the z-axis, applied with electrodes that are not shown, confine the ions to the central region of the trap.

wafer without the need for milling of the substrate itself. There are two general versions of the surface trap electrode geometry, as described in Section 13.6.2 and shown in Figure 13.16. The four-wire geometry has the intrinsic trap axes rotated at 45° to the substrate plane, which allows for efficient laser cooling of the ion. The five-wire geometry has one intrinsic trap axis perpendicular to the surface, which can make that axis difficult to Doppler cool (see Section 13.3.1). To enable Doppler cooling, additional control electrodes can be added to the design to rotate the trap axes away from the intrinsic direction. Alternately, a hybrid between the four- and five-wire designs where the rf strips are of unequal widths (an 'asymmetric' five-wire trap) will rotate the intrinsic axes and enable cooling.

SE traps are relatively new and only a few designs have been demonstrated [15–19]. An SE trap was first demonstrated with charged polystyrene balls using standard PC board fabrication techniques [15]. The first SE trap for atomic ions was

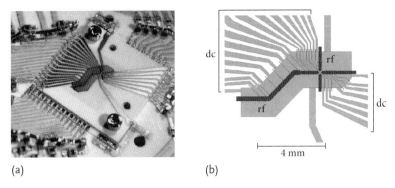

Figure 13.15 (a) Example of a two-wafer trap with an 'X' junction [25]. The trap electrodes are fabricated with evaporated and electroplated gold that is deposited on laser-machined alumina substrates. (b) Detail of the 'X' junction.

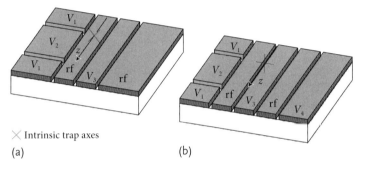

Figure 13.16 (a) Four-wire SE trap geometry and (b) symmetric five-wire SE trap geometry. In practice, the symmetric five-wire geometry is typically not used because of the difficulty of cooling the vertical motion of the trapped ions.

constructed on a fused quartz substrate with electroplated gold electrodes [16, 20] as shown in Figure 13.17. In addition, meander-line resistors were fabricated on the chip as part of the control electrode filtering. Surface-mount capacitors were gap welded to the chip to complete the filters (see Section 13.3.5). The fabrication process sequence is shown in Figure 13.18. The bonding pads and the thin meander-line resistors were formed by liftoff of evaporated gold. Charging of the exposed substrate between the electrodes was a concern, so the trap electrodes were made of 6-μm thick electroplated gold with 8-μm gaps so as to shield the ion somewhat from the charges on the quartz surface.

A similar design was built by a group at MIT for low-temperature testing using 1-μm evaporated silver on quartz [19]. They reported a strong dependence of the anomalous ion heating on temperature (see Section 13.3.6).

The construction of the traps in [16] and [19] was based on adding conducting layers to an insulating substrate. An alternate fabrication method used boron-doped Si wafers anodically bonded to a glass substrate [17] and boron-doped silicon-on-insulator (SOI) wafers [20]. In both cases trenches were etched through the silicon

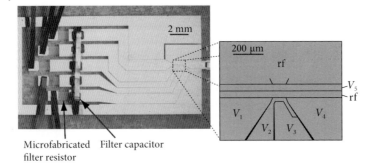

Figure 13.17 An example of a four-wire SE trap constructed of electroplated gold on a quartz substrate [16].

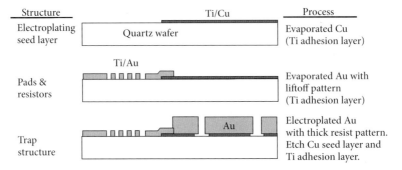

Figure 13.18 Fabrication steps for the example SE trap in Figure 13.17 [16]. The copper seed layer could not be used under the meander-line resistors because the final step of etching the seed layer would fully undercut the narrow meander pattern.

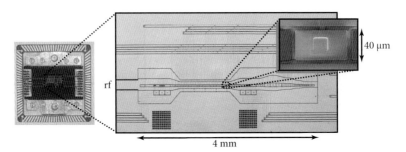

Figure 13.19 Multi-layer, multi-zone, linear SE trap mounted in its carrier and an enlargement of the active region [61].

layer to the glass or embedded insulating layer to define the trap electrodes. The SOI design demonstrated multiple trapping zones in a SE trap and backside loading of ions.

SE traps allow for complex arrangements of trapping zones, but making electrical connections to these electrodes quickly becomes intractable as the complexity

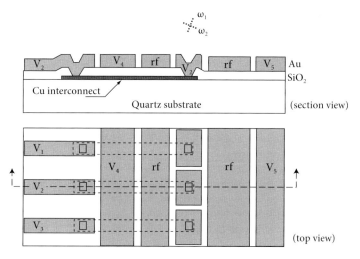

Figure 13.20 Fabrication of an asymmetric five-wire multi-layer SE trap. A CVD oxide insulates the surface electrodes from the second layer of interconnects. Plasma-etched holes in the insulated layer connect the two conducting layers [61].

grows. This problem can be addressed by incorporating multiple conducting layers into the design with only the field from the top layer affecting the ion [6, 61]. An example of such a multi-layer trap fabricated on an amorphous quartz substrate at NIST is shown in Figure 13.19. The metal layers are separated by chemical vapor deposited (CVD) silicon dioxide and connections between metal layers are made by vias that are plasma etched through the oxide, as shown in Figure 13.20. The fabrication process for the surface gold layer is similar to the electroplating shown in Figure 13.18.

In the last three years, micro-fabricated traps have also been produced by Sandia National Laboratory (contact: M. Blain, SNL) and Lucent Technologies (contact: R. Slusher, Georgia Tech Research Institute) and distributed to several ion trap groups in the framework of a "trap foundry" initiated by DTO (now IARPA). Several groups have seen trapping in the Lucent trap, a 17-zone SE trap [62]. The Sandia trap, a 5-zone planar trap where the ions reside in-plane with the electrodes, has also been used to trap ions in two laboratories.

13.8
Future

As ion traps become smaller, trap complexity increases and features such as junctions promise to expand the capabilities of such traps. Two experimentally demonstrated atomic ion traps with junctions have been based on multi-layer designs [25, 27]. The slots and difficulty of alignment and bonding in multi-wafer traps make it difficult to scale such structures.

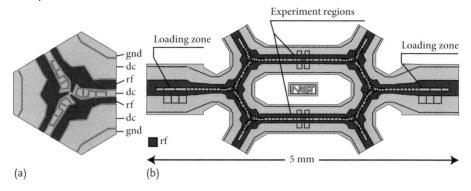

Figure 13.21 (a) Example of a SE trap 'Y' junction and (b) a prototype design using multiple 'Y' junctions to link experimental regions and loading zones [61].

Figure 13.21a shows an example design of a 'Y' version of an SE trap junction. The shape of the RF junction is an example of the optimization that is possible with SE traps because of the efficient methods described in Section 13.6.2 to calculate the fields. The electrode geometry has been optimized to generate a pseudopotential that has minimal axial 'bumps' so that RF micromotion during ion transport will be minimized (see Section 13.3.2). Initial ion transport through such a "Y" junction is reported in [61]. Such components could be assembled into larger structures as shown in Figure 13.21b. SE traps fabricated using standard recipes in a foundry and using standard patterns may eventually make ion traps more accessible to research groups that do not have the resources needed to develop their own.

With increased trap complexity, several other issues arise. One of these is the question of how to package traps and provide all the electrical connections needed to operate them. Another issue is that of corresponding complexity of the lasers used in manipulating the ions. Beyond cooling, state preparation, and detection, lasers are needed to coherently manipulate the internal states of the ions and couple pairs or groups of ions. Multiplexing sets of lasers to address multiple trapping zones for parallel processing will be difficult. Alternatives to laser optical field state manipulation have been proposed [63–67] where magnetic structures, both active wire loops and passive magnetic layers, replace laser beams. If proven to be viable, this would transfer much of the experimental complexity from large laser systems to electronic packages, which can be more reliably engineered and should be scalable [6].

Acknowledgments

This work was supported by the NIST Quantum Information Program and IARPA. This manuscript is a publication of NIST and is not subject to US copyright.

References

1. Cirac, J.I. and Zoller, P. (1995) Quantum computations with cold trapped ions. *Phys. Rev. Lett.* **74**, 4091.
2. Blatt, R. and Wineland, D.J. (2008) Entangled states of trapped atomic ions. *Nature* **453**, 1008.
3. Monroe, C. and Lukin, M. (2008) Remapping the quantum frontier. *Phys. World*, Aug 32.
4. Wineland, D.J., Monroe, C., Itano, W.M., Leibfried, D., King, B.E., and Meekhof, D.M. (1998) Experimental issues in coherent quantum-state manipulation of trapped atomic ions. *J. Res. Natl. Inst. Stand. Technol.* **103**, 259.
5. Kielpinski, D., Monroe, C.R., and Wineland, D.J. (2002) Architecture for a large-scale ion-trap quantum computer. *Nature* **417**, 709.
6. Kim, J., Pau, S., Ma, Z., McLellan, H.R., Gates, J.V., Kornblit, A., Slusher, R.E., Jopson, R.M., Kang, I., and Dinu, M. (2005) System design for large-scale ion trap quantum information processor. *Quantum Inf. Comput.* **5**, 515.
7. Steane, A.M. (2007) How to build a 300 bit, 1 gig-operation quantum computer. *Quantum. Inf. Comput.* **7**, 171.
8. Penning, F.M. (1936) Die Glimmentladung bei niedrigem Druck zwischen koaxialen Zylindern in einem axialen Magnetfeld. *Physica* **3**, 873.
9. Dehmelt, H. (1990) Experiments with an isolated subatomic particle at rest. *Rev. Mod. Phys.* **62**, 525.
10. Paul, W. (1990) Electromagnetic traps for charged and neutral particles. *Rev. Mod. Phys.* **62**, 531.
11. Wineland, D.J. and Houches, L. (2004) *Quantum information processing in ion traps*, in Session LXXIX, 2003, Quantum Entanglement and Information Processing (eds D. Estève, J.-M. Raimond, and J. Dalibard), Elsevier, Amsterdam, pp. 261–293.
12. Grimm, R., Weidemuller, M., and Ovchinnikov, Y.B. (2000) Optical dipole traps for neutral atoms. *Adv. At. Mol. Opt. Phys.* **42**, 95.
13. Dehmelt, H.G. (1967) Radiofrequency spectroscopy of stored ions I – storage. *Adv. At. Mol. Phys.* **3**, 53.
14. Chiaverini, J., Blakestad, R.B., Britton, J., Jost, J.D., Langer, C., Leibfried, D., Ozeri, R., and Wineland, D.J. (2005) Surface-electrode architecture for ion-trap quantum information processing. *Quantum Inf. Comput.* **5**, 419.
15. Pearson, C.E., Leibrandt, D.R., Bakr, W.S., Mallard, W.J., Brown, K.R., and Chuang, I.L. (2006) Experimental investigation of planar ion traps. *Phys. Rev. A* **73**, 032307.
16. Seidelin, S., Chiaverini, J., Reichle, R., Bollinger, J.J., Leibfried, D., Britton, J., Wesenberg, J.H., Blakestad, R.B., Epstein, R.J., Hume, D.B., Itano, W.M., Jost, J.D., Langer, C., Ozeri, R., Shiga, N., and Wineland, D.J. (2006) Microfabricated surface-electrode ion trap for scalable quantum information processing. *Phys. Rev. Lett.* **96**, 253003.
17. Britton, J., Leibfried, D., Beall, J., Blakestad, R.B., Bollinger, J.J., Chiaverini, J., Epstein, R.J., Jost, J.D., Kielpinski, D., Langer, C., Ozeri, R., Reichle, R., Seidelin, S., Shiga, N., Wesenberg, J.H., and Wineland, D.J. (2006) A microfabricated surface-electrode ion trap in silicon. arXiv:quant-ph/0605170.
18. Brown, K.R., Clark, R.J., Labaziewicz, J., Richerme, P., Leibrandt, D.R., and Chuang, I.L. (2007) Loading and characterization of a printed-circuit-board atomic ion trap. *Phys. Rev. A* **75**, 015401.
19. Labaziewicz, J., Ge, Y., Antohi, P., Leibrandt, D., Brown, K.R., and Chuang, I.L. (2008) Suppression of heating rates in cryogenic surface-electrode ion traps. *Phys. Rev. Lett.* **100**, 013001.
20. Britton, J., Leibfried, D., Beall, J.A., Blakestad, R.B., Wesenberg, J.H., and Wineland, D.J. (2009) Scalable arrays of RF Paul traps in degenerate Si. *Appl. Phys. Lett.* **95**, 173102.
21. Rowe, M.A., Ben-Kish, A., DeMarco, B., Leibfried, D., Meyer, V., Beall, J., Britton, J., Hughes, J., Itano, W.M., Jelenkovic, B., Langer, C., Rosenband, T., and Wineland, D.J. (2002) Transport of

quantum states and separation of ions in a dual RF ion trap. *Quantum Inf. Comput.* **2**, 257.

22 Wineland, D.J., Leibfried, D., Barrett, M.D., Ben-kish, A., Bergquist, J.C., Blakestad, R.B., Bollinger, J.J., Britton, J., Chiaverini, J., Demarco, B., Hume, D., Itano, W.M., Jensen, M., Jost, J.D., Knill, E., Koelemeij, J., Langer, C., Oskay, W., Ozeri, R., Reichle, R., Rosenband, T., Schaetz, T., Schmidt, P.O., and Seidelin, S. (2005) Quantum control, quantum control, quantum information processing, and quantum-limited metrology with trapped ions, in *Proc. XVII Int. Conf. on Laser Spectroscopy, Avemore, Scotland, 2005* (eds E.A. Hinds, A. Ferguson, and E. Riis), World Scientific, Singapore, pp. 393–402, (quant-ph/0508025).

23 Huber, G., Deuschle, T., Schnitzler, W., Reichle, R., Singer, K., and Schmidt-Kaler, F. (2008) Transport of ions in a segmented linear Paul trap in printed-circuit-board technology. *New J. Phys.* **10**, 013004.

24 Schulz, S.A., Poschinger, U., Ziesel, F., and Schmidt-Kaler, F. (2008) Sideband cooling and coherent dynamics in a microchip multi-segmented ion trap. *New J. Phys.* **10**, 045007.

25 Blakestad, R.B., Ospelkaus, C., VanDevender, A.P., Amini, J.M., Britton, J., Leibfried, D., and Wineland, D.J. (2009) High-fidelity transport of trapped-ion qubits through an X-junction trap array. *Phys. Rev. Lett.* **102**, 153002.

26 Stick, D., Hensinger, W.K., Olmschenk, S., Madsen, M.J., Schwab, K., and Monroe, C. (2006) Ion trap in a semiconductor chip. *Nat. Phys.* **2**, 36.

27 Hensinger, W.K., Olmschenk, S., Stick, D., Hucul, D., Yeo, M., Acton, M., Deslauriers, L., Monroe, C., and Rabchuk, J. (2006) T-junction ion trap array for two-dimensional ion shuttling, storage and manipulation. *Appl. Phys. Lett.* **88**, 034101.

28 Itano, W.M. and Wineland, D.J. (1982) Laser cooling of ions stored in harmonic and Penning traps. *Phys. Rev. A* **25**, 35.

29 Berkeland, D.J., Miller, J.D., Bergquist, J.C., Itano, W.M., and Wineland, D.J. (1998) Minimization of ion micromotion in a Paul trap. *J. Appl. Phys.* **83**, 5025.

30 Jackson, J.D. (1999) *Classical electrodynamics*, Wiley, New York, 3rd edn.

31 Macalpine, W.W. and Schildknecht, R.O. (1959) Coaxial resonators with helical inner conductor. *Proceedings of the IRE* **47**, 2099.

32 Cohen, M. (1965) Design techniques utilizing helical line resonators. *Microw. J.* **8**, 69.

33 Jefferts, S.R., Monroe, C., Bell, E.W., and Wineland, D.J. (1995) Coaxial-resonator-driven RF (Paul) trap for strong confinement. *Phys. Rev. A* **51**, 3112.

34 Diedrich, F., Bergquist, J.C., Itano, W.M., and Wineland, D.J. (1989) Laser cooling to the zero-point energy of motion. *Phys. Rev. Lett.* **62**, 403.

35 Monroe, C., Meekhof, D.M., King, B.E., Jefferts, S.R., Itano, W.M., Wineland, D.J., and Gould, P. (1995) Resolved-sideband raman cooling of a bound atom to the 3d zero-point energy. *Phys. Rev. Lett.* **75**, 4011.

36 King, B.E., Wood, C.S., Myatt, C.J., Turchette, Q.A., Leibfried, D., Itano, W.M., Monroe, C., and Wineland, D.J. (1998) Cooling the collective motion of trapped ions to initialize a quantum register. *Phys. Rev. Lett.* **81**, 1525.

37 Ozeri, R., Itano, W.M., Blakestad, R.B., Britton, J., Chiaverini, J., Jost, J., Langer, C., Leibfried, D., Reichle, R., Seidelin, S., Wesenberg, J.H., and Wineland, D.J. (2007) Errors in trapped-ion quantum gates due to spontaneous photon scattering. *Phys. Rev. A* **75**, 042329.

38 Turchette, Q.A., Kielpinski, D., King, B.E., Leibfried, D., Meekhof, D.M., Myatt, C.J., Rowe, M.A., Sackett, C.A., Wood, C.S., Itano, W.M., Monroe, C., and Wineland, D.J. (2000) Heating of trapped ions from the quantum ground state. *Phys. Rev. A* **61**, 063418.

39 Deslauriers, L., Olmschenk, S., Stick, D., Hensinger, W.K., Sterk, J., and Monroe, C. (2006) Scaling and suppression of anomalous heating in ion traps. *Phys. Rev. Lett.* **97**, 103007.

40 Leibrandt, D., Yurke, B., and Slusher, R. (2007) Modeling ion trap thermal noise

decoherence. *Quantum Inf. Comput.* **7**, 52.

41 Roos, C., Zeiger, T., Rohde, H., Nägerl, H.C., Eschner, J., Leibfried, D., Schmidt-Kaler, F., and Blatt, R. (1999) Quantum state engineering on an optical transition and decoherence in a Paul trap. *Phys. Rev. Lett.* **83**, 4713.

42 Tamm, C., Engelke, D., and Bühner, V. (2000) Spectroscopy of the electric-quadrupole transition $^2S_{1/2}(F=0)-^2D_{3/2}(F=2)$ in trapped $^{171}Yb^+$. *Phys. Rev. A* **61**, 053405.

43 DeVoe, R.G. and Kurtsiefer, C. (2002) Experimental study of anomalous heating and trap instabilities in a microscopic ^{137}Ba ion trap. *Phys. Rev. A* **65**, 063407.

44 Deslauriers, L., Haljan, P.C., Lee, P.J., Brickman, K.A., Blinov, B.B., Madsen, M.J., and Monroe, C. (2004) Zero-point cooling and low heating of trapped $^{111}Cd^+$ ions. *Phys. Rev. A* **70**, 043408.

45 Home, J. (2006) *Entanglement of two trapped-ion spin qubits*, Ph.D. thesis, University of Oxford.

46 Epstein, R.J., Seidelin, S., Leibfried, D., Wesenberg, J.H., Bollinger, J.J., Amini, J.M., Blakestad, R.B., Britton, J., Home, J.P., Itano, W.M., Jost, J.D., Knill, E., Langer, C., Ozeri, R., Shiga, N., and Wineland, D.J. (2007) Simplified motional heating rate measurements of trapped ions. *Phys. Rev. A* **76**, 033411.

47 Lamoreaux, S.K. (1997) Thermalization of trapped ions: a quantum perturbation approach. *Phys. Rev. A* **56**, 4970.

48 James, D.F.V. (1998) Theory of heating of the quantum ground state of trapped ions. *Phys. Rev. Lett.* **81**, 317.

49 Henkel, C., Potting, S., and Wilkens, M. (1999) Loss and heating of particles in small and noisy traps. *Appl. Phys. B* **69**, 379.

50 Leibfried, D., DeMarco, B., Meyer, V., Lucas, D., Barrett, M., Britton, J., Itano, W.M., Jelenkovic, B., Langer, C., Rosenband, T., and Wineland, D.J. (2003) Experimental demonstration of a robust, high-fidelity geometric two ion-qubit phase gate. *Nature* **422**, 412.

51 Wesenberg, J.H., Epstein, R.J., Leibfried, D., Blakestad, R.B., Britton, J., Home, J.P., Itano, W.M., Jost, J.D., Knill, E., Langer, C., Ozeri, R., Seidelin, S., and Wineland, D.J. (2007) Fluorescence during Doppler cooling of a single trapped atom. *Phys. Rev. A* **76**, 053416.

52 Barrett, M.D., Chiaverini, J., Schaetz, T., Britton, J., Itano, W.M., Jost, J.D., Knill, E., Langer, C., Leibfried, D., Ozeri, R., and Wineland, D.J. (2004) Deterministic quantum teleportation of atomic qubits. *Nature* **429**, 737.

53 Home, J.P. and Steane, A.M. (2006) Electrode configurations for fast separation of trapped ions. *Quantum Inf. Comput.* **6**, 289.

54 Schulz, S., Poschinger, U., Singer, K., and Schmidt-Kaler, F. (2007) *Optimization of segmented linear paul traps and transport of stored particles*, in Elements of Quantum Information (eds W.P. Schleich and H. Walther), Wiley-VCH, p. 45.

55 Reichle, R., Leibfried, D., Blakestad, R.B., Britton, J., Jost, J.D., Knill, E., Ozeri, R., Langer, C.R., Seidelin, S., and Wineland, D.J. (2007) *Transport dynamics of single ions in segmented microstructured paul trap arrays*, in Elements of Quantum Information (eds W.P. Schleich and H. Walther), Wiley-VCH, p. 69.

56 Hucul, D., Yeo, M., K. Hensinger, W., Rabchuk, J., Olmschenk, S., and Monroe, C. (2008) On the transport of atomic ions in linear and multidimensinal ion trap arrays. *Quantum Inf. Comput.* **8**, 501.

57 Oliveira, M.H. and Miranda, J.A. (2001) Biot-Savart-like law in electrostatics. *Eur. J. Phys.* **22**, 31.

58 Wesenberg, J.H. (2008) Electrostatics of surface-electrode ion traps. *Phys. Rev. A* **78**, 063410.

59 Pau, S., Pai, C.S., L. Low, Y., Moxom, J., Reilly, P.T.A., Whitten, W.B., and Ramsey, J.M. (2006) Microfabricated quadrupole ion trap for mass spectrometer applications. *Phys. Rev. Lett.* **96**, 120801.

60 Debatin, M., Kröner, M., Mikosch, J., Trippel, S., Morrison, N., Reetz-Lamour, M., Woias, P., Wester, R., and Weidemüller, M. (2008) Planar multipole ion trap. *Phys. Rev. A* **77**, 033422.

61 Amini, J.M., Uys, H., Wesenberg, J.H., Seidelin, S., Britton, J., Bollinger, J.J.,

Leibfried, D., Ospelkaus, C., VanDevender, A.P., and Wineland, D.J. (2010) Towards scalable ion traps for quantum information processing. *New J. Phys.* **12**, 033031.

62 Leibrandt, D.R., Labaziewicz, J., Clark, R.J., Chuang, I.L., Epstein, R.J., Ospelkaus, C., Wesenberg, J.H., Bollinger, J.J., Leibfried, D., Wineland, D.J., Stick, D., Sterk, J., Monroe, C., Pai, C.S., Low, Y., Frahm, R., and Slusher, R.E. (2009) Demonstration of a scalable, multiplexed ion trap for quantum information processing. *Quantum Inf. Comput.* **9**, 901.

63 Mintert, F. and Wunderlich, C. (2001) Ion-trap quantum logic using long-wavelength radiation. *Phys. Rev. Lett.* **87**, 257904, Erratum: (2003) *Phys. Rev. Lett.* **91**, 029902.

64 Leibfried, D., Knill, E., Ospelkaus, C., and Wineland, D.J. (2007) Transport quantum logic gates for trapped ions. *Phys. Rev. A* **76**, 032324.

65 Chiaverini, J. and Lybarger, Jr., W.E. (2008) Laserless trapped-ion quantum simulations without spontaneous scattering using microtrap arrays. *Phys. Rev. A* **77**, 022324.

66 Ospelkaus, C., Langer, C.E., Amini, J.M., Brown, K.R., Leibfried, D., and Wineland, D.J. (2008) Trapped-ion quantum logic gates based on oscillating magnetic fields. *Phys. Rev. Lett.* **101**, 090502.

67 Johanning, M., Braun, A., Timoney, N., Elman, V., Neuhauser, W., and Wunderlich, C. (2009) Individual addressing of trapped ions and coupling of motional and spin states using RF radiation. *Phys. Rev. Lett.* **102**, 073004.

Index

a
absorption imaging 190, 197, 274, 351, 355
adiabatic approximation 227–228
adiabatic potentials 183, 197, 383, 385
adiabatic splitting of a BEC 19
aluminum nitride 391
analytic solutions 409
anomalous heating 405
array of traps 271
aspect ratio 246
asymmetric double well 19
atom chip 175
atom deflection 141
atom interferometer
    174, 198, 201–202, 204, 207
atom losses 175, 203
atom-optics chip 176
atomic clock 265
atomic fountain 266
atomic interactions 202

b
ballistic expansion 184, 188, 190, 194, 201
beam splitter 259
BEC 173, 317
BEC-BCS crossover 365
BEC splitter 259
Bessel functions 188, 190
Beyond mean-field 350, 352
Biot–Savart 203
blackbody radiation 314
Bloch functions 187
Bose–Einstein condensate 173, 176, 184, 203
    – apparatus 174
    – oscillation 184
bosonic bunching 337, 356

c
canonical ensemble 367
carrier chip 175, 184, 195, 197
Casimir–Polder interaction 155
Casimir–Polder potential 139
center-of-mass oscillation 184, 199
charging energy 259
circuit QED 323
coherence factor 243, 248–249, 251
coherence length 259
coherence time 273
coherent state 235
collective excitations 125
collision shift 266, 268–269
combiner 259
condensate 173
conductor cross 41
conveyor 47
cooperativity factor 324
correlation length 246
correlation time 204
corrugated magnetic field 18
CQED 322
Crank–Nicolson method 203
critical current 312
critical current density 312
critical field
    – dressed potential 229
    – two-wire splitter 214
cryostat 314
current-carrying conductors 11
current-carrying wire 126
current fluctuations 126

d
DC matter wave 257
decoherence 162, 204, 247–249
degenerate Fermi gas 365, 369
degenerate frequencies 400
density-dependent 3-body loss 380
density evolution 191

density fluctuations 337
density imprint 203
density of states 367
density profile dynamics 203
density shift 266, 268–269
density shift in a BEC 272
dephasing 163
DFG 365
Dick effect 274
dielectric charging 402
diffraction 173, 184, 196
 – grating 174
 – pattern 188
 – potential 182
diffraction grating 8
diffraction orders 188, 190, 194, 202, 207
 – coherent superposition 193, 200
 – incoherent superposition 193, 196, 199
diffraction pattern 206
dimple 41, 221
dispenser 175
dispersion forces 139
Doppler cooling 399
double meander 177
double well 385–387, 390
double well potential 259
dressed adiabatic potentials 224, 234
 – effective field 259
 – Hamiltonian 226, 228
 – RF polarization 225, 228, 232
 – theory 225, 231
dressed states 224
Drude permittivity 136
dual-species MOT 374, 390
duty factor 276
dyadic Green function 124

e
Earnshaw's theorem 126, 222, 227
effective magnetic field 226–227
effective 1D coupling constant 346
electric dipole 148
electric near-field spectrum 151
electron-beam lithography 177
electroplating 372
electrostatic mirror 4
evaporation efficiency 378
evaporative cooling 15
exchange symmetry 272
expansion 206

f
fabrication: superconducting chips 312

far-field 205–207
Fermi degeneracy
 366, 370–371, 381–382, 388
Fermi–Dirac distribution 367
Fermi function 368, 371
Fermi gas 365
Fermi pressure 371, 381
Fermi sea 367, 370, 391
Fermi temperature 368
fermions in one dimension 368
Fermi's Golden Rule 127
filters 404
five-wire trap 409
Floquet formalism 230
fluctuation-dissipation theorem 124
fluctuation spectrum 150
fluorescence imaging 274
Fock state 238
fountain clock 265
four-wire trap 409
fragmentation 16, 177
Fresnel reflection coefficients 136
fringes 259
fugacity 367–368, 379

g
global phase 235
Green tensor 124
Gross–Pitaevskii equation 174, 191, 202
guide, magnetic 40

h
H trap 43
healing length 173, 204–205, 338
heating 206, 405–406
heating rate 161
Helmholtz equation 124
hexapole field 214, 216
high performance computing 203
hybrid systems 173
hydrodynamic equations 339
hyperbolic functions 179

i
image charge 153
image dipole 153
imprint strength 196
interaction effects 205
interaction energy 206
interference 173, 181, 184, 193, 199, 204
interference amplitude 252
 – average square 253
 – distribution function 255
 – full counting statistics 255, 257

interference pattern   193, 236
  – contrast   237
  – fringe spacing   236
  – phase   259
  – undulation   248
interferometry   211
  – experimental realizations   238
  – independent condensates   240, 252
  – one-dimensional BEC   246, 257
  – split condensates   240, 246
  – theoretical aspects   234
intrinsic axes   401
Ioffe–Pritchard trap   37, 373
Ioffe trap   175
ion trap   395, 407
ions
  – Doppler cooling   399
  – loading   403
  – motional heating   405
  – spacing   398
  – trapping   396

j
Jakobi–Anger expansion   190
Johnson noise   405
Josephson formalism   242, 251
Josephson junction   321
junction   411, 416

k
Kramers–Kronig relations   123

l
Landau–Zener transition   383, 386
Larmor frequency   213, 226
lattice   173
lattice vector   179, 187
Leib–Liniger gas   332
level shift   135
light-induced atom desorption (LIAD)   374
light scattering   391
linear response   124
local density approximation
  347–349, 356, 369–370
local density of states   137
long-range order   259
low-interaction regime   206
low-pass filter   205
Luttinger liquid   247–248, 250, 253, 366
Luttinger parameter   253

m
macor   373
magic trap   266

magic wavelength   267
magnetic dipole   149
magnetic film   10
magnetic film atom chip   13–14
magnetic lattice   23, 28, 174, 177, 203
  – finite size effects   181
  – potential   177
  – reflectivity   183, 197, 203
  – symmetry   183
magnetic microstructures   10
magnetic mirror   4, 7, 9–11
magnetic near-field spectrum   151
magnetic-trap clock transition   267
magneto-optical recording   12
magneto-optical trap   174
Majorana losses   36
Majorana spin-flips   175, 197
Mathieu equation   397
matter wave decoherence   164
matter wave interferometer   198
matter wave splitter
  – comparison   232
  – electrostatic   222
  – five-wire splitter   216
  – multi-mode operation   220
  – RF-dressed potentials   259
  – stability   216, 233
  – static magnetic   213, 222
  – two-wire splitter   214
  – Y splitter   218
mean-field effects   173, 194, 204
mean-field theories   350, 353
meander pattern   174, 177, 196
Meissner effect   131, 319
merging   245, 258
metastability   147
micro-fabricated lattice   174
micro-fabricated magnetic mirror   10
microfabrication   10
micromotion   397, 401
mirror magneto-optical trap   15
mirror-MOT   51, 316
mode matching   52
modeling   408
modulated potential   185
momentum distribution   205
momentum eigenstates   187
monomode guiding   258
MOT   174–175

n
nanotube   157
NEMS   311

noise spectrum   150
non-destructive clock readout   276
non-equilibrium surface interaction   156
non-interacting theory   205
number state   259

o

on-chip detection   258
one-dimensional   331
one-dimensional Bose gas   246
  – coherence dynamics   247
  – density fluctuations   246
  – effective coupling constant   247, 253
  – phase fluctuations   241, 252
  – quantum and thermal fluctuations   249, 254, 257
  – thermal decoherence   249, 252
one-dimensional magnetic lattice   24, 26
optical clock   267
optical dipole trap   388, 390
optical resonator   274
oxygen-free high-conductivity (OFHC) copper   373

p

parallel programming   203
Paul trap   395
  – geometries   398
  – ion motion   396
Pauli exclusion principle   381
Penning trap   395
periodic potential   177, 184
permanent currents   314
phase diffusion   259
phase dynamics   259
phase evolution   192, 202
phase fluctuations   341
phase fluctuations in one dimension   259
phase imprint   174, 190, 194, 203–204, 207
phase modulation   174
phase-modulation index   190, 196
phase-number uncertainty   238, 242
phonons modes   340
photolithography   372
photonic shot noise   355
Planck spectrum   150
plane waves   187, 190
plasma frequency   259
polar molecules   134, 323
polarizability   148, 222
potassium   365, 373–374, 378–380
projection noise limit   274
pseudopotential   397

q

quadratic Zeeman shift   269
quadrupole trap   37, 175
quantum degeneracy   370
quantum electrodynamics in dielectrics   123
quantum fluctuations   204
quantum noise analysis   257
quantum reflection   143, 156
quasi-condensate   246, 259, 338, 348, 358
quasi-particle excitations   340
qubit   321

r

Rabi frequency   227–228
radial frequency   398
Raman–Nath regime   189
Ramsauer–Townsend effect   380
Ramsey interferometry   21
Ramsey oscillation   132
rapid adiabatic transfer   390
Rayleigh range   388
reciprocity theorem   124
recoil energy   174, 206
recombination   259
recooling   407
reflection of cold atoms   7
relative phase   21, 237
  – coherence   259
  – diffusion   241, 245, 258
  – evolution   239–240
  – locking   240, 251–252
  – reproducibility   259
relative population   259
residual RF noise   126
resonant Casimir–Polder potential   140
resonator   404
RF-dressed potentials   259, 365–366, 382–383, 385, 391
rf induced potentials   183
rf shunt   404
rf spectroscopy   16
rf trap   396
RF wires   224, 231–232
ring trap   232
rotating-wave approximation   227
  – beyond-RWA effects   230
rotational transition   137
Rydberg atoms   325

s

scaling laws for dispersion forces   141
scattering cross-section   379–380
scattering halos   204

scattering length 203, 206
Schrödinger-cat state 322
secular motion 397
semi-classical approximation 369
separation 407
sidebands 401
skin depth 151
soliton 245
Sommerfeld expansion 368
species-selective 383, 385, 390
spin-charge separation 366
spin-flip losses 36
spin-flip rate 127
spin-flips 159
spin squeezing 277
splitter 259
splitting
– adiabaticity 242
– coherence 238–240
– imbalance 218, 220, 231, 237, 240
– longitudinal 221
– optimized ramps 221, 241, 243
– timescales 240–243
– transverse 213, 220, 222
spontaneous emission 158
squeezing 243, 249, 258
– factor 243
SQUID 321
standard quantum limit 265
Stark effect 325
state-selective potential 225
Stern–Gerlach force 7
strong coupling regime 322, 324
superconducting film 312
superconducting materials 312
superconductor 160
surface diffraction 202
surface electrode (SE) trap 399, 409, 411
surface interaction 173, 186
surface interactions: superconductors 320
sympathetic evaporative cooling
    365, 373, 378

t
thermal de Broglie wavelength 369
thermal fluctuations 125
Thomas–Fermi limit 191, 195
Thomas–Fermi radius 370
time-of-flight expansion 236, 240
Tonks–Girardeau gas 254
Tonks–Girardeau regime 334
topological excitations 204
transport 407
trap depth 373–377, 390
trap lifetime 159, 317, 320
trap volume 374–375, 377, 390
trigonometric functions 179
tunneling 240, 242, 251
tunneling energy 259
two-body correlation function 337
two-component BEC 21
two-dimensional magnetic lattice 25
two-fluid model 130
two-point correlation function 253

u
U trap 43
ultrahigh vacuum, UHV 314

v
vacuum fluctuations 125
vacuum Rabi frequency 323
van der Waals interaction 154
vibrational transition 137
vortex (superconductor) 320
vortex flux noise 132

w
which-way information 235

y
Yang–Yang thermodynamics 342

z
Z trap 43
zones 407